VOLUME THREE

INTERNATIONAL REVIEW OF MOVEMENT DISORDERS

The Neurobiology of the Gilles De La Tourette Syndrome and Chronic Tics: Part A

VOLUME THREE

INTERNATIONAL REVIEW OF
MOVEMENT DISORDERS

The Neurobiology of the Gilles De La Tourette Syndrome and Chronic Tics: Part A

Edited by

MARC E. LAVOIE

Département de Psychiatrie et Addictologie, Laboratoire de Psychophysiologie Cognitive et Sociale, Groupe d'étude sur les Troubles tic, d'Accumulation Compulsive et Obsessionnel-Compulsif (GE-tic-tac-toc), Centre de recherche de l'Institut Universitaire en Santé Mentale de Montréal, Université de Montréal, Montréal, Québec, Canada

ANDREA E. CAVANNA

Department of Neuropsychiatry, BSMHFT and University of Birmingham, Birmingham; School of Life and Health Sciences, Aston University, Birmingham; University College London and Institute of Neurology, London, United Kingdom

ELSEVIER

ACADEMIC PRESS

An imprint of Elsevier

Academic Press is an imprint of Elsevier
50 Hampshire Street, 5th Floor, Cambridge, MA 02139, United States
525 B Street, Suite 1650, San Diego, CA 92101, United States
The Boulevard, Langford Lane, Kidlington, Oxford OX5 1GB, United Kingdom
125 London Wall, London EC2Y 5AS, United Kingdom

First edition 2022

ISBN: 978-0-323-91034-7
ISSN: 2666-7878

For information on all Academic Press publications
visit our website at https://www.elsevier.com/books-and-journals

Publisher: Zoe Kruze
Acquisitions Editor: Sam Mahfoudh
Developmental Editor: Jhon Michael Peñano
Production Project Manager: Vijayaraj Purushothaman
Cover Designer: Mark Rogers

Typeset by STRAIVE, India

Contents

Contributors

Elia Abi-Jaoude
Department of Psychiatry, The Hospital for Sick Children; Department of Psychiatry, University Health Network, University of Toronto, Toronto, ON, Canada

Laura Bogut Andersen
Pediatric Department, Herlev University Hospital, Herlev, Denmark

Elinor Baldwin
Child Study Center, Yale University School of Medicine, New Haven, CT, United States

Tracy Bhikram
Department of Psychiatry, University Health Network, University of Toronto, Toronto, ON, Canada

Andrea E. Cavanna
Department of Neuropsychiatry, BSMHFT and University of Birmingham; School of Life and Health Sciences, Aston University, Birmingham; University College London and Institute of Neurology, London, United Kingdom

Nicholas Cothros
Department of Medicine, Queen's University School of Medicine, Kingston, ON, Canada

Hugo D. Critchley
Department of Neuroscience, Brighton & Sussex Medical School; Sussex Partnership NHS Foundation Trust, United Kingdom

Nanette Mol Debes
Pediatric Department, Herlev University Hospital, Herlev, Denmark

Katherine Dyke
School of Psychology, University of Nottingham, University Park, Nottingham, United Kingdom

Susan T. Francis
Sir Peter Mansfield Imaging Centre, School of Physics and Astronomy, University of Nottingham, University Park, Nottingham; Beacon of Excellence in Precision Imaging, University of Nottingham, United Kingdom

Madeleine J. Groom
Academic Unit of Mental Health & Clinical Neurosciences, School of Medicine, Institute of Mental Health, University of Nottingham, Nottingham, United Kingdom

Camilla Groth
Pediatric Department, Herlev University Hospital, Herlev, Denmark

Georgina M. Jackson
Institute of Mental Health, School of Medicine, University of Nottingham, University Park; Academic Unit of Mental Health & Clinical Neurosciences, School of Medicine, Institute of Mental Health, University of Nottingham, Nottingham, United Kingdom

Stephen R. Jackson
School of Psychology; Institute of Mental Health, School of Medicine, University of Nottingham, University Park, Nottingham; Beacon of Excellence in Precision Imaging, University of Nottingham, United Kingdom

Rebecca P. Jordan
Child Study Center, Yale University School of Medicine, New Haven, CT, United States

Marc E. Lavoie
Département de Psychiatrie et Addictologie, Laboratoire de Psychophysiologie Cognitive et Sociale, Groupe d'étude sur les Troubles tic, d'Accumulation Compulsive et Obsessionnel-Compulsif (GE-tic-tac-toc), Centre de recherche de l'Institut Universitaire en Santé Mentale de Montréal, Université de Montréal, Montréal, Québec, Canada

James F. Leckman
Child Study Center, Yale University, New Haven, CT, United States

Julie Leclerc
Département de Psychologie, Groupe d'étude sur les Troubles tic, d'Accumulation Compulsive et Obsessionnel-Compulsif (GE-tic-tac-toc), Centre de recherche de l'Institut Universitaire en Santé Mentale de Montréal, Université du Québec à Montréal, Montréal, Québec, Canada

Davide Martino
Department of Clinical Neurosciences, University of Calgary; Hotchkiss Brain Institute, Calgary, AB, Canada

Simon Morand-Beaulieu
Child Study Center, Yale University School of Medicine, New Haven, CT, United States

Marie-Ange Perrault
Département de Psychologie, Laboratoire de Psychophysiologie Cognitive et Sociale, Groupe d'étude sur les Troubles tic, d'Accumulation Compulsive et Obsessionnel-Compulsif (GE-tic-tac-toc), Centre de recherche de l'Institut Universitaire en Santé Mentale de Montréal, Université de Montréal, Montréal, Québec, Canada

Charlotte L. Rae
School of Psychology, University of Sussex, United Kingdom

Rosa Sanchez Panchuelo
Sir Peter Mansfield Imaging Centre, School of Physics and Astronomy, University of Nottingham, University Park, Nottingham, United Kingdom

Paul Sandor
Department of Psychiatry, University Health Network, University of Toronto; Youthdale Treatment Centers, Toronto, ON, Canada

Elizabeth Shephard
Institute of Psychiatry, Psychology & Neuroscience (IoPPN), King's College London, London, United Kingdom; Department of Psychiatry, Faculdade de Medicina FMUSP, University of Sao Paulo, Sao Paulo, Brazil

Hilmar P. Sigurdsson
Brain and Movement Research Group, Clinical Ageing Research Unit, Translational and Clinical Research Institute, Newcastle University, Newcastle upon Tyne, United Kingdom

Liselotte Skov
Pediatric Department, Herlev University Hospital, Herlev, Denmark

Caitlin M. Smith
School of Psychology, University of Nottingham, University Park, Nottingham; Beacon of Excellence in Precision Imaging, University of Nottingham, United Kingdom

Denis G. Sukhodolsky
Child Study Center, Yale University School of Medicine, New Haven, CT, United States

Natalia Szejko
Department of Neurology; Department of Bioethics, Medical University of Warsaw, Poland

Olivier Walusinski
Lauréat de l'Académie nationale de Médecine, Paris, France

Yulia Worbe
Sorbonne University and Paris Brain Institute, Salpetriere Hospital; Department of Neurophysiology, Saint Antoine Hospital, Paris, France

Julia Zhong
Child Study Center, Yale University School of Medicine, New Haven, CT, United States

Preface

The past 20 years have seen an exponentially growing body of research dedicated to the Gilles de la Tourette syndrome and its underlying neurobiology, with more than 4000 publications. Unfortunately, efforts to clarify the etiology and pathophysiology of this quintessentially neuropsychiatric condition are often fragmented. However, there is reason to believe that we are on the edge of a new phase. Since the turn of the millennium, promising avenues are being explored to shed more light on the neurobiology of the Gilles de la Tourette syndrome. In that sense, this book appears to be a timely effort to provide an essential platform to synthesize and disseminate multilevel and transdisciplinary collaborations. Our shared goal was to condense the best evidence on the neurobiology of the Gilles de la Tourette syndrome with the help of cutting-edge researchers on these complex topics.

Section I of this book presents the syndrome, emphasizing its definition, nosology, and current guidelines. Specifically, in Chapter 1, Dr. Walusinski provides a comprehensive review of this clinical syndrome history, named after Georges Gilles de la Tourette. Under the scrutinizing eyes of Dr. Jean-Martin Charcot, Chair of the first modern clinic combining the study of mental and nervous diseases, Gilles de la Tourette and many other renowned neurologists of the 19th century joined efforts to construct the modern narrative of tic disorders. Chapter 2, by Dr. Leckman, introduces the multifaceted phenomenology of tics. The model of the "self under siege" provides a shared perspective that may lead to better clinical, social, and academic outcomes. An in-depth understanding of the clinical aspects is also a prerequisite for studying fundamental pathophysiology. Chapter 3, by Dr. Cavanna, summarizes the current nosology and highlights the role played by the recent development of national and international guidelines in the assessment and treatment of the Gilles de la Tourette syndrome. It is suggested that the standardization of clinical care and alignment with evidence-based treatment protocols should be combined with skilled clinical observation and individualization of therapeutic approaches. Chapter 4 by Dr. Szejko closes the first section by providing an up-to-date overview of the complex pathophysiology involved in tics and coexisting symptoms. The current picture emerges from the synthesis of multiple lines of research, with particular attention to neuroimaging, neurochemistry, and genetic studies.

Throughout Sections II and III, the book unveils a strategy to integrate brain imaging, electrophysiology, and neuropsychology in exploring the Gilles de la Tourette syndrome brain in action. Single and multimodality functional imaging studies exploring brain activation or oscillatory activity in conjunction with tasks involving motor inhibition and motor preparation advanced our knowledge of the urge–tic complex neural correlates. Chapter 5 by Dr. Cothros and Dr. Martino highlights the computational models that revealed a central role of the cerebellum–basal ganglia connectivity in tic generation, possibly linked to GABAergic anomalies. Furthermore, the increased connectivity of shorter-range connections has been associated with tics, whereas a functional disconnection of the frontoparietal network reflects the immaturity of brain development. Functional imaging and electrophysiology can unlock the door to cerebral activation patterns related to inhibitory deficits involving motor, cognitive, emotional, and behavioral domains. Chapter 6 by Dr. Rae and Dr. Critchley reviews these topics and shows that tics likely arise from aberrant interactions between frontal cortices and the basal ganglia, expressed via the corticostriatal–thalamocortical loops. The authors also highlight the role of the insular cortex in generating the premonitory urge and emphasize that it is vital to consider autonomic influences on tic expressions that may underlie tic exacerbations during heightened affective states. Even if we accept that corticostriatal–thalamocortical pathways are involved in the pathophysiology of the Gilles de la Tourette syndrome, we have to admit that the observed relationship between trajectories of brain development and the evolution of clinical symptoms is still scarce. In order to examine the temporal course of symptoms, Dr. Andersen et al. in Chapter 7, review longitudinal studies and point out developmental aspects from a neuroradiological perspective. They suggest that patients in remission of tics and patients with persistent tics might represent different phenotypical variants and thus correspond to different pathophysiological mechanisms. Cross-sectional studies on pediatric and adult clinical populations have shown differences mainly in the corpus callosum volume.

Interestingly, changes in parallel diffusivities in the thalamus, caudate nucleus, and frontal lobe are most pronounced in patients with persisting tics, compared with both patients in remission and healthy controls. Patients with the Gilles de la Tourette syndrome also exhibit activation changes associated with cognitive control of motor output, which appears to be compensatory. Our understanding of the pathophysiology of the Gilles de la Tourette syndrome comes from noninvasive functional imaging studies, postmortem histology analysis, and animal models. These findings must be interpreted in light of our comprehension of the neural circuitry of movement expression, including insights from other movement disorders. Dysfunction of corticobasal ganglia networks is a primary pathophysiological mechanism of symptoms associated with the Gilles de la Tourette syndrome. In Chapter 8, Dr. Worbe argues that the "how" and the "why" of this dysfunction remain a matter of debate. Conventional magnetic resonance brain imaging failed to identify structural abnormalities.

In contrast, advanced structural imaging techniques detected morphological and volumetric changes in various cortical areas, basal ganglia, cerebellum, and white matter pathways connecting these structures. Moreover, developing new brain imaging techniques such as diffusion tension imaging and structural connectivity allows the evaluation of white matter tracts potentially affected by deep brain stimulation. This approach opens new perspectives on potential neuroimaging markers in response to neurosurgical interventions and the localization of metabolic abnormalities. The prefrontal cortex is the binding site for activating the volitional withholding or suppression of tics by pausing motor outflow from the thalamus. Despite the relative consensus, there are inconsistencies in the findings, which in Chapter 9, Dr. Abi-Jaoude and Dr. Sandor ascribe to heterogeneity, including demographic and clinical characteristics, study design, and data analysis, among other factors. It remains unclear to what extent these findings reflect mechanisms intrinsic to the Gilles de la Tourette syndrome pathophysiology vs compensatory changes. Nevertheless, the overall picture implicates a fundamental role for prefrontal cortical control areas in self-regulation across symptom domains, particularly tic inhibition.

Section III focuses on the structure and function of neural circuits from neuroimaging and electrophysiology data. One of the most complex manifestations of the Gilles de la Tourette Syndrome seems to be the alteration of somatosensory information processing, which is the focus of Chapter 10 by Dr. Smith et al. After considering the potential contribution of altered GABA signaling to sensorimotor dysfunction, these contributors also discuss their original and innovative method for precisely mapping cortical sensorimotor representations. Ultra-high-field functional magnetic resonance imaging and neuro-navigated transcranial magnetic stimulation mapping are two key components of this innovative approach. Other studies have used complementary neurophysiological and brain imaging techniques, such as electroencephalography and functional magnetic resonance imaging to investigate inhibitory control and assess possible associations with mechanisms of tic suppression. Some studies have reported decreased inhibitory control, whereas others have reported intact or enhanced inhibitory function and theorized that this might be a consequence of multiple years of suppressing tics. Chapter 11 by Dr. Morand-Beaulieu et al. reviews the psychophysiological correlates of inhibitory control and tic suppression to examine the similarities and differences between these two processes. Chapter 12, by Dr. Lavoie et al., proposes to define etiological concepts of cerebral motor overactivation, which involve suboptimal functioning in subcortical and cortical structures. Drawing on their expertise, the authors propose combining psychological and physiological knowledge and applying it to therapy. They argue that biofeedback implemented with psychological treatment can significantly improve symptoms. Another challenge is the multidimensional and complex nature of the Gilles de la Tourette syndrome, characterized by significant clinical heterogeneity and points to a similarly composite etiological basis. Therefore, a vital issue is the interpretation of these neurobiological mechanisms in the presence of comorbidities. In Chapter 13, Dr. Bhikram reviews functional neuroimaging research on tics, obsessive-compulsive symptoms, sensory phenomena, and functional connectivity. The proposed neurobiological models emphasize how the neurobiology of tics differs from obsessive-compulsive disorder (OCD) and notably how Gilles de la Tourette syndrome with comorbid OCD may differ from Gilles de la Tourette syndrome and OCD alone. Moreover, chronic tics patients frequently have co-occurring attention-deficit/hyperactivity disorder (ADHD), encompassing inattention, hyperactivity, and impulsivity symptoms. Chapter 14 by Dr. Shephard et al. critically appraises the literature on this comorbidity pattern, explicitly focusing on the co-occurrence of Gilles de la Tourette syndrome and ADHD on electrophysiological and neurocognitive mechanisms of cognitive control.

This collective volume reviews state-of-the-art knowledge of the Gilles de la Tourette syndrome, providing a starting point for researchers and practitioners and introducing innovative transdisciplinary topics. We can undoubtedly assume that the research community has made progress in understanding the mechanisms and improving the treatment of this condition. However, many questions remain unanswered. For example, how do altered neurodevelopmental pathways result in the development of tics and their modulation throughout the lifespan? Do symptoms disappear or become successfully inhibited in adults? Why do many patients fail to respond to current treatment approaches? These are just a few examples of research questions that can only be tackled by adopting a multimodal approach, such as the one proposed in this book. Our hope is that this collective work on the current state of research proves useful in tracing the line from where we are coming to where we are heading in the 21st century and beyond.

The coeditors
Marc E. Lavoie and Andrea E. Cavanna
February 2, 2022

SECTION I

Introduction and clinical definition

CHAPTER ONE

Historical background of the Gilles de la Tourette syndrome

Olivier Walusinski*
Lauréat de l'Académie nationale de Médecine, Paris, France
*Corresponding author: e-mail address: walusinski@baillement.com

Contents

International Review of Movement Disorders, Volume 3
ISSN 2666-7878
https://doi.org/10.1016/bs.irmvd.2021.08.001

1. What is a tic?

Motor tics, or expressive mimic movements without motivation, and vocal tics are manifestations that are repetitive, stereotypical, arythmic, sudden (clonic) or prolonged (dystonia), and simple or complex (series of behaviors), with the potential to be inhibited by will and concentration, often followed by relief through a rebound of explosive overexpression. Tics can persist during sleep. They fluctuate over time, may disappear for a few days or months, then reappear. Suggestion hastens their occurrence. These unwanted motor acts are involuntary, i.e., automatic without conscious effort.

Although described since antiquity, they were not medically recognized before the nineteenth century. The word "tic" itself emerged in the seventeenth century, as a sort of onomatopoeia, a monosyllable interjected in a brief and sudden movement. The first descriptions viewed it as a form of chorea. As part of the "danse de Saint-Guy" (St Vitus' dance), the Czech Joseph Bernt (1770–1842) (Bernt, 1810) and the French researchers Michel Bouteille (1732–1916) (Bouteille, 1810) and Claude Foulhioux (1803–1873) (Foulhioux, 1847) distinguished localized forms in a few muscles of the face or the limbs, which they named false chorea, pseudo-chorea, or irregular chorea; these movements can now be recognized as tics.

In fact, the term "tic" comes from veterinary medicine (Bouley & Reynal, 1892). This term can be tracked to a posthumous work of the horse specialist Carlo Ruyni (1456–1530) from Bologna in Italy. The work was published in Venice in 1598 (Ruyni, 1598). It employs this term in the section on "Spasmo delto tico mortale" to distinguish the contractions secondary to tetanus from dystonic spasms observed in certain horses: "Et en differenza del ticco secco, il quale a più tosto vitio che male." That is, he observes that, unlike tetanus, the dry (isolated) tic is brief, without fatal consequences. This is referred to as a "tire-sec," a brief sharp pulling movement. The French physician Jean Jourdin (16?–17?), translated Ruyni's book in 1647 (Jourdin, 1647): "A tique, according to Ferrare, results from retractions of the nerves originating in the brain and is caused by excessive heat or cold or by an immoderate flow of blood. The signs of this disturbance are when the horse twists its head and its ears stand up, when its eyes roll and its mouth locks and its tail extends and its flanks are sunken, and when it presses its teeth into the feeding trough and damages the trough by arching its neck ..." Ruyni was referring to Giovanni Battista Ferraro (1528–1569) who had described a spasm in a horse with tetanus, said to be suffering from a "tiro mortale" (Ferraro, 1602). When a horse experiences discomfort, it may pull on its halter by arching its neck abruptly, a "tire-*sec*." It is why the origin of the word "tic" does appear to be Italian: "tiro-secco," similar to the French "tire-sec" above, became "ticchio" ("caprice" from "capra" [goat] in Italian) by abbreviation; the French became "tique," "ticq," and finally "tic." The same root of this onomatopoeia can be found in other languages. In German: *Zucken, zugen, tucken, ticken* (to lightly touch), tick. In English: tug, tick, tic. And in Spanish: *tico*. In 1903, Fernand Rudler (1874–1910), a military physician, and Claude Chomel (1860–1946), a military veterinarian, confirmed the existence of tics in animals, particularly in horses, this established a parallel between tics in horses and tics in people (Rudler & Chomel, 1903).

Confusion developed when André Nicolas (1704–1780) described in 1756 the "tic douloureux de la face," designating what is currently known as trigeminal neuralgia (André, 1756). Thereafter numerous authors, notably Alexis Pujol (1739–1804) in 1787 (Pujol, 1787), used the name of "tic douloureux." Only the sudden onset of pain is somewhat comparable to a tic, since the grimace accompanying the pain is variable and inconstant. (The same description had been proposed in 1671 by Johannes Laurentius Bausch (1605–1665) in Germany (Dewhurst, 1957) and by John Fothergill (1712–1780) in 1773 in England (Fothergill, 1773). In the famous Panckoucke dictionary published in 1821, the tic is first distinguished from trismus, a symptom of tetanus, and is then defined as follows: "The name "tic" is given to neuralgia, either of the face, in general, or of certain parts of the face. Since this disease is often accompanied by intense pain, it is known especially as 'tic douloureux.' More often, the name 'tic' is applied to unnatural habits in the movement of body parts, strange attitudes, singular gestures, a vicious manner of speaking, etc., the rectification of which often requires a great deal of treatment as well as perseverance, which may not always suffice in obtaining a cure" (Société de Médecins, 1821). Finally, Armand Trousseau (1801–1867) gave the name of epileptoid neuralgia to trigeminal neuralgia, which he clearly distinguished from the simple tic, "a type of chorea, and moreover clearly distinct from the danse de Saint-Guy [St. Vitus dance]" (Eboli, Stone, Aydin, & Slavin, 2009; Trousseau, 1873).

François Boissier Sauvages de Lacroix (1706–1767), a famous physician in Montpellier and author of the first medical nosography, which he wrote in 1763 in Latin, described what were "spasms for the Greeks" and "convulsions for the Romans" or "better yet, cacokinesis, depraved, local movements of the muscles" (Boissier de Sauvages, 1771). "These movements affect only the upper & lower limbs, the head, the eyes, the tongue, and the jaw." The sonorous word "cacokinesis," from the Greek "kakos" meaning bad and "kinesis" meaning movement, must have included different types of convulsions, notably those characteristics of Jacksonian epilepsy and dystonia. Tics appear in his classification among partial tonic spasms, along with strabismus, torticollis, contractions, cramps and priapism, tetanus, and other abnormal movements. Is a ticcer afflicted with "depraved movements"?

To gauge the state of knowledge on tics prior to the work of Gilles de la Tourette and Charcot, the 1883 "tic" entry of Maurice Letulle (1853–1929) in the *Jaccoud Dictionnaire de Médecine* is a useful source. It distinguished between convulsive tics, which we continue to consider tics, and coordinated tics, which encompassed banal movements automatically and subconsciously repeated (smoothing one's hair, licking one's lips, and so forth), and probably included dystonia (Letulle, 1883). The entry stated that the pathophysiology of tics was completely unknown, but noted with certainty the contagious nature of tics, along with the role of imitation and "bad habits" in their triggering.

2. Chorea

Chorea had a millennial history (Lanska, 2010). For most physicians during the nineteenth century, all abrupt, isolated, or repeated movements, whether psychological or organic in origin, were referred to as "chorea," to which was sometimes added an adjective, as in chorea lasciva, chorea imaginativa, malleatory chorea, chorea festinans, procursive chorea, dancing chorea, or saltatory chorea. Charcot's Tuesday Lesson on January 17, 1888, included this caricature: "What did Dr. de Granville call your disease? The patient: Chorea. Charcot: This word 'chorea' really seems to resolve the entire question. I ask you how the condition, before you, resembles chorea. To make it fit into the category of chorea, it would be necessary, alongside the already existing Sydenham's chorea and rhythmical chorea, to create a ticcing chorea." Indeed, chorea was identified by Thomas Sydenham (1624–1689), and its link with acute rheumatic fever and rheumatic heart had been established since the work of Bouteille (1810) later by Jacques Pierre Botrel (1819–1914) (Botrel, 1850) and Germain Sée (1818–1896) in 1850 (Sée, 1850). Sée coined the term "chorée rhumato-cardiaque." It was confirmed in 1889 by William Osler (1849–1919) (Osler, 1887).

In his 1857 dictionary, Robley Dunglison (1798–1869) gave about 15 synonyms for the danse de Saint-Guy, but confused matters by including chronic chorea, which could be partial, and "tic gesticulatoria" (Dunglison, 1857). Trousseau continued in this vein, calling Sydenham's chorea the danse of Saint-Guy to distinguish it from "writer's cramp, or writer's chorea (chorea scriptorum)," i.e. the "functional spasm" of Guillaume Duchenne de Boulogne (1806–1875) or dystonia (Duchenne de Duchenne, 1872).

In 1850, the Hungarian physician David-Didier Roth (1800–1888), seeking, as Bouteille had, to clarify what the term "tic" covered distinguished "the irresistible muscle activity or abnormal chorea" and divided it into "irresistible muscle activity for locomotion and stationary irresistible muscle activity." For the latter, he assigned "the name of tics, involuntary movements that do not move the entire body and occur in a stationary way." He listed under this term of tic essentially what we call forms of dystonia, either cervical or oromandibulary dystonia. His example of the simple and brief tic was the eye-blinking tick or "nictation." He did not fail to notice that certain tics can be associated with "involuntary ventriloquy": "Imitation more or less perfect of the sounds of living beings, dog barking, cat meowing, etc., or the imitation of the sound of inanimate objects, such as that of the saw, the plain, rain, etc" (Roth, 1850). This depiction is apt insofar as what defines the motor and phonic tics of Gilles de la Tourette syndrome, namely the association of abrupt muscular movements accompanied by a noise, is similar to a horse's sudden movements and eructated sounds, described by Jourdin and Ferraro.

After an observation published in 1856, in which the French Pierre Bosredon (1804–1893) referred to "barking delirium" (Bosredon, 1856), a noisy form of manifested hysteria, Étienne Ancelon (1806–1886) responded to him in 1857, referring to chorea of the larynx. Barking consisted of phonic tics associated with motor tics: "Barking chorea always starts with convulsions of the face which in most cases dissipate, leading to sudden jerks that are frequently repeated"; that is, tics (Roth, 1850). Ancelon in 1857, like Roth (1850), described cases of Gilles de la Tourette syndrome. In his clinical lessons, Trousseau also used the term of "laryngeal or diaphragmatic chorea that may constitute the entire tic. This involves not only a vocal outburst or a strange cry; it is also a singular tendency to always repeat the same word, the same exclamation. The individual also says words out loud that he would rather keep silent."

3. How Gilles de la Tourette ended up studying tics

In 1881, while still a medical student, Gilles de la Tourette translated an article by the American alienist George M. Beard (1839–1883), "Experiments with the 'Jumpers' of Maine" (Beard & Gilles de la Tourette, 1881) which had been published in 1880 in *The Popular Science Monthly*, a periodical for the general public (Beard, 1880a). When he began his third year of residency in 1884 with Charcot, Charcot knew of this translation. He encouraged Gilles de la Tourette to study abnormal movements and, in particular, to clarify those classified without distinction under the term of "chorea." Beard's writings were the starting point for this work.

3.1 Jumpers of Maine

Beard's name remains associated with the description of neurasthenia in 1879 (Beard, 1879). In 1878, Beard told his colleagues at the fourth annual meeting of the American Neurological Association that French-Canadian woodsmen from northern Maine would jump or strike when excited. The corresponding report was 24 lines long in the *Journal of Nervous and Mental Disease* (Beard, 1878). In his presentation, Beard described an exchange with his friend William Augustus Croffut (1835–1915). While traveling through northern Maine, Croffut observed lumberjacks who, when taken by surprise, began to jump around excessively, hence the name of "Jumpers." "It is a psychological and not a pathological state. It is not a nervous disease. It is the result of a psychical condition among a people of a low order of intellectual development." As they jumped, they hit themselves or their entourage violently. They seemed incapable of refraining from this type of sudden, vehement reaction. Since many subjects reacted this way in this region, Beard viewed the phenomenon as a surviving example of "an epidemic of habits" from the Middle Ages. Gilles de la Tourette never mentioned this 1878 presentation and its summary report, also published in London (Anonymous, 1878). In fact, 20 years before Beard's depiction, the *Boston Daily Advertiser* published on February 15, 1861, an article titled "Jumping Frenchmen of Madawaska," referring to French immigrants in Québec. Madawaska is located in Aroostook County in northern Maine on the border with New Brunswick (Canada). The article's author, Rev. M. R. Keep, estimated that from one-fourth to one-third of the French-speaking men in the Madawaska region were "jumpers" or "Jumping Frenchmen." When startled, men with this affliction would lose control of themselves and would obey sudden commands, such as to throw whatever was in their hands, including tools like axes, occasionally injuring bystanders. Beard subsequently traveled to the Moosehead Lake region of Maine to observe firsthand the "Jumpers" or "Jumping Frenchmen." Beard presented his findings on this syndrome in a paper at the sixth annual meeting of the American Neurological Association in 1880, and they were published in both the Journal of *Nervous and Mental Disease* (Beard, 1880b) and *Popular Science Monthly* (Beard, 1881).

Many authors refer to an article published by Beard in the *Journal of Nervous and Mental Disease* when they cite the source used by Gilles de la Tourette. In fact, Gilles de la Tourette himself is the cause of this error. In a note in his seminal 1885 article, he wrote: "Beard's report was published in *The Journal of Nervous and Mental Disease*, vol. VII, p. 487. We published the translation in No. 5, vol. II, of *Les Archives de Neurologie*, 1881, pp. 146–150." But as he had indicated in his 1881 translation, the source text appeared in *The Popular Science Monthly*! Some have contended that Gilles de la Tourette did not translate faithfully and that he considerably embellished the material. The reality is that Gilles de la Tourette did in fact translate the first-person text written by Beard in *The Popular Science Monthly*, adding a French literary style, cutting out a few sentences, and omitting several proper names given by Beard. The only one that Gilles de la Tourette included is misspelled: Beard wrote Dr. Edward Steese, which Gilles de la Tourette transcribed as Dr. E. Stewe. The article published in the *Journal of Nervous and Mental Disease* is a third-person account of Beard's findings. It reads like a condensed version of the previously cited article but was published in a prestigious medical journal. Gilles de la Tourette was thus more or less faithful to the original, but not the text commonly attributed to Beard, a text that Beard probably did not write himself, even though it is the most often cited.

Beard described young men who startled in an excessive way at the slightest stimulation but behaved normally and appropriately outside of surprising situations. They would jump and cry out at "any sound from behind that was sharp and unexpected... These phenomena suggest epilepsy, particularly in their explosive character and in the nature of the cry. The hands strike or throw with a quick, impulsive movement, which is very hard to imitate artificially. They go off like a piece of machinery; it is more like the explosion of a gun than the movement of the limbs of even an angry man; and the cry suggests that which we hear in hysteria and in epilepsy." Aside from their violently emotional startle reflex, these men were calm and well-balanced; Beard described them as "modest, quiet, and retiring." The complete description associated an exaggerated startle reflex in the Jumpers with raised arms, a piercing scream, and violent hitting, most often of themselves, with the hand directed against the head or chest. This movement was then repeated several times. This sequence occurred most often after receiving an order with sudden violence or hearing an unexpected noise. The Jumpers were lumberjacks with little education or capacity to verbalize their emotions. Beard noted similarities with hysteria: the psycho-contagious and familial aspects of the phenomenon. He also suggested a hereditary aspect. Beard did not explicitly highlight the idea of immediate and involuntary imitation, aside from the repetition of the order before its execution (Lanska, 2018). It is very strange that Gilles de la Tourette did not highlight Beard's idea that the symptoms observed were probably hysterical in nature. Several American journals contained Beard's work and accounts of the jumpers appeared for about 10 years in the United States. Mark Paul Richard has suggested that these articles highlighted the significance of Canadian immigrants, French in origin, to the United States after 1870 (Richard, 2018).

3.2 Latah of Malaysia

At the Académie des Sciences, Charcot made the acquaintance of Armand de Breau de Quatrefages (1810–1892), a physician, zoologist, and anthropologist in 1884. He said to Charcot: "Look, I have something for you!" Quatrefages showed him an article, then gave it to Gilles de la Tourette. It was written by a businessman and traveler, H. A. O'Brien, who described the "Latah" of Malaysia. O′Brien's text was published in 1883 (O'Brien, 1883), with additions made in 1884 in an obscure journal founded by a group of British colonial administrators. Gilles de la Tourette translated this revised document (O'Bren, 1884). In a new article (Gilles de la Tourette, 1884), he introduced extracts from O′Brien's writings and from a very recent publication by an American neurologist and former U.S. surgeon general, William Hammond (1828–1900): "Today, we have the account of a similar state [to the Jumpers of Maine] which exists not in America, but in Asia, specifically Malaysia. The narrator, Mr. O'Brien, is not a physician, just someone acting on curiosity; nevertheless, he gives a good report of what he observed, abstaining from commentary. Instead of the American 'Jumping,' the term used in the Malay language is 'Latah' which may designate either the affected individual or his special state."

Gilles de la Tourette highlighted the similarities between the Jumpers of Maine and the Latah described by O'Brien: "Persons of a peculiar nervous organization, ranging from those who, from their mental constitution, seem absolutely subservient to another's will; to those who appear merely of a markedly excitable temperament." The word "Latah" can be translated by "ticklish." Upon hearing a loud or unexpected noise, a Latah will startle violently and exclaim in a manner that O'Brien characterized as always obscene. He or she often exteriorizes a state of panic after a sudden unforeseen stimulation. According to O'Brien, these individuals imitate the words, sounds, or movements of those around them, without being asked, all while enjoying a perfectly regular mental state in the intervals between fits. This propensity for involuntary imitation emerges as the dominant trait in O'Brien's descriptions, unlike the accounts of Beard who never mentioned imitation. "I have met a man several times lately who is a very strong Latah subject. He is cook on board a local steamer, and is naturally (alas, for human nature!) the butt of all the crew, who daily and almost hourly exercise their clumsy wit—the wit of sailors and Orientals—at his expense. All of this skylarking, however, had a tragical ending the other day, which illustrates the point of which I am speaking. This cook was dandling his child forward one day; one of the crew came and stood before him with a billet of wood in his arms, which he began nursing in the same way as the Latah was nursing his baby. Presently he began tossing the billet up to the awning and the cook tossed his child up also, time for time. At last, the sailor opened his hands wide apart and let the wood fall upon the deck, and the cook immediately spread out his hands away from the descending child, who never moved again after striking the boards." While the Jumping in Maine seemed only to occur in men, O'Brien told of women subjects, and most often older women, who suffered from Latah. He wrote that without any effort on his part, they "completely abandoned themselves to [his] will and powers of direction."

3.3 Miryachit in Siberia

Gilles de la Tourette recopied most of Hammond's article: "These singular facts are not limited to America or Malaysia, countries where the climate is warm; they can also be observed in Siberia, as Dr. Hammond relates with examples from the accounts of American officers." Hammond was at that time a professor of nervous and mental diseases at New York University and is still known for his seminal description of athetosis in 1871 (Hammond, 1871). Gilles de la Tourette referred to the text published in Paris in the journal *L'Union Médicale* (Hammond, 1884c), entitled "Miryachit, nouvelle maladie du système nerveux." The translator is not named, but apparently translated an Italian version (Hammond, 1884b) of the original article, which was published in English in the *New York State Journal of Medicine* (Hammond, 1884a). Whereas Hammond spelled the name of the condition "Miryachit," Gilles de la Tourette wrote "Myriachit," which means "to act stupidly." The description of the condition is based on an account, published in 1883 (Buckingham, Foulk, & McLean, 1883), of a voyage undertaken in 1882 by three American Navy officers, the lieutenant Benjamin H. Buckingham[a] (Zapffe, 1978), the ensign George Clayton Foulk (1856–1893),[b] and Walter McLean (1855–1930).[c] These three officers observed the unusual propensity of certain individuals in the Amur River basin in eastern Siberia to imitate the motor behavior or speech of those around them, in an involuntary manner. Hammond seemed unaware that he was not the first to describe the Myriachit. It was, in fact, the German physician and botanist Georg Wilhelm Steller (1709–1746) (Nuttal, 2004) who first wrote of the phenomenon, after observations during a northern expedition from 1729 to 1731 (Steller, 1774). He referred to "hysteria" and cited the various appellations used by the peoples of the Russian Far East where these involuntary imitation behaviors were observed: "amurakh" (Yakuts), "olan" (Tougouses), "irkunii" (Yukaghirs/Yukagirs), "menkeiti" (Koryaks/Koriaks), and "imu" (Ainus/Aynus). Hammond drew parallels between these accounts and Beard's observations: "From this description it will at once, I think, be perceived that there are striking analogies

[a] Lieutenant Benjamin H. Buckingham was part of an American military delegation that came to test the qualities of iron-nickel alloys developed by the Schneider-Creusot steel mill in 1891 and later used in American weapons due to their superiority over the alloys available up to that time.

[b] George Clayton Foulk (1856-1893), a navy officer and diplomat, first served in Asia from 1877 to 1883, then represented the U. S. at the American legation in Korea from 1883 to 1885. He went on to serve as chargé d'affaires until 1887. After being discharged from the army, he settled in Japan where he represented American companies from 1888 to 1890. From 1890 until his death, he was a professor of mathematics at Doshisha University in Kyoto (The United States in Asia: A Historical Dictionary).

[c] Walter McLean (1855-1930) would serve as the commander of the Norfolk Naval Shipyard from Novembe 25, 1915, to February 4, 1918. He is known for having inspected several German warships and for holding them prisoner in American ports until the end of World War I (Wikipedia).

between 'Miryachit' and this disorder of the 'Jumping-Frenchmen' of Maine. Indeed, it appears to me that if the two affections were carefully studied, it would be found that they were identical, or that, at any rate, the phenomena of the one could readily be developed into those of the other... There is another analogous condition known by the Germans as *Schlaftmukenheit* [sic] ..." For this condition, Hammond gave examples of confusional states, occurring when the person was purposely awakened, and for which we see no link with the stress-related startle reflex and the imitation behaviors reported above. His descriptions suggest a sort of parasomnia categorized as "confusional arousals" of which the subject has no memory; such phenomena are frequent in children but may persist in 1–2% of adults (Mainieri, Loddo, & Provini, 2021). Hammond concluded: "They all appear to be due to the fact that a motor impulse is excited by perceptions without the necessary concurrence of the volition of the individual to cause the discharge. They are, therefore, analogous to reflex actions and especially to certain epileptic paroxysms due to reflex irritations. It would seem as though the nerve-cells were very much in the condition of a package of dynamite or nitroglycerin, in which a very slight impression is competent to effect a discharge of nerve-force. They differ, however, from the epileptic paroxysms, in the fact that the discharge is consonant with the perception—which is in these cases an irritation—and is hence an apparently logical act; whereas, in epilepsy, the discharge is more violent, is illogical, and does not cease with the cessation of the irritation."

Gilles de la Tourette continued: "We will not attempt to interpret this bizarre state: Jumping in Maine, Latah in Malaysia, Myriachit in Siberia, and perhaps Schlaftrunkenheit in Germany. We prefer to make our own contribution to this series of related observations. In the department of our teacher, Professor Charcot, there is currently a case in many ways similar to this singular state. The subject is a 15-year-old boy, intelligent, perfectly capable of reasoning, of sound constitution, who suffers from extreme hyperexcitability, specific tics, jolting movements of the head and trunk, after which he almost invariably and loudly pronounces the word of Cambronne [shit]. Moreover, if we speak in front of him, he faithfully echoes the last two or three words of the sentence just pronounced. This is a source of observations nearly unexplored and whose interpretation cannot be attempted until a great many facts, carefully studied, without preconceived ideas, have been collected." Gilles de la Tourette added this note: "Since the writing of this article, we have had the good fortune to come across another subject; the corresponding observation and three others from Mr. Charcot will be described in a forthcoming text on this matter" (Gilles de la Tourette, 1885).

4. Beard, O'Brien, and Hammond: Valid introductory material?

Gilles de la Tourette's personal behavior, as described by his contemporaries, suggests that he himself may have identified with these hyper-reactive individuals who jump at the slightest stimulation. Incidentally, Gilles de la Tourette indicated that he went to London to question two famous English neurologists, Sir William Henry Broadbent (1835–1907) and John Hughlings Jackson (1835–1911), and that he met "numerous physicians of diverse nationalities" at the international conference of medical sciences in Copenhagen in 1884. They "told us they had never observed similar cases, nor read descriptions of this disease."

Turning to the articles of Beard, O'Brien, and Hammond, was Gilles de la Tourette correct in viewing them as a valid introduction to his own publication? Jumping, in the sense of startling, is a universal, stereotyped, phylogenetically old behavior that can be observed in all vertebrates. It is a reflex response to a sudden, intense, and unexpected stimulation and serves the purpose of protection or preparation for fight or flight (Hale et al., 2002). In humans, this natural mechanism can be modified through cultural learning. Anxiety states, withdrawal from alcohol or other legal or illegal psychotropic substances, post-traumatic stress disorder, and certain affective states may increase the startle reflex, but this does not constitute a specific disease. In 1918, the Austrian pediatrician Ernst Moro (1874–1951) described a reflex present in newborns that disappears before the age of 4 months. Known as the "reflexe de Moro," the clasping reflex, Moro reflex, or "Umklammerungs reflexe," it entails an extension of the four limbs, followed by their sudden flexion, after a noise, a sensation of falling, or air blown abruptly onto the face or belly (Moro, 1918, 1920). In 1936, Carney Landis (1897–1962) and William Hunt (1902–1981) described the "startle response," which appears in humans as the clasping reflex and eventually disappears (Hunt & Landis, 1936, 1937a, 1937b). These reflexes, said to be archaic, are physiological reactions and resemble, in their manifestation, the response of the Jumpers in America, Asia, and Siberia (Goldstein et al., 1938).

Reuben Rabinovitch (1909–1965), a neurologist in Montréal, wrote a personal account of his childhood in 1965 that is also a medical account. In it, he described the arrival of lumberjacks in his childhood village in the Laurentides region north of Montréal. These men would go into the forest to cut down trees with 30 or so horses, which would sometimes kick in response to violent yelling. The lumberjacks would overcome their shock by jumping, which became a strange and comic habit. Rabinovitch agreed with physicians in the region accustomed to this behavior that it was a cultural form of hysteria (Rabinovitch, 1965). In 1967, Charles Kunkle examined 15 Jumpers entirely comparable to those described by Beard. All of them underwent a neurological examination and were found to be normal (Kunkle, 1967). His analysis highlighted the closed, masculine communities where they lived and their lack of education. Kunkle's diagnosis was collective neurosis.

In 1986, a team of Montréal neurologists studied eight Jumpers from the Beauce region of Québec who worked in the winter as lumberjacks in the nearby forests of Maine. Although all of them described themselves as shy and ticklish, they felt quite violent by nature. In particular, they would startle when teased by others, and this reflex was always accompanied by aggressive movements directed at the tormentor. Fatigue and anxiety were found to accentuate the symptoms. They lived in groups, isolated in the forest for several months and without other social contacts, in very primitive conditions without individual beds, without changing their clothes or washing. They had little in the way of schooling. The "jumping" had started in adolescence. Several of the Jumpers had ascendants who were recognized Jumpers, but none of their own children were affected. The authors filmed the Jumpers during a "fit" and the very same behavior described by Beard can be observed. No signs of echolalia, echopraxia, or coprolalia were observed. All of them underwent a neurological examination and were found to be normal. Today, the "Jumpers of Maine" seem to have completely disappeared. This is apparently due to the fact that the poor socio-economic conditions under which the lumberjacks once worked no longer exist. Quebecer neurologists believe that psychological and material poverty, along with a rudimentary lifestyle, explain the phenomenon of jumping and its related manifestations. They consider it a form of conversion hysteria, a form of psychocultural conditioning in response to stress (the hysteria of Charcot and Gilles de la Tourette!). No neurological organicity has been found (Saint-Hilaire & Saint-Hilaire, 2001; Saint-Hilaire, Saint-Hilaire, & Granger, 1986). Jumpers have been identified in different regions and in other socio-cultural groups of various ethnic origins in the U.S. which nonetheless have similar lifestyles (Kunkle, 1965). The phenomenon has nothing to do with motor tics; rather, it is a psychopathological condition based on cultural factors (Bakker & Tijssen, 2010).

In 1980, the American Joseph Hardison recalled that in his youth, the term "goosey" was applied to people who were ultra-sensitive to being tickled or teased. "Goosey" is a familiar English word but is not without special significance. "To goose" means to push, shake, tickle, incite, or tease. Hardison drew a parallel with the Jumpers of Maine (Hardison, 1980; Massey, 1984). In 1992, a review of the medical literature examined all of the psychiatric pathologies in which excessive startle reactions may occur. The list included schizophrenia, post-traumatic stress disorder, drug withdrawal, and hysteria (Howard & Ford, 1992).

Holman Francis Day (1865–1935), a writer and filmmaker born in Vassalboro, Maine, penned a short story in 1902 fictionalizing the life of the jumping lumberjacks in Maine. Their miserable living conditions were made unrealistically appealing, but their behavior during jumping fits and their violent outbursts, often involving their spouses, were faithfully portrayed (Day, 1902).

5. Differential diagnosis

Startle is a universal and phylogenetically ancient stereotyped reflex response to sudden stimulation, which can be exaggerated in a wide variety of neuropsychiatric disorders, including culture-bound syndromes, e.g., jumping, Miryachit, and Latah. There is also a rare form of epilepsy characterized by an abnormal startle reflex. Hereditary hyperekplexia, hyperekplexia (Suhren, Bruyn, & Tuynman, 1966), Kok disease (Kok & Bruyn, 1962), familial startle disease, or stiff baby syndrome is characterized by "an exaggerated startle reflex." About 150 cases have been identified (Thomas, 2015). This pathology does not apply to the Jumpers.

William James West (1794–1848) s described in 1841 a syndrome in his son, James Edwin West (1840–1860), who showed the first symptoms at the age of about 4 months. The syndrome he described was called West syndrome (West, 1840-1841). The name initially given in France, "Salaam tic," may be the cause of possible confusion with common tics (André, 1892). In fact, it is a form of myoclonus epilepsy with onset in infancy, characterized by seizures involving the muscles of the neck, trunk, and limbs, with nodding of the head and flexion and abduction of the arms as typical features. Mental retardation is associated in most cases. The condition has nothing in common with true tics.

In 1955, Théophile Alajouanine (1890–1980) and Henri Gastaut (1915–1995) described "startle epilepsy" (Alajouanine & Gastaut, 1955).

It occurs in children with brain damage in the motor cortex, most often subsequent to neonatal anoxia, and the corresponding symptomatology is stereotyped (Alajouanine & Gastaut, 1958). Again, this has nothing to with Jumpers.

6. Historical and current interpretations of Jumping, Latah, and Myriachit

The word "Latah" seems to have appeared for the first time in Malay literature in the fifteenth century, as a metaphorical form interpreted as either "love sickness" or "tickle"! (Winzeler, 1984) Many cultures share the idea of a startle, and the language for it expresses the agreeable-disagreeable ambivalence of games and contact, the first steps toward the physical closeness of lovers: "Imu" in Japan, "Ikota" in Siberia, "Mali-mali" in the Philippines, "Yaun" in Vietnam, "Orgamci" or "Belenci" in Mongolia, and so forth (Yap, 1967).

Knowledge of Latah by Europeans dates from the middle of the nineteenth century and is based on the accounts of John Turnbull Thomson (1821–1884), an English engineer involved in infrastructure projects in Malaysia (Thomson, 1864), and of Frederik Johannes van Leent (1829–1895), a physician and head of the Navy medical department in the Dutch East Indies (van Leent, 1867), one of the first to attribute beriberi to a nutritional deficiency (van Leent, 1880). A more precise description was written in 1885 by Henry Ogg Forbes (1851–1932), a Scottish ornithologist and explorer who observed "that curious affection called by the natives 'Latah.'" His servant would jump like a jack-in-the-box upon seeing a caterpillar. Terrorized, cursing loudly, he would go into a "trance." Forbes attributed the nervousness of the inhabitants to the climate in order to explain his observations (Forbes, 1885). By the end of the nineteenth century, there were many medical descriptions, such as that of John Desmond Gimlette (1867–1934), who highlighted a mix of superstition and religious fervor. He saw Latah as a manifestation of hysteria expressed with local cultural acquired acceptance: "In both cases, self-control being lost for the time being, the attention is occupied mainly by a single idea, in the 'Latah' woman by an uncontrollable desire to imitate, in which the servile portion of human nature is unconsciously displayed; in the amok man, by a reckless drive to kill, in which man's wild, animal side dominates" (Gimlette, 1897). In the twentieth century, authors underlined colonialism's role in inciting this behavioral reaction (Murphy, 1982). Robert L. Winzeler, of the

University of Nevada in Reno, published in 1995 a complete ethnographical and medical study on Latah, which stands as the reference work on this subject (Winzeler, 1995).

An assistant to Russian neuropsychiatrist Sergei Korsakoff (1854–1900), Ardalion Ardalionowicz Tokarski (Ардалион Ардалионович Токарс кий, 1859–1901), after having observed numerous cases of "meriatschenje" in Siberia, published two observations of pathological imitation behavior in 1890 (Tokarski, 1890). In addition to his own observations, he included that of a Russian military physician, Nicolai Ivanowich Kaschin (1825–1872), who in 1868, around Lake Baikal, was confronted with a military corps that simultaneously repeated all of the orders received, illustrating behavioral contagion. For Tokarski, myriachit had nothing in common with the echolalia and echopraxia of what came to be known as Gilles de la Tourette syndrome but were included with other conditions under the terms "artic hysteria" or "Siberian hysteria" (Tokarski, 1891). Nikolaĭ Nikolaevich Bazhenov (1857–1923), chef physician at the Preobrazhenskaya psychiatric hospital in Moscow, submitted a review to *Les Archives de Neurologie* in 1911 entitled "Myriatchenié et Klikouschisme," which the authors qualified as psychoneurosis described by Kaschin and Tokarski (Bagenoff, 1911). The "Klikouschis," meaning those who cry out—because they are "possessed by the demon" or "have been cursed"—exhibited attacks of *grande hystérie* (hysteria major) as described by Charcot, which Bazhenov likened to the epidemics of demonopathic convulsionaries, exemplified by the Ursulines of Loudun.

An ethological analysis of Jumping, Latah, and Myriachit describes these involuntary and automatic behaviors as a reaction to a visual, auditory, or tactile stimulus, which can be schematically described as "startle and imitate" with the addition, in some cases but not all, of echopraxia and echolalia. No form of tic can be likened to these startle reflexes (Joseph & Saint-Hilaire, 1998). This jumping, or inversely a frozen immobility or behaviors of avoidance or aggression, is an immediate and single reaction to a stimulus, whereas tics occur without an exterior triggering factor and are by definition repetitive. The activity of mirror neurons, a substrate for imitating behaviors and decoding intentionality that was discovered by Giacomo Rizzolatti's team at the University of Parma, explains this innate and involuntary capacity in Latahs and Myriachits to mime motor action (Rizzolatti, Fogassi, & Gallese, 2001; Rizzolatti & Sinigaglia, 2016). The imitation behaviors observed in a Latah fit are accompanied by cursing directed at the tormentor. Latah seems to affect shy and introverted people whose cognitive, motor, and social development is nonetheless normal, especially middle-aged women who reproduce the movements and gestures of their mothers (Kenny, 1990). Latah can thus be likened to hysteria, with specific manifestations that reflect the subjects' culture. Myriachit is similar because of the behavioral imitation, but there is no cursing. It seems currently accepted to view all of these behaviors as psychiatric disturbances where the acceptance and tolerance of the social group in which they occur acts as reinforcements and favors social integration in the group. These historical documents are considered particularly valuable as anthropological studies of psychiatric conditions and how culture influences their expression (Simons, 2013).

In 1965, Harold Stevens noted that strangely, publications on the Jumpers disappeared after 1912. He raised the issue of translation errors in various articles that could have led Gilles de la Tourette to consider the jumping in Maine, Latah, and Myriachit as the same disease he was trying to isolate, whereas Charcot drew a clear distinction (Stevens, 1965). In none of the cases do these brief and isolated startle reflexes, secondary to a triggering factor outside the individual, resemble motor and phonic tics. Is the cursing the same thing as coprolalia? It is indeed, in the medical and anthropological opinion of Ronald C. Simons (Simons, 1994). The imitation behaviors are harder to analyze in the accounts, but there are many situations in normal life where bodily attitudes are imitated, such as a yawn or laughter, without this necessarily constituting the echopraxia and echolalia present in Gilles de la Tourette syndrome (Chapel, 1970).

It is unfortunate that Gilles de la Tourette was unfamiliar with the original thesis of the alienist Prosper Lucas (1805–1855), defended in 1833, which subtly analyzes all of the phenomena of "contagious imitation" and in which Gilles de la Tourette could have found ideas relevant to the behaviors of the Jumpers, Latahs, and Myriachits: "Acquired similarities result from the habit of the same acts, the same feelings, the same sensations, the same ideas, in a word from one's upbringing, which over time imprints the entire organism with this character that Lavater called, in the traits and movements, the 'secondary physiognomy.' In these sympathetic propagations of nervous affection, women and those individuals they pass through like an electric shock are almost always not only controlled by the same organic conditions, but also under the yoke of the same opinions, the same superstitions, the same daily impressions, and all of the same life circumstances" (Lucas, 1833). A physician in Lyon, Claude Sigaud (1862–1921), defended in 1889 his thesis entitled *De l'échomatisme*, which was subsequent to Gilles de la Tourette's seminal article, with a meaning close to Charcot's echokinesis. He gave a psychophysiological explanation of these automatic imitation phenomena, seeing them as a cerebral automatism secondary to "psychological poverty." He drew on the work of multiple authors, including Paul Sollier (1861–1933), Alfred Binet (1857–1911), Pierre Janet (1859–1947), and Théodule Ribot (1839–1916).

7. The predecessors

At the beginning of his seminal article, Gilles de la Tourette mentioned the innovative book of Étienne-Michel Bouteille, "*Traité de la chorée ou Danse de Saint-Guy*" (Bouteille, 1810). This physician, trained at the Faculty of Medicine of Montpellier (in southern France), clearly distinguished between Sydenham's chorea and "pseudo-chorea or false chorea," the latter predominated by "involuntary movements of different body parts and grimace-inducing convulsions of the face." Gilles de la Tourette noted that "the disease we are going to study would have fallen into this complexus," which was to say that the disease he was going to describe was chorea. He went on to indicate that the first description of involuntary movements accompanied by involuntary grunting and cursing was given in 1825 by Jean-Gaspard Itard (1775–1838) (Itard, 1825). Itard, who was appointed physician in 1799 at the National Institution for Deaf Mutes, was concerned essentially with the hearing organ, its diseases, and related investigations. His reputation throughout Europe at the time was related to his attempt to educate, from 1801 to 1806 the "Wild Boy of Aveyron," who likely had autism (Itard, 1801). In any case, the article that interests us is devoted to various abnormal behaviors and movements that today seem to lack any specific link, aside from a possible psychic cause. It was his tenth case that interested Gilles de la Tourette, who presented it as the first observation of his article, highlighting it as a prototypical case of the disease he wanted to describe. Before Gilles de la Tourette, aside from David-Didier Roth, already mentioned, Stanislas Sandras (1802–1856), a physician at Hôpital Hôtel-Dieu in Paris and professor *agrégé* at the Faculté de Médecine, wrote in 1851 a book entitled *Traité pratique des Maladies Nerveuses*, one of the first practical treatises on nervous diseases. He also copied Itard's observation, classifying it among the "partial forms of chorea" (Sandras, 1851). Sandras never used the word "tic," and this chapter of his book only seems relevant for having helped disseminate Itard's work, but with no new information.

8. The first described clinical case of Gilles de la Tourette syndrome

Itard's patient was Mme de D., who had suffered since the age of seven from movements that "were involuntary and convulsive, with participation of the shoulder, neck, and face muscles. The result was extraordinary contortions and grimaces. The disease had continued to progress and, the spasm affecting the organs of the voice and speech, this young person would utter bizarre cries and words devoid of meaning, but never in a state of delirium, without any disturbance of the mental faculties." After a few years of progression, she spent nearly a year in Switzerland for treatment and returned much improved, with only the occasional tic. "She was then married. But the marriage, rather than strengthening and completing her cure as had been hoped, caused her disease to return quite rapidly. Mme de D... did not have any children and was thus deprived of the favorable opportunities of the physical and moral transformation ordinarily produced by motherhood." Itard was especially struck by the phonic tics: "Among the continual and chaotic movements resulting from these morbid contractions, those affecting the organs of the voice and speech are the only one worthy of our attention, as they represent an exceptional phenomenon, and constitute a most disagreeable inconvenience that deprives the patient of all of the joys of society. The disturbance they create is due to the pleasure the subject takes in social interaction. For example, in the middle of a conversation in which the subject is most interested, she suddenly, without being able to restrain herself, interrupts what she is saying or hearing with bizarre cries and words that are even more extraordinary and contrast deplorably with her intellect and her distinguished manners. These words are mostly vulgar curses, obscene epithets, and, what is no less embarrassing for her and others listening, blunt expressions of judgment or unfavorable opinions of those present in society." Itard concluded with the idea that this pathology could be classified as chorea: "Between this neurosis and the one at the center of this dissertation, there is certainly no essential difference justifying a separate category amongst the numerous diseases of the brain." The observation of Mme de D. was reproduced in theses on chorea as the prototype of "partial chorea," for example in the 1857 thesis of Émile Quantin (Quantin, 1857).

In 1859, Pierre Briquet (1796–1881), in his *Traité clinique et thérapeutique de l'hystérie*, mentioned "a remarkable disturbance of phonation in hysterics," referring to Itard's observation in which he likened it to fits of involuntary laughter (Briquet, 1859). Armand Trousseau told the story of Mme de D. in 1862 in his medical department at Hôtel-Dieu hospital, considering her condition a form of epilepsy: "Everyone here has heard of a society lady who, when out in the world—at the theatre, at church, or while taking a walk—would suddenly utter the most serious insults, or the most obscene words, behavior of which she seemed unaware. She was a respectable woman in every way and of great intelligence." He saw this as a case of the sudden and irresistible impulses that epileptics seemed to experience: "The society lady I spoke of just now, and who, under the control of singular, irresistible influences, would utter the strangest words, of which she seemed unaware, would express out loud during her attack witty ideas that, according to convention, were usually kept silent. Even though in this case the impulse was irresistible, the veracity of the response or statement could make it seem, to those unaccustomed to the phenomena of epilepsy, that these words were intentionally uttered" (Trousseau, 1862).

A young Italian, Giulio Melotti (1857–19?), who graduated from the University of Bologna in July 1882, spent 2 years in Paris from 1884 to 1886. Melotti transcribed one of Charcot's lessons on tics and jumping, the *Lezione quattordicesima*, published in Milan in 1887 (Charcot & Melotti, 1887). There is no trace of this lesson in Charcot's complete works, and it is likely that Melotti, rather than transcribing an actual lesson, compiled several he had heard into one, using information from Gilles de la Tourette himself to give a detailed clinical picture of the tic disease to his Italian readers. He wrote, among other things, that Charcot had examined Mme de D., giving credence to this oft-repeated error. Here is what Charcot did say, on Tuesday, December 13, 1887: "In Parisian high society, there was a person of the most aristocratic circles who was known for uttering filthy words. I did not have the honor of knowing her, though I met her one day on my way up the stairs from the Salon and I was surprised to hear her suddenly say 'shit' and 'fucking pig'" (Charcot, 1887).

9. Madame D. or Madame La Comtesse Picot de Dampierre

Strangely, the true identity of Mme de D., known throughout the nineteenth century as the Marquise de Dampierre, has remained a mystery until 2020. None of the authors mentioned above explicitly named this patient.

In their journal dated January 27, 1860, Jules de Goncourt (1830–1870) and Edmond de Goncourt (1822–1896) were apparently familiar with the singularities of speech of this aristocratic woman, referring to the fine account by the painter Paul Gavarni, a pseudonym of Sulpice-Guillaume Chevalier (1804–1866): "As to the way in which Mme de Dampierre barks and spits out her thoughts, he is frightened by these utterances that toss thought out before the mediation of the will: 'It is like a fearsome leakage'" (Goncourt de & Goncourt de, 1891).

Articles in the Parisian daily press in July 1884, after the death of La Marquise de Dampierre, provide some clues. On 17 July 1884, Gil Blas reported anecdotes on the unusual personality of the "Comtesse Picot de Dampierre": "During high mass at the Madeleine, it was not uncommon to hear such perfect barking that one might have believed a pack of hounds belonging to the Duchess of Uzès was present in the church. It was, in fact, Madame de Dampierre, during the Elevation. This good, witty, charitable woman, well-bred by all accounts, could not keep her condition silent at certain times. Having a remarkable talent for producing artificial flowers and decorative frames, cut from cordovan leather, for example, she was awarded a first-class gold medal at the horticulture exhibit which took place at the Champs Élysées and the Carré Marigny under the Second Empire. The Emperor presided over the awards ceremony. When she heard her name called, Mme de Dampierre stepped up to receive her medal. But as an old Bourbon royalist, finding herself for the first time in the presence of the nephew of the "Corsican Ogre," she totally forgot the good Catholic history of Father Lariquet that she had learned as a child. Instead of receiving a large medal from the Marquis de Bonaparte's nephew, Lieutenant-General of the Royal Armies, she saw only the usurper, the despoiler of Orléans properties ["Orléans" suggesting the king and the royalty], and in response to the Emperor's courtesy, stupefying all present, she proffered a volley of insults and barking that none of the spectators were likely to forget, even twenty-five years later" (Anonyme, 1884).

So, La Marquise de Dampierre was in fact "Madame La Comtesse Picot de Dampierre." Who was she? Her birth name was Ernestine Émilie Prondre de Guermantes, born on August 22, 1800 ("4 Fructidor an VIII" according to the Revolutionary calendar) at the Guermantes family castle in Seine et Marne, east of Paris. The public registry indicates a name stripped of any link to the French nobility, owing to the French Revolution. On 02 March 1798, her father, Emmanuel Paulin Prondre de Guermantes

(1775–1800), married Eulalie de Brisay (1779–1866). The young couple had their first daughter, Albertine Adélaïde (1799–1819), on June 02, 1799. In poor health (tuberculosis?), the father died at only 25 years old, on November 03, 1800, when his daughter Ernestine was only 3 months old. A widow at age 24, Eulalie de Brizay remarried Jean-Baptiste de Tholozan, Marquis de Vernon, on November 21, 1802 and had two more children, Eulalie de Tholozan (1804–1889) and Ernest René de Tholozan (1808–1890).

Ernestine was a source of worry for her mother and step-father due to her homeliness and her behavioral problems, which Itard described as follows: "At age 7, she was overcome with convulsive contractions in the muscles of her hands and arms, especially when she was practicing her writing and would abruptly move her hand away from the letters she was forming. After shifting her hand, its movements became regular and could be controlled voluntarily, until another jolt once again interrupted her work. At first this was considered a sign of vivacity and harmless mischief, but since the behavior was repeated over and over, adults around the girl began to respond with reprimands and punishment. Soon, however, they became convinced that these movements were involuntary and convulsive, involving the shoulder, neck, and face muscles. The result was extraordinary contortions and grimaces. The disease continued to progress, and with the spasm spreading to the organs of the voice and speech, this youngster would make bizarre cries and utter words devoid of meaning, but never in a state of delirium, without any disturbance of the mental faculties." Physicians offered the hope that her condition might improve with puberty, but this did not happen. After a stay in Switzerland that seemed to bring temporary improvement, "either because of the spa treatments or because she enjoyed her time there in the mountains," she relapsed after returning home.

Her physicians advised her mother to find her a husband, since marriage was an ancestral treatment for hysteria. Eulalie de Brisay-Tholozan selected a first cousin, Louis Augustin Picot de Dampierre (1780–1841), as a potential son-in-law capable of accepting this young woman as his wife, despite her natural defects. A brave soldier under Napoleon who was wounded at the Battle of Friedland and survived the retreat from Russia, he was made "Baron d'Empire" and, at 36, aspired to a peaceful life. Thus, the fortune of the Prondre de Guermantes was nothing if not attractive to him, given his mountain of debt. Since he was Ernestine's third-degree relative, he had to request a dispensation for the marriage from the Bishop of Meaux, in these terms: "The most pressing reason is this young woman's health:

for several years, she has suffered from an acute nervous and convulsive disease. All the medical remedies used by physicians in Paris, Brittany, and Switzerland have not improved her constitution. Her doctors agree that only a change in her civil status can bring an end to the accidents she experiences. In this situation, it is improbable that she should rapidly find a husband appropriate to her rank and birth, and the idea of a delay in the consummation of the union proposed to her and approved by her, or of some other obstacle, may worsen her condition" (Archives de Seine-et-Marne, 1817). Authorization was granted and the marriage contract approved by King Louis XVIII. The marriage was celebrated at the Guermantes castle on 07 December 1817.

Ernestine's older sister died on March 23, 1819, leaving her as the sole heiress to the Guermantes fortune. Gradually she established her dominion over the castle and its occupants, being the only one to possess a master key to every door, including the gates to the grounds. The Count of Dampierre was only too willing to grant his wife this authority, a role that tied her to the castle and left him free to frequent the Parisian salons, without her and her uncontrollable speech problem. He died in Paris, on February 11, 1841. Ernestine would survive him by more than 40 years. She died in her Paris residence, at 191 Boulevard Saint-Germain (less than 200 m from Charcot's home) on July 08, 1884, at age 83, in the presence of her nephew, the Baron Henri Baillardel de Lareinty (1824–1901), a French senator from 1876 to 1901.

Ernestine's coprolalia consisted extreme outbursts of language, shocking for an aristocratic woman born during the French Revolution. Her vocabulary was at times filthy, borrowed from working class speech. Within the family, it was seen as a transgression, a comic form of the caustic judgments heard in Parisian salons. In 1847, Eugène Billod (1819–1886) provided a number of details not found in the writings of Itard. Among others, he reported that Mme de D. was sometimes able, with great effort, to restrain her compulsion to curse: "She would sometimes interrupt herself and cut off a sentence to direct an epithet at some society person, always something unsuitable and often obscene. These words of course came with a struggle, evidenced by the shame and reddening of her face, her lowered eyes, and an embarrassed, speechless demeanor in this poor lady. The onslaught of these words, for a moment withheld by an effort of will, came rushing out like an arrow released from a bow's supple cord. Thus, after having said, "*Vous êtes un foutu cochon*," [you're a fucking pig] she remained for some time in this struggle without pronouncing the injurious words which were soon deflated like a surge of feeling [...]. It is natural to suppose that this ability to only express ideas considered proper, this control of the intelligence over phonation, was impaired in this lady. The result of this impairment was that the expression of certain ideas was forced upon her; her will, this force subjected to other forces, rose up in insubordination, something that the patient wanted in spite of herself. Furthermore, this condition was not limited to the expression of certain words she could not withhold. One sometimes sees Mme de D. could at times be seen shouting, imitating the barking of a dog, or pinching or kicking her neighbor" (Billod, 1847). Gilles de la Tourette omitted to mention Billod's quotation in his seminal article.

Actually, some of her reported behaviors would not be regarded as coprolalic but would correspond to what has been described as non-obscene socially inappropriate symptoms (NOSIS) (Eddy & Cavanna, 2013; Kurlan et al., 1996). There are significant social and emotional sequelae to living with tics and NOSIS, which can adversely affect the quality of life. However, these inconveniences do not seem to have hindered Ernestine's social life, in particular after her husband's death. She may have suffered, however, from physical consequences such as discomfort from the repetitive movements and the stigma of her severe, violent, or socially inappropriate movements, vocalizations, or perhaps actions, not reported by contemporaries. Some ticcers have low self-esteem or poorer psychosocial functioning. It seems that this was not the case with Ernestine. Unfortunately, we did not find any personal handwritten testimony that could inform us about her psychological state and her own perception of her illness.

At any rate, it may seem surprising that in the accounts of Ernestine de Guermantes' eruptions in high-society circles, there is no mention of a medical diagnosis. This reality is confirmed in the *Journal inédit* of the Count Alfred of Gramont (1856–1915), a diplomat made famous by the Ems Dispatch: "This morning, General de Vacquières came for breakfast. He is very agreeable and told us many amusing stories, especially about the famous Countess of Dampierre, whom everyone used to know before her death in 1884. Deaf and always spitting, she was high-spirited and couldn't keep herself from saying what she thought out loud." Alfred de Gramont also related the exclamation "Tambour! Tambour!" that Ernestine uttered to Madame Santerre, recalling the order of Antoine Joseph Santerre (1752–1809) for a drumroll to drown out the voice of King Louis XVI on the scaffold. There was also Ernestine's introduction of three priests to her guests in her salon: "Un cu..., deux cu..., trois cu...rés de village!", where "curé" means priest, but "cu" hints at "cul", slang for ass. Other exclamations seem to lay bare an inner thought. To the Baroness of Rothschild, her neighbor who was hosting her along with other guests at her castle in Ferrières, she said: "I have an entire castle; this is merely a boutique!" (Mension-Rigau, 2011).

The Madame D. observed by Itard was thus not the Marquise de Dampierre, but a countess. Her real name was Ernestine Émilie Prondre de Guermantes, and she married the Count Augustin Louis Picot of Dampierre. This toponymic addition to her name resulted in the confusion that has surrounded her identity since 1825 (Walusinski & Féray, 2020).

10. Trousseau's depiction: 12 years before Gilles de la Tourette

Gilles de la Tourette mentioned very clearly that Trousseau had described a clinical case while treating "several types of chorea" in 1862, referring specifically to tics: "Once again, I was recently consulted by a young Englishman sent to me from Dieppe whose tic consisted in convulsive, violent movements of the head and right shoulder… These tics are in some cases accompanied by a cry, the voice bursting forth more or less loudly, which is characteristic. While on this subject, I shall mention the story that I have recounted several times, of one of my old classmates, whom I once recognized, after twenty years of not seeing him, while he was walking behind me, by the bark-like noise I had heard him make when we were students together. This cry, this yelp, this burst of voice, a veritable laryngeal or diaphragmatic chorea, may constitute the entire tic. In a few cases, it is not only a bursting forth of the voice, a singular cry; it is also an irresistible tendency to always repeat the same word, the same exclamation, and the individual may even reiterate words aloud that it would be better to hold back. These tics are quite often hereditary" (Trousseau, 1862). Trousseau precisely described a chronic disease associating multiple motor tics, and stereotyped phonic tics, given that he was able to recognize them 20 years later. He noted their familial and hereditary character, but also that this condition did not reduce life expectancy. Gilles de la Tourette vehemently criticized the diagnosis of chorea proposed by Trousseau: "Laryngeal chorea, diaphragmatic chorea—the false interpretation is too prejudicial for a description that, as we shall see, is accurate in many details." Charles Handfield Jones (1819–1890) of Liverpool also copied Trousseau's description in 1870 without adding any new elements (Handfield-Jones, 1870).

11. Gilles de la Tourette's observations

Among the other eight clinical cases described by Gilles de la Tourette, the second case was an observation by Pierre Marie (1853–1940), at that time *chef de clinique* (senior house officer or resident) in the La Salpêtrière department (1883–1884). The eighth case was observed by a former Charcot resident who became a professor in Bordeaux, Albert Pitres (1848–1928). The ninth case was recounted by Charles Féré (1852–1907), a Charcot *interne* 3 years before Gilles de la Tourette. In other words, Gilles de la Tourette prepared five observations himself. The article reveals a surprising fact concerning the geographical origin of the patients. Five lived in Le Havre, and one in Evreux, both cities in the French region of Normandy. Gilles de la Tourette noted that the patients described in observations II and VI were treated by Dr. Gibert in Le Havre. Joseph Gibert (1829–1899), founder of the journal *La Normandie Médicale,* would publish seven articles describing attacks of hysteria, between 1891 and 1893, using expressions such as "passionate attitudes"; he came to be known as the "Charcot of Normandy." His was regularly in contact with Charcot. When Dr. Gibert's son became a student of the philosopher Pierre Janet (1849–1947), Janet began to spend time at Gibert's hospital where he examined Léonie Leboulanger, a "somnambulant hysteric." The study of Leboulanger was an important part of Janet's thesis, *L'automatisme psychologique, essai de psychologie expérimentale sur les formes inférieures de l'activité humaine.* It was also Gibert who presented Janet to Charcot, which led to a fruitful collaboration. Rather than indicating a higher incidence of the disease in Normandy, this geographical concentration is more likely due to the fact that Gibert, a subtle and attentive clinician, played the discreet role of referring physician, and sent or recommended sending these ticcers to the master at La Salpêtrière. In his lesson on January 17, 1888, Charcot noted with irony: "*I don't know why the ticcers are almost always from the seaside. We have had several from Le Havre and Rochefort.*" Gilles de la Tourette went to Le Havre himself, on July 15, 1884, to evaluate the progression of the disease in a young male patient examined by Pierre Marie (obs. II) and released from La Salpêtrière a year prior. Three days earlier, during the same voyage, he had re-examined a boy who only had simple tics (obs. V). Probably on the way home, on July 20, 1884, he saw another boy already seen by Dr. Gibert, who had advised rest at the boy's grandfather's home in Rouen. This boy would not see Charcot at La Salpêtrière until October 1884.

12. "Motor incoordination"

Gilles de la Tourette's writings give the impression that he avoided using the word "tic," preferring the strange terms of "motor incoordination" or "contortion": "Muscular jerks limited to muscular groups of the face and one or both upper limbs." Was the term tic "too trivial for the rather extravagant character of Gilles de la Tourette"? As Georges Guinon (1859–1932) would note, "the involuntary movements of these patients are neither absurd nor incoordinated. On this matter, we thus cannot share the opinion of Gilles de la Tourette. What, precisely, is motor incoordination? It is still called ataxia, that is, impaired arrangement of movement... Do we find anything of the sort in our patients and can the muscular jerks that we see in them be compared in any way to these highly characteristic motor problems? No, the patients are never awkward in their voluntary movements. They flex and extend their legs forcefully, hit the ground squarely with their feet, are perfectly solid on their legs, and never miss a step. While it is true that they may stop walking to execute a tic, they do not throw their legs out to the sides, and they do not entangle them to the point of falling to the ground... The patient's movements are not incoordinated; furthermore, they are not even illogical. Overall, and in their invariable repetition, they present an arrangement that one can truly call systematic" (Guinon, 1886). Guinon's criticisms are perfectly well founded. There can be no doubt that Gilles de la Tourette proposed this regrettable term of "motor incoordination" to bring his description closer to the cases of Jumping, Latah, and Myriachit. For example, in observation II, which included all of the clinical signs of the disease, Gilles de la Tourette used the word "movements" with special emphasis (referring only once to "tics"), while highlighting the fact that "all of the emotions act very keenly on [the patient]. He jumps when called." In other words, he based his framework on the articles of Beard, O'Brien, and Hammond, and he wanted his descriptions to fit this framework as precisely as possible. Gilles de la Tourette employed the word "tic" only four times in his comments on the nine observations—a paradox, as tics were the subject of his paper.

13. Evaluating the initial description

Do the nine observations proposed by Gilles de la Tourette describe the syndrome named for him? (Kushner, 1995). In fact, no! Observation IV was a description of obsessive problems (the patient, a 24-year-old male, felt compelled to repeat a word or phrase when he attended a lecture and relieved this compulsion by repeating the word or words aloud when alone). The patient made a few involuntary movements (Gilles de la Tourette referred to jumping), but had no real tics, nor any echolalia or coprolalia. Observation V described simple isolated tics that tended to disappear spontaneously. Observation VII told the story of a young man who had felt anxious since the traumatic events around the siege of Paris in 1871 and who exhibited a few transient tics. He would startle when surprised, similar to the Jumpers of Maine, but nothing more. As to the ninth and last observation, it involved delayed pubescence, a case of "infantilism" according to Gilles de la Tourette, with rhythmical movements that did not resemble tics and with no echolalia. Sometimes this male patient would utter curse words, but without real coprolalia.

The first three observations, particularly the one recounted by Itard concerning Mme D., perfectly described the pathology that Gilles de la Tourette was attempting to introduce in the neurological nosology, with all clinical signs present: motor and phonic tics, associated echolalia, echokinesis, and coprolalia. For Itard, the symptoms he described did not merit a separate disease classification and he assumed that, with his general framework, the case of Mme D. expressed a clinical form of chorea. Although Gilles de la Tourette appropriated Itard's depiction of Mme D. as the first example of the disease he wished to construct, his conclusions were the opposite of Itard's. He argued that Mme D. merited a distinct classification from chorea.

The observation taken from Pitres was also entirely characteristic, but oddly enough, Gilles de la Tourette spoke of choreoid tics and thus muddled, involuntarily of course, the demonstration by which he hoped to isolate "his disease" from chorea. For observation VI, Gilles de la Tourette used the word "tic" twice in his description for eye blinking. The child would say "oh, oh!" but without any other symptoms, and the progression seemed spontaneously favorable. So, this case may have been nothing more than simple transient tics.

In total, four of the observations were indisputably cases of Gilles de la Tourette syndrome. In his discussion, Gilles de la Tourette perfectly described the childhood onset: between the ages of 6 and 16. He highlighted the "utterly slow and insidious" progression, with its periods of remission and exacerbation. He also clearly reported the capacity of patients to temporarily inhibit the exteriorization of their tics, and the subsequent rebound: "He could refrain from all his abrupt movements and stop them completely. But, immediately following this cessation, which required of him the most violent constraint of mind, the movements would return with unusual violence." Gilles de la Tourette insisted on the echolalia: "This is one of the most constant symptoms of their condition, and one that, in particular, most impressed the foreign authors. It is one of the first symptoms to appear after the motor incoordination and seems to be one of the last to persist during quiescent periods.... This faculty of imitation, this echolalia of movements and actions is thus, in certain cases, very complete in these patients." For the involuntary imitation of a movement, Charcot coined the term "echokinesis." Pierre Marie spoke of "echomatisme." A clinician from Lyon, Claude Sigaud (1862–1921) drew on the work of the Russian Ivan Mikhaïlovitch Setchenoff (1829–1905) "on the reflex actions of the brain (Сеченов Setchenov, 1871), and focused in this thesis defended in 1889 on showing that echomatism was an automatic brain function (Sigaud, 1889). He compared it to a reflex in the motor domain, thereby presaging the recent discovery of motor neurons and the function of decoding intentionality that they enable (Hurley, 2005).

Gilles de la Tourette insisted on the pathognomonic nature of coprolalia (his coinage) stating that "it is never observed in other conditions... Not only do patients utter obscene words; it also seems that they may also exhibit echolalia, which they are often reliant on. However, "coprolalia is not found in all subjects, in the same way that echolalia is not always present." Finally, "the tics are often hereditary," implying "degeneration," which caused neurological diseases according to the Salpêtrière School (Walusinski, 2020). For the prognosis, Gilles de la Tourette expressed a pessimistic view, but he could not, at that time, understand the progression over a lifetime: "We have never seen the disease disappear completely; it may improve, but patients are not cured, and furthermore, an improvement may merely be the prelude to a new exacerbation."

14. Championed by Charcot

On July 19, 1885, Paul Le Gendre (1854–1936), a colleague and friend of Gilles de la Tourette, gave readers of the journal *L'Union Médicale* a complete description of the disease, entitling his piece "Gilles de la Tourette disease." He began as follows: "As the ancient proverb says, 'It is not for every man to go to Corinth'... and it is even less the common fate to bring a new disease to the baptismal font. While we do not know whether our colleague Gilles de la Tourette has been favored in the first case, in the second, he has most certainly deserved his good fortune, by his skillful historical evaluation and his wise clinical studies in which he pulled from the chaos of chorea a distinct morbid entity, claiming for it a place of its own in the nosography. Mr. Charcot—and there is no better judge in such matters—declared in a clinical lesson this year that the nervous disease described by Gilles de la Tourette in *Les Archives de Neurologie* has special characteristics definite enough to legitimize the naming of this disease after our friend. It would be disrespectful on our part not to adopt the master's position, regardless of whether one is of the opinion that naming diseases for the person who first described them, or who appears to have done so, may be problematic, and we are among those who hold this opinion" (Le Gendre, 1885).

15. Georges Guinon, the forgotten co-author

Georges Guinon (1859–1932) took Gilles de la Tourette's place as Charcot's resident for 1885 (Walusinski, 2021a). As early as January 1886, just 1 year after his predecessor's seminal publication, Guinon published four new observations in *La Revue de Médecine*. Charcot's fundamental role, albeit hidden, cannot be denied; the master sought to enhance the description of this new entity and situate it precisely in the neurological nosology: "We wish to thank Mr. Charcot, who was the inspiration for this work and helped us with his generous advice." Two cases were seen by Charcot during 1885 and were reported by Melotti.

Who was Guinon? His career was surprising. The meeting with Charcot was crucial, as it was for Gilles de la Tourette. Léon Bernard (1872–1934) wrote in his 1932 homage to Guinon: "In all aspects of his person and his emotional as well as his intellectual being, Guinon had been absorbed by the famous personality, whose esteem he considered as the honor of his life. Its memory remained alive for him, like the memory of a great love. Guinon was a simple and straightforward soul. When Charcot suddenly passed away, he faltered. Everything he had planned, all high hopes for his life came crashing down; he left" (Bernard, 1932). The deep bereavement gave his career an unexpected and unique turn, which was unusual for a brilliant student of Charcot. Shortly after his teacher's death, he left for Normandy where he worked as a family physician. Later he set up his practice near the Douarnenez harbor in Brittany, where he was known as a discreet and extremely devoted physician until 1919, comforting families that had lost their loved ones during World War I. In 1919, he joined Léon Bernard and the fight against tuberculosis by becoming a medical inspector for the office of public hygiene in the Paris region; the new Léon-Bourgeois law had set up such offices in every French département. Guinon had defended his thesis on the agents that provoke hysteria in 1889, with Charcot presiding over the jury (Guinon, 1889). He wrote up the master's lessons from 1889 to 1893, succeeding Gilles de la Tourette as Charcot's personal secretary. In 1887, he wrote the article on convulsive tics in the monumental *Dictionnaire Encyclopédique des Sciences Médicales* of Amédée Dechambre (1812–1886) (Guinon, 1887a). In it, he noted that Gilles de la Tourette "did in fact describe the convulsive tic disease, as we have shown in a work on this condition"; he did not refer to motor incoordination. Guinon clearly stated: "What is found, most often in isolation, are firstly eye blinking and then the alternating opening and closing movement of the mouth." He then insisted on the systematic nature of involuntary movements: "In everyone, we find a few identical tics, such as touching one's beard and spitting."

16. "Idea tics"

Acting on Charcot's wise clinical advice, Guinon added to Gilles de la Tourette's description the frequent presence of "idées fixes," which Charcot called "idea tics": "Mr. Charcot called our attention to the existence in our patients of a series of psychic phenomena that we have not found noted in other, similar observations and that, if we refer to the cases we have studied, must be quite frequent in the serious forms of the tic disease. These phenomena are *idées fixes*. In absolute terms, outside the condition that interests us, *idées fixes* constitute a chapter of mental pathology, as is well known. The extent to which they can vary is common knowledge, and to cite just a few examples, we mention *la folie du pourquoi* where the patients cannot resist asking the reasons for absolutely insignificant things; *la folie du doute avec délire du toucher*, which differs slightly from simple *idées fixes* in that it leads to truly delirious ideas; arithmomania or the mania of numbers and calculations; and onomatomania, recently described by Mr. Charcot and Mr. Magnan" (Charcot & Magnan, 1885). In 1838, Jean-Étienne Esquirol (1772–1840) had included these symptoms together with delirium in a large group: *monomanie raisonnante* or intellectual monomania (Esquirol, 1838). *La folie du doute* was isolated from "monomania" by Henri Legrand du Saulle (1830–1886) in 1875 (Legrand du Saulle, 1875) and corresponds to modern-day obsessive-compulsive disorder. Guinon intuited a common pathophysiological mechanism for the disease's various symptoms: "It is now easy to understand the link between all of these phenomena—tics, involuntary words, *idées fixes*. All three appear to be of the same family. The cause is the same; it is merely a question of degree: (1) simple convulsive or impulsive motor representation producing facial grimaces or tics in the limbs; (2) a more complex convulsive motor representation resulting in utterance of a sound or pronunciation of a word or phrase; and (3) ordinary convulsive or impulsive ideas, either resulting in more or less complicated actions, such as counting and arranging objects on a table, or remaining limited to a conscious fact without external manifestations, such the perpetual feeling of unmotivated fear, which we have noted in one of our patients." Through and with the help of Guinon, who added to the initial work of Gilles de la Tourette, Charcot built the complete clinical picture of the disease as we know it today. When he was finished, Guinon proposed a new term: "Convulsive tic disease, the name which, according to Professor Charcot, would be most appropriate for this condition." This brief remark is colored by the thinly veiled spite that Guinon felt after Charcot initially named the disease for Gilles de la Tourette.

Guinon would discuss the disease again the following year, but this time he formulated a personal opinion, inspired by Briquet, as a sort of challenge to Charcot (Guinon, 1887b). Tic disorder and hysteria could be associated and could be difficult to distinguish, given the points they had in common: "Hysteria, among its numerous manifestations, also includes involuntary movements, in which we may include coughing, hiccupping, barking, and involuntary exclamations. These, as we have shown elsewhere, are nothing other than slightly more complex motor phenomena ... If we compare these phenomena to those observed in the convulsive tic disease, it is easy to see that objectively, the difference between them is far from distinct. Furthermore, as with tics, hysterical involuntary movements recur in fits.... It is not the movements themselves that will provide the nosological characteristic for classifying each case appropriately, but rather the phenomena of all sorts that accompany them most of the time. These phenomena are all different depending on whether the subject is a hysteric or suffering from convulsive tics. Echolalia, observed in the two diseases, constitute an exception. And in the tic disease, coprolalia is often noted, along with *idées fixes, folie du doute*, etc., which characterize delirium in hereditary degenerates. On the contrary, in hysteria, the well-known mental state of the patients is entirely different; most especially, neurosis will have left its mark on the patient in more or less of a lasting way. The observation of hysterical stigmata is thus critically important." Guinon presented three new observations, including one given to him by Charcot (Guinon (1887b). He prepared the two others during his year as a resident under Émile Troisier (1844–1919) at Hôpital Saint-Antoine in 1886. The first involved a hysteric girl with anarchical movements followed by prolonged lipothymia. The second and third cases were clearly convulsive ticcers in whom Guinon observed inappropriate emotional states and hemianaesthesia accompanied by a reduced visual field, which to him indicated hysteria. The age of onset, much earlier for tics, together with the fact that they progress over a much longer period, are arguments for differential diagnosis. "Here we have the convulsive tic disease in hysterical subjects. And why couldn't these two neuroses be associated? Heredity, which creates in and of itself the tic disease, plays no small role in the etiology of hysteria. Moreover, don't we see similar morbid associations every day? Epilepsy often combines with hysteria to form the disease called 'hysteroepilepsy with separate crises.' Similarly, there are cases in which hysteria or the convulsive tic disease develops where the other already exists."

17. Later publications in France

Maurice Lannois (1856–1942), a native of Lyon and founder of the Société française d'Oto-Rhino-Laryngologie, published his thesis on the nosography of chorea in 1886 (Lannois, 1886). In a chapter entitled "pseudo-chorea," he gave a complete, accurate clinical picture of the disease and added an observation from La Salpêtrière of a little girl whose explosive movements started at age 7: "She is constantly crying out with a little, contained 'hum!' During fits, there is a moment where she cannot utter anything, then suddenly a barrage of words ushers forth and she has to cry out, unable to prevent herself, such that she is heard throughout the house. At this point, she sometimes cannot stop vulgar language from slipping out, the word 'shit,' for example." Lannois wrote very clearly that it was Charcot who enriched the initial work of Gilles de la Tourette by his description of "idea tics," making no mention of Guinon.

18. Clinical cases presented during Charcot's lessons

The *Leçons de Charcot*, published in 1893 and compiled by Guinon, included an "appendix" transcribing two lessons from 1886 that were omitted from the third volume of the master's works. One of them, on "hysterical sounds," included a clarification on the differential diagnosis of hysteria, convulsive tic disorder, chorea, and paramyoclonus (Charcot,

1893). The two briefly described cases of convulsive tic disorder, already presented to students on an earlier occasion, were probably those that Melotti referred to in his article. Shortly thereafter, Charcot incorporated young patients with the disorder in his Tuesday lessons. On December 13, 1887, Charcot presented "a young fellow aged 12 or 13," noting that he had a tic. "In front of you, he is holding back, but from time to time, he blinks and has contractions in the lower limbs which result in certain involuntary movements. Now and again, he strikes the ground with his foot. If he only did that, it would be nothing serious, but sometimes he makes a sort of grunting noise: hugh! hugh! hugh! And he presents the phenomenon of coprolalia. What is coprolalia? If you are a little ticklish regarding the value of certain words, cover your ears. Coprolalia is the irresistible urge to pronounce the word that Victor Hugo put in the mouth of Cambronne at Waterloo and which, according to him, should take the place of the heroic phrase popularized by the legend; that is to say, the word: ... [shit]. What does this mean? you ask. Hasn't this child been raised to be polite? But of course, he has been raised as children should be. This word, he has heard it pronounced, but then, in the street one hears so many words not of one's vocabulary! Indeed! He utters it continually, in spite of himself, by impulsion. We have often seen this condition in children. Mr. Gilles de la Tourette and Mr. Guinon have dealt with the question; this is the tic disease. When someone suffers from this disorder, he develops a series of phenomena—some of them psychic, others physical—that we have observed to combine with each other" (Charcot, 1887).

On January 17, 1888, Charcot examined a 17-year-old boy—another native of Normandy! Making the boy write and execute several movements, Charcot lectured on the differential diagnosis between "vulgar" chorea and convulsive tic disorder. His treatment consisted in hydrotherapy, "*teinture de Mars,*" (*teinture de Mars de Zwefler* or aromatic tincture of iron acetate), and "*vin de Colombo.*" On Tuesday, January 24, 1888, Charcot briefly presented those in attendance with two other adolescents. On February 21, 1888, Charcot began by telling a 21-year-old man who suffered from tics, "Moreover, you know that having tics in a family is not without consequences. Tics are a special mark." What followed was a thorough interrogation on personal and family antecedents. The patient, it turned out, also had a *folie du doute*, characterized by the fear of rabies at the sight of an animal. He had two brothers and a sister with tics: "When there is one ticcer in a family, it is rarely an isolated case."

In his Tuesday lesson on October 23, 1888, Charcot clearly contradicted Gilles de la Tourette: "The movements of ticcers, however complex and bizarre they may be, are not always and essentially disorganized, incoordinated, or contradictory, as is too often believed. On the contrary, they are generally systematic, to the extent that they always recur in the same way in a given subject. Moreover, very often at least, they reproduce, albeit exaggeratedly, certain automatic physiological movements that have a purpose." During this lesson, Charcot credited Gilles de la Tourette with the coinage of the word coprolalia: "This is the remarkable phenomenon, among all of the others, that Gilles de la Tourette, in his interesting work on the tic disease, ingeniously designated using the term 'coprolalia.'" At the Tuesday lesson on June 4, 1889, Charcot refined the semiological description: "The complex tic movement is not absurd in itself. It is absurd and illogical because it is executed outside of a context, without an apparent reason. The act of scratching oneself occurs without an itch, the eyes blink in the absence of any foreign object, and so forth. In addition, tic movements are abrupt, rapid, momentary, and do not have, for example, the slowness of choreic gesticulations. They are not continuous but supervene by fits repeated at greater or lesser frequencies and lasting for shorter or longer periods, fits that patients can often, for a time, stop by an effort of will. It is also often the case that while jerking and grimacing, patients utter exclamations and entire words, quite often filthy words. You will have accumulated such specific clinical characteristics that you will be able to distinguish with certainty the tic disease from all other types of convulsive conditions" (Charcot, 1889).

The last Tuesday lesson during which Charcot mentioned "convulsive tic disease" took place on June 4, 1889. He was once again teaching the difference between tics and chorea by presenting the students with a ticcer alongside a chorea patient (Huntington's chorea), both female. By way of introduction, he declared: "You see to what extent the electric jerks of the ticcer differ profoundly from the slow and permanent gesticulations of subjects with Sydenham's chorea. Between tics and chorea, the distance is vast: do not forget this since these conditions are sometimes wrongly given the same name and the prognosis is quite different." Charcot showed the opposition between the speed of the movements and highlighted the constancy of choreic writhing, whereas tics stop intermittently. The chorea patient made no sounds, whereas the ticcer "makes, from time to time, expressive laryngeal noises." The movements of the latter were chaotic and unpredictable, whereas in the former, everything was rhythmical, regular, and systematic. The ticcer started having tics at the age of 12, whereas the other patient was "over 51" at the time, with an age of onset at 33. "Chronic chorea is considered an incurable disease; its outcome is fatal, and the therapies are unable to stop its relentless progress, not even for a moment. This is not quite the case with the tic disease. It is also one of the most difficult conditions, to the extent that we cannot say if we ever actually cure it, but we can count on periods of cessation, often quite long, and on a certain waxing and waning, either spontaneous or brought about by appropriate therapy. Hydrotherapy, rational gymnastics, and isolation are among the approaches that may be helpful in such cases. You can now recognize the value of learning to distinguish between these two morbid states, for which the prognosis is so different."

19. Breitman introduces degeneration theory

On November 28, 1888, Grégoire Breitman (1859–1914), a physician from the city of Ananyiv, a part of Moldavia at the time, currently in the Ukraine (160 km north of Odessa), defended the first thesis on Gilles de la Tourette syndrome since the seminal article's publication, with Charcot presiding over his jury: "Alienists such as Magnan and his students Legrain and Saury see echolalia, coprolalia, and echokinesis as an episodic syndrome of *la folie des dégénérés*" (Breitman, 1888). Paul-Maurice Legrain (1860–1939) and Honoré Saury (1854–1906), disciples of Valentin Magnan (1835–1916), introduced the concept of *bouffée délirante* (brief psychotic disorder) into the psychiatric nosology. This was after both had done work on *la folie des dégénérés* (insanity in degenerates), Legrain writing his thesis (Legrain, 1886) on the subject and Saury a book (Saury, 1886). Breitman, a student alienist at the Vaucluse asylum near Paris (currently the Hôpital de Santé mentale de Perray-Vaucluse, in Epinay sur Orge), wanted to study the heredity of involuntary imitation symptoms in order to clarify the role of "dégénération" in their genesis. This had been the accepted etiology of neuropsychiatric diseases since the work of Bénédicte-Augustin Morel (1809–1873): "As the semiology of hereditary degenerates grows more detailed, exact, and manifest, we are relegating the search for the original flaw in the various forms of alienation to a secondary position"; given that "the aim is to find out whether, by studying the semiology of each psychic state and by grouping the signs into syndromes, we can, by deduction, demonstrate the manifestly degenerative hereditary origin." He explained that "under the combined influence of peripheral and central excitations, we will see degenerates, these future candidates for mental alienation, executing a series of bizarre actions, designated as 'Latah, Jumping, and Myriachit' by foreign authors." Breitman showed, based on several observations he had collected, that echolalia, coprolalia, and echokinesis can be seen in various types of mental alienation, especially in delirium. He coined *égoécholalie* and *égoéchokinésie* for when a patient would respectively speak to himself in front of a mirror and control his movements while watching himself. For Breitman, these symptoms were related "by the fact that they were impulses arising from an obsession." And he categorically concluded that "echolalia, coprolalia, and echokinesis with tics can constitute a perfectly independent morbid syndrome with a determined course, that is, tic disease or Gilles de la Tourette disease. Patients with this disease are degenerate" (Walusinski, 2020).

20. "Enhanced Habit Formation"

In 1888, Ferdinand Burot (1846–1921), a Navy physician, member of the Académie de Médecine, professor at the École de Médecine de Rochefort, published his essay on the treatment of convulsive tic disorder by persuasion. He treated the young female patient studied and described by Pitres, Charcot, and Gilles de la Tourette (obs. VIII). This was the beginning of the psychiatrization of the disease. Burot concluded: "Convulsive tic disease is the result of a specific cerebral excess. It is a psychic condition almost exclusively in the domain of moral treatment. This entails training the brain to operate using special instruction and brain exercises. Persuasion, which is nothing other than suggestion without sleep, is an important therapeutic agent, albeit a methodical one. It acts by increasing tenfold the will of the subject, who is then able to moderate and dominate his impulses" (Burot, 1888a) A few months later, Burot reported on the effect of his treatment, which he considered very successful, with the near-complete disappearance of all symptoms: "The purpose of moral gymnastics is to destroy these vicious associations, these organized habits, to moderate the activity of the automatic sphere, and to shift the balance to benefit the psycho-intellectual sphere, which should predominate" (Burot, 1888b).

On November 16, 1888, in Bordeaux, Jean Urbain-Joseph Chauvreau (1863–?), a student at the École de Médecine de la Marine in Rochefort, defended his thesis, focused on Burot and Pitres, the latter presiding over the thesis jury (Chauvreau, 1888). He started by describing Pitres's observation, then presented the case of a youngster he had personally observed. The child's multiple motor tics began at age 3 and his echolalia and coprolalia, entirely characteristic, at age 5. Chauvreau's third observation had been presented in Toulouse, at the 1887 conference of the Association française pour l'Avancement des Sciences, during the discussion following Burot's presentation on his attempts to use persuasion treatment. During this conference, Émile Duplouy (1856–1896), a Navy physician, later director of the Navy health department in Rochefort, reported on the case of a distinguished navy officer whom he had treated for 4 years. "Even before entering the École Navale, this officer had a convulsive tic characterized by the sudden and involuntary projection of the upper limb, with convulsions in the pharynx muscles and a guttural cry in all ways similar to a bark. These phenomena would occur when the officer approached one of his superiors, which led them to slap him frequently, at which point inappropriate words would escape the officer's lips. Hydrotherapy completely failed against this convulsive tic. The patient, always very energetic, was only able to overcome it with time. Without wishing to discourage Mr. Burot in the treatment by suggestion he is pursuing with praiseworthy perseverance that has already been rewarded by notable improvement, it is Mr. Duplouy's opinion that the aforementioned officer's cure is above all due to his advancing age. Very pronounced in adolescence, the disease, which Mr. Duplouy readily linked to one of these most bizarre manifestations of hysteria that are sometimes observed in men, improved as the subject grew closer to the middle period of his life, and ended as soon as he had passed it" (Burot & Duplouy, 1887). Chauvreau then included two observations provided by Gilles de la Tourette, and that of Itard concerning the Marquise de Dampierre. After reviewing the clinical methods developed by Gilles de la Tourette, Guinon, and Charcot, he discussed at greater length echolalia, also observed "during minor nervous accidents" and in hysteria, accompanied by cries and "barks."

21. Catrou, Gilles de la Tourette's student

On March 6, 1890, with Charcot presiding over the jury, Jacques Catrou (1865–?), a former non-resident medical student, defended his thesis on convulsive tic disorder, in which he included Jumping, Latah, and

Myriachit (Catrou, 1890). He started by summarizing (nearly copying) the entire beginning of the seminal article, then cited Charcot's lesson "recorded by Mr. Melotti of Bologna" from June 1885: "In lesson XIV, after reviewing the history as we have already done, Mr. Charcot validated the word 'coprolalia' coined by Mr. Gilles de la Tourette and replaced the term 'motor incoordination' by 'convulsive tic.' 'There are,' he stated, 'three main elements in Gilles de la Tourette disease: the tic, echolalia, and coprolalia'... Mr. Charcot also introduced in the description, a symptom of prime importance, involving the mental state of the subject. In these patients, one very often observes what he calls an 'idea tic,' such as *la folie du doute, le délire du toucher*, misophobia, arithmomania, and so forth." Catrou added 28 unpublished observations that Charcot and Gilles de la Tourette had sent him. He then proceeded to analyze all of the 45 observations published to date. His first observation was taken from the letter of Yankowsky in the Russian medical journal *Vratch* ("the physician"). Yankowsky noted that he had worked in eastern Siberia. There he read an article on the "Miryatschenye." "The word 'miryachit' or 'meryajet' means in ordinary language 'to act foolishly.' The person who acts in this manner is called 'meryascha.'" He recounted that one evening, he examined 14 soldiers "in bizarre positions. Some were walking, others were lying down or sitting, and all of them continued doing so, even in my presence. I asked them, 'What's wrong with you?' and all of them said in unison, 'What's wrong with you?' I asked them, 'Are you sick?' Response: 'Are you sick?' and so on. All of them answered each question together, repeating the same question. The commander of the company then arrived and noted that all of these soldiers had eaten potatoes with hempseed oil [hempseed oil is thought to contain only very small quantities of euphoria-inducing substances] purchased from an inhabitant of Korea. Hearing the word 'oil,' all of them began to repeat 'oil, oil, oil,' in different tones. Neither exhortations nor orders could prevent the patients from repeating the words pronounced by anyone present." In addition to the general hilarity, Yankowsky noted either mydriasis or miosis in the patients' pupils. He began to suspect the oil, then learned that four soldiers, who ate the same oil but had no symptoms, did not see the oil seller, who, he was informed, was a 'meryascha.'" This explained for him the epidemic of behavioral problems, "which proves that 'miryachenie' is contagious." This account confirms the hysterical nature of Myriachit (spelling used by Gilles de la Tourette). Catrou then reviewed all of the symptoms described by Gilles de la Tourette, but his thesis took a highly original turn when he argued against the theories attributing the etiology to "degeneration": "We do not subscribe in any way to the current opinion whereby all nervous pathology falls under mental degeneration, aside from conditions where a physical lesion has been demonstrated."

22. Publications outside France

Thomas Carlton Railton (1844–1922), in Manchester, published an observation in April 1886 of a man approximately 40 years old with abnormal movements since the age of seven, making him an entirely typical case of the condition: "When the lower extremity was attacked, he often kicked the toe of his boot vigorously into the ground, and usually tapped his forehead with his right hand and turned up his eyes at the same time. On other occasions he would perform the following series of movements: first, holding out his right hand, he supinated and pronated the forearm two or three times, digging his elbow into his side with each supination; then he raised his right knee and patted it several times; and lastly, he struck his nose, forehead, or the top of his head with his hand, opening his mouth and turning up his eyes, and accompanying each tap with an inarticulate cry.... During this period, the cry was perfectly inarticulate: sometimes it was almost a squeak, sometimes like ahem, as in a clearing of the throat. It never occurred except at the time of one of his involuntary movements. The movements varied greatly, but a particular set seemed to predominate for a certain time, then give way in turn to another series. For some time, he was in the habit of making a peculiar noise with his tongue and lips, like the sound of spitting out some small thing from between the lips. Probably the disease had slowly reached this stage in the course of over 20 years, and it remained much the same for the succeeding 3 years of my acquaintance with him, sometimes seeming worse, the movements being larger, more perceptible, and more frequent in their occurrence; sometimes they improved, amounting to only to a slight occasional jerk of the hand, or a tapping of the forehead.

"In 1882, I observed that a new phenomenon had supervened, rendering his complaint much more seriously inconvenient. Instead of the inarticulate cry, the movements were accompanied at times by a muffled sound, the gist of which could not be distinguished when it was first heard, but which became resolvable by its repetition into words never used in polite society. He would be painfully conscious of the effect upon his audience of the word or words he used and would endeavor to turn off the obnoxious sound by continuing it either into a humming of a bar of music, or as a strenuous cough... Such attacks formed a most painful contrast to his surroundings and to his manners, which were those of a cultivated gentleman, who, in conversation, never made use of obscene or improper language. Mr. de la Tourette, who has collected nine cases of this disease, has applied the term 'coprolalia' to this symptom, from copro = filth and lalia = to talk; and he considers that it is pathognomonic of the disease" (Railton, 1886).

In September 1885, Dimitri Yankowsky published a note in the journal *Vratch*. Jacques Roubinovitch (1862–1950), born in Odessa and a resident in asylums around Paris at the time, translated the note for Gilles de la Tourette, as Jacques Catrou noted in his thesis. In July 1886, two American physicians, C.L. Dana and W.P. Wilkin, reported the first American case (Dana & Wilkin, 1886). Incidentally, they referred to another Russian publication: "In March 1886, Dr. I. I. Pautynkhoff reported (*Russkaya Meditizina*, March 16) 'a case of imitative disease,' resembling that of Tourette's, the patient showing echolalia and echokinesis (imitation of speech and movement). The patient had deficient intelligence." They started with a discussion of echolalia and coprolalia: "It was then noticed that when sitting quietly he would suddenly and involuntarily burst out into expressions of the most profane and obscene nature; repeating them rapidly for a few movements and then stopping. A surprise or sudden noise of any kind was liable to bring on an outburst. He did not seem to understand the significance of the words used... The word he used oftenest, repeating it rapidly and automatically, was the equivalent of the French word 'merde,' which is the word quoted by Tourette as most frequently expressed by his patients. He used to be much worse mornings, and before or during his toilet would regale the house with his automatic blasphemies so loudly as to disturb the neighbors." They continued with a description of the abnormal movements: "When examined by us he was observed to be anemic, but fairly well grown and nourished. He has choreic movements in the face, and slight convulsive movement in legs and arms. When he is making a strong effort to repress the obnoxious words, or when he is under any restraint, the movements are more prominent. The choreic movements, as stated, affect his face and shoulders most. He had also a kind of expiratory spasm, producing a noise like a half-developed cough. If more marked, it might be put down as an illustration of laryngeal chorea." The authors then gave the complete clinical picture of the disease as Gilles de la Tourette described it in his article and repeated one of his key points: "The coprolalia is pathognomonic of the disease, according to Tourette." As early as 1886, Gilles de la Tourette's name was being amputated by North American authors!

On July 27, 1889, at the Friedrich-Wilhelms-Universität in Berlin, Otto Müller (1865–?), a student of Hermann von Helmholtz (1821–1894) and Emil du Bois-Raymond (1818–1896) defended his thesis on tics, with as jury president Cölestin Slawyk, from Spandau, a pioneer in lung surgery (Müller, 1889). He noted that the work of Gilles de la Tourette had generated little interest in Germany, aside from a thesis defended in 1885 by Theodor von Färber, a student of Hermann Oppenheim (1857–1919) (Färber, 1885). Müller presented four cases of motor and phonic tics. He elaborated on Guinon's discussion of the link with hysteria and considered tics as a disease of the will, for which he recommended the use of morphine and arsenic.

Heymann Hirschfeld (1860–?) described three patients with motor and phonic tics associated with echolalia and echokinesis in his thesis defended in Berlin in 1891 (Hirschfeld, 1891). His discussion centered on the differential diagnosis of chorea, hysteria, and paramyoclonus multiplex, a pathology described in 1881 in Germany by Nikolaus Friedreich (1825–1882) and analyzed by Pierre Marie (1853–1940) in *Le Progrès Médical* in 1886. Parmyoclonus multiplex is a pathology similar to Morvan's fibrillary chorea (myokimias explained by the presence of antibodies disturbing the functioning of potassium channels) (Walusinski & Honnorat, 2013).

Richard Rawdon Stawell (1864–1935) studied at the National Hospital for Diseases of the Nervous System, Queen's Square, and the Great Ormond Street Hospital for Sick Children in London. In 1895, Stawell used the expression "habit spasm" coined by Richard Gowers (1845–1915) to described two cases of convulsive tics (Stawell, 1895): "These movements are not choreic in nature." He maintained that "correction of errors of refraction and the removal of post-nasal growths have completely been successful in putting a stop to the disordered movements." For him, it was important to stop these types of movement disorders as soon as possible because following late treatment "the prognosis of this affection is uncertain; it does tend, however, to be bad."

In 1899, Georg Köster (1867–1932) provided two entirely characteristic observations (Köster, 1899). And in the 1899 medical treatise directed by the Englishman Sir Thomas Clifford Allbutt (1836–1925, inventor of the clinical thermometer), James Samuel Risen Russell (1863–1939), born in French Guiana, paid homage to Charcot, Gilles de la Tourette, and Guinon in his chapter on tics, completely supporting their point of view (Risen Russell, 1899).

23. North American contribution

The Canadian William Osler (1849–1919) gave a lesson at Johns Hopkins Hospital in Baltimore, published on October 11, 1890, during which he presented a remarkable observation of a 13-year-old girl with all of the symptoms described by Gilles de la Tourette, including coprolalia, but not obsessional ideas. Osler, nonetheless, described them using a word coined by Charcot, arithmomania, and suggested *la folie du toucher*. He pointed out the difficulties of translating expressions that are apparently similar but have different meanings in French and English: "There is a curious disease or perhaps, more correctly, symptom group, met with chiefly in children, to which attention has been called of late by French writers, which is characterized by irregular, spasmodic movements, the utterance of involuntary explosive sounds or words, and mental defects of various sorts. It is not a very common affection in this country, and I take this opportunity to bring to your notice a case which we have been studying for the past few weeks. The cases have usually been described as chorea, or 'habit-spasm,' both of which conditions are simulated very closely by the irregular movements; certain instances also have been reported as hysteria. Unfortunately, Charcot and his pupils, Guinon and Gilles de la Tourette, have given to this affection the name "Maladie des tics convulsifs." I say unfortunately, for here and in England we use the term Convulsive Tic to characterize a totally different affection, involving usually the facial muscles and of either central or peripheral origin, but not necessarily coming on in childhood and not characterized by the other features presented by the disease of which we are at present speaking; and thus it happens that if we turn to the recent editions of French books we find under "tic convulsif" a disease very different from that described by the same name in English and American works" (Osler, 1890).

In 1897, James Cornelius Wilson (1847–1934) from Philadelphia published the case of a school-boy 15 years old who presented a series of movements recurring at short intervals: spasmodic twitch of the muscles of the face, rapid opening and closing of his, "bending forward, grasping a convenient chair or table for support, or more frequently resting his hands upon his knees. His head is drawn forward, his body strongly curved, his thighs and knees flexed, his feet separated. His hands tightly grasp the object upon which he rests, or the lower part of his thighs just above his knees. His head and the upper part of his body are violently moved backward and forward. After a series of such oscillations, numbering from five to eight, each accompanied with an inarticulate sound h-m, h-m-m, h-m-m-m and during which respiration is very shallow and incomplete, he straightens up, regains his breath, and the paroxysm ends with a long, sighing inspiration" (Wilson, 1897). Wilson's conclusions supported the demonstration proposed by Gilles de la Tourette: "This case belongs to the group of nervous disorders which include the "Jumpers" described by Beard, the Latah of Malays, the Myriachit of Siberia and which were demonstrated by de la Tourette in 1884 to be one and the same affection."

Otto Lerch (1855–1948) of New Orleans (Louisiana) after comprehensive definitions of what were new words at the time (convulsive tics, echokinesis, echolalia, and coprolalia) submitted an observation of a German emigrant, unusual due to the age of the disease's onset. This man, 55 years old, residing in the U.S. for 21 years, had demonstrated "spasm" for 3 years: "The patient opens and closes his eyes and rolls his eyeballs. His strongly drawn backward to return with ease to the natural position. Now and then the movement is so violent that the whole trunk becomes involved and is twisted to the right or the left. The motions are frequently accompanied by loud, noisy, belching, especially morning and night" (Lerch, 1901).

24. Elsewhere in France

Grégoire André (1844–19?), chief physician at the Toulouse hospital, noted in 1891, "Charcot added an important chapter to the symptomatology of convulsive tics, showing that in severe forms there were specific psychic phenomena, or *idées fixes*. These are what Mr. Grasset calls the 'psychic stigmata of the tic disease' or 'psychic tics.'" André described a man of 45 with motor tics in one arm, which he did not complain about, but he was very disturbed by his obsessional ideas, by "searching for accurate names" or onomatomania [a term coined by Charcot and Valentin Magnan (Charcot & Magnan, 1885)], and by hypochondriacal ideas. The clinical picture was clearly and primarily psychiatric. While Guinon was cited, Gilles de la Tourette was not mentioned (André, 1891).

In 1892, Raoul Brunon (1854–1929) from the city of Rouen, a friend of Charles Féré, proposed five observations of ticcers in the journal *La Normandie Médicale*. He wanted to show that many cases could be cured using a treatment that involved, every morning, "cold lotion all over the body then rubbing with a massage glove and pills containing belladonna extract and reduced iron (in increasing doses, up to 12 centigrams per day); a laxative every three days; physical exercise out of doors; and fewer hours of study at school." However, his succinct descriptions leave one to suspect that none of his patients had Gilles de la Tourette syndrome (Brunon, 1892).

The alienist Jacques Roubinovitch published an observation in 1893 of a female patient hospitalized in the department of Auguste Voisin (1829–1898) at La Salpêtrière who exhibited the complete clinical picture of Gilles de la Tourette syndrome. She had a large callus on her forehead from striking herself involuntarily. Her obsession was to break objects, driven as she was by irresistible clastic impulses. Roubinovitch named her condition "krouomania." He subscribed to Tokarski's idea whereby compulsive tic disorder and Myriachit were two distinct pathologies (Roubinovitch, 1893).

Julien Noir (1866–1948), a student of Bourneville and, like him, an editor for *Le Progrès Médical*, defended his thesis in 1893 with Victor Cornil (1837–1908) presiding over the jury. The thesis examined tics in children considered to be "degenerates, imbeciles, and idiots" and contained numerous illustrations and photographs (Noir, 1893). Noir gave 73 observations of youngsters with mental disabilities, some of whom had tics, with echolalia, coprolalia, and echokinesis in certain cases, but these three mimetic behaviors sometimes existed without tics. Using a purely descriptive nosographical approach, he proposed four families of tics: simple motor tics; coordinated tics observed in cases of mental retardation, that is, repeated rhythmical movements (rocking, *krouomania*, etc.); Gilles de la Tourette syndrome, Jumping, Latah, and Myriachit (including echolalia, echokinesis, and coprolalia); and finally, "purely psychic tics," that is, *idées fixes*, obsessions, impulses, and so forth. None of his observations were characteristic of Gilles de la Tourette syndrome. Aside from the descriptive interest of this thesis and its underlying intellectual approach, indicative of the humanity Noir and Bourneville showed in caring for these poor patients, the work is a manifesto by colleagues of the Salpêtrière School in support of the recognition and isolation of Gilles de la Tourette syndrome, which faced opposition from detractors such as Magnan and his students, André Lemoine and Nicolas Lemaire, who considered "convulsive and psychic tics as simple stigmata of degeneration," refusing to give them their own place as a morbid entity, contrary to Gilles de la Tourette, whose ideas Catrou defended in his thesis. Noir also presented novel ideas in the area of etiological description and research concerning repetitive and rhythmical movements having no purpose, frequently observed in people with mental disabilities. Paul Sollier (1861–1933) had initiated this study in his 1890 thesis also directed by Bourneville, with Cornil presiding over the jury (Sollier, 1890). The Franco-American Édouard Séguin (1812–1880) (Séguin, 1846), a student of Itard and Esquirol (1772–1840), had not mentioned tics in his pioneering work on the treatment of "idiots." By contrast, Paul Moreau from the city of Tours (1844–1908), in one of the first pedopsychiatric treatises, entitled *La folie chez l'enfant*, had described the simple and isolated tic form, citing anecdotes from Prosper Lucas and Charles Darwin (1809–1882) and agreeing with them on the role of heredity as a predisposing factor (Moreau de Tours, 1888).

In 1893, Louis Chabbert (1862–1945) from Toulouse published four observations of patients with motor tics in *Les Archives de Neurologie*. The complete clinical picture of Gilles de la Tourette syndrome, with the association of phonic tics, echolalia, and coprolalia, was only present in the third observation. Chabbert observed Charcot's echokinesis, but preferred the term "echocinesis." Apparently acknowledging the grounds on which Gilles de la Tourette syndrome had been described, he not only argued, in case after case, for the role of heredity, but also sought to highlight all possible links with hysteria (Chabbert, 1893). His approach is reminiscent of that of Guinon.

Defended in 1894 with Jacques-Joseph Grancher (1843–1907) presiding over the jury, the thesis of Casimir Janowicz (1859–1918), born in Kamyanets-Podilsky in what was then Poland, covered the differential diagnosis of convulsive tics, paramyoclonus multiplex, and electric chorea (Janowicz, 1891). After describing two cases of simple transient tics, Janowicz explained the difficulties posed by the inaccurate descriptions of paramyoclonus and electric chorea—simple syndromes, still poorly defined, involving muscular spasms—whereas convulsive tic disorder had a well-established clinical picture. For Janowicz, paramyoclonus and electric chorea had a psychic origin suggesting a form of hysteria, whereas Augustin Morvan (1819–1897), in his 1890 description of fibrillary chorea ("myokimias"), explicitly described a neurological disease that is now recognized as an autoimmune channelopathy known as Morvan's syndrome (Walusinski & Honnorat, 2013).

Ezio Sciamanna (1850–1905), an Italian alienist and student of Charcot and Moriz Benedikt (1835–1920), constructed a pathophysiology in 1898 around 10 ticcing patients that combined tics and paranoia. Tics, for Sciamanna, were impulsive movements aimed at removing some imaginary danger. These pointless, non-adaptive actions became, by his definition, a delirium, influencing all of the individual's behaviors, preparing the way for paranoia (Sciamanna, 1898).

In 1899, Constantin Oddo (1860–1926) in Marseille set out to help practitioners differentiate tic disorder from Sydenham's chorea, but the resulting clinical picture had atrocious implications, for the poor patients. While chorea required hereditary nervous problems to develop, "the same cannot be said for ticcers, whose hereditary past is burdened in another manner, whose flaws are more pronounced. Theirs is not only a hereditary neuropathy, but degeneration, in its most severe form... If the ticcer is not yet a degenerate, he will be, he is marked to become one... Chorea patients are almost always cured; the ticcer is never cured." The entire article is in this same vein (Oddo, 1899).

25. Gilles de la Tourette's final contribution

Aside from the observations he gave Catrou for his thesis, Gilles de la Tourette never published any other data on the disease named for him during Charcot's lifetime. It was not until 1899 that the transcript of one of his lessons at Hôpital Saint-Antoine was published (Gilles de la Tourette, 1899). A 22-year-old woman came "to ask for treatment for chorea, which seemed to explain, at least in appearance, the agitation of the muscles of her face and the left side of her body." The abnormal movements began when she was 80 years old. Periods lasting a few weeks to a few months alternated with calm periods, without "muscular incoordination," but with a few residual tics in her face, nonetheless. Gilles de la Tourette once again employed this awkward expression, as if not to back down, all while disowning it in the lines that followed! All of the physicians consulted had diagnosed chorea and prescribed arsenic and antipyrine. In addition to her motor tics, she had phonic tics (hum! hum! or oh ah!). "When she grows irritated with her little boy, who is very agitated, her jerks worsen and she calls him 'pig, ass,' often including the word of Cambronne [shit] in her interjections... I was now sure; the unanimous diagnosis of chorea should be replaced by a diagnosis of convulsive tic disease with coprolalia." This young woman also exhibited some phobias. In his commentary, Gilles de la Tourette then recognized the contributions of Guinon and Charcot to his initial description and gave this recommendation: "If you wish to read a good survey of the question, I urge you to consult the thesis Mr. Catrou, my student, prepared in 1890 on the convulsive tic disease, based on twenty-six unpublished observations I sent to him." Gilles de la Tourette admitted at this point that the expression "motor incoordination" was unsuitable and acknowledged the validity of Guinon's "convulsive tics": "Jerks, or tics, to use what has become the established expression." There followed a long section on coprolalia and echolalia, "which can truly be deemed psychic stigmata." While the patients could be considered to belong "to the unbalanced group," Gilles de la Tourette said he "did not care for this expression." For him, the use of arsenic had no place in their treatment. After discussing differential diagnosis, especially regarding chorea, Gilles de la Tourette vigorously contested the view of Brissaud (see below), expressed in the thesis of his student Georges Patry (1869–?) describing "the variable chorea of degenerates": "In my opinion, I shall say so immediately, this condition cannot be differentiated from the convulsive tic disease." Fourteen years after the initial article, this clarification in 1899 presented the disease's clinical picture and course in a comprehensive fashion that remains relevant today. It was an occasion for Gilles de la Tourette to repeat the aphorism, "Once a ticcer, always a ticcer." In 1900, when he published his last paper in *La Revue Neurologique*, he revisited the nosology of chorea in pregnant women, denying the existence of an isolated pathology; at that time, abortions were performed in such cases to save the mother and avoid the birth of a disabled child. For Gilles de la Tourette, these cases of chorea were not chorea, but either hysteria or convulsive tic disorder. However, he erroneously reaffirmed once again that Sydenham's chorea did not exist after puberty and had no link with rheumatism.

26. Theory of Meige and Feindel

In 1902, Henry Meige (1866–1940) and Eugène Feindel (1862–1930) published their work on tics and their treatment (Meige & Feindel, 1902). A translation in German appeared in 1903 (Meige & Feindel, 1903) and in English in 1907 (Meige & Feindel, 1907). In the introduction, Brissaud was categorical: "The authors of this book have resolutely identified the pathogenesis of tics as a mental process. The psychology of the ticcer can be easily penetrated in certain cases where the tic is equivalent to or superposed on 'other episodic stigmata of degeneration,' but the task becomes infinitely more complex when the incorrectness of psychic acts is limited to tics alone. Even under these circumstances, in examining the mental state, one always observes the insufficiency of inhibition. This weakness allows countless unfortunate habits to form and grow entrenched over time. Such is the role of habit in the genesis of tics, revealing their similarity with all functional acts. A tic is often merely a function performed in an untimely or unsuitable manner."

Brissaud spoke of Gilles de la Tourette syndrome as "a lamentable neurosis": "Its only special feature is the somewhat systematic regularity of its progression. It is merely the superlative expression of a neuropathic and psychopathic disposition, in all ways similar to the disposition that gives rise to the most benign tics (Brissaud, 1896). The disease, in its initial manifestations, presents like a simple convulsive oddity that appears insignificant. But when one considers that this oddity can go on indefinitely in time, spreading gradually to all limbs, accompanied by serious mental disorders that often lead to dementia, one realizes that this little symptom, this little premonitory tic, always deserves to be viewed with defiance and energetically fought." This book laid the groundwork for the interference of psychoanalysis in the treatment of ticcers. The authors contested the etiopathogenic view, in all of its foresight, that René Cruchet (1875–1959) from Bordeaux proposed in 1901 (Cruchet, 1904) and 1902 (Cruchet, 1902; Walusinski, 2021b).

Cruchet based his reasoning on the work of Constantin von Monakow (1853–1930) and Wladimir Aleksandrowicz Muratoff or Muratow (Владимир Александрович Муратов, 1865–1916) concerning athetosis and chorea. Similarly to these authors, he proposed a functional disorder "of the cortex and basal ganglia of the brain." For Meige and Feindel the tic, a coordinated movement, was evidence of a "reflex" activity of the cerebral cortex. They deduced it could be overcome by education (a movement oriented toward a goal) and therefore the will. They compared the process to a baby's sucking, an innate automatism that becomes mastication, a learned automatism. They thus conceived of tics as "the frequent repetition of a primitively voluntary movement." In their view, the premonitory sensation associated with tics was proof that they were conscious acts, justifying their position based on Brissaud: "The tic is a coordinated automatic act and thus of cortical origin." They went so far as to compare tics with micturition and defecation, noting that their execution "is preceded by a need and followed by satisfaction." They considered the tic "as a functional disturbance representing an act driven by an exaggerated, untimely need, followed by excessive, inappropriate satisfaction." Meige and Feindel used the incomplete observations reported in the seminal article of Gilles de la Tourette to posit that alongside "convulsive tic disease," there was a different clinical entity, neglected by Gilles de la Tourette but described by Brissaud: "the variable chorea of degenerates." This chorea "has no uniformity in its current manifestations, or regularity in its progression, or constancy in its duration… It is a neurosis that lacks symptomatic rigor, as it were... Do the involuntary manifestations of the syndrome warrant inclusion in the pathologies listed in the chapter on myoclonia?" It is surprising to read the following: "Gilles de la Tourette refused to consider this condition as a chorea. For him, it is simply one of the forms of the convulsive tic disease; indeed, variable chorea does sometimes entail the utterance of explosive syllables and even coprolalia." Brissaud seemed unaware of, or unconcerned by, Charcot's efforts to demonstrate the difference between chorea and tics. In any case, posterity has forgotten variable chorea and recognized Gilles de la Tourette syndrome!

Meige and Feindel listed all of the medications already tried to reduce tics and stated their inefficacy, adding: "Since ticcers have a mental state that favors all types of mania, one must always avoid using ether, morphine, or cocaine." While electrotherapy and bathing in the sea were advised against, hydrotherapy was recommended, along with rigorous rubbing with a massage glove. Only "re-education treatment" found favor with Meige and Feindel: "Immobilization of movements and movements of immobilization... The first of these procedures consists in training the ticcer to keep the limbs and face absolutely still, photographically still, over progressively increasing time periods... Parallel to imposing this discipline of immobility on patients, an effort will be made to teach them discipline in their movements. This involves making them execute slow, regular, correct movements, on command, that exercise the muscles in the region where the tic is located... The adults responsible for young ticcers often lack sufficient firmness and authority to correct them, if the need arises... After treatment of a tic, the task of the physician is not finished. Once the untimely movement is gone, there is still the mental state of the ticcer, which makes him easy prey for another offensive of the disease. If the mental state is sufficiently modified by re-education, then it is truly possible to hope for a definitive cure of the tic, through a firming of the will, where the congenital debility had previously allowed the tics to set in. But for some time, it is wise to speak only of improvement... Firmness, patience, goodness, and good sense—these are the all-powerful weapons that the physician must wield to wage war on tics. Docility, trust, and perseverance—these are the auxiliary means that the patient must provide to emerge victorious from this combat. Once this alliance is formed, the fight will begin, and will be without respite or mercy. It must be pursued unfailingly against all of the untimely habits, motor or otherwise."

After Morel, Ribot, a professor at the prestigious Collège de France and director of the influential *Revue Philosophique*, continued to develop the concept of degeneration ("degeneration, that is, instability and psychological incoordination") and the associated notion of heredity, thereby providing an acceptable conceptual basis for explaining various social and asocial behaviors, along with all sorts of problems, from alcoholism to depression to sterility (Pinell, 2001; Walusinski, 2020). The directive and authoritarian ideology developed by Meige and Feindel for treating tics was based on Billod's article (Billod, 1847) and Ribot's, 1883 book on diseases of the will, a best-seller in the psychological literature with its 37 editions from 1883 to 1936 (Ribot, 1883). Ribot declared: "The voluntary faculty is closely tied to the involuntary faculty; the former rests on, and takes its force from the latter and is, by comparison, quite fragile. The education of the attention basically consists in nothing more than arousing and developing feigned perceptions and endeavoring to make them stable through repetition." This explains why the will should make it possible to prevent outbursts of involuntary movements through sustained attention. The repetition of the adaptive movements should prevent involuntary movements because fundamentally, "the brain is a motor organ, which is to say that many of its elements produce movement and there is not a single state of consciousness that does not contain, to some degree, motor elements." According to Ribot, Gilles de la Tourette syndrome, like aboulia, depression, or hysteria, was "a state of incoordination, lost equilibrium, anarchy, and moral ataxia that results from a constitutional impotence of the will." It is easy to see why Meige and Feindel spoke of a fight "without respite or mercy"! These clinical, pathogenic, and therapeutic considerations would become the accepted rule for more than half a century, making the book of Meige and Feindel the reference for diagnosing and treating tics. The Englishman Samuel Alexander Kinnier Wilson (1878–1937) (Broussolle, Trocello, Woimant, Lachaux, & Quinn, 2013), known for his description of the eponymous disease involving excess copper in the organism due to a metabolic defect (Wilson, 1912), adopted the theories of Meige and Feindel on tics in 1927 (Wilson, 1927). However, he noted the difficulty of giving a pathophysiological interpretation based on the psychopathology causing the abnormal movements, in particular tics, which were present in the sequelae of the encephalitis epidemic (Constantin von Economo, 1876–1931). A recent publication casts new light on these theories based on a new pathophysiological paradigm: "Aberrant reinforcement signals to the sensorimotor striatum may be fundamental for the formation of stimulus-response associations and may contribute to the habitual behavior and tics of this syndrome" (Delorme et al., 2016).

In 1909, Cruchet wrote a volume for general practitioners in the "Consultations médicales françaises" collection entitled *La tiquose* (Cruchet, 1909). "Any simple or complex movement that is sudden and frequently repeated without any apparent reason or purpose at irregular intervals constitutes a tic." For him, any normal movement could be transformed into a tic. Cruchet held to the principle that there was no continuum between the "habitual tic" and the "convulsive tic" or dystonia. For him, the first "was an action movement" whereas the second was an attitude or a "fixed movement." And yet he used the word "tic" for both of them! Without any discussion, he cited the popular belief that a blinking ocular tic may occur following the presence of a foreign body under the eyelid or after making a game of voluntary blinking or imitating another ticcer. In a popular book published in 1911, with a second edition in 1930, Cruchet the pediatrician advised parents and teachers on "bad habits"; that is, behaviors in children that, for him, caused lasting neuropsychiatric pathologies (Cruchet, 1911). He listed the descriptions of various localizations of tics in the face and limbs, including nail-biting and trichotillomania as tics. He distinguished "visceral tics," including snoring, sniffing, yawning, sneezing, coughing, and laryngeal and phonatory sounds, with coprolalia falling in this category. It should be noted that Cruchet considered thumb-sucking, rhythmic movement disorder, stuttering, polydipsia, bed-wetting, sleepwalking, and masturbation as "bad habits" in children that needed to be vigorously corrected. His explanation went as follows: "Any bad habit, that is, one that damages either the body's outer conformity or its physical or moral heath, can only be explained in two ways. The first entails excessive emotionalism, by virtue of which an act, once it enters one's consciousness, tends to be reproduced solely because of the initial first impression it causes. The second explanation lies in an insufficient will, which, in the presence of a sensation experienced normally, is unable to drive it away, even while recognizing it as dangerous." Everything thus came down to how a child was raised, which should hold any spontaneous impulses in check: "Tics are bad habits due to a lack of authority and control." Cruchet railed against "Freudism," an approach far removed from his principles: "Tics are attributed to narcissism, that is, this form of autoerotism characterized by the fact that subject adores themselves. A tic occurs in the part of their body where there is an erogenous zone, a sensation of pleasure… One could never have imagined that the child's legendary innocence was in fact concealing native lust and perversity." The therapeutic methods he suggested ranged from a straight-jack type restraint to application of an eye ointment or wash containing cocaine or camphor to gymnastic exercises that he called the Brissaud method and the Pitres method (Walusinski, 2021b).

27. In medical treatises

In 1894, in the *Traité des Maladies du Système nerveux* by Joseph Grasset (1849–1918) and Georges Rauzier (1862–1920), the tic chapter used elements developed by Letulle in the Jaccoud dictionary without presenting Gilles de la Tourette syndrome as an isolated entity (Grasset & Rauzier, 1894).

In 1902, Henri Triboulet (1864–1920) wrote the tic chapter in the *Traité de Médecine* of Brouardel and Augustin Gilbert (1858–1927) (Triboulet, 1902). He did not distinguish transient childhood tics, leaving the reader with the impression that all ticcers have Gilles de la Tourette syndrome: "The prognosis is always serious, due to the etiological principle: a transmittable hereditary flaw." The discussion of differential diagnosis was more soundly structured and highlighted how to distinguish tics from chorea, hysteria, and "paramyoclonus." He recommended treating patients—as early as possible, he emphasized—with the methods of Meige and Feindel.

In the so-called *Charcot-Bouchard Traité de Médecine*, actually the work of Brissaud and published in 1894, there was no chapter on tics (Charcot et al., 1894). By contrast, the *Pratique Médico-Chirurgicale* of Brissaud, Pinard, and Reclus Brissaud, Pinard, & Reclus (1907) contained a long chapter on the subject, written by Meige (Meige, 1907). The descriptions of tics were detailed, and their etiology was linked to "neuropathic and psychopathic" heredity. Gilles de la Tourette syndrome was reduced to banal tics associated with coprolalia, the nature of which was contested in keeping with Magnan's opinion.

28. Psychoanalytic inquiry

The era of psychoanalytic iatrogenesis began in 1921 with the publication of a tic theory by the Hungarian psychoanalyst Sandor Ferenczi (1873–1933), who prior to that point had never examined any patients, basing his ideas on the account of one ticcer, which formed the long first chapter of the Meige and Feindel book and was entitled "The Confessions of a Victim to Tic" (Ferenczi, 1921). For him, tics were a stereotyped equivalent of onanism; ticcers were thus frustrated masturbators! Sigmund Freud (1856–1939) wrote little on tics. He was in Paris from October 1885 to February 1886, where he met Gilles de la Tourette and Guinon, and thus must have been aware of their articles on "convulsive tics." In his study on hysteria with Josef Breuer (1842–1925), the second case was that of Frau Emmy von N., who may have been misdiagnosed by Freud and in fact suffered from Gilles de la Tourette syndrome (Freud, 1962; Pappenheim, 1980). Of course, it is not possible to establish a retrospective diagnosis with certitude. It seems probable that Freud eliminated Gilles de la Tourette syndrome as a diagnosis because his patient's tics seemed to disappear, although temporarily, under hypnosis whereas "convulsive tics" were not modified. Freud was of Guinon's opinion that links existed between the two pathologies or that they were frequently associated (Kushner, 1998). After 1893, Freud never proposed any psychoanalytic explanations of the pathophysiology of tics. As for Charles-Louis Trepsat (1879–1937) in 1922, he recounted in *Le Progrès Médical* his psychoanalytic interpretation of the confession of a ticcer, a poor victim of onanism in childhood. Trepsat supported his case by noting that the diagnosis of tics was made by Joseph Babiński (1857–1932), whereas the clinical picture resembles dystonia (Trepsat, 1922).

The Swiss psychiatrist and psychoanalyst Raymond de Saussure (1894–1971), Freud's student, stands out honorably among other psychoanalysts. He discussed the etiological role of encephalitis lethargica, described by von Economo, and presented an observation of Gilles de la Tourette syndrome in which the tics often took the form of repeated yawning (Gilles de la Tourette, Huet, & Guinon, 1890): "Despite the similarities between Gilles de la Tourette's description and Blanche's symptoms, we still hesitate to consider her yawning fits as hysterical symptoms." For Saussure, the abnormal movements of the young patient he had treated at the Clinique de Céry, where Gilles de la Tourette died, were organic in origin, like the Parkinsonian symptoms of other victims of this epidemic, and not psychological (Saussure de, 1923).

The psychoanalyst Serge Lebovici (1915–2000) was at the other end of the spectrum, and in 1951 wrote that tics were the "symptom of an obsessional psychoneurosis"; hence the need for early treatment, which entailed "prophylactic mental hygiene," to avoid the development of a neurosis in adulthood (Lebovici, 1951).

29. Early modern concepts

In 1941, André DeWulf (1903–2000) and Ludo van Bogaert (1897–1989) confirmed the validity of the conclusions of Charcot and his students Gilles de la Tourette and Guinon on the neurological organicity of the disease (DeWulf & von Bogaert, 1941). The clinical picture for their patient was atypical, with onset at age 18. When they began treating him at age 30, his general condition was very impaired by advanced osseous tuberculosis. Aside from the abnormal movements, the patient had no psychiatric comorbidity. Shortly after his death following status epilepticus, they performed an autopsy: "We undertook a detailed histological study, exploring in particular the two motor zones, the cerebellar apparatus, the red nucleus, the basal ganglia, the olivary bodies, and the anterior horns of the cervical spine. Aside from what appears to be cellular sclerosis in layers IV-V-VI in the frontal, motor, and parietal regions, we did not observe any noteworthy details.... What we find important in this negative examination is that it permits the conclusion that Gilles de la Tourette disease has no lesional substratum, which in turn permits us to include it among the systematized extra-pyramidal diseases." The fact that this 1941 paper was written in Belgium under Nazi occupation and published in Switzerland may explain its lack of dissemination after the war among psychoanalysts. Nonetheless, the work gave the authors an opportunity to produce one of the first films of motor and phonic tics, 1 month before the patient's death. The photos of the patient that illustrated the article were reprinted in 1949 in a medical treatise published in Paris (Lemierre, Lenormant, & Pagniez, 1949).

The publications in German on this subject include the 1892 paper by Friedrich Jolly (1804–1904), a professor of psychiatry and director of the neuropsychiatric clinic at the Berlin Charité Hospital (Jolly, 1892); the 1896 paper by Johannes Bresler (1866–1942), who worked in an asylum in Kreuzburg (Upper Silesia), then part of Germany, currently Kluczbork (Opole Voivodeship) in Poland (Bresler, 1896); the 1898 paper by Hermann Wille of Münsterlingen, son of and assistant to Ludwig Wille (1834–1912), a German psychiatrist who worked in Basel, Switzerland (Wille, 1898); the 1899 paper of Georg Köster (1867–1932), a professor of psychiatry and neurology in Leipzig (Köster, 1899); the 1927 paper by Erwin Straus (1891–1975), a neuropsychiatrist and philosopher, assistant to Karl Bonhoeffer (1868–1948) at Berlin Charité Hospital (immigrated to Baltimore in 1938), also a critic of psychoanalysis and friend of Eugène Minkowski (1885–1972) (Straus, 1927); the 1927 work of Joseph Wilder (1895–1976), assistant to Otto Pötzl (1877–1962) at the university neuropsychiatric clinic of Vienna (fled to New York in 1937) (Wilder, 1927); and the 1932 paper of Werner Runge (1882–?), a neurologist in Chemnitz (Runge, 1932).

The full history of Gilles de la Tourette syndrome after World War II, especially the importance of psychoanalytic theory, particularly in the U.S., is well presented in Howard Kushner's book, which stands as the reference work on this subject (Kushner, 1999).

Gilles de la Tourette disease disappeared almost totally from the medical literature from the early 20th century until the late 1950s. In the 1960s, the disease became a syndrome, firmly reassociated with Gilles de la Tourette's name.

30. Arthur and Elaine Shapiro

In a 1965 review of the literature, Diane Kelman found 44 authentic descriptions of Gilles de la Tourette syndrome in all medical publications worldwide between 1906 and 1964. She found slightly higher incidence in males than in females, onset before age 10, and normal intelligence without any particular risk of "degeneration" (she used the nineteenth-century term). She confirmed the absence of any link between Jumping, Latah, Myriachit, and Gilles de la Tourette syndrome. Finally, she noted that neuroleptics seemed to reduce the symptomatology, but that proper perspective was lacking to accurately assess their impact (Kelman, 1965). Indeed, it had only been recently, in 1961, that Jean-Noël Seignot had succeeded for the first time in effectively treating one of his patients, through experimental use of haloperidol (Seignot, 1961). His publication did not generate many immediate echoes. Then in 1968, Arthur K. Shapiro (1923–1995) and his wife Elaine, both American psychiatrists, successfully replicated Seignot's observation. In their article, which American medical journals refused to publish, they demonstrated the efficacy of haloperidol in calming the symptomatology of Gilles de la Tourette syndrome (Shapiro & Shapiro, 1968). Their work did have a large impact, first in the community of American psychiatrists, and later among American neurologists. Shapiro thus became the instigator of a radical paradigm shift. This disease was henceforth seen as organic; psychotherapy, and to an even greater degree psychoanalysis, have no role in its treatment, aside from helping patients and their families tolerate the psycho-affective repercussions of the disease. The Shapiros estimated the number of patients in the U.S. to be approximately 100,000 in their excellent 1978 book, which can be considered the first work since the early twentieth century to comprehensively address the subject (Shapiro et al., 1978). One of their colleagues, Ruth Bruun of Cornell University, described Arthur Shapiro as "a revolutionary, willing to challenge the prevailing dogma, dynamic, charming, and relentlessly stubborn when fighting for what he thought was right and more honest than psychoanalysis then fashionable, an engaging speaker and a man of diverse interests and enthusiasms. It is extremely unusual for a couple of researchers to completely change the prevailing view of a disease, but this is exactly what they did. We salute the tenacity and perseverance of the Shapiros, who brought about the 'renaissance' in work on Gilles de la Tourette syndrome and improved treatment for patients" (Brunn & Bruun, 1994). Twenty years after their landmark publication, the Shapiros were able to attend a large meeting held on May 2 and 3, 1985, at Hôpital La Salpêtrière to celebrate the 100-year anniversary of the seminal article.

Gilles de la Tourette had highlighted the descriptions of some of his predecessors, notably Itard and Trousseau. However, according to Shapiro, the first description of the disease appeared in the book *Malleus Maleficarum* (often translated as *Hammer of the Witches*), written by two monks, Jakob Sprenger (1436–1495) and Henricus Institoris (1430–1505), and published in 1487 with the approval of Pope Innocent VIII (Institoris & Sprenger, 1487). There were several successive editions (Sprenger & Institoris, 1520). These authors described a priest with motor and phonic tics which, at the time, were considered a sign of satanic possession, to be addressed by exorcism. The 1973 American horror film *The Exorcist*, directed by William Friedkin and based on William Peter Blatty's 1971 novel of the same name, was largely inspired by the accounts in *Malleus Maleficarum*. The mother of a 12-year-old girl with "spasms" and coprolalia, very probably a true case of Gilles de la Tourette syndrome, thinks her daughter is the victim of a possession and calls on two priests for an exorcism… in our day and age (Kushner, 1995).

Statement of ethics

This work required no approval from an institutional review board and was prepared in accordance with ethical guidelines of the publisher.

Conflict of interest statement

The author has no conflicts of interest to declare.

Funding sources

No funding was obtained for this work.

References

Alajouanine, T., & Gastaut, H. (1955). La syncinésie-sursaut et l'épilepsie-sursaut à déclenchement sensoriel ou sensitif inopiné. *Revue Neurologique (Paris)*, *93*, 29–41.

Alajouanine, T., & Gastaut, H. (1958). La syncinésie-sursaut et l'épilepsie-sursaut à déclenchement sensoriel ou sensitif inopiné: Considérations sur les épilepsies dites réflexes. In *Bases physiologiques et aspects cliniques de l'épilepsie* (pp. 199–231). Paris: Masson.

André, N. (1756). *Observations pratiques sur les maladies de l'urèthre, et sur plusieurs faits convulsifs.* Paris: chez Delaguette.
André, G. (1891). Une observation de maladie des tics convulsifs avec onomatomanie. *Le Mercredi Médical, 2*, 325–326.
André, G. (1892). *Les nouvelles maladies nerveuses.* Paris: Octave Doin.
Anonyme. (1884). Nouvelles & échos. *Gil Blas, 6*(1703), 1.
Anonymous. (1878). Jumping Frenchmen. *The London Medical Record, 5*, 326.
Archives de Seine-et-Marne. (1817). *Chartrier du château de Guermantes (Pièce "À Monseigneur l'évêque de Meaux".)* (dossier 157 J 28, pp. 1–1). Melun: Archives de Seine et Marne.
Bagenoff, P. R. (1911). Myriatchénié et Klikouchisme. *Archives de Neurologie, 2*, 273–274.
Bakker, M. J., & Tijssen, M. A. (2010). Jumping Frenchman of maine. In K. Kompoliti, & L. Verhagen-Metman (Eds.), *The encyclopedia of movements disorder.* Oxford: Academic Press.
Beard, G. M. (1878). Remarks upon 'Jumpers or Jumping Frenchmen'. *The Journal of Nervous and Mental Disease, 5*, 526.
Beard, G. M. (1879). Neurasthenia (nerve exhaustion): With remarks on treatment. *The St. Louis Medical and Surgical Journal, 36*, 345–365.
Beard, G. M. (1880). Experiments with the 'jumpers' of maine. *Popular Science Monthly, 18*, 170–178.
Beard, G. M. (1880). Experiments with the "jumpers" or "jumping Frenchmen" of Maine. *Journal of Nervous and Mental Disease, 7*, 487–490.
Beard, G. M. (1881). Experiments with the "jumpers" of Maine. *Popular Science Monthly, 18*, 170–178.
Beard, G., & Gilles de la Tourette, G. (1881). Les sauteurs du Maine (États-Unis). *Archives de Neurologie, 2*, 146–150.
Bernard, L. (1932). Georges Guinon. *La Presse Médicale, 39*, 541–542.
Bernt, J. (1810). *Monographia Choreae Sti. Viti.* Pragae: Sumptibus J. G. Calve.
Billod, E. (1847). Maladies de la volonté. *Annales Médico-Psychologiques, 10*, 15–35. 170-202, 317-347.
Boissier de Sauvages, F. (1771). *Nosologie méthodique dans laquelle les maladies sont rangées par classes, suivent le système de Sydenham, & l'ordre des botanistes.* Paris: Hérissant le fils.
Bosredon, P. (1856). Note sur la nature de l'affection désignée sous le nom de délire des aboyeurs. *Gazette des Hôpitaux civils et militaires, 29*, 564.
Botrel, J. P. (1850). *De la chorée considérée comme une affection rhumatismale.* Thèse Paris n 79 Imp. Rignoux.
Bouley, H. M., & Reynal, J. (1892). *Nouveau dictionnaire pratique de médecine, de chirurgie et d'hygiène vétérinaires.* Paris: Asselin & Houzeau.
Bouteille, E. M. (1810). *Traité de la chorée ou Danse de Saint-Guy.* Paris: Vinçard.
Breitman, G. (1888). *Contribution à l'étude de l'écholalie, de la coprolalie et de l'imitation des gestes chez les dégénérés et les aliénés.* Thèse Paris n°28 Ollier-Henry.
Bresler, J. (1896). Beitrag zur Lehre von der Maladie des Tics convulsif (mimische Krampfneurose). *Neurologisches Centralblatt, 15*, 965–972.
Briquet, P. (1859). *Traité clinique et thérapeutique de l'hystérie.* Paris: J.B. Baillière.
Brissaud, E. (1896). La chorée variable des dégénérés. *Revue Neurologique (Paris), 4*, 417–431.
Brissaud, E., Pinard, A., & Reclus, P. (1907). Tic. *Vol. 6. Pratique médico-chirurgicale* (1st, pp. 483–498). Paris: Masson.
Broussolle, E., Trocello, J. M., Woimant, F., Lachaux, A., & Quinn, N. (2013). Samuel Alexander Kinnier Wilson. Wilson's disease, Queen Square and neurology. *Revue Neurologique (Paris), 169*, 927–935.
Brunn, R. D., & Bruun, B. (1994). *A mind of its own: Tourette's syndrome: A story and a guide.* New York: Oxford University Press.
Brunon, M. (1892). Tics et Tiqueurs. *La Normandie médicale, 8*, 169–174.

Buckingham, B. H., Foulk, G. C., & McLean, W. (1883). *Observations upon the Korean Coast, Japanese-Korean ports and Siberia, made during a journey from the Asiatic Stations to the United States through Siberia and Europe, June 3 to September 8, 1882*. Information from Abroad, General Information Series Number 1 Washington: Office of Naval intelligence, Navy Department, Bureau of Navigation. Government Printing Office.
Burot, F., & Duplouy, E. (1887). De la maladie des tics convulsifs et de son traitement par l'hypnotisme. Seizième congrès de l'Association française pour l'avancement des Sciences. *Bulletin général de Thérapeutique Médicale et Chirurgicale, 113*, 518–519.
Burot, F. (1888). Un cas de la maladie des tics convulsifs, traité et amélioré par la persuasion. *Revue de l'Hypnotisme, 2*, 141–144.
Burot, F. (1888). Sur un cas de tics convulsifs avec écholalie et coprolalie. *Revue de l'Hypnotisme, 2*, 330–334.
Catrou, J. (1890). *Étude sur la maladie des tics convulsifs (Jumping, Latah, Myriachit)*. Thèse Paris n °129 impr H. Jouve.
Chabbert, L. (1893). De la maladie des tics (tics, chorée, hystérie: diagnostic). *Archives de Neurologie, 25*, 12–41.
Chapel, J. L. (1970). Gilles de la Tourette's syndrome, Latah, Myriachit, and Jumpers revisited. *New York State Journal of Medicine, 70*, 2201–2204.
Charcot, J. M. (1887). *Leçons du Mardi à La Salpêtrière. Policlinique 1887-1888*. Paris: Bureaux du Progrès médical et A. Delahaye & E. Lecrosnier.
Charcot, J. M. (1889). *Leçons du Mardi à La Salpêtrière. Policliniques 1888-1889*. Paris: Bureaux du Progrès Médical et E Lecrosnier & Babé.
Charcot, J. M. (1893). *Clinique des Maladies du Système Nerveux. Leçons du professeur, Mémoires, Notes et Observations*. Paris: Bureaux du Progrès Médical et Félix Alcan.
Charcot, J. M., & Magnan, V. (1885). De l'onomatomanie. *Archives de Neurologie, 10*, 157–168.
Charcot, J. M., & Melotti, G. (1887). Lezione quattordicesima: Intorno ad alcuni casi di tic convulsivo con coprolalia ed ecolalia. Analogia col Jumping di Beard, il Latah della Malesia ed il Miriachit di Hammond. In *Nuove lezioni sulle malattie del sistema nervoso ed in modo particolare sull'isterismo nell'uomo: fatte alla Salpetrière negli anni scolastici 1884-85 ed 1885-86*. Milano: Dottor Francesco Vallardi.
Charcot, Bouchard, & Brissaud. (1894). *Traité de Médecine*. Paris: Masson.
Chauvreau, J. U. J. (1888). *Les tics coordonnés avec émission brusque & involontaire de cris et de mots articulés*. Ch Thèze: Rochefort sur mer.
Cruchet, R. (1902). *Étude critique sur le tic convulsif et son traitement par la gymnastique*. Thèse Bordeaux n°60 impr Gounouilhou.
Cruchet, R. (1904). Le tic convulsif et ses variétés cliniques (mémoire pour la quatrième année d'internat) 1901. In *Titres et Travaux scientifiques*. Bordeaux: impr Gounouilhou.
Cruchet, R. (1909). *La Tiquose. Consultations médicales françaises*. Paris: A. Poinat. Fascicle 8.
Cruchet, R. (1911). *Les mauvaises habitudes chez les enfants*. Paris: Expansion scientifique française.
Dana, C. L., & Wilkin, W. P. (1886). On convulsive tic with explosive disturbances of speech (so-called Gilles de la Tourette's disease). *The Journal of Nervous and Mental Disease, 13*, 407–412.
Day, H. F. (1902). *Pine tree ballads: rhymed stories of unplanned human natur' up in Maine*. Boston: Small, Maynard & Co.
Delorme, C., Salvador, A., Valabrègue, R., Roze, E., Palminteri, S., Vidailhet, M., et al. (2016). Enhanced habit formation in Gilles de la Tourette syndrome. *Brain, 139*(Pt. 2), 605–615.
Dewhurst, K. (1957). A symposium on trigeminal neuralgia with contributions by Locke, Sydenham, and other eminent seventeenth century physicians' disease. *Journal of the History of Medicine and Allied Sciences, 12*, 21–36.

DeWulf, A., & von Bogaert, L. (1941). Études anatomo-cliniques des syndromes hypercinétiques complexes. Une observation anatomo-clinique de maladie des tics (Gilles de la Tourette). *Monatsschrift für Psychiatrie und Neurologie, 104*, 53–61.
Duchenne de Boulogne, G. (1872). *De l'électrisation localisée et de son application à la pathologie et à la thérapeutique*. Paris: J.B. Baillière.
Dunglison, R. (1857). *A dictionary of medical science*. Philadelphia: Blanchard and Lea.
Eboli, P., Stone, J. L., Aydin, S., & Slavin, K. V. (2009). Historical characterization of trigeminal neuralgia. *Neurosurgery, 64*, 1183–1186.
Eddy, C. M., & Cavanna, A. E. (2013). On being your own worst enemy: an investigation of socially inappropriate symptoms in Tourette syndrome. *Journal of Psychiatric Research, 47*, 1259–1263.
Esquirol, J. E. (1838). *Des maladies mentales considérées sous les rapports médical, hygiénique et médico-légal*. Paris: J.B. Baillière.
Färber, T. (1885). *Über atypische Formen und Complicationen der Chorea*. Berlin: von Gustav Schade.
Ferenczi, S. (1921). Psycho-analytical observation on tic. *The International Journal of Psycho-Analysis, 2*, 1–30.
Ferraro, G. B. (1602). *Del Tiro reproduced in Cavallo Frenato by Pirro-Antonio Ferraro (son of Ferraro GB)*. Napoli: Antonio Pace.
Forbes, H. O. (1885). *A naturalist's wanderings in eastern archipelago*. New York: Harper and Brothers.
Fothergill, J. (1773). Of a painful affection of the face. *Medical Observations and Inquiries by a Society of Physicians London, 5*, 129–142.
Foulhioux, C. (1847). *Recherches sur la nature et sur le traitement de la danse de Saint-Guy*. Lyon: L. Perrin.
Freud, S. (1962). *Studies in hysteria. "Frau Emmy von N." The standard edition of the complete psychological works of Sigmund Freud*. London: The Hogarth Press. The Institute of Psycho-Analysis.
Gilles de la Tourette, G., Huet, E., & Guinon, G. (1890). Contribution à l'étude des bâillements hystériques. *Nouvelle Iconographie de La Salpêtrière, 3*, 97–119.
Gilles de la Tourette, G. (1884). Jumping, Latah, Myriachit. *Archives de Neurologie, 8*, 68–74.
Gilles de la Tourette, G. (1885). Étude sur une affection nerveuse caractérisée par l'incoordination motrice, accompagnée d'écholalie et de coprolalie. *Archives de neurologie, 9*, 19–42. 158-200.
Gilles de la Tourette, G. (1899). La maladie des tics convulsifs. *La Semaine Médicale, 19*, 153–156.
Gimlette, J. D. (1897). Remarks on the etiology, symptoms and treatment of Latah, with report of two cases. *The British Medical Journal, 2*, 455–457.
Goldstein, K., Landis, C., Hunt, W. A., & Clarke, F. M. (1938). Moro reflex and startle pattern. *Archives of Neurology and Psychiatry, 40*, 322–327.
Goncourt de, J., & Goncourt de, E. (1891). *Journal: mémoires de la vie littéraire (volume 3, Vendredi 27 janvier 1860)*. Monaco: Ed de l'imp. nationale.
Grasset, J., & Rauzier, G. (1894). *Traité Pratique des Maladies du Système Nerveux*. Montpellier, Paris: Camille Coulet, Masson.
Guinon, G. (1886). Sur la maladie des tics convulsifs. *Revue de Médecine, 6*, 50–80.
Guinon, G. (1887). Tic convulsif. *Dechambre A. Dictionnaire encyclopédique des Sciences médicales série, 3*, 17555–17593.
Guinon, G. (1887). Tics convulsifs et hystérie. *Revue de Médecine, 7*, 509–519.
Guinon, G. (1889). *Les agents provocateurs de l'hystérie*. Thèse Paris n°140 Aux Bureaux du Progrès médical et A. Delahaye & Lecrosnier.
Hale, M. E., Long, J. H., McHenry, M. J., & Westneat, M. W. (2002). Evolution of behavior and neural control of the fast-start escape response. *Evolution, 56*, 993–1007.

Hammond, W. (1871). *A treatise on diseases of the nervous system*. New York: D. Appleton and Co.

Hammond, W. (1884). Miryachit: a newly described disease of the nervous system and its analogues. *British Medical Journal, 1*, 758–759.

Hammond, W. (1884). Miryachit, nuova malattia del sistema nervoso. *Medicina Contemporânea, 1*, 126–130.

Hammond, W. (1884). Miryachit, nouvelle maladie du système nerveux. *L'Union Médicale, 38*, 706–708.

Handfield-Jones, C. (1870). *Studies on functional nervous disorders*. London: John Churchill and sons.

Hardison, J. E. (1980). Are the jumping Frenchmen of Maine goosey? *JAMA, 244*, 70.

Hirschfeld, H. (1891). *Über Maladie des Tics convulsifs*. Inaugural-Dissertation n°362 Berlin: Buchdruckerei von Gustav Schade.

Howard, R., & Ford, R. (1992). From the jumping Frenchmen of Maine to post-traumatic stress disorder: The startle response in neuropsychiatry. *Psychological Medicine, 22*, 695–707.

Hunt, W. A., & Landis, C. (1936). Studies of the startle pattern. *The Journal of Psychology, 2*, 201–219. 339-352.

Hunt, W. A., & Landis, C. (1937a). Studies of the startle pattern. *The Journal of Psychology, 3*, 339–343. 487-490.

Hunt, W. A., & Landis, C. (1937b). Studies of the startle pattern. *The Journal of Psychology, 4*, 199–206.

Hurley, S. (2005). The shared circuits Hypothesis: A unified functional architecture for Control, Imitation and Simulation. In S. Herley, & N. Chater (Eds.), *Perspectives on imitation: From neuroscience to social science*. Cambridge, Massachusetts: A Bradford Book the MIT Press.

Institoris, H., & Sprenger, J. (1487). *Malleus maleficarum*. Speyer: Peter Drach.

Itard, J. G. (1801). *De l'éducation d'un homme sauvage, ou des premier développements physiques et moraux du jeune sauvage de l'Aveyron*. Paris: Goujon fils.

Itard, J. G. (1825). Mémoire sur quelques fonctions involontaires des appareils de la locomotion, de la préhension et de la voix. *Archives Générales de Médecine, 3*, 385–407.

Janowicz, C. (1891). *Tic convulsif des enfants: paramyoclonus multiplex et chorée électrique, étude critique et comparative*. Thèse Paris n°356 impr Henri Jouve.

Jolly, F. (1892). Über die Sogen. Maladie des Tics convulsifs. *Charité Annalen, 17*, 740–753.

Joseph, A. B., & Saint-Hilaire, M.-H. (1998). Startle syndromes. In A. B. Joseph, & R. Young (Eds.), *Movement disorders in neurology and neuropsychiatry* (2nd ed., pp. 457–462). Blackwell Science.

Jourdin, J. (1647). *La Vraye cognoissance du cheval, ses maladies et remèdes, avec l'anatomie de Ruyni, contenant 64 tables en taille-douce par le moyen desquelles on pourra facilement cognoitre toutes les parties du cheval*. Paris: chez Thomas de Ninville.

Kelman, D. H. (1965). Gilles de la Tourette disease in children: A review of the literature. *Journal of Child Psychology and Psychiatry, 6*, 219–226.

Kenny, M. G. (1990). Latah: The logic of fear. In W. J. Karim (Ed.), *Emotions of culture: A Malay perspective* (pp. 123–141). Singapor: Oxford University Press.

Kok, O., & Bruyn, G. (1962). An unidentified hereditary disease. *The Lancet, 279*, 1359.

Köster, G. (1899). Über die Maladie des Tics impulsifs (mimische Krampfneurose). *Deutsche Zeitschrift für Nervenheilkunde, 15*, 147–158.

Kunkle, E. C. (1965). The "jumpers" of Maine: Past history and present status. *The Journal of the Maine Medical Association, 56*, 191–193.

Kunkle, E. C. (1967). The "jumpers" of Maine: A reappraisal. *Archives of Internal Medicine, 119*, 355–358.

Kurlan, R., Daragjati, C., Como, P. G., Mcdermott, M. P., Trinidad, K. S., Roddy, S., et al. (1996). Non obscene complex socially inappropriate behavior in Tourette's syndrome. *The Journal of Neuropsychiatry and Clinical Neurosciences, 8*, 311–317.

Kushner, H. (1995). Medical fictions: The case of the cursing marquise and the (re)construction of Gilles de la Tourette's s syndrome. *Bulletin of the History of Medicine, 69*, 224–254.

Kushner, H. (1998). Freud and the diagnosis of Gilles de la Tourette's illness. *History of Psychiatry, 9*, 1–25.

Kushner, H. (1999). *A cursing brain. the histories of Tourette syndrome*. Cambridge, USA: Harvard University Press.

Lannois, M. (1886). *Nosographie des chorées*. Paris: J.B. Baillière.

Lanska, D. J. (2010). The history of movement disorders. *Handbook of Clinical Neurology, 95*, 501–546.

Lanska, D. J. (2018). Jumping Frenchmen, Miryachit, and Latah: Culture-specific hyperstartle-plus syndromes. *Frontiers of Neurology and Neuroscience, 42*, 122–131.

Le Gendre, P. (1885). La maladie de Gilles de la Tourette. *L'Union Médicale, 40*, 109–114.

Lebovici, S. (1951). *Les Tics chez l'enfant*. Paris: Presses Universitaires de France.

Legrain, P. M. (1886). *Du délire chez les dégénérés*. Thèse Paris n°199 impr. A. Parent.

Legrand du Saulle, H. (1875). *La folie du doute (avec délire du toucher)*. Paris: Adrien Delahaye.

Lemierre, A., Lenormant, C., & Pagniez, P. (1949). *Traité de médecine. XVI, Maladies du système nerveux*. Paris: Masson.

Lerch, O. (1901). Convulsive tics. *American Medicine (Philadelphia), 1*, 694–695.

Letulle, M. (1883). Tics. In S. Jaccoud (Ed.), *Vol. 35. Nouveau Dictionnaire de Médecine et de Chirurgie pratiques*. Paris: J.B. Baillière.

Lucas, P. (1833). *De l'imitation contagieuse ou de la propagation sympathique des névroses et des monomanies*. Paris: Didot Jeune.

Mainieri, G., Loddo, G., & Provini, F. (2021). Disorders of arousal: A chronobiological perspective. *Clocks Sleep, 3*, 53–65.

Massey, E. W. (1984). Goosey patients: Relationship to jumping Frenchmen, Myriachit, Latah and tic convulsif. *North Carolina Medical Journal, 45*, 556–558.

Meige, H. (1907). *Les tics. Pratique Médico-Chirurgicale*. Paris: Masson.

Meige, H., & Feindel, E. (1902). *Les tics et leur traitement*. Paris: Masson.

Meige, H., & Feindel, H. (1903). *Der Tic, sein Wesen und seine Behandlung; nebst einer Vorrede von Professor Brissaud; Deutsche autorisierte Ausgabe von O. Giese* (p. 1903). Leipzig: Deuticke.

Meige, H., & Feindel, H. (1907). *Tics and their treatment*. New York: William Wood and Company.

Mension-Rigau, E. (2011). *L'ami du prince: Journal inédit d'Alfred de Gramont (1892-1915)*. Paris: Fayard.

Moreau de Tours, P. (1888). *La folie chez les enfants*. Paris: J.B. Baillière.

Moro, E. (1918). Das erste Trimenon. *Münchener medizinische Wochenschrift, 65*, 1147–1159.

Moro, E. (1920). Das erste Trimenon. *Münchener medizinische Wochenschrift, 67*, 360.

Müller, O. (1889). *Über die als "Maladie des tics convulsifs" beschriebene Krankheit*. Berlin: Buchdruckerei von Gustav Schade.

Murphy, H. B. M. (1982). *Comparative psychiatry: The international and intercultural distribution of mental illness*. Berlin: Springer Verlag.

Noir, J. (1893). *Étude sur les tics chez les dégénérés, les imbéciles et les idiots*. Thèse Paris n°171 aux Bureaux du Progrès Médical et Félix Alcan.

Nuttal, M. (2004). *Encyclopedia of the arctic*. New York: Routledge.

O'Brien, H. A. (1883). Latah. *Journal of the Royal Asiatic Society, 11*, 143–154.

O'Brien, H. A. (1884). Latah. *Journal of the Royal Asiatic Society, 12*, 283–284.

Oddo, C. (1899). Le diagnostic différentiel de la maladie des tics et de la chorée de Sydenham. *La Presse Médicale*, 7, 189–191.

Osler, W. (1887). On the general etiology and symptoms of chorea, based on the records of 410 cases at the Infirmary for Nervous Diseases. *Philadelphia Medical News, 51*, 437–441. 465-470.

Osler, W. (1890). On the form of convulsive tic associated coprolalia, etc. *The Medical News, 57*, 645–647.

Pappenheim, E. (1980). Freud and Gilles de la Tourette. Diagnostic speculations on 'Frau Emmy von N'. *International Review of Psycho-Analysis*, 7, 267–277.

Pinell, P. (2001). Degeneration theory and heredity patterns between 1850 and 1900. In J. P. Gaudillière, & I. Löwy (Eds.), *Heredity and Infection. The history of disease transmission* (pp. 245–260). London & New York: Routledge.

Pujol, A. (1787). *Essai sur la maladie de la face nommée tic douloureux*. Paris: Chez Théophile Barrois.

Quantin, E. (1857). *De la chorée*. Thèse Paris n°54 imprimerie Rignoux.

Rabinovitch, R. (1965). An exaggerated Startle Reflex resembling a kicking Horse. *Canadian Medical Association Journal, 93*, 130.

Railton, T. C. (1886). Notes of a case of involuntary muscular movements accompanied by coprolalia. *The Medical Chronicle of Manchester, 4*, 24–29.

Ribot, T. H. (1883). *Les maladies de la volonté*. Paris: Germer Baillière.

Richard, M. P. (2018). A peculiar condition: A history of the Jumping Frenchmen Syndrome in scientific and popular accounts. *Journal of the History of the Neurosciences, 27*, 355–374.

Risen Russell, J. S. (1899). The tics. In A. Y. Clifford (Ed.), *Vol. 7. A system of medicine* (1st). New York: Macmillan and Co.

Rizzolatti, G., Fogassi, L., & Gallese, V. (2001). Neurophysiological mechanisms underlying the understanding and imitation of action. *Nature Reviews. Neuroscience, 2*, 661–670.

Rizzolatti, G., & Sinigaglia, C. (2016). The mirror mechanism: A basic principle of brain function. *Nature Reviews. Neuroscience, 17*, 757–765.

Roth, D. D. (1850). *Histoire de la musculation irrésistible ou de la chorée anormale*. Paris: J.B. Baillière.

Roubinovitch, J. (1893). *A propos d'un cas de maladie des tics convulsifs avec mouvements par obsession*. Paris: impr Cour d'Appel.

Rudler, F., & Chomel, C. (1903). Tics et stéréotypies de léchage chez l'homme et chez le cheval. *Nouvelle Iconographie de La Salpêtrière, 16*, 369–390.

Runge, W. (1932). Beitrag zum Ticproblem. *Deutsche Zeitschrift für Nervenheilkunde, 127*, 96–113.

Ruyni, C. (1598). *Dell'anatomia, e dell'infermità del cavallo*. Venetia: Appresso Gasparo Bindoni.

Saint-Hilaire, M. H., Saint-Hilaire, J. M., & Granger, L. (1986). Jumping Frenchmen of Maine. *Neurology, 36*, 1269–1271.

Saint-Hilaire, M. H., & Saint-Hilaire, J. M. (2001). Jumping Frenchmen of Maine. *Movement Disorders, 16*, 530.

Sandras, S. (1851). *Traité pratique des Maladies Nerveuses*. Paris: Germer Baillière.

Saury, H. (1886). *Étude clinique sur la folie héréditaire*. Paris: A. Delahaye et E. Lecrosnier.

Saussure De, R. (1923). Discussion sur l'étiologie d'un tic survenu quinze mois après une encéphalite léthargique atypique. *Archives suisses de Neurologie et de Psychiatrie, 12*, 292–317.

Sciamanna, E. (1898). Sulla malattia del tic. *Rivista quindicinale di psicologia psichiatria neuropatologia, 99*, 16–17.

Sée, G. (1850). De la chorée, rapports du rhumatisme et des maladies cœur avec les affections nerveuses convulsives. *Mémoires de l'Académie impériale de Médecine, 15*, 374–525.

Séguin, E. (1846). *Traitement moral, hygie'ne et e'ducation des idiots et autres enfants arrie're's* (1st ed., Vol. 1). Paris: J.B. Baillière.

Seignot, J. N. (1961). Un cas de maladie de Gilles de la Tourette guéri par le R1625. *Annales Médico-Psychologiques*, *119*, 578–579.
Сеченов, И.М., & Setchenov, I. M. (1871). *Рефлексы головнаго мозга (Les réflexes du cerveau)*. С.-Петербургъ (S-Petersburg): типографія К. Б. Трубникова.
Shapiro, A. K., Shapiro, E., Bruun, R. D., & Sweet, R. D. (1978). *Gilles de la Tourette syndrome*. New York: Raven Press.
Shapiro, A. K., & Shapiro, E. (1968). Treatment of Gilles de la Tourette's Syndrome with haloperidol. *The British Journal of Psychiatry*, *114*, 345–350.
Sigaud, C. (1889). *De l'échomatisme, essai de pathogénie psycho-physiologique: zoandrie, échokinésie, écholalie*. Thèse Lyon: Pitrat ainé.
Simons, R. C. (1994). The interminable debate on the nature of Latah. *The Journal of Nervous and Mental Diseases*, *182*, 339–341.
Simons, R. C. (2013). *The culture-bound syndromes: Folk illnesses of psychiatric and anthropological interest*. Dordrecht, Boston, Lancaster: D. Reidel Publishing Company.
Société de Médecins et de Chirurgiens. (1821). *Dictionnaire des sciences médicales Tét-Tric*. Paris: Panckoucke.
Sollier, P. (1890). *Psychologie de l'idiot et de l'imbécile: essai de psychologie morbide*. Thèse Paris n° 66 Félix Alcan.
Sprenger, J., & Institoris, H. (1520). *Malleus maleficarum, maleficas et earum haeresim, ut phramea potentissima conterens*. Köln: Anno XX Coloniae excudebat Joannes Gymnicus.
Stawell, R. R. (1895). Habit spasm in children. *The Australian Medical Journal*, *17*, 153–158.
Steller, G. W. (1774). *Beschreibung Von Dem Lande Kamtschatka*. Frankfurt und Leipzig: JG. Flaischer.
Stevens, H. (1965). "Jumping frenchmen of Maine" Myriachit. *Archives of Neurology*, *12*, 311–314.
Straus, E. (1927). Untersuchungen über die postchoreatischen Motilitätsstörungen, insbesondere die Beziehungen der Chorea minor zum Tic. *Monatschrift für Psychiatrie und Neurologie*, *66*, 281–300.
Suhren, O., Bruyn, G. W., & Tuynman, A. (1966). Hyperekplexia, a hereditary startle syndrome. *Journal of the Neurological Sciences*, *3*, 577–605.
Thomas, R. H. (2015). Hyperekplexia: Overexcitable and underdiagnosed. *Developmental Medicine and Child Neurology*, *57*, 313.
Thomson, J. T. (1864). *Some glimpses into life in the Far East*. London: Richardson.
Tokarski, A. (1890). Meriatschenje und maladie des tics convulsifs. *Neurologisches Central-Blatt*, *9*, 662–663.
Tokarski, A. (1891). Myriachit et maladie des tics convulsifs. *La Semaine Médicale*, *11*, 184.
Trepsat, C. (1922). Traitement d'un tiqueur par la psychanalyse. *Le Progrès Médical*, *35*, 182–184.
Triboulet, H. (1902). Tics, Maladie des tics convulsifs. In P. Brouardel, & A. Gilbert (Eds.), *Vol. 10. Traité de Médecine et de Thérapeutique*. Paris: J.B. Baillière.
Trousseau, A. (1862). *Clinique Médicale de l'Hôtel-Dieu de Paris*. *Vol. 2* (1st pp. 23–23). Paris: J.B. Baillière.
Trousseau, A. (1873). Névralgie épileptiforme. *Vol. 2. Clinique me´dicale de l'Ho^tel-Dieu de Paris* (4th, pp. 156–167). Paris: J.B. Baillière.
van Leent, F. J. (1867). Contributions à la géographie médicale. *Archives de Médecine Navale*, *8*, 172–173.
van Leent, F. J. (1880). Communication sur le béri-béri. *Congrès périodique international des Sciences médicales*, *6*, 170–194.
Walusinski, O. (2020). The concepts of heredity and degeneration in the work of Jean-Martin Charcot. *Journal of the History of the Neurosciences*, *29*, 299–324.
Walusinski, O. (2021a). *Les 33 internes de Jean-Martin Charcot à La Salpêtrière de 1862 à 1893*. Brou: Oscitatio.

Walusinski, O. (2021b). René Cruchet (1875–1959), beyond encephalitis lethargica. *Journal of the History of the Neurosciences, 30*, 1–19. https://doi.org/10.1080/0964704X.2021.1911913. Epub ahead of print. PMID: 34241573.

Walusinski, O., & Féray, J. C. (2020). The Marquise de Dampierre identified at last, the first described clinical case of Gilles de la Tourette syndrome. *Revue Neurologique (Paris), 176*, 754–762.

Walusinski, O., & Honnorat, J. (2013). Augustin Morvan (1819-1897), a little-known rural physician and neurologist. *Revue Neurologique (Paris), 169*, 2–8.

West, W. J. (1840-1841). On a peculiar form of infantile convulsions. Letter. *Lancet, 1*, 724–725.

Wilder, J. (1927). Ein Fall von Maladie des Tics. In J. Wilder, & J. Silbermann (Eds.), *Beiträge zum Ticproblem*. Berlin: Karger.

Wille, H. (1898). Über einen Fall von Maladie des Tics impulsifs. *Monatschrift für Psychiatrie und Neurologie, 4*, 210–226.

Wilson, J. C. (1897). A case of tic convulsive. *Archives of Pediatrics, 14*, 881–887.

Wilson, S. A. K. (1912). Progressive lenticular degeneration. A familial nervous disease associated with cirrhosis of the liver. *Brain, 34*, 295–507.

Wilson, S. A. K. (1927). The tics and allied conditions. *The Journal of Neurology and Psychopathology, 8*, 93–108.

Winzeler, R. L. (1984). The study of Malayan Latah. *Indonesia, 37*, 77–104.

Winzeler, R. L. (1995). *Latah in Southeast Asia: The history and ethnography of a culture-bound syndrome*. New York: Cambridge University Press.

Yap, P. (1967). Classification of the culture-bound reactive syndromes. *Australian and New Zealand Journal of Psychiatry, 1*, 172–179.

Zapffe, C. A. (1978). A brief history of alloy steel. *Metal Progress, 18*, 459–467.

CHAPTER TWO

The neurobiology of Gilles de la Tourette syndrome and chronic tics

James F. Leckman*
Child Study Center, Yale University, New Haven, CT, United States
*Corresponding author: e-mail address: james.leckman@yale.edu

Contents

1. Introduction

It has been a fascinating and inspiring journey to learn from individuals with tics. They have taught us a great deal from "premonitory" urges to the fact that their tics typically disappear when they are engaged in a physical activity that requires both focused attention and motor control. However, the more we learn, the more questions arise concerning the underlying neurobiology of tics and related behaviors, as well as the sensory and mental states that often accompany them.

International Review of Movement Disorders, Volume 3
ISSN 2666-7878
https://doi.org/10.1016/bs.irmvd.2021.11.005

In this chapter, we focus on the phenomenology of tics and related behaviors, beginning with brief descriptions of the amazing variety of simple and complex motor and phonic tics. We then turn to the premonitory urges and the fleeting and sense of relief that often follows the completion of a series, or a "bout" of tics. Remarkably, many individuals with tics also experience a heightened sensitivity to other sensations in their bodies and perceptions of the world around them, including for many individuals, the need for things to be "just right."

The natural history of tics is highly variable. Interestingly, a majority of individuals experiencing simple motor tics in childhood do not go on to experience a full range of motor and vocal tics later in life. Fortunately, only a relatively small percentage of individuals go on to experience severe tics in adulthood. In addition to tics, many individuals with tics have a diverse range of other neuropsychiatric symptoms, ranging from attentional difficulties to obsessive-compulsive symptoms (Cavanna et al., 2020). Some individuals are also on the autism spectrum. Remarkably, at present and with so many children and adolescents connecting with one another through social media, there has also been an increase in the number of individuals with functional movement disorders that can closely resemble tics. In the age of COVID-19, a substantial number of parents are concerned about their child having an immune-related condition (Robertson, Eapen, Rizzo, Stern, & Hartmann, 2020).

Future research that integrates clinical and neurobiological data, including neuroimaging and genetics, is expected to reveal the pathogenesis of tic disorders at the neural circuit level, which hopefully will lead to targeted and effective interventions (Robertson et al., 2017).

We end the chapter by considering the "unlikely strengths" that frequently accompany tics, and the potential value of an individual's "acceptance" of their tics.

2. What are tics?

When Georges Albert Édouard Brutus Gilles de la Tourette (1885) first described the syndrome that bears his name, he referred to it as the *maladie des tics*. So, what are tics? Tics are a nearly unlimited number of sudden and abrupt movements and/or sounds. Often more easily recognized than precisely defined, tics are nonrhythmic motor movements or sounds that occur and recur over the course of time. Tics can be easily mimicked, but they usually have a "stereotypic" character, which simply means that a specific tic looks or sounds more or less the same each time it happens. Given their sudden and recurrent nature, the observer may think that specific tics are being done "on purpose." With few exceptions including Janet (1903) and Freud (1913), descriptions from the 19th century onward, including those of Itard (1825) and Gilles de la Tourette (1885), have largely focused on cataloguing and classifying tics as viewed from the outside.

In this chapter, we first describe "tics" before considering the internal sensory states that are often associated with them. Tics are characterized on their anatomical location, frequency, duration, and impact on the individual's day-to-day life. Another useful descriptor is the "forcefulness" of the tic, as some tics call attention to themselves simply by virtue of their forceful character. Some tics, as barely noticeable (a slight shrug or a hushed guttural noise from the throat) to powerful movements and loud sounds, can be frightening as well as exhausting. Tics also vary in terms of their "complexity." They range from brief, meaningless, abrupt movements or sounds (simple tics), to tics that are longer in duration and seemingly more purposive in character (complex tics).

2.1 Motor tics

Motor tics usually begin with transient tics involving the face or head. A typical history describes repeated eye blinking or other sudden eye movements and/or facial and head movements of variable intensity, beginning in kindergarten or in the early elementary school years. These initial tic movements come and go, and often entirely disappear only to recur at a later point in time. In a typical case, there is often a rostral–caudal progression of motor tics, with tics of the face, head, and shoulders appearing earlier and in a higher proportion of cases than motor tics involving the arms, legs, or the torso (Fig. 1; Jagger et al., 1982; Leckman, King, & Cohen, 1998; Shapiro, Shapiro, Young, & Feinberg, 1988). The observed range of motor tics is extraordinary, so that virtually any voluntary movement can emerge as a motor tic. Table 1 presents a brief listing of some of the common motor tics.

Tics are usually described as being either "simple" or "complex." *Simple motor tics* are sudden, brief (usually less than <1 or −2 s in duration), meaningless movements that involve only a single set of muscles. Common examples include eye blinking, facial grimacing, mouth movements, head jerks, shoulder shrugs, and arm and leg jerks. Early on, many children are totally unaware of their simple motor tics. Over time, many individuals with a tic disorder develop *complex motor tics*, which are more purposive appearing

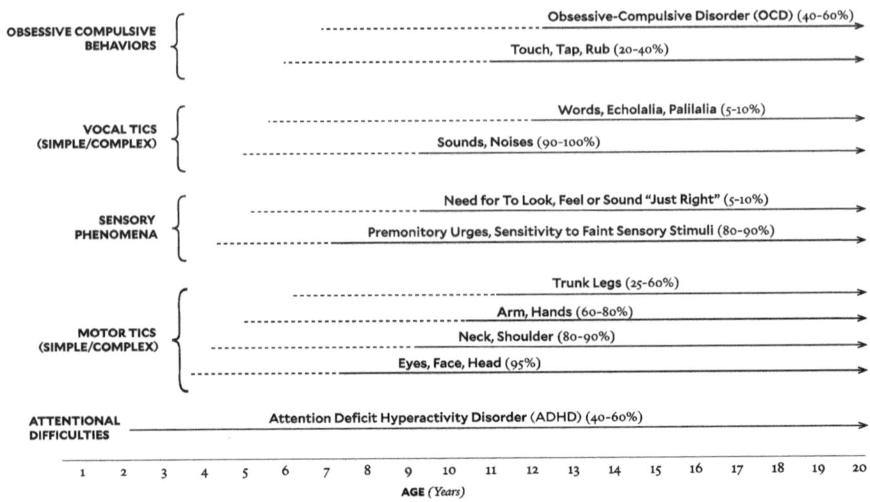

Fig. 1 Natural history of tics and associated behaviors. The occurrence of motor and vocal tics and related behaviors progressing at different ages from transient episodes (broken lines) to sustained periods (solid lines). *Data from Oxford University Press, Jagger, J., Prusoff, B. A., Cohen, D. J., Kidd, K. K., Carbonari, C. M., & John, K. (1982). The epidemiology of Tourette's syndrome: A pilot study.* Schizophrenia Bulletin, *8(2), 267–278. doi:10.1093/schbul/8.2.267, Copyright © 1982.*

Table 1 Simple and complex motor and vocal tics.

Tic symptoms	**Examples**
Simple motor tics: Sudden, brief, meaningless movements	Eye blinking, eye movements, grimacing, nose twitching, mouth movements, jaw snaps, lip pouting, head jerks and turns, shoulder shrugging, arm jerks, abdominal tensing, kicks, finger movements, rapid jerking of any part of the body
Complex motor tics: Slower, longer, more "purposeful" movements	Sustained "looks," facial gestures, biting, touching objects or self, throwing, banging, thrusting arms, gestures with hands, gyrating and bending, twisting dystonic postures, copropraxia (obscene gestures)
Simple phonic tics: Sudden, meaningless, sounds or noises	Throat clearing, coughing, sniffling, spitting, screeching, barking, grunting, gurgling, clacking, hissing, sucking, and innumerable other sounds
Complex phonic tics: Sudden, more "meaningful" utterances	Syllables, words, phrases, statements such as "oh, okay," "I've got to," "stop that," "shut up," "what makes me do this?" "how about it," or "now you've seen it," speech atypicalities (usually rhythms, speed, tone, accents, intensity of speech); echo phenomenon (echolalia, i.e., the immediate repetition of one's own or another's words or phrases); and coprolalia (obscene, inappropriate, and aggressive words and statements)

movements that consist of a sequence of movements that resemble voluntary motor acts (e.g., facial gestures, shoulder shrugging, repetitive touching) (McGuire et al., 2013). These movements are referred as *tonic tics* to distinguish them from *dystonic tics*. Dystonic tics are slower and can involve gyrating, bending, and twisting movements that can result in an abnormal posture. Complex motor tics are rarely seen in the absence of simple motor tics. Paroxysms, or continuous orchestrated displays of simple and complex motor tics, can occur in more severely affected individuals. Fortunately, obscene gestures (*copropraxia*) and self-injurious acts (hitting the face and biting a hand or wrist) are only observed in a small number of cases. At times, it may be difficult to distinguish complex tics from motor dyskinesias, chorea, and other hyperkinetic movement disorders, and some individuals may have more than one movement disorder. For example, motor tics and motor stereotypies are often seen together in individuals with both autism spectrum disorder and a tic disorder (Martino & Hedderly, 2019). In some cases, it can be difficult to distinguish tics from motor stereotypies.

The impact and the level of impairment and behavioral disruption associated with motor tics is highly variable. In addition to the frequency, intensity, complexity, and duration of specific motor tics, estimates of global impairment also need to include the impact on the individual's self-esteem, family relationships, social acceptance, school, or job functioning, and as well as their physical well-being. Patients are often acutely aware of their complex motor tics and their potential negative impact on how they are perceived by their peers and their own self-perception. For example, a recurrent, but dramatic copropraxic gesture can be quite problematic. Physical injuries, including injuries to the spinal cord and blindness from a retinal detachment, also occur in a small minority of adolescent and adult cases, secondary to repetitive, severe self-injurious tics.

2.2 Phonic and/or vocal tics

For many individuals, their vocal and/or phonic tics begin 1–2 years after the onset of their motor tics. Initially, similar to the motor tics, they are usually "simple" in character (e.g., throat clearing, grunting, and squeaks) (see Fig. 1). Vocal tics usually show a similar progression to that seen in motor tics, from transient bouts of tics to the tics occurring over a more sustained period of time (Jagger et al., 1982; Shapiro et al., 1988). Again, the range of possible phonic or vocal symptoms is extraordinary in that any noise or sound a person is capable of making can emerge as a tic (see Table 1).

As with motor tics, phonic symptoms can be characterized by their frequency, duration, intensity (volume), complexity (noises vs. syllables or words), and their level of interference.

Simple vocal-phonic tics are fast, meaningless sounds, or noises. Sniffing, throat clearing, grunting, barks, and high-pitched squeaks are some of the most common simple vocal-phonic symptoms. *Complex vocal-phonic tics* are quite diverse and include syllables, words, or phrases, as well as odd patterns of speech in which there are sudden changes in rate, volume, and/or rhythm. Cavanna, Ali, Leckman, and Robertson (2010) reported the re-occurrence of episodic and contextually inappropriate outbursts of laughter in eight adults with TS. Not surprisingly, this rare complex vocal tic was invariably described as distressing and socially disabling. Immediate echo phenomena, such as repeating words and phrases are common in some patients [*echolalia* (repeating words or phrases said by another person) and *palilalia* (repeating words or phrases said by the individual)]. In a minority of patients, *coprolalia* is present in which socially inappropriate syllables, words, or phrases are repeated, at times in a loud, explosive manner. Shapiro et al. (1988) reported that among the 666 cases with TS evaluated in their specialty clinic, 30% endorsed the presence of one or more of these vocal echo phenomena. Although coprolalia is commonly associated with TS in movies and social media, it is a rare symptom. Indeed, even among adults with tics who seek treatment with specialists, only 20–35% have a history of coprolalia (Freeman et al., 2009). The Freeman study was based on approximately 600 consecutive new patients with TS at 15 clinical sites in seven countries. In that study, the age at onset for coprolalia was 11 years. Over a course of days, complex vocal-phonic tics are rarely, if ever, present in the absence of simple vocal tics as well as motor tics.

2.3 Clusters of tic symptoms

Tic symptoms are classically separated along dimensions of simple versus complex tics. Several studies have sought to identify possible tic symptom dimensions (Alsobrook & Pauls, 2002; Kircanski, Woods, Chang, Ricketts, & Piacentini, 2010; Mathews et al., 2007; McGuire et al., 2013; Robertson, Althoff, Hafez, & Pauls, 2008; Robertson & Cavanna, 2007). For example, McGuire et al. (2013) reported the results of an investigation of tic symptom clusters, using a sample of 239 youth and adults diagnosed with TS or another chronic tic disorder, across two university-based outpatient clinics (Fig. 2). Their hierarchical analysis of the 40 tics included in the

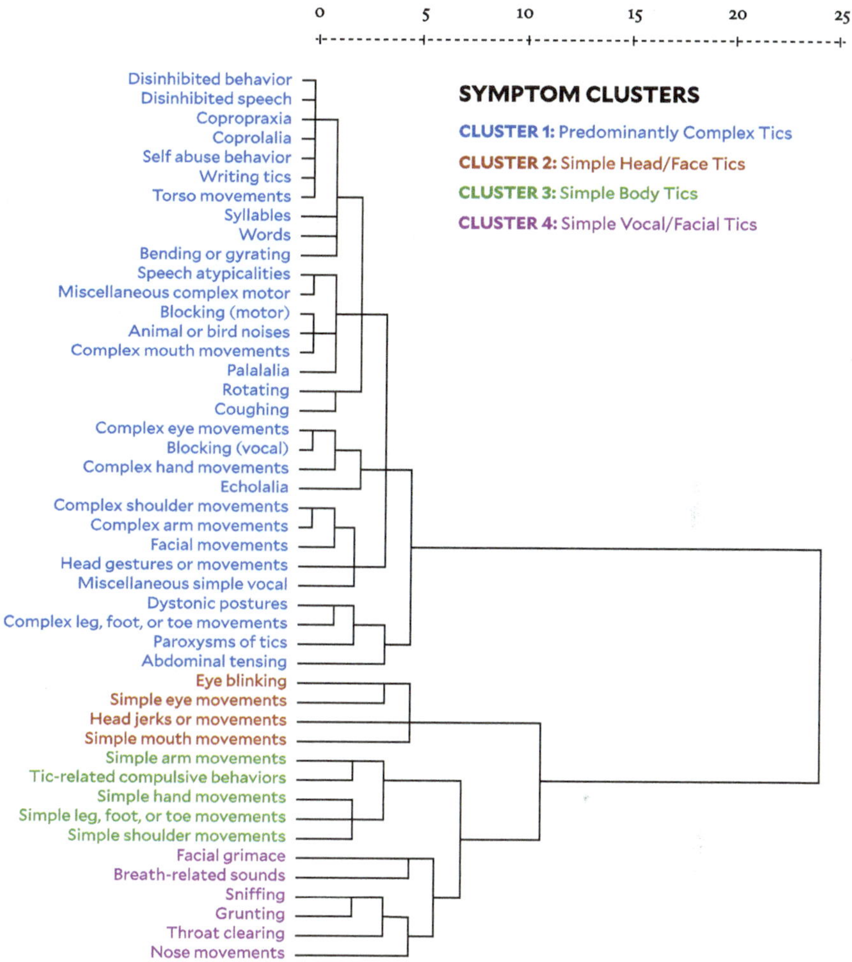

Fig. 2 Tic symptom dimensions. This figure presents the rescaled dendrogram of agglomerative hierarchical cluster analysis using Ward's method of the 40 YGTSS symptom checklist items. Text colors indicate clustered symptoms. *Reprinted by permission from Springer Nature Customer Service Centre GmbH: Nature/Springer/Palgrave, Kircanski, K., Woods, D. W., Chang, S. W., Ricketts, E. J., & Piacentini, J. C. (2010). Cluster analysis of the Yale global tic severity scale (YGTSS): Symptom dimensions and clinical correlates in an outpatient youth sample.* Journal of Abnormal Child Psychology, 38*(6), 777–788. doi:https://doi.org/10.1007/s10802-010-9410-5, Copyright © 2010.*

Yale Global Tic Severity Scale (YGTSS) symptom checklist identified four clusters: (1) complex phonic tics, and disinhibited behaviors and vocalizations; (2) complex motor tics; (3) simple motor and/or vocal tics involving the head region; and (4) primarily, simple motor tics. The identified clusters were similar for both the adults and the adolescents, and largely replicated the results of an earlier cluster analysis (Kircanski et al., 2010). Interestingly, there was no association with any of the clusters with comorbid conditions, including Attention-deficit/hyperactivity disorder (ADHD), obsessive-compulsive disorder (OCD), or an anxiety disorder. Unfortunately, an individual's cluster membership scores were not associated with their response to either cognitive–behavioral interventions or pharmacological treatments. Further studies of symptom dimensions in tic disorders are needed. They may lead to the identification of tic symptom clusters that are differentially associated with an individual's clinical long-term outcome (Groth, Debes, & Skov, 2017; Rizzo, Gulisano, Calì, & Curatolo, 2011), as well as greater clarity concerning the neurocircuitry underlying their disorder (Bhikram, Crawley, Arnold, Abi-Jaoude, & Sandor, 2020).

3. Premonitory urges

Although motor and vocal-phonic tics are the defining feature of TS, many individuals report that some of their tics are preceded by a localized bodily sensation and/or an urge to tic (Cavanna, Black, Hallett, & Voon, 2017; Cohen & Leckman, 1992; Leckman, Walker, & Cohen, 1993; Woods, Piacentini, Himle, & Chang, 2005). For many individuals, it is often difficult to find the right words to accurately describe these sensations and urges. Interestingly, there is often a momentary feeling of "relief" following the performance of a tic or a series of tics. Consequently, the tics associated with these *premonitory sensations and urges* are often experienced as not being completely "involuntary" in that the urge and the associated tic can be consciously resisted. The premonitory urges usually are focal in character and limited to a specific anatomical location (Fig. 3). Tics involving the head, neck, shoulders, or the mid-line abdomen were most frequently preceded by these urges, while simple tics, such as eye blinking and mouth movements, were less likely to be preceded by urges. Some individuals report a "sense of inner tension" (Chee & Sachdev, 1997; Kurlan, Lichter, & Hewitt, 1989). Many individuals report having both focal and more generalized antecedent urges and sensations. In some instances, these premonitory sensations and urges are experienced as being more troublesome than the tics themselves (Kane, 1994).

Initial research on premonitory urges stemmed from detailed personal narratives provided by individuals with TS. For example, Bliss, Cohen, and Freedman (1980) described the sensory signals that preceded his tics

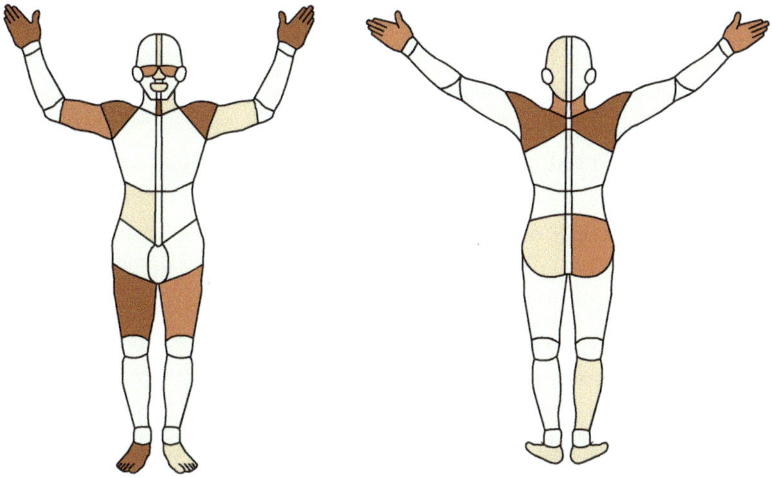

Fig. 3 Density of premonitory sensations and urges. The densities of premonitory urges for each of 89 anatomical regions are depicted indicates the proportion of cases that identified a specific body region as a site where they experienced a premonitory urge. The highest density (intensity of the red color) on the scale represents the total premonitory urges per region per person, the lowest "0" urges per region per person and the midpoint "20" urges per region per person. These data are based on self-reported data ($N = 135$). *Reprinted from Leckman, J. F., Walker, D. E., & Cohen, D. J. (1993). Premonitory urges in Tourette's syndrome.* The American Journal of Psychiatry, 150*(1), 98–102. doi:10.1176/ajp.150.1.98 (Copyright © 1993). American Psychiatric Association. All Rights Reserved.*

as well as "a very rapidly escalating desire to satisfy the sensations with movements intended to free oneself from the insistent feeling" (p. 37). Kane (1994) confirmed this description and added, "These sensations are not mere precursors to tics...They precipitate tics more than providing a signal of imminence, the pre-tic sensation acts as the aversive stimulus toward which tics are directed" (p. 806). These urges often involve a feeling of pressure or tingling localized in the muscles involved in the performance of the tics. Some individuals with tics perceive these urges and other sensory phenomena as being the "core" of TS (Hollenbeck, 2001). As with tics, the occurrences of urges vary in their frequency, intensity, and duration (Woods et al., 2005).

Developmentally, many individuals with a tic disorder report that they first became aware of premonitory urges 2–3 years after the onset of their tics. A majority of older children and adults confirm the presence of premonitory urges associated with one or more of their tics (Banaschewski, Woerner, & Rothenberger, 2003; Kwak, Dat Vuong, & Jankovic, 2003).

Advances in this field include the development of the Premonitory Urge for Tics Scale (PUTS) (Woods et al., 2005). The PUTS was found to have excellent psychometric properties for children older than age 10 years, and the PUTS scores were significantly correlated with overall tic severity. More recently, McGuire et al. (2016) developed the Individualized Premonitory Urge for Tics Scale (I-PUTS), which clinicians administer to assess urge severity across multiple dimensions (number, frequency, and intensity). In this study 75 youth aged 6–17 years described experiencing premonitory urges approximately 50% of the time when they had a tic. In a majority of cases, these urges were localized in the head/face, neck/throat, and arm regions. Although other regions had lower urge frequency, whole body/other urges had the greatest intensity compared to all other body regions. Urges experienced in the neck/throat region also exhibited an elevated urge intensity.

The development of the PUTS and I-PUTS has made it possible to track changes in the intensity of premonitory urges in response to pharmacological treatments (Jankovic, Jimenez-Shahed, & Brown, 2010; Rath, Tavy, Wertenbroek, van Woerkom, & de Bruijn, 2010). For example, the severity of premonitory urges has been shown to improve as the tics improve with some medications, including topiramate and Botulinum toxin.

An increased awareness of the importance of premonitory urges has also led to significant advances in cognitive–behavioral interventions for tics. More specifically, the Comprehensive Behavioral Intervention for Tics (C-BIT) is based on enhancing the individual's awareness of their premonitory urges and then working with the individual to develop and practice a "competing response" that is specific to a specific premonitory urge (Houghton, Capriotti, Conelea, & Woods, 2014; Piacentini et al., 2010; Pringsheim et al., 2019). C-BIT has been identified both by the American Academy of Neurology (see: https://www.aan.com/GuidelineDetail/958) and the European Society for the Study of Tourette Syndrome (ESSTS; Müller-Vahl et al., 2021) as a first line treatment for individuals with tics, as CBIT is effective in both children and adults and has no major side effects (Woods et al., 2011).

Moving forward, it will be important to gain a deeper understanding of the neurobiological basis of premonitory urges and their interconnection with the circuitry underlying the tics and the momentary relief that is associated with the actually performing of a tic (Evers & van de Wetering, 1994; Jackson, Sigurdsson, Dyke, Condon, & Jackson, 2021; Woods et al., 2005).

4. Heightened sensitivity to sensory perceptions

In addition to premonitory urges, many individuals with tic disorders are often times remarkably sensitive to sensory perceptions arising from their bodies as well as the external world. As first noted by Gilles de la Tourette (1885) and described previously, individuals may unconsciously mirror the behavior (echopraxia) and speech (echolalia) of others as well as of themselves (palilalia): They do and say what they have just seen or heard. Other examples include *interoception*, *site sensitization*, and the need for things to be "just-right," based on visual, tactile, and/or auditory perceptions.

4.1 Interoception

In addition to the premonitory urges, there is a growing interest in *interception*—how perceptions of internal body sensations and the feeling states that they engender may contribute to tic severity and comorbid symptoms. Both adults and children with tic disorders have been shown to have a lower ability to accurately perceive internal body sensations, compared with healthy control subjects (Ganos et al., 2015; Pile, Lau, Topor, Hedderly, & Robinson, 2018; Rae, Larsson, Garfinkel, & Critchley, 2019). Some studies, but not all, have documented that the level of interoceptive awareness was significantly associated with the intensity of their premonitory urges (Ganos et al., 2015; Rae et al., 2019). More recently, Pile et al. (2018) found that individuals with tic disorders demonstrated lower interoceptive accuracy than controls, and sadly, higher levels of interoceptive accuracy were associated with higher anxiety and lower quality of life, but not with premonitory urge (Pile et al., 2018). There is also evidence that interoceptive accuracy can be altered in response to targeted interventions (Isaacs & Riordan, 2020; Pile et al., 2018).

Investigators have sought to advance our understanding of the processing of sensory stimuli in individuals with tic disorders through the use of behavioral and neurophysiological tests. For example, the Sensory Gating Inventory (SGI) is a self-report instrument that includes many items relevant to the sensory phenomena associated with tic disorders (Hetrick, Erickson, & Smith, 2012; Sutherland Owens, Miguel, & Swerdlow, 2011). Another useful tool for assessing and monitoring sensory phenomena in subjects with tic disorders is the University of São Paulo Sensory Phenomena Scale (USP-SPS) (Miguel et al., 2000; Rosario et al., 2009).

Initially developed to focus primarily on individuals with OCD, the authors found that sensory phenomena were present in two-thirds of the individuals with OCD and that the prevalence was significantly higher in early-onset OCD patients. In addition, tics were present in 20% of the subjects. There were no significant differences based on the co-occurrence of tics.

Likewise, investigators have used physiological procedures, including pre-pulse inhibition (PPI) and event-related potential (ERP), in an effort to gain a more complete understanding of the hypothesized abnormalities in cortical-subcortical circuits involved in sensory processing. For example, PPI is a simple behavioral measure of inhibition of the startle blink reflex, referring to reduction in startle blink magnitude when a stimulus (prepulse) occurs 30–500 ms before a startle stimulus. The prepulse is believed to activate automatic brain mechanisms that protect or "gate" the processing of that sensory stimulus for a brief window of time. Several studies have shown reduced PPI in TS subjects, compared to healthy controls (Buse, Beste, Herrmann, & Roessner, 2016; Castellanos et al., 1996; Smith & Lees, 1989; Swerdlow, 2013; Zebardast et al., 2013). PPI may emerge as an endophenotype of TS that can be utilized in both human and animal studies. The electrophysiological underpinnings of tic disorders are addressed more fully in subsequent chapters "Integrating electrophysiology into cognitive-behavioral treatment for the Gilles de la Tourette Syndrome" by Lavoie; "Neurotransmitters involved in the pathophysiology of GTS" by Hartmann; "Control of actions and tics: A psychophysiological perspective" by Morand-Beaulieu; "Integrating electrophysiology into cognitive-behavioral treatment for the Gilles de la Tourette Syndrome" by Lavoie; "Network oscillatory activity in chronic tics and Tourette syndrome" by Jimenez-Shahed; "Co-occurring ADHD symptoms and electrophysiological correlates of cognitive control in young people with Tourette syndrome" by Shephard.

4.2 Site sensitization

Individuals with tic disorders are often acutely aware of, distracted by, and at times distressed by faint sensory stimuli. Examples include faint sensations associated with feeling tags on shirts or pants, elastic bands on socks, and the sound of others chewing food. Some individuals find it very difficult to attend to more salient stimuli (Cohen & Leckman, 1992). Belluscio, Jin, Watters, Lee, and Hallett (2011) found that the most bothersome sensory stimuli were those that were faint, repetitive, or constant and non-salient. Investigators have also documented that those individuals with

TS do not differ from control subjects with regard to their ability to detect specific olfactory and tactile stimuli. These findings have been independently replicated using olfactory and tactile stimuli (Schunke et al., 2016; Weisman et al., 2018). The presence of neuropsychiatric comorbidities in TS is associated with even more profound sensory processing dysfunction as well as a significant reduction in the individual's quality of life (Isaacs & Riordan, 2020; Weisman et al., 2018).

4.3 "Just-right" phenomena

Another set of sensory phenomena frequently encountered in individuals with tic disorders involves a need for things to feel, look, or sound "just right" (Leckman, Walker, Goodman, Pauls, & Cohen, 1994). In our second study of more than 130 subjects with tic disorders (aged 9–71 years), 59 (44%) reported the presence of "just right" phenomena. Most individuals could readily distinguish these "just right" sensations from the premonitory urges associated with tic behavior. Frequently, this took the form of pointing out that the "just right" perception was more of a mental phenomenon than a bodily sensation. The "just right" awareness, most commonly referred to, was for visual (31%) or tactile (25%), as opposed to auditory (10%) perceptions. These symptoms were far more common in individuals with comorbid OCD (81%) compared to individuals with sub-clinical OCD (61%). Worbe et al. (2010), using a semi-structured interview, also found that 30% of 166 consecutive patients with TS (aged 15–68 years) endorsed the presence of "just right" perceptions. When the cases with TS were stratified, according to whether or not they had repetitive behaviors and thoughts that were "tic-like" verses "OCD-like," the tic-like group had significantly higher rates of the "just right" perceptions.

As mentioned earlier, the USP-SPS (Miguel et al., 2000; Rosario et al., 2009) was developed to characterize the sensory abnormalities frequently encountered in individuals with OCD, as well as with tic disorders. In the initial study, Miguel et al. (2000) performed in-depth interviews with 62 adults (aged >17 years), including 21 adult subjects with TS without OCD, 20 cases with TS with comorbid OCD, and 20 with OCD alone. They found that 90% of the cases with TS plus OCD, versus 48% of those with TS alone, reported the presence of one or more "just right" (visual > tactile > auditory) perceptions. The rate in the cases with OCD alone was just 35%. Although "just right" sensations appear to be distinct from premonitory urges, one study did report a correlation between PUTS and USP-SPS scores (Sutherland Owens et al., 2011).

5. Timing, course, and context of tics

5.1 Waxing and waning of tics

Virtually all descriptions of tic disorders describe their *waxing and waning course*. This simply means that over any period of time (minutes, hours, days, weeks, months, and years), an individual's tics get better or worse. We still do not understand why this happens. Is it due to changes in stress levels or the seasons of the year? Or perhaps it is due to some intrinsic properties of the neurons and neural circuits that underlie tics (Leckman, Vaccarino, Kalanithi, & Rothenberger, 2006). One of the few empirical studies on this topic suggested the presence of a "fractal process" in the tic time-series (Chappell et al., 1994; Peterson & Leckman, 1998). Fractals are infinitely complex patterns that are "self-similar" across different time scales. If true, this simply means that for any temporal dimension (minutes, hours, days, weeks, months, or years), the same pattern of tics is evident.

More specifically, individual tics are known to occur in discrete bursts or *bouts* with very brief inter-tic intervals (measured in seconds, or fractions of seconds). Similarly, for longer time scales measured in hours, these bouts of tics also occur in bouts so that on any given day there are both tic-free periods and periods of intense ticcing. For long time scales measured in weeks or months, a waxing and waning pattern is also evident, although the timing of tics over longer time intervals has not been vigorously evaluated (Fig. 4).

In one project, we videotaped more than 20 medication-free subjects with TS (aged 8–49 years) for extended periods of time (Peterson & Leckman, 1998). The time intervals between temporally adjacent tics were measured, and the statistical properties of these intervals were assessed through autoregressive integrated moving average (ARIMA) modeling, and the construction of first return maps. ARIMA is a statistical analysis model that uses time series data to determine if earlier time intervals predict future time intervals. Remarkably, the frequency distribution of tic interval durations followed an inverse power law of temporal scaling, which is typical of fractal processes (Fig. 5). Spectral analyses similarly demonstrated that the spectral power density of tic interval duration was related inversely to frequency. The first return maps revealed the "burst-like" behavior and short-term periodicity in tics (Peterson & Leckman, 1998). Consequently, the "bouts of bouts" of tics are not haphazard events, but instead demonstrate the properties observed in non-linear dynamical systems (Hashemiyoon, Kuhn, & Visser-Vandewalle, 2017). Given the time-intensive nature of this

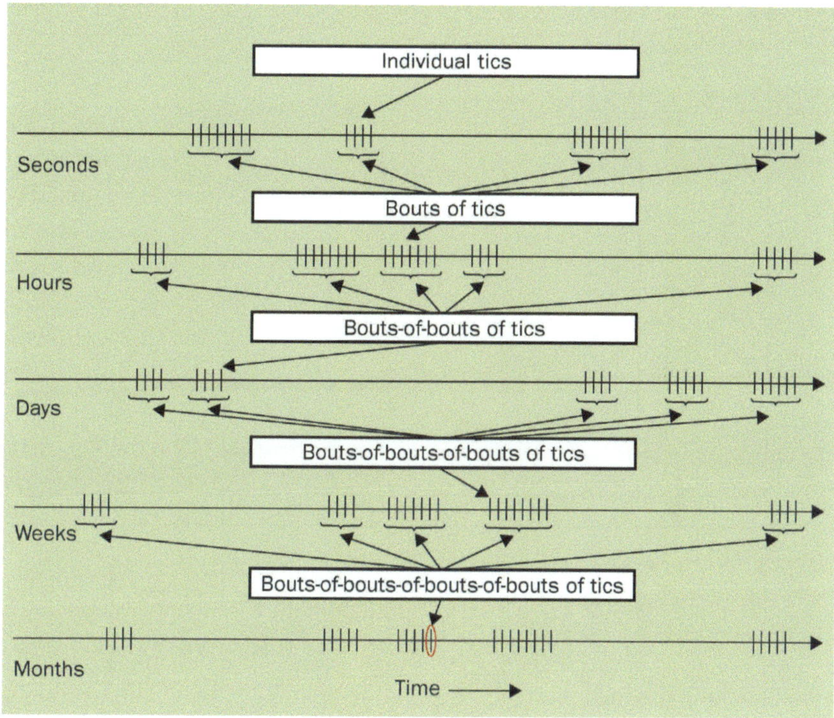

Fig. 4 Fractal character of the temporal occurrence of tics. Progressively longer time scales (seconds to months) are depicted in this figure (A–E). Tics occur in bouts (A) and bouts of tics occur in bouts (B). We predict that regardless of the temporal dimension, the bout-like appearance of tics (or higher order combinations of bouts and bouts-of-bouts of tics) will be present. This fractal quality may well underlie the waxing and waning of tics observed over weeks to months (E) as well as other features of the natural history of Tourette syndrome. *Reprinted from Leckman, J. F. (2002). Tourette's syndrome.* The Lancet (British Edition), 360 *(9345), 1577–1586. doi:https://doi.org/10.1016/S0140-6736(02)11526-1, Copyright © 2002, with permission from Elsevier. Data from Peterson, B. S., & Leckman, J. F. (1998). The temporal dynamics of tics in Gilles de la Tourette syndrome.* Biological Psychiatry, 44 *(12), 1337–1348. doi:10.1016/S0006-3223(98)00176-0, Copyright © 1998.*

research, the development, and refinement of automated tic counting devices may be of great value (Bernabei et al., 2010; Brandt, Patalay, Bäumer, Brass, & Münchau, 2016).

If these studies are replicated, tics then might be best understood as being the product of processes that emerge over multiple time scales, from milliseconds to minutes, to days, weeks, months and years, and perhaps even decades. As such, they may be exemplars of a dynamical system at work

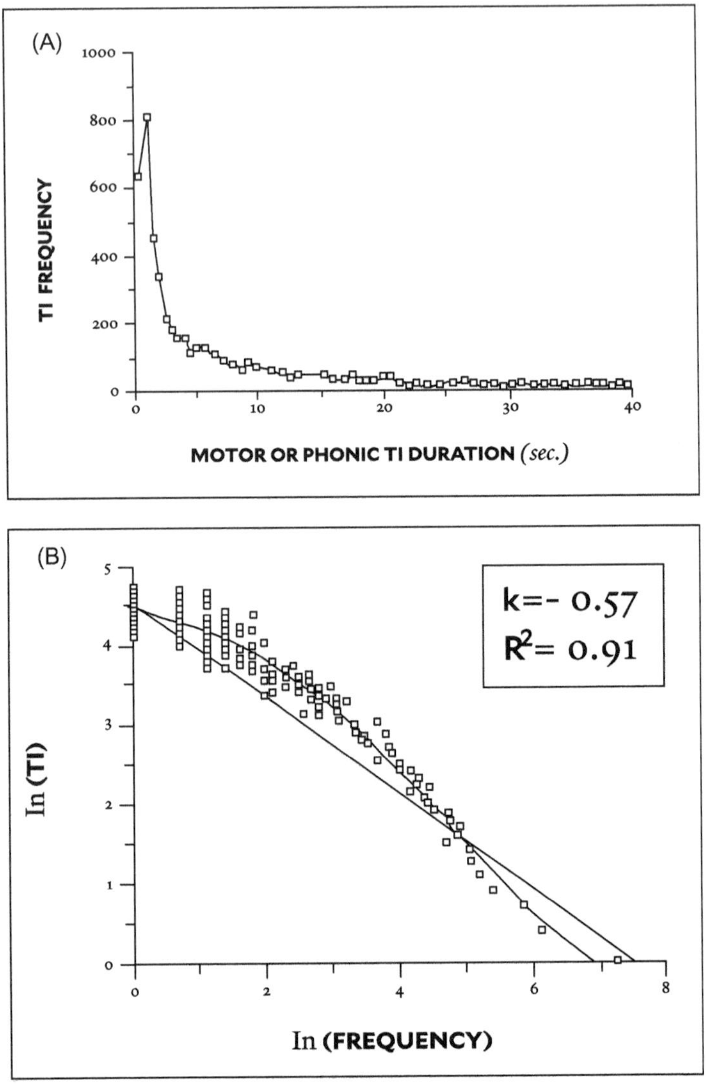

Fig. 5 Tics in time, evidence for the inverse Power law. (A) Group composite inverse power law frequency distributions. Tic interval (TI) frequency count plotted as a function of TI duration (in seconds) for time series in which both motor and phonic tics were regarded as "events." (B) The log-log transforms of the frequency distributions produce a straight-line relationship between TI duration and TI frequency. The slope (*k*) of this descending line segment was estimated using simple linear regression; R^2 is the amount of variance in the distribution explained by the regression equation. *Reprinted from Peterson, B. S., & Leckman, J. F. (1998). The temporal dynamics of tics in Gilles de la Tourette syndrome.* Biological Psychiatry, 44 *(12), 1337–1348. doi:10.1016/S0006-3223 (98)00176-0, Copyright © 1998, with permission from Elsevier.*

in neurodevelopment (Smith & Thelen, 2003). It is possible that an examination of the underlying neural mechanisms of this fractal phenomenon may identify related events that can be measured in milliseconds that are associated with the generation of tics. Surely, the timing of the firing of sets of interconnected neurons may also provide important clues to clarifying the underlying neurobiology of tics (Bortolato & Pittenger, 2017).

5.2 Course of tic disorders

Tics typically have an onset between the ages of 4 and 6 years and reach their worst-ever severity between the ages of 10 and 12 years (Fig. 6). On average, tic severity declines during adolescence. By early adulthood, roughly three-quarters of individuals with TS will have greatly diminished tic symptoms, and over one-third will be virtually tic free. However, as discussed later in this chapter, individuals with TS often have symptoms associated with attentional difficulties as well as obsessions and compulsions.

5.3 Tic exacerbations

Although there is compelling data to support the inherent fractal nature of the timing of tics, the description of tics as simply intermittent trains of

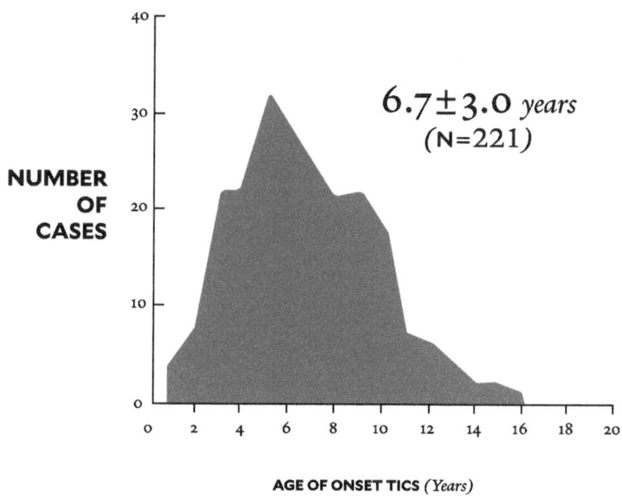

Fig. 6 Age of onset distribution. This figure presents the age of onset of tics in 221 individuals with Tourette syndrome evaluated at the Yale Child Study Center TS/OCD Clinic. *Reprinted from Leckman, J. F., & Cohen, D. J. (1998).* Tourette syndrome: Tics, obsessions, compulsions. Developmental psychopathology and clinical care. *John Wiley and Sons, Inc., Wiley Books, Copyright © 1998, with permission from Elsevier.*

semi-involuntary motor or phonic acts, is incomplete. The intensity and severity of tics is highly variable. There can be a sudden exacerbation of symptoms. In an effort to establish an objective, prospective, and quantitative method for identifying and monitoring symptom exacerbations, Lin et al. (2002) modeled monthly consecutive YGTSS scores in 64 youth (aged 7–16 years) for periods ranging from 3 to 39 months. Exacerbation thresholds were estimated by using state-of-the-art bootstrap methods. Not surprisingly, the severity of tic symptoms was quite variable. When both tic and OC symptoms were present, there was a significant degree of covariation in the severity of the tic and OC symptoms.

5.4 Contextual factors influencing tic severity

Tic expression can be influenced, both worsened and attenuated, by *antecedent environmental events* including *sensory cues* as well as the individual's *emotional state*. A list of many of these contextual factors are presented in Table 2. Information supporting each of these factors has come from a broad range of sources ranging from personal diaries (O'Connor et al., 1994; O'Connor et al., 2003) as well as anecdotal accounts from clinicians (Bornstein et al., 1990; Eapen et al., 2004; Robertson et al., 2002; Silva et al., 1995). Structured observational and experimental studies have also been conducted (Barnea et al., 2016; Conelea et al., 2011; Meidinger et al., 2005; Piacentini et al., 2006; Scahill et al., 2001; Steinberg, Shmuel-Baruch, Horesh, & Apter, 2013; Watson et al., 2005; Wood et al., 2003; Woods et al., 2001). Large cross-sectional and prospective longitudinal studies have also been completed (Hoekstra et al., 2004; Lin et al., 2007). More recently, in a cross-sectional study of 132 children with TS, Horesh et al. (2018) found that a higher quantity of negative major life events and a lower quantity of positive major life events were associated with more severe expression of tics and had a negative impact on the child's self-esteem and social participation. Indeed, a compelling case can be made that an individual's current level of psychosocial stress is associated with future levels of tic severity (Bornstein et al., 1990; Cohen & Leckman, 1992; Eapen et al., 2004; Hoekstra et al., 2004; Lin et al., 2007, 2010; Robertson et al., 2002; Silva et al., 1995).

Another potentially related phenomenon is the emergence of inappropriate, and at times even dangerous responses to sensory cues in the environment. In an initial study, 10 (36%) of the 28 subjects with TS reported examples of *reflexive tics* (Cohen & Leckman, 1992). Examples included the urge to make a loud vocal tic in a quiet library immediately upon seeing

Table 2 What makes tics better or worse?[a]

Better	• Sleep (almost always the tics will disappear)
	• Doing something that requires focused attention and motor control, like riding a bicycle or playing ping-pong or playing a musical instrument (almost always)
	• Relaxation (variable)
	• Verbal instructions to suppress tics, especially with incentives
	• During and after physical exercise (usually the tics will diminish or be less forceful)
Worse	• Stressful life events (50–90%)
	• Returning to school (60–70%)
	• Being anxious or upset or noticing stress in a loved one (80–90%)
	• Excitement, like going off to Disney World for the first time, birthdays, holidays especially with presents (60–70%)
	• Watching exciting movies (in the theater) or television shows (almost always)
	• Fatigued, being tired, but not quite ready to fall asleep (almost always) (50–75%)
	• Being alone (40–50%)
	• Quiet reading tasks
	• Talking about tics, tics are "suggestible" (almost always)
	• Menstruation (only for some women)
	• Eating (only for some individuals)
	• Drinking coffee (only for some individuals)
	• Being too hot, hot weather (only for some individuals)

[a]Based in part on the review of this field by Conelea and Woods (2008a). The self-report and anecdotal data are drawn from descriptive studies (Bornstein, Stefl, & Hammond, 1990; Eapen, Fox-Hiley, Banerjee, & Robertson, 2004; Robertson, Banerjee, Eapen, & Fox-Hiley, 2002; Silva, Munoz, Barickman, & Friedhoff, 1995; Zinner, Conelea, Glew, Woods, & Budman, 2012); Self report diaries (O'Connor, Brisebois, Brault, Robillard, & Loiselle, 2003; O'Connor, Gareau, & Blowers, 1994); experimental studies (Conelea, Woods, & Brandt, 2011; Meidinger et al., 2005; Piacentini et al., 2006; Scahill et al., 2001; Watson, Dufrene, Weaver, Butler, & Meeks, 2005; Wood et al., 2003; Woods, Watson, Wolfe, Twohig, & Friman, 2001); and prospective longitudinal studies (Hoekstra, Steenhuis, Kallenberg, & Minderaa, 2004; Lin et al., 2007).

the sign, "Quiet Please." Others described the "need" to touch a hot iron or to put the car in reverse gear while driving down a highway or touch the foot of an adjacent person. In another instance, a child had load vocal tic just when the teacher was trying to get the attention of the class. Another extreme example was related by a physicist, who during World War II, had to give up a job in a high energy physics laboratory, because whenever he saw the sign "Danger High Voltage," he had the strong urge to touch the apparatus. Cavanna et al. (2010) noted that two of the eight individuals with laughing tics reported that their tic was cued by hearing certain words or phrases. In one instance, this tic only occurred when an adolescent subject heard the word "tree." When his classmates realized this, he was subjected to unrelenting teasing in the school environment that only subsided through the active intervention of another student and the school personnel.

Experimental studies have also documented that tic-related conversations can prompt transient worsening of tics in some individuals. For example, Woods et al. (2001) performed a functional analysis of the effects of "tic-related" talk on the frequency of vocal and motor tics in two boys (aged 6 and 16 years) with TS. Using the A-B-A-B withdrawal designs, these two boys were alternately exposed to conditions with and without talk of their tics. They found that vocal tics markedly increased when the content pertained to tics and decreased when talk did not pertain to tics. Contextual distractions do not appear to affect tic frequency (Conelea & Woods, 2008b).

Interestingly, Wood et al. (2003) conducted a detailed study of four children (aged 8–14 years) as they watched a movie known to elicit specific emotional responses. They found that the frequency of the children's tics was highest during periods associated with anticipation and the resolution of emotional charged situations, and lowest during periods depicting moments of anger and happiness, and intermediate during periods of sadness and fear. These investigators also monitored the activity of the children's autonomic nervous systems while they watched the movie. Although the frequency of the child's tics was influenced differentially by various emotional states, no effect was evident regarding the activity of their autonomic nervous system. This finding is perhaps consistent with the observation by Belluscio et al. (2011) that individuals with tic disorders show a greater sensitivity to faint versus more intense sensory stimuli.

Tics are also *suggestible*. A classic example regularly occurs during clinical consulting when the individual is asked about specific tic symptoms. In this context, some patients and other family members with a lifetime history of tics will display a particular type of tic just after that type of tic is mentioned.

Remarkably, some of these tics are not part of the individual's current repertoire of tics.

Likewise, tics can be *contagious*, such that when an individual is in a group of individuals with severe tics, one individual's tics can rapidly spread across the entire group (thoughts of group meetings hosted by the Tourette Association of America where when one individual made a loud vocal tic, others spontaneously repeated the very same tic even though it had never been part of their repertoire of vocal tics).

Another reality in this age of social media, is that tics and tic-like behaviors can be contagious and "go viral." Müller-Vahl, Pisarenko, Jakubovski, and Fremer (2021) recently described the "outbreak" of a new type of mass sociogenic illness in Germany. The group in Germany have labeled this phenomenon "mass social media-induced illness" (MSMI). The outbreak of MSMI was initiated by a "virtual" index case that was posted on YouTube. Affected teenagers present with similar or identical functional "Tourette-like" behaviors, which can be clearly differentiated from tics in Tourette syndrome.

A team in Canada, led by Davide Martino, is leading a study to expand our knowledge of this phenomenon by pooling data from a large group of experts from around the world (see: https://www.essts.org/news/survey-on-functional-tic-like-behaviours; https://cumming.ucalgary.ca/resource/tourette-ocd/children-and-adults/disorder-specificresources/tourette-syndrome-and-0; https://www.medscape.com/viewarticle/949882). Remarkably, seeing tics and tic-like behaviors on interactive internet sites, including TikTok, has led to a substantial number of individuals developing complex tic-like behaviors. Adolescent females appear to be at greater risk of developing tic-like behaviors. The group in Germany has suggested that these "Tourette-like" symptoms can be viewed as the 21st century expression of a culture bound stress reaction of our post-modern society that celebrates unique individuals. Current estimates are that these attention-seeking behaviors are present in large numbers of young people across the globe, since spread via social media is no longer restricted to specific locations, such as local communities or school environments. This has the potential to have a considerable impact on health care systems, and society as a whole.

In contrast, there are also circumstances where tics virtually disappear. In one of the few studies to date, Bodeck, Lappe, and Evers (2015) assessed the impact of engaging in musical activity, both listening and performing, on tic frequency in a group of 29 individuals. A within-subject repeated measures design was also conducted in eight individuals. Five experimental conditions were tested: baseline, during a musical performance, versus just listening to music. Tics were counted using videotape recordings. Consistent with our clinical experience, the tics almost completely disappeared during an actual musical performance, and the tics were also less frequent when the individual was listening to music.

5.4.1 Comorbid conditions

It is also clear that for many children and adults, their tics are just part of the story. As discussed by several other chapters in both Parts A & B of this volume, there are the range of other difficulties frequently encountered by individuals with chronic tic disorders including OCD, ADHD, Mood Disorders, anger control problems, and in a minority, Autism Spectrum Disorders, and Learning Disabilities (see Fig. 1). These comorbid conditions occur more frequently in individuals with TS compared to those with just a chronic motor or vocal tic disorder (Khalifa & von Knorring, 2006) and often are of greater emotional and prognostic significance than are the tics themselves, given their impact on self-esteem, family and peer relationships, academic performance, peer acceptance, early mortality, as well as suicidal ideation and attempts (Fernández de la Cruz et al., 2017; Gorman et al., 2010; Johnco et al., 2016; Meier, Dalsgaard, Mortensen, Leckman, & Plessen, 2017; Sukhodolsky et al., 2003; Weingarden et al., 2018). In addition to these neuropsychiatric disorders, children and adults with chronic tic disorders are also at higher risk to develop traumatic spinal cord and brain injuries (Chen, Su, Wang, Hsu, & Shen, 2019; Patterson et al., 2016), migraine headaches (Ghosh et al., 2012; Kwak et al., 2003) and sleep disorders (Ghosh et al., 2014; Lee et al., 2017; Martino et al., 2017; Ricketts et al., 2018).

As reviewed elsewhere in this volume, the presence of comorbid ADHD and OCD may be differentially associated with specific tic symptoms. Thus far, there have been at least 10 studies that have examined symptom factors/clusters in cohorts of well characterized cases with TS. For example, Cavanna et al. (2011) analyzed symptom data from a sample of 639 patients with TS and identified three factors that in total accounted for approximately 45% of the phenotypic variance. These factors included: (1) complex motor tics and echo-phenomena (echolalia, echopraxia, and palilalia) and OC behaviors; (2) complex vocal tics and coprolalia and copropraxia; and (3) attention-deficit and hyperactivity symptoms plus aggressive behaviors and OC behaviors.

A better characterization of the TS phenotypes may help identify the molecular and neural pathways underlying this etiologically heterogeneous condition. For example, Darrow et al. (2017) examined the data from more than 3000 individuals recruited for genetic studies and completed symptom-level factor and latent class analyses that were then replicated in an independent sample of approximately 900 individuals. Heritability and polygenic load associated with TS, OCD, and ADHD were estimated. Using these genetic data, they identified two cross-disorder symptom-based phenotypes. One related to a subset of OC symptoms (a need for symmetry, evening up, checking obsessions; ordering, arranging, counting, writing–rewriting compulsions; and repetitive writing tics). The second related to a range of symptoms associated with disinhibition (uttering syllables/words, echolalia/palilalia, coprolalia/copropraxia, and obsessive urges to offend/mutilate/be destructive).

Although tics are the sine qua non of TS, they are often not the most enduring or impairing symptoms in children with TS. Measures used to enhance self-esteem, such as encouraging strong friendships and the exploration of interests, are crucial to ensuring positive adulthood outcome in TS.

6. Conclusions and questions for future research

As briefly summarized in Box 1, for more than 150 years, a growing literature has focused on the phenomenology of tics and tic disorders. There has been increased attention to efforts to understand and describe the sensory, subjective, and contextual experiences associated with TS. Indeed, when clinicians, parents, teachers, and peers understand the sensory aspects of TS, they "understand" at some level "why" the child is doing what they are doing. This increased level of understanding has also led to refinements of existing cognitive behavioral interventions.

One important focus of attention has been to explore and understand the contextual factors that influence tic suppression. Other major advances concern efforts to understand the timing of tics and their inherent waxing and waning character. Although tics, in some form, persist into adulthood, it has also become clear that tic severity typically peaks early in the second decade with many patients showing a marked reduction in tic severity by the age of 19 or 20 years. Another important area of research has been focused on understanding the neural basis of co-occurring OCD and ADHD symptoms. For better or worse, tics alone are the exception rather than the rule. Indeed, the presence of chronic motor and/or vocal tics *alone* in the absence of other difficulties usually heralds a positive outcome—especially in the presence of other strengths.

BOX 1 Highlights

- Tics often arise from a *heightened and selective sensitivity* to cues from within the body or from the outside world. This sensitivity may be due in part to a loss in the normal "automatic" ability to suppress or "gate" or habituate irrelevant information in sensory, motor, and cognitive domains
- Motor and phonic tics occur in "bouts" over the course of a day and wax and wane in severity over the course of weeks to months. Less well known is the "self-similarity" of these patterns across different time scales. Understanding the dynamical patterns of tic expression may be useful to families, clinicians and teachers
- Tic expression, the nature of tics as well as their severity, can be influenced (both exacerbated and attenuated) by environmental factors involving both internal and external stimuli. Understanding the contextual factors that influence tic expression is key to refining and developing novel behavioral interventions
- Tics are worse during periods of fatigue, stress, and excitement and better during periods of goal-directed behavior that requires motor control, e.g., playing a musical instrument
- In most cases, motor and vocal tic severity peaks early in the second decade with many patients showing a marked reduction in tic severity by the age of 19 or 20 years
- In the majority of cases, tics in some form, persist well into adulthood
- The presence of chronic motor and vocal tics *alone* in the absence of other difficulties, often heralds a positive outcome – especially in the presence of other strengths – *however tics alone are the exception rather than the rule*
- Social, emotional, and academic outcomes in adulthood are *not* synonymous with tic outcomes
- The education of clinicians, teachers, parents, and peers concerning the key phenomenological features of Tourette syndrome and related disorders is needed to ensure the best outcomes
- Future progress in this field depends on the close interdisciplinary collaboration among *teams* of scientists, parents, teachers, and advocacy groups working together

Efforts are continuing to develop and refine intervention programs. These efforts include scientists from a broad range of disciplines: neuroimagers, developmental and behavioral neuroscientists, geneticists, and immunologists, as well as representatives of advocacy organizations (Robertson et al., 2017). Many of the recent scientific advances are the result initiatives funded by organizations, such as the Tourette Association of

America, European Society for the Study of Tourette Syndrome, and affiliated organizations across the globe (Müller-Vahl, Szejko, et al., 2021). Box 2 points to some of the remaining unanswered questions regarding the "enigmatic world" of TS.

BOX 2 Questions for future research

1. **Why are males more commonly affected with Tourette syndrome (TS) than females?** What endocrine, molecular, and neural mechanisms play a role and at what point in neural development do these events unfold?
2. **What determines tic onset?** What are the underlying neurobiological processes that lead tics to appear during a specific phase of neural development?
3. **What is the neurobiological origin of premonitory urges and what functional role do they play in tic expression?** Are they causally related or are they simply non-functional responses which become strengthened and maintained by automatic negative (the urges themselves) and positive (the momentary relief) reinforcement? How can we diminish the intensity of tic urges?
4. **How can we better understand the nonlinear dynamics of the waxing and waning of tics, particularly with regard to its neural origins, its potential prognostic value as well as a foundation for novel treatment interventions?** Do certain classes of neurons show a similar temporal pattern of neural activity? If it were possible to monitor in real time patient-specific variables, would that provide us with more accurate tools to assess tic severity? Is it also possible, if we can help individuals regulate the activity of specific neurons, that this will have therapeutic value?
5. **What genetic, environmental, and neurobiological factors influence the natural history of tic disorders leading in some cases to a marked attenuation or remission of tics in early adulthood?** Is there an endophenotype of TS, such as diminished prepulse inhibition that remains even if the tics largely remit? Is remission a return to a "normal" state of neural function or does it reflect a progressive adaptation to the anomalous motor and sensory states associated with tics and associated behaviors? Prospective longitudinal studies are needed to clarify and further define the patterns of neural interconnectivity seen in individuals with TS
6. **How can we best understand the range of comorbid conditions associated with TS?** Why is the rate of these disorders higher in cases with TS compared to individuals with a chronic tic disorder?
7. **Are there unique etiological subtypes of TS that can be identified based on their clinical presentation?** Are tic disorders on a continuum? Why are certain clusters of tic symptoms differentially associated with premonitory urges and co-occurring ADHD and OCD symptoms? Will these phenotypes be useful in clarifying the vulnerability genes and other aspects of the developmental pathobiology of TS?

References

Alsobrook, J. P., & Pauls, D. L. (2002). A factor analysis of tic symptoms in Gilles de la Tourette's syndrome. *The American Journal of Psychiatry*, *159*(2), 291–296. https://doi.org/10.1176/appi.ajp.159.2.291.

Banaschewski, T., Woerner, W., & Rothenberger, A. (2003). Premonitory sensory phenomena and suppressibility of tics in Tourette syndrome: Developmental aspects in children and adolescents. *Developmental Medicine and Child Neurology*, *45*(10), 700–703. https://doi.org/10.1017/S0012162203001294.

Barnea, M., Benaroya-Milshtein, N., Gilboa-Sechtman, E., Woods, D. W., Piacentini, J., Fennig, S., et al. (2016). Subjective versus objective measures of tic severity in Tourette syndrome—The influence of environment. *Psychiatry Research*, *242*, 204–209. https://doi.org/10.1016/j.psychres.2016.05.047.

Belluscio, B. A., Jin, L., Watters, V., Lee, T. H., & Hallett, M. (2011). Sensory sensitivity to external stimuli in Tourette syndrome patients: Sensory sensitivity in TS. *Movement Disorders*, *26*(14), 2538–2543. https://doi.org/10.1002/mds.23977.

Bernabei, M., Andreoni, G., Mendez Garcia, M. O., Piccini, L., Aletti, F., Sassi, M., et al. (2010). Automatic detection of tic activity in the Tourette syndrome. In *2010 Annual International Conference of the IEEE Engineering in Medicine and Biology* (pp. 422–425). https://doi.org/10.1109/IEMBS.2010.5627374.

Bhikram, T., Crawley, A., Arnold, P., Abi-Jaoude, E., & Sandor, P. (2020). Examining the functional activity of different obsessive–compulsive symptom dimensions in Tourette syndrome. *NeuroImage Clinical*, *26*, 102198. https://doi.org/10.1016/j.nicl.2020.102198.

Bliss, J., Cohen, D. J., & Freedman, D. X. (1980). Sensory experiences of Gilles de la Tourette syndrome. *Archives of General Psychiatry*, *37*(12), 1343–1347. https://doi.org/10.1001/archpsyc.1980.01780250029002.

Bodeck, S., Lappe, C., & Evers, S. (2015). Tic-reducing effects of music in patients with Tourette's syndrome: Self-reported and objective analysis. *Journal of the Neurological Sciences*, *352*(1), 41–47. https://doi.org/10.1016/j.jns.2015.03.016.

Bornstein, R. A., Stefl, M. E., & Hammond, L. (1990). A survey of Tourette syndrome patients and their families: The 1987 Ohio Tourette survey. *The Journal of Neuropsychiatry and Clinical Neurosciences*, *2*(3), 275–281. https://doi.org/10.1176/jnp.2.3.275.

Bortolato, M., & Pittenger, C. (2017). Modeling tics in rodents: Conceptual challenges and paths forward. *Journal of Neuroscience Methods*, *292*, 12–19. https://doi.org/10.1016/j.jneumeth.2017.02.007.

Brandt, V. C., Patalay, P., Bäumer, T., Brass, M., & Münchau, A. (2016). Tics as a model of over-learned behavior-imitation and inhibition of facial tics. *Movement Disorders*, *31*(8), 1155–1162. https://doi.org/10.1002/mds.26607.

Buse, J., Beste, C., Herrmann, E., & Roessner, V. (2016). Neural correlates of altered sensorimotor gating in boys with Tourette syndrome: A combined EMG/fMRI study. *The World Journal of biological Psychiatry: The Official Journal of the World Federation of Societies of Biological Psychiatry*, *17*(3), 187–197. https://doi.org/10.3109/15622975.2015.1112033.

Castellanos, F. X., Fine, E. J., Kaysen, D., Marsh, W. L., Rapoport, J. L., & Hallett, M. (1996). Sensorimotor gating in boys with Tourette's syndrome and ADHD: Preliminary results. *Biological Psychiatry (1969)*, *39*(1), 33–41. https://doi.org/10.1016/0006-3223(95)00101-8.

Cavanna, A. E., Ali, F., Leckman, J. F., & Robertson, M. M. (2010). Pathological laughter in Gilles de la Tourette syndrome: An unusual phonic tic. *Movement Disorders*, *25*(13), 2233–2239. https://doi.org/10.1002/mds.23216.

Cavanna, A. E., Black, K. J., Hallett, M., & Voon, V. (2017). Neurobiology of the premonitory urge in Tourette's syndrome: Pathophysiology and treatment implications. *The Journal of Neuropsychiatry and Clinical Neurosciences*, *29*(2), 95–104. https://doi.org/10.1176/appi.neuropsych.16070141.

Cavanna, A. E., Critchley, H. D., Orth, M., Stern, J. S., Young, M.-B., & Robertson, M. M. (2011). Dissecting the Gilles de la Tourette spectrum: A factor analytic study on 639 patients. *Journal of Neurology, Neurosurgery and Psychiatry*, *82*(12), 1320–1323. https://doi.org/10.1136/jnnp.2010.225029.

Cavanna, A. E., Ganos, C., Hartmann, A., Martino, D., Pringsheim, T., & Seri, S. (2020). The cognitive neuropsychiatry of Tourette syndrome. *Cognitive Neuropsychiatry*, *25*(4), 254–268. https://doi.org/10.1080/13546805.2020.1760812.

Chappell, P. B., McSwiggan-Hardin, M. T., Scahill, L., Rubenstein, M., Walker, D. E., Cohen, D. J., et al. (1994). Videotape tic counts in the assessment of Tourette's syndrome: Stability, reliability, and validity. *Journal of the American Academy of Child and Adolescent Psychiatry*, *33*(3), 386–393. https://doi.org/10.1097/00004583-199403000-00013.

Chee, K. Y., & Sachdev, P. (1997). A controlled study of sensory tics in Gilles de 1a Tourette syndrome and obsessive-compulsive disorder using a structured interview. *Journal of Neurology, Neurosurgery, and Psychiatry*, *62*(2), 188–192. https://doi.org/10.1136/jnnp.62.2.188.

Chen, S.-F., Su, Y.-C., Wang, L.-Y., Hsu, C.-Y., & Shen, Y.-C. (2019). Tourette's syndrome is associated with an increased risk of traumatic brain injury: A nationwide population-based cohort study. *Parkinsonism & Related Disorders*, *63*, 88–93. https://doi.org/10.1016/j.parkreldis.2019.02.033.

Cohen, A. J., & Leckman, J. F. (1992). Sensory phenomena associated with Gilles de la Tourette's syndrome. *The Journal of Clinical Psychiatry*, *53*(9), 319–323.

Conelea, C. A., & Woods, D. W. (2008a). The influence of contextual factors on tic expression in Tourette's syndrome: A review. *Journal of Psychosomatic Research*, *65*(5), 487–496. https://doi.org/10.1016/j.jpsychores.2008.04.010.

Conelea, C. A., & Woods, D. W. (2008b). Examining the impact of distraction on tic suppression in children and adolescents with Tourette syndrome. *Behaviour Research and Therapy*, *46*(11), 1193–1200. https://doi.org/10.1016/j.brat.2008.07.005.

Conelea, C. A., Woods, D. W., & Brandt, B. C. (2011). The impact of a stress induction task on tic frequencies in youth with Tourette syndrome. *Behaviour Research and Therapy*, *49*(8), 492–497. https://doi.org/10.1016/j.brat.2011.05.006.

Darrow, S. M., Hirschtritt, M. E., Davis, L. K., Illmann, C., Osiecki, L., Grados, M., et al. (2017). Identification of two heritable cross-disorder endophenotypes for Tourette syndrome. *The American Journal of Psychiatry*, *174*(4), 387–396. https://doi.org/10.1176/appi.ajp.2016.16020240.

Eapen, V., Fox-Hiley, P., Banerjee, S., & Robertson, M. (2004). Clinical features and associated psychopathology in a Tourette syndrome cohort. *Acta Neurologica Scandinavica*, *109*(4), 255–260. https://doi.org/10.1046/j.1600-0404.2003.00228.x.

Evers, R. A. F., & van de Wetering, B. J. M. (1994). A treatment model for motor tics based on a specific tension-reduction technique. *Journal of Behavior Therapy and Experimental Psychiatry*, *25*(3), 255–260. https://doi.org/10.1016/0005-7916(94)90026-4.

Fernández de la Cruz, L., Rydell, M., Runeson, B., Brander, G., Rück, C., D'Onofrio, B. M., et al. (2017). Suicide in Tourette's and chronic tic disorders. *Biological Psychiatry*, *82*(2), 111–118. https://doi.org/10.1016/j.biopsych.2016.08.023.

Freeman, R. D., Zinner, S. H., Müller-Vahl, K. R., Fast, D. K., Burd, L. J., Kano, Y., et al. (2009). Coprophenomena in Tourette syndrome. *Developmental Medicine and Child Neurology*, *51*(3), 218–227. https://doi.org/10.1111/j.1469-8749.2008.03135.x.

Freud, S. (1913). The disposition to obsessional neurosis. In J. Strachey, A. Strachey, & A. Tyson (Eds.), *Vol. XII. The complete psychological works of Sigmund Freud* Hogarth Press.

Ganos, C., Garrido, A., Navalpotro-Gómez, I., Ricciardi, L., Martino, D., Edwards, M. J., et al. (2015). Premonitory urge to tic in Tourette's is associated with interoceptive awareness. *Movement Disorders*, *30*(9), 1198–1202. https://doi.org/10.1002/mds.26228.

Ghosh, D., Rajan, P. V., Das, D., Datta, P., Rothner, A. D., & Erenberg, G. (2012). Headache in children with Tourette syndrome. *The Journal of Pediatrics*, *161*(2), 303–307.e6. https://doi.org/10.1016/j.jpeds.2012.01.072.

Ghosh, D., Rajan, P. V., Das, D., Datta, P., Rothner, A. D., & Erenberg, G. (2014). Sleep disorders in children with Tourette syndrome. *Pediatric Neurology*, *51*(1), 31–35. https://doi.org/10.1016/j.pediatrneurol.2014.03.017.

Gilles de la Tourette, G. (1885). Etude sur une affection nerveuse caracte ´rise'e par de l'incoordination motrice accompagne'e de l'e´cholalie et de la coprolalie. *Archives of Neurology*, *9*(19–42), 158–200.

Gorman, D. A., Thompson, N., Plessen, K. J., Robertson, M. M., Leckman, J. F., & Peterson, B. S. (2010). Psychosocial outcome and psychiatric comorbidity in older adolescents with Tourette syndrome: Controlled study. *British Journal of Psychiatry*, *197*(1), 36–44. https://doi.org/10.1192/bjp.bp.109.071050.

Groth, C., Debes, N. M., & Skov, L. (2017). Phenotype development in adolescents with Tourette syndrome: A large clinical longitudinal study. *Journal of Child Neurology*, *32*(13), 1047–1057. https://doi.org/10.1177/0883073817729917.

Hashemiyoon, R., Kuhn, J., & Visser-Vandewalle, V. (2017). Putting the pieces together in Gilles de la Tourette syndrome: Exploring the link between clinical observations and the biological basis of dysfunction. *Brain Topography*, *30*(1), 3–29. https://doi.org/10.1007/s10548-016-0525-z.

Hetrick, W. P., Erickson, M. A., & Smith, D. A. (2012). Phenomenological dimensions of sensory gating. *Schizophrenia Bulletin*, *38*(1), 178–191. https://doi.org/10.1093/schbul/sbq054.

Hoekstra, P., Steenhuis, M., Kallenberg, C., & Minderaa, R. (2004). Association of small life events with self reports of tic severity in pediatric and adult tic disorder patients: A prospective longitudinal study. *The Journal of Clinical Psychiatry*, *65*(3), 426–431. https://doi.org/10.4088/JCP.v65n0320.

Hollenbeck, P. J. (2001). Insight and hindsight into Tourette syndrome. *Advances in Neurology*, *85*, 363–367.

Horesh, N., Shmuel-Baruch, S., Farbstein, D., Ruhrman, D., Milshtein, N. B. A., Fennig, S., et al. (2018). Major and minor life events, personality and psychopathology in children with Tourette syndrome. *Psychiatry Research*, *260*, 1–9. https://doi.org/10.1016/j.psychres.2017.11.016.

Houghton, D. C., Capriotti, M. R., Conelea, C. A., & Woods, D. W. (2014). Sensory phenomena in Tourette syndrome: Their role in symptom formation and treatment. *Current Developmental Disorders Reports*, *1*(4), 245–251. https://doi.org/10.1007/s40474-014-0026-2.

Isaacs, D., & Riordan, H. (2020). Sensory hypersensitivity in Tourette syndrome: A review. *Brain & Development*, *42*(9), 627–638. https://doi.org/10.1016/j.braindev.2020.06.003.

Itard, J. M. G. (1825). Mémoire sur quelques fonctions involontaires des appareils de la locomotion, de la préhension et de la voix. *Archives Generales de Médecine*, *8*, 385–407.

Jackson, S. R., Sigurdsson, H. P., Dyke, K., Condon, M., & Jackson, G. M. (2021). The role of the cingulate cortex in the generation of motor tics and the experience of the premonitory urge-to-tic in Tourette syndrome. *Journal of Neuropsychology*, *15*(3), 340–362. https://doi.org/10.1111/jnp.12242.

Jagger, J., Prusoff, B. A., Cohen, D. J., Kidd, K. K., Carbonari, C. M., & John, K. (1982). The epidemiology of Tourette's syndrome: A pilot study. *Schizophrenia Bulletin*, *8*(2), 267–278. https://doi.org/10.1093/schbul/8.2.267.

Janet, P. (1903). *Les obsessions et la Psychiasthenie. Vol. 1*. Paris: Alcan (reprinted in New York, Arno, 1976).

Jankovic, J., Jimenez-Shahed, J., & Brown, L. W. (2010). A randomised, double-blind, placebo-controlled study of topiramate in the treatment of Tourette syndrome. *Journal of Neurology, Neurosurgery and Psychiatry, 81*(1), 70–73. https://doi.org/10.1136/jnnp.2009.185348.

Johnco, C., McGuire, J. F., McBride, N. M., Murphy, T. K., Lewin, A. B., & Storch, E. A. (2016). Suicidal ideation in youth with tic disorders. *Journal of Affective Disorders, 200*, 204–211. https://doi.org/10.1016/j.jad.2016.04.027.

Kane, M. J. (1994). Premonitory urges as "attentional tics" in Tourette's syndrome. *Journal of the American Academy of Child and Adolescent Psychiatry, 33*(6), 805–808. https://doi.org/10.1097/00004583-199407000-00005.

Khalifa, N., & Von Knorring, A. L. (2006). Psychopathology in a Swedish population of school children with tic disorders. *Journal of the American Academy of Child and Adolescent Psychiatry, 45*(11), 1346–1353. https://doi.org/10.1097/01.chi.0000251210.98749.83.

Kircanski, K., Woods, D. W., Chang, S. W., Ricketts, E. J., & Piacentini, J. C. (2010). Cluster analysis of the Yale global tic severity scale (YGTSS): Symptom dimensions and clinical correlates in an outpatient youth sample. *Journal of Abnormal Child Psychology, 38*(6), 777–788. https://doi.org/10.1007/s10802-010-9410-5.

Kurlan, R., Lichter, D., & Hewitt, D. (1989). Sensory tics in Tourette's syndrome. *Neurology, 39*(5), 731–734. https://doi.org/10.1212/WNL.39.5.731.

Kwak, C., Dat Vuong, K., & Jankovic, J. (2003). Premonitory sensory phenomenon in Tourette's syndrome. *Movement Disorders, 18*(12), 1530–1533. https://doi.org/10.1002/mds.10618.

Leckman, J. F., King, R. A., & Cohen, D. J. (1998). Tics and tic disorders. In J. F. Leckman, & D. J. Cohen (Eds.), *Tics, Obsessions, Compulsions. Developmental Psychopathology and Clinical Care* (pp. 23–42). John Wiley and Sons.

Leckman, J. F., Vaccarino, F. M., Kalanithi, P. S., & Rothenberger, A. (2006). Tourette syndrome: A relentless drumbeat—Driven by misguided brain oscillations. *Journal of Child Psychology and Psychiatry, 47*(6), 537–550. https://doi.org/10.1111/j.1469-7610.2006.01620.x.

Leckman, J. F., Walker, D. E., & Cohen, D. J. (1993). Premonitory urges in Tourette's syndrome. *The American Journal of Psychiatry, 150*(1), 98–102. https://doi.org/10.1176/ajp.150.1.98.

Leckman, J. F., Walker, D. E., Goodman, W. K., Pauls, D. L., & Cohen, D. J. (1994). "Just right" perceptions associated with compulsive behavior in Tourette's syndrome. *The American Journal of Psychiatry, 151*(5), 675–680. https://doi.org/10.1176/ajp.151.5.675.

Lee, W.-T., Huang, H.-L., Wong, L. C., Weng, W.-C., Vasylenko, T., Jong, Y.-J., et al. (2017). Tourette syndrome as an independent risk factor for subsequent sleep disorders in children: A nationwide population-based case–control study. *Sleep, 40*(3), 1–7. https://doi.org/10.1093/sleep/zsw072.

Lin, H., Katsovich, L., Ghebremichael, M., Findley, D. B., Grantz, H., Lombroso, P. J., et al. (2007). Psychosocial stress predicts future symptom severities in children and adolescents with Tourette syndrome and/or obsessive-compulsive disorder. *Journal of Child Psychology and Psychiatry, 48*(2), 157–166. https://doi.org/10.1111/j.1469-7610.2006.01687.x.

Lin, H., Williams, K. A., Katsovich, L., Findley, D. B., Grantz, H., Lombroso, P. J., et al. (2010). Streptococcal upper respiratory tract infections and psychosocial stress predict future tic and obsessive-compulsive symptom severity in children and adolescents with Tourette syndrome and obsessive-compulsive disorder. *Biological Psychiatry, 67*(7), 684–691. https://doi.org/10.1016/j.biopsych.2009.08.020.

Lin, H., Yeh, C.-B., Peterson, B. S., Scahill, L., Grantz, H., Findley, D. B., et al. (2002). Assessment of symptom exacerbations in a longitudinal study of children with Tourette's syndrome or obsessive-compulsive disorder. *Journal of the American Academy of Child & Adolescent Psychiatry*, *41*(9), 1070–1077. https://doi.org/10.1097/00004583-200209000-00007.

Martino, D., & Hedderly, T. (2019). Tics and stereotypies: A comparative clinical review. *Parkinsonism & Related Disorders*, *59*, 117–124. https://doi.org/10.1016/j.parkreldis.2019.02.005.

Martino, D., Pringsheim, T. M., Cavanna, A. E., Colosimo, C., Hartmann, A., Leckman, J. F., et al. (2017). Systematic review of severity scales and screening instruments for tics: Critique and recommendations. *Movement Disorders*, *32*(3), 467–473. https://doi.org/10.1002/mds.26891.

Mathews, C. A., Jang, K. L., Herrera, L. D., Lowe, T. L., Budman, C. L., Erenberg, G., et al. (2007). Tic symptom profiles in subjects with Tourette syndrome from two genetically isolated populations. *Biological Psychiatry*, *61*(3), 292–300. https://doi.org/10.1016/j.biopsych.2006.02.009.

McGuire, J. F., McBride, N., Piacentini, J., Johnco, C., Lewin, A. B., Murphy, T. K., et al. (2016). The premonitory urge revisited: An individualized premonitory urge for tics scale. *Journal of Psychiatric Research*, *83*, 176–183. https://doi.org/10.1016/j.jpsychires.2016.09.007.

McGuire, J. F., Nyirabahizi, E., Kircanski, K., Piacentini, J., Peterson, A. L., Woods, D. W., et al. (2013). A cluster analysis of tic symptoms in children and adults with Tourette syndrome: Clinical correlates and treatment outcome. *Psychiatry Research*, *210*(3), 1198–1204. https://doi.org/10.1016/j.psychres.2013.09.021.

Meidinger, A. L., Miltenberger, R. G., Himle, M., Omvig, M., Trainor, C., & Crosby, R. (2005). An investigation of tic suppression and the rebound effect in Tourette's disorder. *Behavior Modification*, *29*(5), 716–745. https://doi.org/10.1177/0145445505279262.

Meier, S. M., Dalsgaard, S., Mortensen, P. B., Leckman, J. F., & Plessen, K. J. (2017). Mortality risk in a nationwide cohort of individuals with tic disorders and with Tourette syndrome. *Movement Disorders*, *32*(4), 605–609. https://doi.org/10.1002/mds.26939.

Miguel, E. C., do Rosário-Campos, M. C., Prado, H. S., do Valle, R., Rauch, S. L., Coffey, B. J., et al. (2000). Sensory phenomena in obsessive-compulsive disorder and Tourette's disorder. *The Journal of Clinical Psychiatry*, *61*(2), 150–157. https://doi.org/10.4088/jcp.v61n0213.

Müller-Vahl, K. R., Pisarenko, A., Jakubovski, E., & Fremer, C. (2021). Stop that! It's not Tourette's but a new type of mass sociogenic illness. *Brain*. https://doi.org/10.1093/brain/awab316, awab316. Advance online publication.

Müller-Vahl, K. R., Szejko, N., Verdellen, C., Roessner, V., Hoekstra, P. J., Hartmann, A., et al. (2021). European clinical guidelines for Tourette syndrome and other tic disorders: Summary statement. *European Child & Adolescent Psychiatry*. https://doi.org/10.1007/s00787-021-01832-4. Advance online publication.

O'Connor, K., Brisebois, H., Brault, M., Robillard, S., & Loiselle, J. (2003). Behavioral activity associated with onset in chronic tic and habit disorder. *Behaviour Research and Therapy*, *41*(2), 241–249. https://doi.org/10.1016/S0005-7967(02)00051-7.

O'Connor, K. P., Gareau, D., & Blowers, G. H. (1994). Personal constructs associated with tics. *British Journal of Clinical Psychology*, *33*(2), 151–158. https://doi.org/10.1111/j.2044-8260.1994.tb01106.x.

Patterson, A. L., Choudhri, A. F., Igarashi, M., McVicar, K., Shah, N., & Morgan, R. (2016). Severe neurological complications associated with Tourette syndrome. *Pediatric Neurology*, *61*, 99–106. https://doi.org/10.1016/j.pediatrneurol.2016.05.008.

Peterson, B. S., & Leckman, J. F. (1998). The temporal dynamics of tics in Gilles de la Tourette syndrome. *Biological Psychiatry*, *44*(12), 1337–1348. https://doi.org/10.1016/S0006-3223(98)00176-0.

Piacentini, J., Himle, M. B., Chang, S., Baruch, D. E., Buzzella, B. A., Pearlman, A., et al. (2006). Reactivity of tic observation procedures to situation and setting. *Journal of Abnormal Child Psychology*, *34*(5), 647–656. https://doi.org/10.1007/s10802-006-9048-5.

Piacentini, J., Woods, D. W., Scahill, L., Wilhelm, S., Peterson, A. L., Chang, S., et al. (2010). Behavior therapy for children with Tourette disorder: A randomized controlled trial. *Journal of the American Medical Association*, *303*(19), 1929–1937. https://doi.org/10.1001/jama.2010.607.

Pile, V., Lau, J. Y. F., Topor, M., Hedderly, T., & Robinson, S. (2018). Interoceptive accuracy in youth with tic disorders: Exploring links with premonitory urge, anxiety and quality of life. *Journal of Autism and Developmental Disorders*, *48*(10), 3474–3482. https://doi.org/10.1007/s10803-018-3608-8.

Pringsheim, T., Holler-Managan, Y., Okun, M. S., Jankovic, J., Piacentini, J., Cavanna, A. E., et al. (2019). Comprehensive systematic review summary: Treatment of tics in people with Tourette syndrome and chronic tic disorders. *Neurology*, *92*(19), 907–915. https://doi.org/10.1212/WNL.0000000000007467.

Rae, C. L., Larsson, D. E. O., Garfinkel, S. N., & Critchley, H. D. (2019). Dimensions of interoception predict premonitory urges and tic severity in Tourette syndrome. *Psychiatry Research*, *271*, 469–475. https://doi.org/10.1016/j.psychres.2018.12.036.

Rath, J. G., Tavy, D. L., Wertenbroek, A. A. A. C. M., van Woerkom, T. C. A. M., & de Bruijn, S. F. T. M. (2010). Botulinum toxin type a in simple motor tics: Short-term and long-term treatment-effects. *Parkinsonism & Related Disorders*, *16*(7), 478–481. https://doi.org/10.1016/j.parkreldis.2009.11.011.

Ricketts, E. J., Rozenman, M., Choy, C., Goldberg, H. B., Kim, J. S., Colwell, C. S., et al. (2018). Sleep sufficiency in pediatric and adolescent Tourette's disorder: National survey of children's health. *Journal of Developmental and Behavioral Pediatrics*, *39*(1), 72–76. https://doi.org/10.1097/DBP.0000000000000518.

Rizzo, R., Gulisano, M., Calì, P. V., & Curatolo, P. (2011). Long term clinical course of Tourette syndrome. *Brain & Development*, *34*(8), 667–673. https://doi.org/10.1016/j.braindev.2011.11.006.

Robertson, M. M., Althoff, R. R., Hafez, A., & Pauls, D. L. (2008). Principal components analysis of a large cohort with Tourette syndrome. *British Journal of Psychiatry*, *193*(1), 31–36. https://doi.org/10.1192/bjp.bp.107.039909.

Robertson, M. M., Banerjee, S., Eapen, V., & Fox-Hiley, P. (2002). Obsessive compulsive behaviour and depressive symptoms in young people with Tourette syndrome: A controlled study. *European Child & Adolescent Psychiatry*, *11*(6), 261–265. https://doi.org/10.1007/s00787-002-0301-3.

Robertson, M. M., & Cavanna, A. E. (2007). The Gilles de la Tourette syndrome: A principal component factor analytic study of a large pedigree. *Psychiatric Genetics*, *17*(3), 143–152. https://doi.org/10.1097/YPG.0b013e328015b937.

Robertson, M. M., Eapen, V., Rizzo, R., Stern, J. S., & Hartmann, A. (2020). Gilles de la Tourette syndrome: Advice in the times of COVID-19. *F1000 Research*, *9*, 257. https://doi.org/10.12688/f1000research.23275.1.

Robertson, M. M., Eapen, V., Singer, H. S., Martino, D., Scharf, J. M., Paschou, P., et al. (2017). Gilles de la Tourette syndrome. *Nature Reviews. Disease Primers*, *3*, 16097. https://doi.org/10.1038/nrdp.2016.97.

Rosario, M. C., Prado, H. S., Borcato, S., Diniz, J. B., Shavitt, R. G., Hounie, A. G., et al. (2009). Validation of the university of São Paulo sensory phenomena scale: Initial psychometric properties. *CNS Spectrums*, *14*(6), 315–323. https://doi.org/10.1017/S1092852900020319.

Scahill, L., Lombroso, P. J., Mack, G., Van Wattum, P. J., Zhang, H., Vitale, A., et al. (2001). Thermal sensitivity in Tourette syndrome: Preliminary report. *Perceptual and Motor Skills*, *92*(2), 419–432. https://doi.org/10.2466/pms.2001.92.2.419.

Schunke, O., Grashorn, W., Kahl, U., Schöttle, D., Haggard, P., Münchau, A., et al. (2016). Quantitative sensory testing in adults with Tourette syndrome. *Parkinsonism & Related Disorders*, *24*, 132–136. https://doi.org/10.1016/j.parkreldis.2016.01.006.

Shapiro, A. K., Shapiro, E. S., Young, J. G., & Feinberg, T. E. (1988). *Gilles de la Tourette syndrome* (2nd ed.). Raven Press.

Silva, R. R., Munoz, D. M., Barickman, J., & Friedhoff, A. J. (1995). Environmental factors and related fluctuation of symptoms in children and adolescents with Tourette's disorder. *Journal of Child Psychology and Psychiatry*, *36*(2), 305–312. https://doi.org/10.1111/j.1469-7610.1995.tb01826.x.

Smith, S. J., & Lees, A. J. (1989). Abnormalities of the blink reflex in Gilles de la Tourette syndrome. *Journal of Neurology, Neurosurgery and Psychiatry*, *52*(7), 895–898. https://doi.org/10.1136/jnnp.52.7.895.

Smith, L. B., & Thelen, E. (2003). Development as a dynamic system. *Trends in Cognitive Sciences*, *7*(8), 343–348. https://doi.org/10.1016/S1364-6613(03)00156-6.

Steinberg, T., Shmuel-Baruch, S., Horesh, N., & Apter, A. (2013). Life events and Tourette syndrome. *Comprehensive Psychiatry*, *54*(5), 467–473. https://doi.org/10.1016/j.comppsych.2012.10.015.

Sukhodolsky, D. G., Scahill, L., Zhang, H., Peterson, B. S., King, R. A., Lombroso, P. J., et al. (2003). Disruptive behavior in children with Tourette's syndrome: Association with ADHD comorbidity, tic severity, and functional impairment. *Journal of the American Academy of Child and Adolescent Psychiatry*, *42*(1), 98–105. https://doi.org/10.1097/00004583-200301000-00016.

Sutherland Owens, A. N., Miguel, E. C., & Swerdlow, N. R. (2011). Sensory gating scales and premonitory urges in Tourette syndrome. *The Scientific World Journal*, *11*, 736–741. https://doi.org/10.1100/tsw.2011.57.

Swerdlow, N. R. (2013). Studies of prepulse inhibition of startle, with particular relevance to the pathophysiology or treatment of Tourette syndrome. *Neuroscience and Biobehavioral Reviews*, *37*(6), 1150–1156. https://doi.org/10.1016/j.neubiorev.2012.09.002.

Watson, T. S., Dufrene, B., Weaver, A., Butler, T., & Meeks, C. (2005). Brief antecedent assessment and treatment of tics in the general education classroom: A preliminary investigation. *Behavior Modification*, *29*(6), 839–857. https://doi.org/10.1177/0145445505279252.

Weingarden, H., Scahill, L., Hoeppner, S., Peterson, A. L., Woods, D. W., Walkup, J. T., et al. (2018). Self-esteem in adults with Tourette syndrome and chronic tic disorders: The roles of tic severity, treatment, and comorbidity. *Comprehensive Psychiatry*, *84*, 95–100. https://doi.org/10.1016/j.comppsych.2018.04.008.

Weisman, H., Parush, S., Apter, A., Fennig, S., Benaroya-Milshtein, N., & Steinberg, T. (2018). A study of sensory dysregulation in children with tic disorders. *Journal of Neural Transmission*, *125*(7), 1077–1085. https://doi.org/10.1007/s00702-018-1858-4.

Wood, B. L., Klebba, K., Gbadebo, O., Lichter, D., Kurlan, R., & Miller, B. (2003). Pilot study of effect of emotional stimuli on tic severity in children with Tourette's syndrome. *Movement Disorders*, *18*(11), 1392–1395. https://doi.org/10.1002/mds.10552.

Woods, D. W., Piacentini, J., Himle, M. B., & Chang, S. (2005). Premonitory urge for tics scale (PUTS): Initial psychometric results and examination of the premonitory urge phenomenon in youths with tic disorders. *Journal of Developmental and Behavioral Pediatrics*, *26*(6), 397–403. https://doi.org/10.1097/00004703-200512000-00001.

Woods, D. W., Piacentini, J. C., Scahill, L., Peterson, A. L., Wilhelm, S., Chang, S., et al. (2011). Behavior therapy for tics in children: Acute and long-term effects on psychiatric and psychosocial functioning. *Journal of Child Neurology*, *26*(7), 858–865. https://doi.org/10.1177/0883073810397046.

Woods, D. W., Watson, T. S., Wolfe, E., Twohig, M. P., & Friman, P. C. (2001). Analyzing the influence of tic-related talk on vocal and motor tics in children with Tourette's syndrome. *Journal of Applied Behavior Analysis*, *34*(3), 353–356. https://doi.org/10.1901/jaba.2001.34-353.

Worbe, Y., Mallet, L., Golmard, J.-L., Béhar, C., Durif, F., Jalenques, I., et al. (2010). Repetitive behaviours in patients with Gilles de la Tourette syndrome: Tics, compulsions, or both? *PLoS One*, *5*(9), e12959. https://doi.org/10.1371/journal.pone.0012959.

Zebardast, N., Crowley, M. J., Bloch, M. H., Mayes, L. C., Wyk, B. V., Leckman, J. F., et al. (2013). Brain mechanisms for prepulse inhibition in adults with Tourette syndrome: Initial findings. *Psychiatry Research: Neuroimaging*, *214*(1), 33–41. https://doi.org/10.1016/j.pscychresns.2013.05.009.

Zinner, S. H., Conelea, C. A., Glew, G. M., Woods, D. W., & Budman, C. L. (2012). Peer victimization in Yyuth with Tourette syndrome and other chronic tic disorders. *Child Psychiatry and Human Development*, *43*(1), 124–136. https://doi.org/10.1007/s10578-011-0249-y.

CHAPTER THREE

Tourette syndrome: Current nosology and guidelines

Andrea E. Cavanna[a,b,c,*]
[a]Department of Neuropsychiatry, BSMHFT and University of Birmingham, Birmingham, United Kingdom
[b]School of Life and Health Sciences, Aston University, Birmingham, United Kingdom
[c]University College London and Institute of Neurology, London, United Kingdom
*Corresponding author: e-mail address: a.e.cavanna@bham.ac.uk

Contents

1. Tourette syndrome and the nosology of tic disorders

Both medical terminology and classification systems evolve over time. The field of Tourette syndrome (TS) and other tic disorders is no exception, as nosology has been progressively refined, reflecting the state-of-the-art of scientific advances. Similarly, since the advent of evidence-based medicine and its developments during the last few decades, clinical guidelines have been produced to assist clinicians in the assessment and treatment of patients with tics (Evidence Based Medicine Working Group, 1992; Sackett, Rosenberg, Gray, Haynes, & Richardson, 1996).

International Review of Movement Disorders, Volume 3
ISSN 2666-7878
https://doi.org/10.1016/bs.irmvd.2022.01.002

For classification purposes, tics are currently defined as sudden, repetitive, nonrhythmic movements (motor tics) or vocalizations (vocal tics) (American Psychiatric Association (APA), 2013). Arguably, vocal tics should be more appropriately referred to as phonic tics, because the vocal cords are not involved in all tics that produce sound by moving air through the nose, mouth, or throat (Martino, Madhusudan, Zis, & Cavanna, 2013). It has also been suggested that tics are not strictly speaking involuntary, but semi-voluntary or unvoluntary (Tourette Syndrome Classification Study Group, 1993): they tend to be subjectively perceived as a voluntary response (willful capitulation) to a build-up of premonitory urges, or sensory experiences of mounting inner tension (Cox, Seri, & Cavanna, 2018). The theoretical debate on whether such intermediate states more closely resemble voluntary or involuntary actions dates back at least to Aristotle: "It is generally held that actions are involuntary when done under compulsion or through ignorance [...] But there is some doubt about actions done through fear of a worse alternative [...] It is open to question whether such actions are voluntary or involuntary. [...] Acts of this kind, then, are mixed or composite, but they approximate rather to the voluntary class [...] Such acts therefore are voluntary apart from circumstances—for no one would choose to do any such action in and for itself" (Nicomachean Ethics, book III, 2–6). In most cases, tics are experienced as physical urges that must eventually be expressed: as such, they are often described as suppressible yet irresistible (Cavanna & Nani, 2013).

The current edition of the Diagnostic and Statistical Manual for Mental Disorders (DSM-5) was published by the American Psychiatric Association in 2013 (American Psychiatric Association (APA), 2013), updating the previous edition (DSM-IV-TR), which had been published in 2000 (American Psychiatric Association (APA), 2000). A few changes affected the core concept of tics. The word "stereotyped" was removed from the definition of tics to help distinguish between stereotypies (repetitive behaviors more commonly seen in patients with autism spectrum disorders) and tics. This wording adjustment was aimed at reducing the risk that individuals with a stereotyped movement disorder could be misdiagnosed as having a tic disorder (Cavanna, 2014; Termine, Grossi, Anelli, Derhemi, & Cavanna, 2021). Moreover, the new definition of tics was made consistent across all tic disorders. Separate diagnostic categories capture the different tic disorders in the current classification system. Individual tic disorders differ from each other in terms of the types of tics (motor, vocal, or a combination of both), as well as their duration.

Primary tic disorders are currently classified as movement disorders within the "Neurodevelopmental disorders" section (Robertson & Eapen, 2014). Specifically, the DSM-5 lists four diagnostic criteria for TS (American Psychiatric Association (APA), 2013): presence of both motor and vocal/phonic tics at some time during the illness (not necessarily concurrently); persistence of tics for more than 1 year since first tic onset, irrespective of the duration of tic-free periods; onset before the age of 18 years; finally, the requirement that tics are not attributable to the physiological effects of a substance (e.g., cocaine) or a general medical condition (e.g., Huntington's disease or post-viral encephalitis) (Badenoch, Searle, Watson, & Cavanna, 2021). This is the final step of a process of diagnostic refinement which began in 1885, when Georges Gilles de la Tourette first described his eponymous syndrome as a medical condition characterized by the clinical trial of motor tics, echolalia, and coprolalia (Plessen, 2013; Rickards & Cavanna, 2009).

Contrary to the DSM-IV-TR, the DSM-5 includes the diagnosis of TS for patients with tics who experienced a 3-month or longer remission since their first tic, as long as their first tic developed at least 1 year before the assessment. This change reflects the common clinical observation that the retrospective differentiation of tic-free intervals is often difficult and mostly neglected in everyday practice for diagnostic purposes (Roessner et al., 2014). Although problems in social functioning and tic-related difficulties in multiple areas of life are frequently reported by patients with TS, the requirement of tics to cause clinically significant impairment was removed over two decades ago, during the transition between the DSM-IV (1994) and the DSM-IV-TR (2000). The underlying issue about the removal of the impairment criterion is the unresolved question if the classification of TS should follow clinical and/or research purposes. On the one hand, there is evidence that often co-morbid conditions such as obsessive-compulsive disorder and attention-deficit and hyperactivity disorder are associated with more significant social and academic impairment than tics themselves. On the other hand, for research purposes (e.g., genetic studies) it is important to be able to capture under the TS diagnosis also patients without a significant clinical impairment from the tics whose presentation clearly meets the symptom criteria for TS.

Persistent (chronic) motor or vocal tic disorder is diagnosed when either motor or vocal tics (but not both motor and vocal tics) are present for at least 12 months during the illness, and the diagnostic criteria for TS are not fully met (American Psychiatric Association (APA), 2013). As with TS, tics start before the age of 18 years and are not due to the physiological effects of a substance or a general medical condition. In the DSM-5, a specifier was added to distinguish between patients with motor tics only or with vocal tics only that are chronic in nature. The DSM-5 diagnostic category of provisional tic disorder replaced the DSM-IV-TR category of transient tic disorder to capture patients with motor and/or vocal tics that have been present for less than 12 months in a row since their onset (American Psychiatric Association (APA), 2013). The age at onset is before 18 years and the tics are not due to the physiological effects of a substance or a general medical condition. The rationale for renaming the diagnostic category of "transient" tic disorder as "provisional" tic disorder is that it is difficult to know whether the presentation of tics is "transient," as assumed by its name: at the time of the diagnosis, it cannot be ruled out that tics will continue to develop into a persistent (chronic) tic disorder or indeed TS—and in any case tics were "present" (rather than "transient") at that point in time. The change promotes a more systematic epidemiological approach to tic disorders, but does not have relevant clinical implications. It is important to highlight that the term "provisional" should not be taken to suggest less need for treatment: this might not be the case, even though the patient does not meet the temporal criterion for TS.

The received wisdom among clinicians (largely based on biased and incomplete data) is that for most children their tics are temporary, disappearing within a few months. However, prospective research on provisional tic disorder has shown newly developed tics to persist past a few months more often than previously thought, although usually at minimal severity (Black, Black, Greene, & Schlaggar, 2016). The surprising results of a recent study showed that at least 90% of children who had experienced tics for only a mean of 3 months still had tics at the 12-month anniversary of their first tic, although in some cases tics were seen only with remote video observation of the child sitting alone (Black, Kim, Schlaggar, & Greene, 2020; Black, Kim, Yang, & Greene, 2021; Kim et al., 2019).

The "other specified tic disorder" category should be applied when there is strong evidence from the history, physical examination and/or laboratory results to suggest a plausible underlying cause for the tics, when the age at onset is above 18 years, and in general when tics are present, but full criteria for a tic disorder or for a specific neurodevelopmental disorder are not met (American Psychiatric Association (APA), 2013). Finally, "the unspecified tic disorder" category can be used in situations where there is insufficient information to make a more specific diagnosis in patients with tics who do not meet the full criteria for a tic disorder or for a specific neurodevelopmental disorder (American Psychiatric Association (APA), 2013).

It has been highlighted that there is no genetic or phenotypic evidence supporting the DSM-5 diagnostic categorization of the primary tic disorders into distinct nosological entities. A retrospective chart analytic study on a large sample of patients with chronic tic disorders of all ages showed that the demographic and clinical characteristics of patients with a diagnosis of chronic motor tic disorder were similar to those of patients with TS (Müller-Vahl, Sambrani, & Jakubovski, 2019). The only significant differences were that patients with chronic motor tic disorder had lower tic severity, lower prevalence of complex motor tics (including copropraxia and echopraxia), and lower number of co-morbid conditions, than patients with TS. Based on these findings, it was suggested that both disorders exist along a symptom severity continuum, of which chronic motor tic disorder constitutes a less severe form and TS a more severe form. The authors therefore proposed the introduction of the term "tic spectrum disorders," instead of using the different DSM-5 diagnostic categories (Müller-Vahl et al., 2019). The hypothesis that persistent motor or vocal tic disorder is a *forme fruste* of TS was recently confirmed by the findings of a comprehensive study conducted by the Tourette Association of America International Consortium for Genetics and the OCD Collaborative Genetics Association (Claudio-Campos et al., 2021). Persistent motor or vocal tic disorder and TS occur along a clinical spectrum in which TS is a more severe and persistent motor or vocal tic disorder a less severe manifestation of a continuous neurodevelopmental tic spectrum disorder. The spectrum perspective is consistent with the use of the same definition of tics for all DSM-5 diagnostic categories, with the basic unit being similar throughout all tic disorders.

There is a considerable degree of overlap between the outlined DSM criteria and the diagnostic criteria proposed by the Tourette Syndrome Classification Study Group (TSCSG), despite minor differences (e.g., age at tic onset under 21 years rather than 18 years according to the TSCSG criteria) (Tourette Syndrome Classification Study Group, 1993). Finally, the 11th revision of the World Health Organization's International Statistical Classification of Diseases and Related Health Problems (ICD-11) (World Health Organization (WHO), 2019), which officially came into effect in 2022 to replace the ICD-10 (World Health Organization (WHO), 1993), stands as the global standard for coding health information and causes of death. The ICD-11 codes for TS and other primary tic disorders closely match their DSM-5 categorization (Table 1). Somewhat controversially, despite the

Table 1 Diagnostic codes for primary tic disorders in the current versions of the American Psychiatric Association—Diagnostic and Statistical manual of Mental Disorders (DSM) and World Health Organization—International Classification of Diseases (2022).

DSM-5 (2013)		ICD-11 (2022)	
Code	**Diagnosis**	**Code**	**Diagnosis**
307.23	Tourette's disorder	8A05.00	Tourette syndrome
307.22	Persistent (chronic) motor or vocal tic disorder (specify motor or vocal)	8A05.01	Chronic motor tic disorder
		8A05.02	Chronic phonic tic disorder
307.21	Provisional tic disorder	8A05.03	Transient motor tics
307.20	Other specified tic disorder (specify reason)	8A05.0Y	Other specified primary tics or tic disorders
307.20	Unspecified tic disorder	8A05.0Z	Primary tics or tic disorders, unspecified

evidence that psychiatric disorders are present in the vast majority of patients with TS, tic disorders have been shifted to the "Movement disorders" subgroup of the "Diseases of the nervous system" ICD-11 category, a group of conditions characterized as being in or associated with the nervous system (Woods & Thomsen, 2014).

Over the last few years there have been significant developments toward a standardized approach to the assessment and treatment of TS and other tic disorders. The publication of dedicated sets of guidelines has been made possible by an increasingly more widespread consensus on the nosology of TS (Yang et al., 2016, 2018).

2. Current clinical guidelines on Tourette syndrome

2.1 Canada (2012)

In 2012, a group of Canadian experts published a set of two treatment guidelines for TS in the Canadian Journal of Psychiatry (Pringsheim et al., 2012; Steeves et al., 2012). The two documents provided evidence-based recommendations on the pharmacotherapy and the non-pharmacological treatment interventions (behavioral therapy, deep brain stimulation, and transcranial magnetic stimulation), respectively.

The pharmacotherapy guideline sought to provide clinicians with guidance on the pharmacological management of both children and adults with tic disorders (Pringsheim et al., 2012). The authors combined the results of a systematic literature review with multidisciplinary clinical expertise. Specifically, a group of 14 experts in psychiatry, child psychiatry, neurology, pediatrics, and psychology took part in a consensus meeting where the available evidence was presented and discussed, and nominal group techniques were employed to arrive at consensus on recommendations. At the time of extracting the available evidence, the authors noted the limitations in the strength of the available evidence, as many of the reviewed trials were relatively small and included clinically heterogeneous samples. The authors also highlighted that the reviewed studies performed before 1990 used a wide variety of outcome measures for the measurement of tic severity. Moreover, these studies were often conducted according to crossover study protocols, with poor reporting of results. As a result, it was not possible to perform a meta-analysis of study results for most pharmacological agents. A decrease in the YGTSS total tic score of 8 points (out of 50) was considered clinically meaningful. The authors adopted the Grading of Recommendations, Assessment, Development and Evaluations (GRADE) system to assign a rating to the body of evidence for each pharmacological agent. The grading had four possible ratings: high (we are very confident that the true effect lies close to that of the estimate of the effect), moderate (we are moderately confident in the effect estimate: the true effect is likely to be close to the estimate of the effect, but there is a possibility that it is substantially different), low (our confidence in the effect estimate is limited: the true effect may be substantially different from the estimate of the effect), or very low (we have very little confidence in the effect estimate: the true effect is likely to be substantially different from the estimate of the effect).

A classification scheme based on the GRADE system was also used to make formal recommendations for the pharmacotherapy of tics. A strong recommendation was made when the benefits of treatment were found to clearly outweigh the risks and burdens, and it was felt that the pharmacological agent could be prescribed to most patients in most circumstances without reservation. A weak recommendation was made when the benefits, risks, and burdens were found to be more closely balanced, and it was felt that the best action might differ depending on circumstances. Finally, the authors created a third category for pharmacological agents for which insufficient evidence existed to make a formal recommendation. The 2012 Canadian guidelines on the pharmacotherapy of tics made strong recommendations for the use of alpha-2 adrenergic agonists (Clonidine and Guanfacine) in children. There was evidence supporting the efficacy of several antidopaminergic medications for the treatment of tics; however, the high rates of adverse effects associated with these pharmacological agents resulted in downgrading of recommendations to the weak category, whereby the benefits have to be closely balanced with the risks and adverse effects. For example, the use of a medication with a weak recommendation is not warranted is not indicated in situations where tics are not perceived as severe or disabling. A number of pharmacological agents received weak recommendations: these included Aripiprazole, Fluphenazine, Haloperidol, Metoclopramide (children only), Olanzapine, Pimozide, Quetiapine, Risperidone, Tetrabenazine, Topiramate, and Ziprasidone. The Canadian consensus group determined that Clonidine and Guanfacine should be considered first-line medications for tics, especially in young patients with a co-morbid diagnosis of attention-deficit and hyperactivity disorder, as evidence supports their efficacy for treating the behavioral co-morbidity as well. Aripiprazole and Risperidone were indicated as second-line pharmacological agents, whereas Fluphenazine, Haloperidol, Pimozide, and Ziprasidone were considered third-line options. A recommendation was formulated about avoiding Olanzapine, Quetiapine, and Risperidone in overweight patients, because of the risk of further weight gain as an adverse effect. Based on the available evidence, it was acknowledged that the ability of clinicians to predict which pharmacological agent has the greatest chance of success for a given patient is limited.

The second clinical guideline provided recommendations for non-pharmacological treatments for tic disorders (Steeves et al., 2012). The methodology for the systematic review, consensus group meeting, and generation of treatment recommendations was the same as the one that was used in the guideline on the pharmacotherapy of tic disorders. Based on the available evidence, the authors made strong recommendations for the use of habit reversal therapy and exposure and response prevention, preferably embedded within a supportive psychoeducational program, and with the option of combining either of these approaches with pharmacotherapy. The quality of the evidence for the efficacy of deep brain stimulation was rated as poor and the authors recommended that this intervention should continue to be considered as an experimental treatment for severe, treatment-refractory tics that impose severe limitations on the health-related quality of life of adult patients (Macerollo et al., 2016). According to the guideline, deep brain stimulation should only be performed by experts within the context of research studies. Finally, the authors found no evidence to support the use of transcranial magnetic stimulation in the treatment of tics, however, they advocated further studies within research protocols, in consideration of the low rate of known complications.

2.2 United States (2013)

In 2013, a group of North American experts in TS and the American Academy of Child and Adolescent Psychiatry (AACAP) Committee on Quality Issues published a Practice Parameter for the assessment and treatment of children and adolescents with tic disorders in the Journal of the American Academy of Child and Adolescent Psychiatry (Murphy, Lewin, Storch, & Stock, 2013). The AACAP Practice Parameter was based on a systematic literature review and proposed a set of nine evidence-based recommendations on the assessment (four recommendations) and treatment (five recommendations) of young patients with tic disorders. The strength of the underlying empirical and/or clinical support was rated for each recommendation according to four categories: Clinical Standard (recommendations based on rigorous empirical evidence—e.g., meta-analyses, systematic reviews, randomized controlled trials—and/or overwhelming clinical consensus), Clinical Guideline (recommendations based on strong empirical evidence—e.g., non-randomized controlled trials, cohort studies, case-control studies—and/or strong clinical consensus), Clinical Option (recommendations based on emerging empirical evidence—e.g., uncontrolled trials or case series/reports—or clinical opinion, in the absence of strong empirical evidence/consensus), and Not Endorsed (practices known to be ineffective or contraindicated).

The four recommendations about the assessment of young patients with tic disorders were assigned to the Clinical Standard category of support. According to the first recommendation, the child and adolescent psychiatric assessment should involve routine screening for unusual movements, stereotypies, tics, and family history of tic disorders. The second recommendation advocated a more thorough assessment for tic disorders in case of positive screening. The third recommendation stated that the assessment for tic disorders should involve a careful examination for general medical condition or substance etiologies. Moreover, the assessment for tic disorders should involve a careful examination for co-morbid psychiatric conditions, according to the fourth recommendation.

The final five recommendations focused on the treatment of young patients with tic disorders and were assigned to more heterogenous categories of support. According to the fifth recommendation (Clinical Standard), education regarding chronic tic disorders should be provided regarding expectations for course and prognosis, and treatment planning should consider classroom-based accommodations. The sixth recommendation (Clinical Standard) stated that treatment for chronic tic disorders should address the levels of impairment and distress caused by the tics, as well as any co-morbid conditions. According to the seventh recommendation (Clinical Guideline), behavioral interventions for chronic tic disorders should be considered when tics cause impairment, are moderate in severity, or if behavioral-responsive psychiatric co-morbidities are present. The eight recommendation (Clinical Guideline) dealt with the use of pharmacotherapy. The AACAP Practice Parameter stated that medications for chronic tic disorders should be considered for moderate to severe tics causing severe impairment in quality of life or when medication-responsive psychiatric co-morbidities are present that target both tic symptoms and co-morbid conditions. The authors acknowledged the widespread off-label use of second generation antidopaminergic medications before first generation antidopaminergic agents of Haloperidol and Pimozide, because of concerns about tolerability. It was also highlighted that some prescribers prefer to use alpha-2 adrenergic agonists as first-line agents over antidopaminergic medications because of the better adverse effect profile, in addition to their efficacy in ameliorating tics. The authors found evidence that the use of antidopaminergic medications as augmentation therapy may result in additional benefit for young patients with both tics and obsessive-compulsive symptoms. Likewise, they highlighted that for young patients with co-morbid attention-deficit and hyperactivity disorder, tics are not necessarily increased by stimulant medication, although the presence of tics appeared to limit the maximum dose achieved. For patients with co-morbid anxiety and affective disorders, the recommendation was to use evidence-based treatment for anxiety and affective conditions, in the absence of dedicated studies in patients with tic disorders. Disruptive behavior disorder (including explosive anger and rage outbursts) was found to be relatively common among youngsters with tics. Behavioral therapies addressing antecedents (including sensory issues and poor frustration tolerance) and anger management were deemed to be useful. The authors found preliminary positive findings for second generation antidopaminergic agents, as well as for selective serotonin reuptake inhibitors, however, caution was advocated in consideration of significant design limitations, small sample sizes, relatively weak effects, and associated risks. Finally, according to the ninth recommendation (Not Endorsed), deep brain stimulation, repetitive magnetic stimulation, special diets, and dietary supplements lacked empirical support for the treatment of tic disorders and were not recommended.

2.3 United Kingdom (2016)

In 2016 a group of experts based in the United Kingdom published a Health Technology Assessment on different treatment strategies for tics in children and adolescents with TS (Hollis et al., 2016). This document contained two parts, combining the results of a systematic literature review published in 2016 in the Journal of Child Psychology and Psychiatry (Whittington et al., 2016) and a qualitative analysis on perceptions of treatment for tics among youngsters with TS and parents published in 2015 in BMC Psychiatry (Cuenca et al., 2015), respectively.

The first part consisted of a systematic review and meta-analysis of the studies investigating benefits and risks of pharmacological, behavioral, and physical interventions for the treatment of tics in children and adolescents with TS or other chronic tic disorders. The authors included in their systematic review clinical studies in adults or mixed populations, which were considered as supporting evidence. The Cochrane risk of bias tool was used for the risk of bias assessment and the GRADE approach for assessing the overall quality of the evidence. The number and quality of the reviewed clinical trials was low, resulting in downgrading of the strength of both the evidence and conclusions. The authors acknowledged the need for larger and better-conducted trials addressing important clinical uncertainties. Although the quality of the evidence was generally low, the main findings of the systematic literature review suggested that there is clear evidence that both alpha-2 adrenergic agonists and antidopaminergic agents produce improvements in tics that may be clinically meaningful in children and adolescents with TS. Moreover, the available evidence suggested that significant differences in tic reduction between alpha-2 adrenergic agonists and antidopaminergic agents are unlikely. The overall findings of this systematic review and evidence synthesis showed that when pharmacotherapy is considered appropriate for the treatment of tics, the balance of clinical benefits to harm favors alpha-2 adrenergic agonists (Clonidine and Guanfacine) as first-line options. Antidopaminergic agents were regarded as likely to be useful as second-line options (when alpha-2 adrenergic agonists are either ineffective or poorly tolerated), in consideration of their tolerability profiles. Topiramate and Metoclopramide were included among other agents with relatively weak evidence suggesting effectiveness in reducing tics. Finally, the evidence that medications for co-morbid obsessive-compulsive disorder and attention-deficit and hyperactivity disorder (selective serotonin reuptake inhibitors and central nervous system stimulants/atomoxetine, respectively) significantly exacerbate or worsen tics in the short-term was rated as very low quality.

The second part of the Health Technology Assessment consisted in a qualitative study aimed at exploring the experience of services and treatment and at understanding which outcomes are most valued from the perspective of youngsters with TS and parents of young patients with TS. According to the results of the online survey, just over half of young patients with TS had received pharmacotherapy for their tics. The most commonly used pharmacological agents were Aripiprazole, Clonidine, and Risperidone. Specifically, the parents of young patients with TS perceived Aripiprazole as being most helpful for tic control with least troublesome adverse effects. Both young patients and parents reported that pharmacotherapy could be helpful in reducing tics, but often expressed concerns about tolerability issues and lack of provision of relevant information explaining the rationale for using medication for tics and possible adverse effects. In addition to tic frequency and intensity, young patients with TS (and—to a lesser extent—parents) highlighted the importance of recognizing and managing co-morbid anxiety symptoms and stress associated with tics. Other key themes from the qualitative study included difficulties in access to specialist care and diagnostic delay.

2.4 American Academy of Neurology (2019)

In 2019 a multidisciplinary panel convened by the American Academy of Neurology (AAN) guideline committee published two influential documents in the journal Neurology: a systematic literature review and a set of guideline recommendations on the assessment and treatment of tics in both children and adults with TS and other chronic tic disorders (Pringsheim, Holler-Managan, et al., 2019; Pringsheim, Okun, et al., 2019). This evidence-based practice guideline was completed following the methodologies outlined in the 2011 edition of the AAN guideline development process manual. In their review process, the authors calculated the effect size, expressed as a measure of the size of the intervention effect relative to the variability observed in each study (standardized mean difference). A standardized mean difference of 0.20 was considered the minimal clinically meaningful difference for reduction in tic severity, whereas effect sizes smaller than 0.10 were considered clinically unimportant. There was moderate confidence that a few of the most commonly used medications for tics, including Aripiprazole, Clonidine, Haloperidol, Risperidone, and Tiapride, were probably more likely than placebo to reduce tics. There was low confidence that other commonly used agents, such as Guanfacine, Metoclopramide, Pimozide, Topiramate, and Ziprasidone, were possibly more likely than placebo to reduce tics. The authors acknowledged that several treatment interventions had only been studied in a single randomized controlled trial of short duration, with modest sample size. The guideline document contained a total of 46 recommendations regarding the assessment and management of tics. In addition to pharmacotherapy, the topics included counseling recommendations on the natural history of tic disorders, psychoeducation for teachers and peers, assessment for co-morbidities, and periodic re-assessment of treatment needs. Watchful waiting was included among the available treatment options, and recommendations were provided on how to offer and monitor different interventions, including pharmacotherapy.

The guideline development panel assigned each treatment intervention to one of three designations, based on the level of recommendation strength. Level A was associated with the strongest recommendation level: these recommendations, denoted by the use of the verb "must," were rare because they are based on high confidence in the evidence and require both a high magnitude of benefit and a low risk. Level B was denoted by the verb "should": these recommendations were more common because the requirements were less stringent, despite being still based on the clinical evidence and benefit-risk ratio. Level C was denoted by the verb "may": the lowest allowable recommendation deemed to be useful according to the AAN guideline development process manual within the scope of clinical practice, in line with the highest degree of practice variation. Finally, a few key non-evidence-based factors were transparently and systematically considered when formulating the recommendations: the relative value of the benefit compared with the risk, the feasibility of complying with the intervention (for instance, because of its limited availability), the cost of the intervention, and the expected variation in patient preferences relative to risks, burdens, and benefits.

Based on these principles, it was recommended that clinicians must inform patients and their caregivers about the natural history of tic disorders and evaluate functional impairment related to tics from the patient's (and, if applicable, the caregiver's) perspective (Level A). It was noted that as a result of partial or complete remission during the natural course of the tic disorder, pharmacotherapy for tics in young patients might no longer be required over time.

Clinicians prescribing medications for tics must periodically re-evaluate the need for ongoing medical treatment (Level A) should inform patients and caregivers that watchful waiting is an acceptable approach in patients who do not experience functional impairment from their tics (Level B). Clinicians should refer patients with TS to resources for psychoeducation for both teachers and peers (Level B) and might measure tic severity using a valid scale to assess treatment effects (Level C). Clinicians must also counsel patients that treatments for tics, including medications, infrequently result in complete cessation of tics (Level A).

Evidence from trials mostly conducted in children indicated that patients with tics receiving alpha-2 adrenergic agents are probably (Clonidine) or possibly (Guanfacine) more likely to experience reduced tic severity than those receiving placebo. Moreover, Clonidine and Guanfacine demonstrated beneficial effects on both tics and co-morbid attention-deficit and hyperactivity symptoms. Based on these findings, the AAN guideline recommended that clinicians should counsel patients with tics and co-morbid attention-deficit and hyperactivity disorder that alpha-2 adrenergic agents may provide benefit for both conditions (Level B). With regard to adverse effects, it was recommended that clinicians should prescribe alpha-2 adrenergic agonists for the treatment of tics when the benefits of treatment outweigh the risks (Level B). Clinicians must also counsel patients with tics regarding the adverse effects of alpha-2 adrenergic agents, including sedation (Level A); clinicians must monitor heart rate and blood pressure, as well as QTc interval in patients on Guanfacine extended release with a history of cardiac conditions, patients taking other QT-prolonging medications, or patients with a family history of long QT syndrome (Level A); clinicians discontinuing alpha-2 adrenergic agents must gradually taper them to avoid rebound hypertension (Level A).

With regard to antidopaminergic medications, there was evidence that Aripiprazole, Haloperidol, Risperidone, and Tiapride are probably more likely to reduce tic severity than placebo, whereas Metoclopramide, Pimozide, and Ziprasidone are possibly more likely to reduce tic severity than placebo. Evidence from the systematic literature review of studies in patients with tics showed a higher risk of medication-induced movement disorders with Haloperidol, Pimozide, and Risperidone, a higher risk of weight gain with Aripiprazole and Risperidone, a higher risk of somnolence with Aripiprazole, Risperidone, and Tiapride, a higher risk of QT prolongation with Pimozide, and a higher risk of elevated prolactin with Haloperidol, Metoclopramide, and Pimozide. Adverse effects were found to be medication-specific and often dose-dependent. Based on these lines of evidence on the efficacy and tolerability of antidopaminergic medications, it was recommended that clinicians may prescribe antidopaminergic agents for the treatment of tics when the benefits of treatment outweigh the risks (Level C). In case of prescription of these medications, clinicians must counsel patients on the relative propensity for extrapyramidal, metabolic, and hormonal adverse effects to inform decision-making on which antidopaminergic agent should be prescribed (Level A). Clinicians must also prescribe the lowest effective dose of antidopaminergic medication for tics to decrease the risk of adverse effects (Level A). Clinicians should monitor patients for medication-induced movement disorders and for metabolic and hormonal adverse effects, using evidence-based monitoring protocols (Level B). Electrocardiography and measurement of the QTc interval is mandatory before and after starting Pimozide or Ziprasidone, or in case of co-administration with other medications that can prolong the QTc interval (Level A). Finally, it was recommended that when attempting to discontinue antidopaminergic pharmacotherapy for tics, clinicians should gradually taper medications over weeks to months to avoid withdrawal dyskinesias (Level B).

Topiramate was identified as a potentially useful alternative options for patients with mild but troublesome tics who are not obtaining a satisfactory response or experience adverse effects from other pharmacological agents. Specifically, it was recommended that clinicians should prescribe Topiramate for the treatment of tics when the benefits of treatment outweigh the risks, based on the evidence that this medication is possibly more likely to reduce tic severity than placebo (Level B). When opting for this medication, clinicians must counsel patients regarding common adverse effects, including cognitive and language problems, somnolence, weight loss, and an increased risk of kidney stones (Level A).

The AAN guideline also covered patients with tics and co-morbid behavioral problems. It was recommended that clinicians should ensure that patients with tics are assessed for co-morbid obsessive-compulsive disorder and attention-deficit and hyperactivity disorder, evaluating the burden/functional impairment of behavioral symptoms and ensuring that appropriate treatment is provided (Level B). Specifically, the role of cognitive behavioral therapy as first-line intervention for co-morbid obsessive-compulsive disorder was highlighted. In children with tics and co-morbid attention-deficit and hyperactivity disorder, Clonidine (with or without concomitant use of psychostimulants) and Guanfacine were considered probably more likely to reduce both tic severity and attention-deficit and hyperactivity symptoms than placebo. Finally, clinicians must ensure that appropriate screening for anxiety, affective, and disruptive behavior disorders is performed in patients with tics (Level A). It was recommended that clinicians must inquire about suicidal thoughts and suicide attempts in patients with TS and refer to appropriate resources if present (Level A).

2.5 European Society for the Study of Tourette Syndrome (2022)

In 2022 the leading European experts and members of the European Society for the Study of Tourette Syndrome (ESSTS), published a set of four guidelines on the assessment and treatment of TS in the journal European Child and Adolescent Psychiatry (Andrén et al., 2022; Roessner et al., 2022; Szejko, Robinson, et al., 2022; Szejko, Worbe, et al., 2022). The ESSTS guidelines included four documents covering the following topics: Assessment (Part I), Psychosocial interventions (Part II), Pharmacological treatment (Part III), and Deep brain stimulation (Part IV). The authors combined their literature review with the assessment of the current prescribing practices of expert clinicians who routinely treat patients with TS, in order to formulate treatment recommendations that are adherent to real-life clinical practice settings and constraints. This work built on the previous version of the ESSTS guidelines, which had also been published in four parts, in 2011 (Cath et al., 2011; Müller-Vahl et al., 2011; Roessner et al., 2011; Verdellen, van de Griendt, Hartmann, Murphy, & ESSTS Guidelines Group, 2011). Finally, the current ESSTS guidelines updated the practical parameters of the TS Practice Guidelines, published by the AAN in 2019, which represented a synthesis of studies conducted prior to 2018 (Pringsheim, Holler-Managan, et al., 2019; Pringsheim, Okun, et al., 2019).

After reviewing the DSM-5 diagnostic criteria, the ESSTS guideline on the assessment of TS incorporated the newly implemented ICD-11 criteria (Müller-Vahl et al., 2022). Contrary to the DSM-5 nosology, in the ICD-11 tic disorders were removed from the ICD-10 category "Behavioral and emotional disorders with onset usually occurring in childhood and adolescence" to be reassigned to the "Movement disorders" section (World Health Organization (WHO), 2019). According to the ESSTS experts, this change was not in line with the growing body of both genetic and clinical evidence that tic disorders are related to neurodevelopmental and psychiatric disorders. Recommendations for the assessment of tics and psychiatric co-morbidities were produced in the context of both routine practice and research.

The ESSTS experts also argued that there is no scientific evidence to support the introduction of "Infectious or post-infectious tics" as a subcategory of "Secondary tics." Consistently, they did not recommend assessment of children for the proposed diagnosis of Pediatric Autoimmune Neuropsychiatric Disorders Associated with Streptococcal infections (PANDAS)-related TS—or for PANDAS-derived conditions. The ESSTS guideline on the assessment of TS introduced newly developed tic rating scales and addressed the differential diagnosis between tics and functional tic-like behaviors (Szejko, Robinson, et al., 2022).

In addition to complementing the previous version of the guidelines with a review of the English-language literature since 2011, the updated ESSTS guidelines included the results of a survey among ESSTS experts to formulate new European recommendations for the pharmacotherapy of TS (Roessner et al., 2022). The survey conducted by the ESSTS Guidelines Group gathered the prescription practices of the ESSTS experts for the treatment of children and adults with TS. The experts were asked which medication they would consider as first, second, third, and subsequent choices, provided absence of contraindications to the available agents and absence of co-morbid conditions. The selection of relevant pharmacological agents was conducted by combining the medications with at least moderate or low evidence according to the AAN guidelines (Pringsheim, Holler-Managan, et al., 2019) with those mentioned in the ESSTS survey. In the 2011 survey (22 experts), the top six medications were, in order of preference, Risperidone, Clonidine, Aripiprazole, Pimozide, Sulpiride, Tiapride (Roessner et al., 2011). In the survey published in 2022 (50 experts), the top positions were occupied by Aripiprazole, Clonidine, Tiapride, Haloperidol, Guanfacine, Risperidone (Roessner et al., 2022). Overall, a comparison of the results of the ESSTS surveys published in 2011 and in 2022 documented a clear shift from the use of Risperidone, Pimozide, and Sulpiride to the use of Aripiprazole for the treatment of tics. Contrary to their prior survey from 2011, in the updated version the EESTS experts were asked to give their recommendations separately for children/adolescents and adults. In general, choices in children/adolescents did not differ from choices in adults and pointed to a preference for Aripiprazole in both age groups. The main difference between the age groups was that Haloperidol was much more commonly considered in adults, while Tiapride was mentioned more often in children/adolescents.

The ESSTS survey indicated that Aripiprazole had become the most commonly used agent for the pharmacological treatment of TS across all age groups in Europe. The authors listed several factors as possible reasons, including its unique pharmacological profile as a dopamine partial agonist, the availability of multiple randomized controlled trials documenting a favorable benefit-risk ratio, and its positive profile of adverse effects (Cox & Cavanna, 2021; Cox, Seri, & Cavanna, 2016). Tiapride was reported as the second most commonly prescribed agent in children and adolescents with tics, especially in Germany. This substituted benzamide has a similar working mechanism as aripiprazole, showing a maximum of 80% of dopamine receptor occupation even in the presence of excess concentrations, and a good tolerability (very low potential for extrapyramidal adverse effects compared to Haloperidol). Another recommended antidopaminergic agent, Risperidone, has a good evidence base, but is associated with weight gain and metabolic adverse effects. The results of a recently published review of the comparative efficacy and safety of newer antidopaminergic medications for tic disorders confirmed that Aripiprazole and Risperidone are the most robust evidence-based options and suggested that Quetiapine may be a promising therapy, whereas for Ziprasidone and Olanzapine the evidence is lacking (Yang et al., 2019). The alpha-2 agonist agent Clonidine was rated as the second most preferred medication for children/adolescents with TS (third for adults) in the ESSTS guideline. The authors acknowledged that alpha-2 agonists seem to be characterized by relatively better effectiveness in patients with TS and co-morbid attention-deficit and hyperactivity disorder, for the treatment of both conditions. Accordingly, alpha-2 agonists were recommended as first line treatment of mild to moderate tics in patients with co-morbid attention-deficit and hyperactivity disorder. Moreover, it was stated that in some patients with mild tics only, alpha-2 agonists might be more acceptable than antidopaminergic agents, based on a more favorable tolerability profile. Topiramate, an antiepileptic drug included in the AAN guidelines, was listed among the pharmacological options that might be considered for treatment resistant cases, based on limited evidence base and less frequent prescription rates by ESSTS experts (Cavanna, 2018).

There is a substantial degree of overlap between the 2022 ESSTS guidelines and the 2019 AAN guidelines, as these sets of documents are largely consistent with one another in terms of both content and scope. There are, however, a few minor differences due to the fact that additional studies have been published since the systematic literature review conducted by the authors of the AAN guidelines. Moreover, it is widely known that there are differing availabilities of and indications for pharmacological agents in Europe compared to the United States. Finally, the ESSTS guidelines have a clear focus on actual current clinician practice as a benchmark (Mathews, 2022). The authors pay more attention to the assessment and treatment of psychiatric co-morbidities, as well as to the psychiatric effects of pharmacological agents typically used to treat tics (Cavanna, 2020). The actual recommendations for the pharmacotherapy of tics differ somewhat in the order of choice, with Aripiprazole presented as the first line treatment for tics in the ESSTS guidelines, whereas the AAN guidelines favor alpha-2 agonists.

A few differences between the ESSTS guidelines and the AAN guidelines on psychological and behavioral interventions are also apparent (Andrén et al., 2022). While the AAN guidelines focused on Habit Reversal Therapy, mainly as a key component of the Comprehensive Behavioral Intervention for Tics, the ESSTS guidelines also discussed additional data suggesting the efficacy of exposure and response prevention. In Europe, access to Habit Reversal Therapy and implementation of Comprehensive Behavioral Intervention for Tics are less widespread than in North America, and the ESSTS members might be more likely to be familiar with the principles and practices of exposure and response prevention, a gold standard treatment for anxiety and obsessive-compulsive symptoms, which are frequently co-morbid in patients with TS. Moreover, the ESSTS guidelines addressed group treatment strategies as a potentially effective form of delivering behavioral interventions to patients with TS. Over the last few years, especially since the onset of the covid-19 pandemic, internet-based modules of established behavioral treatments were developed to improve the accessibility of such interventions. Finally, adaptations were made to broaden the focus of behavioral treatments from reduction of tic severity to improvement of the patient's health-related quality of life.

The contents of the ESSTS guideline on deep brain stimulation reflected the increasing knowledge about its efficacy in patients with severe and refractory TS (Szejko, Worbe, et al., 2022). Reported effects were modest and somewhat contradictory, at least partly due to the limited number and sample size of randomized controlled trials: an ongoing issue since the development of early national statement papers (Cavanna et al., 2011). The heterogeneous results on deep brain stimulation were interpreted in the light of the European contributions to a meta-analysis (Baldermann et al., 2016) and data of the International Tourette syndrome DBS Public Database and Registry (Martinez-Ramirez et al., 2018). Although a standardized definition of failure of medical treatments for tics help clinicians to determine when to offer less established treatments, such definition is currently lacking. In order to fill this gap, an international expert consensus paper was recently published (Martino et al., 2022).

For the first time, the ESSTS guidelines included a statement by patient representatives in Europe who conducted a world-wide survey among patient advocacy groups, describing the patient perspective on different research topics and reflecting the fruitful collaboration between advocacy groups and experts in the ESSTS community (Anderson & Tics and Tourette Around the Globe TTAG representing Tic and Tourette Syndrome TS patient associations around the world, 2022). Overall, the revised ESSTS guidelines contain updated information about the recent developments on assessment and treatment of TS, including an updated decision tree providing a practical therapeutic algorithm (Müller-Vahl et al., 2022).

3. Expert consensus on the treatment of Tourette syndrome

3.1 Europe and United Kingdom (2012)

In 2012, the results the first large-scale survey on prescribing habits for the pharmacological management of TS in Europe were published in the European Journal of Pediatric Neurology (Cavanna et al., 2012; Rickards, Cavanna, & Worrall, 2012). All ESSTS members actively prescribing for pediatric and/or adult patients with TS were invited to complete an online questionnaire covering the pharmacological treatment of five symptom domains: tics, attention-deficit and hyperactivity symptoms, obsessive-compulsive symptoms, anxiety, and depression.

The results of this systematic assessment of treatment practices in TS across Europe-based clinician revealed a variety of pharmacological agents being prescribed as first-line options. Antidopaminergic medications were widely used, and Risperidone was the most commonly prescribed medication for the treatment of tics. There was evidence of high variability in both medication choices and dosages. Medical prescribers with a background in psychiatry more often prescribed Risperidone as first-line option for tics, whereas prescribers with a background in neurology tended to choose Clonidine more frequently. Clinicians prescribing for children and clinicians prescribing for adults opted for similar pharmacological classes, although their individual medication choices were different. Most clinicians working with adult patients opted for Aripiprazole as first-line option for the treatment of tics, despite the lack of evidence from double-blind controlled trials at the time of the survey. It is possible that their choice was related to the perception of a more favorable tolerability profile for Aripiprazole compared to other antidopaminergic agents. Overall, the results of the ESSTS survey contradicted previous findings that first generation antidopaminergic agents were the most commonly prescribed pharmacotherapy for tics, documenting a change in prescribing practices.

3.2 Japan (2019)

In 2019, the results of a questionnaire-based survey conducted in Japan and modeled on the 2012 ESSTS survey were published in the journal Brain and Development (Hamamoto et al., 2019). The aim of the survey was to collate the consensus of Japanese experts based on their pharmacotherapy practice for tic disorders and co-morbid conditions. The results were compared with available international evidence to improve the quality of pharmacotherapeutic approaches to tic disorders and co-morbid conditions in Japan. The authors conducted a sequential two-step survey: in the first step (2015), national experts were asked to fill a newly developed questionnaire; in the second step (2016), a revised version of the questionnaire based on feedback of the findings from the first survey was circulated. As part of the initial survey, the authors devised a questionnaire referring to the ESSTS guidelines (Roessner et al., 2011) and expert consensus (Cavanna et al., 2012; Rickards et al., 2012) on pharmacotherapy for patients with TS. The participating clinicians were members of the Japanese Society of Tourette Syndrome Research or had published research articles on tic disorders or childhood obsessive-compulsive disorder during the 5-year period preceding the survey.

The results of the initial survey revealed variability in the use of pharmacological agents among the experts in Japan, and formed the basis for the revision of the questionnaire. The Delphi methodology was used to determine the extent of agreement among the experts. The authors were able to build a general consensus on pharmacotherapy for tic disorders and co-morbid conditions based on the second step of the survey. Antidopaminergic agents were prescribed more frequently than alpha-2 adrenergic agonists, despite their widespread use in the treatment of tic disorders internationally. Specifically, the Japanese experts indicated Aripiprazole and Risperidone as first- and second-line pharmacological options for tic disorders, respectively. There was no evidence of significant group differences between prescribing psychiatrists and pediatricians in the selection of pharmacological agents.

3.3 China (2020)

In 2020, the Chinese Child Neurology Society published the results of an Expert Consensus on Diagnosis and Treatment of Tic Disorders in China in the Child and Adolescent Psychiatry section of the journal Frontiers in

Psychiatry (Liu et al., 2020). This consensus statement encompassed the clinical diagnosis and evaluation of tic disorders and their co-morbidities, focusing on treatment interventions, including traditional Chinese medicine and acupuncture, as well as prognosis in children with TD in China. The recommendations based on experts' clinical experience aimed at improving the understanding of tic disorders for pediatricians and other health professionals, and at optimizing their diagnosis and treatment in China. Recommended medications for the treatment of pediatric patients with tic disorders were based on the authors' clinical experience and availability of the drugs in China, and were assigned to two tiers: first-line options and second-line options. The first-line tier included two antidopaminergic agents (Aripiprazole and Tiapride) and the alpha-2 adrenergic agent Clonidine (with a specific indication for patients with TS and co-morbid attention-deficit and hyperactivity disorder). The group of first-line options also included two proprietary polyherbal Chinese medicines (Changma Xifeng tablets and Jiuwei Xifeng granules) that had been approved by the Chinese National Administration of Traditional Chinese Medicine and had been endorsed by the National Guideline of Traditional Chinese Medicine for Pediatric Diagnosis and Treatment through a Tic Disorders Expert Committee (Rong, Ma, Han, & Wu, 2019). The second-line tier included two antidopaminergic agents (Risperidone and Haloperidol) and the anticonvulsant agent Topiramate.

There had been a standardized national guideline for the diagnosis and treatment of tic disorders using traditional Chinese medicine in China since 2012, with an updated version published in the Journal of Pediatric Traditional Chinese Medicine in 2019 (Rong et al., 2019). Moreover, the results of recent meta-analyses supported the efficacy and safety of acupuncture, traditional Chinese medicine alone, and traditional Chinese medicine plus Western pharmacotherapy, in treating patients with TS and other tic disorders in China (Qi et al., 2020; Zhao, Xin, Wang, & Sun, 2020). Finally, it was acknowledged that the total number and sample sizes of the reported trials on the use of acupuncture and traditional Chinese medications for the treatment for tics were still relatively small, compared to pharmacological trials of Western medicine.

4. Conclusions

The literature reporting guidelines and expert consensus statements on the treatment of patients with TS and other tic disorders has expanded considerable over the last decade (Table 2).

Table 2 Published guidelines and expert consensus on pharmacotherapy for patients with Tourette syndrome and other tic disorders.

Authors	Year	Journal	Country	Population	SLR/MA/ EC
Pringsheim et al.	2012	Can J Psychiatry	Canada	Children + Adults	SLR (2 parts)
Rickards et al.	2012	Eur J Paediatr Neurol	EU (+UK)	Children + Adults	EC
AACAP	2013	J Am Acad Child Adolesc Psychiatry	USA	Children	SLR
Hollis et al.	2016	Health Technol Assess	UK	Children	SLR + MA
Hamamoto et al.	2019	Brain Dev	Japan	Children + Adults	EC
AAN	2019	Neurology	USA	Children + Adults	SLR + MA
Liu et al.	2020	Front Psychiatry	China	Children	EC
ESSTS	2022	Eur J Child Adolesc Psychiatry	EU (+UK)	Children + Adults	SLR (4 parts)

Abbreviations: AACAP, American Academy of Child and Adolescent Psychiatry; AAN, American Academy of Neurology; EC, expert survey and consensus; ESSTS, European Society for the Study of Tourette Syndrome; MA, meta-analysis; SLR, systematic literature review.

The Tic Disorders and Tourette syndrome Study Group members of the Movement Disorder Society recently published the results of a survey on clinical practice in tic disorders (Ganos et al., 2021). These results mapped well to the available evidence from the current guidelines. In general, differences in regional pharmacotherapeutic preferences are reflected in the treatment guidelines from Europe and North America (Bhikram, Elmaghraby, Abi-Jaoude, & Sandor, 2021).

Evidence-based guidelines are notoriously difficult to apply to clinical conditions characterized by complex co-morbidities. Moreover, statistically significant benefits observed in randomized controlled trials of pharmacotherapy for tics may be marginal in day-to-day practice. Finally, placebos have often demonstrated substantial reduction in tic and tic-related impairments: it has been estimated that the placebo effect can account for 30% of improvement in tic severity and 40% of improvement in overall impairment in about 20% of patients (Cavanna, Strigaro, & Monaco, 2007; Cubo et al., 2013; Ferreira, Trenkwalder, & Mestre, 2018). The role of placebos as tic suppressors implies that there might be therapeutic effects on tics due to the very act of intervening: this highlights the importance of conducting placebo-controlled trials to determine the efficacy of pharmacological agents for tics. Standardization of clinical care and alignment with evidence based treatment protocols have to be combined with skilled clinical observation and individualization of therapeutic approaches (Accad & Francis, 2018). TS is a quintessentially neuropsychiatric condition, and individualized management plans for patients with tic disorders should be developed based on a number of factors, including symptom severity, clinical presentation, patient preference, and resource availability. The combination of the principles of evidence-based medicine with the subtleties of clinical judgment can promote a patient-centered approach focused on improving patients' health-related quality of life.

Acknowledgments

The author is grateful to Tourettes Action UK and Tourette Association of America for their ongoing support.

Funding

The present work was invited and did not receive any specific funding.

Conflicts of interest

The author declares that he has no conflicts of interest about the present work.

References

Accad, M., & Francis, D. (2018). Does evidence based medicine adversely affect clinical judgment? *British Medical Journal, 362*, k2799.

American Psychiatric Association (APA). (2000). *Diagnostic and statistical manual of mental disorders, fourth edition, text revision (DSM-IV-TR)*. Washington, DC: American Psychiatric Association.

American Psychiatric Association (APA). (2013). *Diagnostic and statistical manual of mental disorders, fifth edition (DSM-5)*. Washington, DC: American Psychiatric Association.

Anderson, S. M., & Tics and Tourette Around the Globe (TTAG) representing Tic and Tourette Syndrome (TS) patient associations around the world. (2022). European clinical guidelines for Tourette syndrome and other tic disorders: patients' perspectives on research and treatment. *European Child and Adolescent Psychiatry*, in press.

Andrén, P., Jakubovski, E., Murphy, T. L., Woitecki, K., Tarnok, Z., Zimmerman-Brenner, S., et al. (2022). European clinical guidelines for Tourette syndrome and other tic disorders-version 2.0. Part II: Psychological interventions. *European Child and Adolescent Psychiatry*, in press.

Badenoch, J., Searle, T., Watson, I., & Cavanna, A. E. (2021). Tics in patients with encephalitis. *Neurological Sciences*, *42*(4), 1311–1323.

Baldermann, J. C., Schüller, T., Huys, D., Becker, I., Timmermann, L., Jessen, F., et al. (2016). Deep brain stimulation for Tourette syndrome: A systematic review and meta-analysis. *Brain Stimulation, 9*(2), 296–304.

Bhikram, T., Elmaghraby, R., Abi-Jaoude, E., & Sandor, P. (2021). An international survey of health care services available to patients with Tourette syndrome. *Frontiers in Psychiatry, 12*, 621874.

Black, K. J., Black, E. R., Greene, D. J., & Schlaggar, B. L. (2016). Provisional tic disorder: What to tell parents when their child first starts ticcing. *F1000Research, 5*, 696.

Black, K. J., Kim, S., Schlaggar, B. L., & Greene, D. J. (2020). The new tics study: A novel approach to pathophysiology and cause of tic disorders. *Journal of Psychiatry and Brain Science, 5*, e200012.

Black, K. J., Kim, S., Yang, N. Y., & Greene, D. J. (2021). Course of tic disorders over the lifespan. *Current Developmental Disorders Reports, 8*(2), 121–132.

Cath, D. C., Hedderly, T., Ludolph, A. G., Stern, J. S., Murphy, T., Hartmann, A., et al. (2011). European clinical guidelines for Tourette syndrome and other tic disorders. Part I: Assessment. *European Child and Adolescent Psychiatry, 20*(4), 155–171.

Cavanna, A. E. (2014). Improved criteria for the diagnosis of tic disorders in DSM-5. *Future Neurology, 9*, 251–253.

Cavanna, A. E. (2018). *Behavioural neurology of antiepileptic drugs: A practical guide*. Oxford: Oxford University Press.

Cavanna, A. E. (2020). *Pharmacological treatment of tics*. Cambridge: Cambridge University Press.

Cavanna, A. E., Eddy, C. M., Mitchell, R., Pall, H., Mitchell, I., Zrinzo, L., et al. (2011). An approach to deep brain stimulation for severe treatment-refractory Tourette syndrome: The UK perspective. *British Journal of Neurosurgery, 25*(1), 38–44.

Cavanna, A. E., & Nani, A. (2013). Tourette syndrome and consciousness of action. *Tremor and Other Hyperkinetic Movements, 3*, (tre-03-181-4368-1).

Cavanna, A. E., Rickards, H., Worrall, R., Hoekstra, P. J., Plessen, K. J., & Roessner, V. (2012). From ipse dixit to evidence-based guidelines: On the optimal management of Tourette syndrome. *European Journal of Paediatric Neurology, 16*(3), 310–311.

Cavanna, A. E., Strigaro, G., & Monaco, F. (2007). Brain mechanisms underlying the placebo effect in neurological disorders. *Functional Neurology, 22*(2), 89–94.

Claudio-Campos, K., Stevens, D., Koo, S. W., Valko, A., Bienvenu, O. J., Budman, C. B., et al. (2021). Is persistent motor or vocal tic disorder a milder form of Tourette syndrome? *Movement Disorders, 36*(8), 1899–1910.

Cox, J. H., & Cavanna, A. E. (2021). Aripiprazole for the treatment of Tourette syndrome. *Expert Review of Neurotherapeutics, 21*(4), 381–391.

Cox, J. H., Seri, S., & Cavanna, A. E. (2016). Safety and efficacy of aripiprazole for the treatment of pediatric Tourette syndrome and other chronic tic disorders. *Pediatric Health, Medicine and Therapeutics, 7*, 57–64.

Cox, J. H., Seri, S., & Cavanna, A. E. (2018). Sensory aspects of Tourette syndrome. *Neuroscience and Biobehavioral Reviews, 88*, 170–176.

Cubo, E., González, M., Singer, H., Mahone, E. M., Scahill, L., Müller-Vahl, K. R., et al. (2013). Impact of placebo assignment in clinical trials of tic disorders. *Movement Disorders, 28*(9), 1288–1292.

Cuenca, J., Glazebrook, C., Kendall, T., Hedderly, T., Heyman, I., Jackson, G., et al. (2015). Perceptions of treatment for tics among young people with Tourette syndrome and their parents: A mixed methods study. *BMC Psychiatry, 15*, 46.

Evidence Based Medicine Working Group. (1992). Evidence based medicine: A new approach to teaching the practice of medicine. *Journal of the American Medical Association, 268*, 2420–2425.

Ferreira, J. J., Trenkwalder, C., & Mestre, T. A. (2018). Placebo and nocebo responses in other movement disorders besides Parkinson's disease: How much do we know? *Movement Disorders*, *33*(8), 1228–1235.

Ganos, C., Sarva, H., Kurvits, L., Gilbert, D. L., Hartmann, A., Worbe, Y., et al. (2021). Clinical practice patterns in tic disorders among Movement Disorder Society members. *Tremor and Other Hyperkinetic Movements*, *11*, 43.

Hamamoto, Y., Fujio, M., Nonaka, M., Matsuda, N., Kono, T., & Kano, Y. (2019). Expert consensus on pharmacotherapy for tic disorders in Japan. *Brain and Development*, *41*(6), 501–506.

Hollis, C., Pennant, M., Cuenca, J., Glazebrook, C., Kendall, T., Whittington, C., et al. (2016). Clinical effectiveness and patient perspectives of different treatment strategies for tics in children and adolescents with Tourette syndrome: A systematic review and qualitative analysis. *Health Technology Assessment*, *20*(4).

Kim, S., Greene, D. J., Bihun, E. C., Koller, J. M., Hampton, J. M., Acevedo, H., et al. (2019). Provisional tic disorder is not so transient. *Scientific Reports*, *9*(1), 3951.

Liu, Z. S., Cui, Y. H., Sun, D., Lu, Q., Jiang, Y. W., Jiang, L., et al. (2020). Current status, diagnosis, and treatment recommendation for tic disorders in China. *Frontiers in Psychiatry*, *11*, 774.

Macerollo, A., Martino, D., Cavanna, A. E., Gulisano, M., Hartmann, A., Hoekstra, P. J., et al. (2016). Refractoriness to pharmacological treatment for tics: A multicenter European audit. *Journal of the Neurological Sciences*, *366*, 136–138.

Martinez-Ramirez, D., Jimenez-Shahed, J., Leckman, J. F., Porta, M., Servello, D., Meng, F. G., et al. (2018). Efficacy and safety of deep brain stimulation in Tourette syndrome: The international Tourette syndrome deep brain stimulation public database and registry. *JAMA Neurology*, *75*(3), 353–359.

Martino, D., Madhusudan, N., Zis, P., & Cavanna, A. E. (2013). An introduction to the clinical phenomenology of Tourette syndrome. *International Review of Neurobiology*, *112*, 1–33.

Martino, D., Malaty, I., Müller-Vahl, K., Nosratmirshekarlou, E., Pringsheim, T. M., Shprecher, D., et al. (2022). Treatment failure in persistent tic disorders: An expert clinicians' consensus-based definition. *European Child and Adolescent Psychiatry*, in press.

Mathews, C. A. (2022). Treating Tourette syndrome and other chronic tic disorders: Updated guidelines by the European society for the study of Tourette syndrome (ESSTS). *European Child and Adolescent Psychiatry*, in press.

Müller-Vahl, K. R., Cath, D. C., Cavanna, A. E., Dehning, S., Porta, M., Robertson, M. M., et al. (2011). European clinical guidelines for Tourette syndrome and other tic disorders. Part IV: Deep brain stimulation. *European Child and Adolescent Psychiatry*, *20*(4), 209–217.

Müller-Vahl, K. R., Sambrani, T., & Jakubovski, E. (2019). Tic disorders revisited: Introduction of the term 'tic spectrum disorders'. *European Child & Adolescent Psychiatry*, *28*(8), 1129–1135.

Müller-Vahl, K. R., Szejko, N., Verdellen, C., Roessner, V., Hoekstra, P. J., Hartmann, A., et al. (2022). European clinical guidelines for Tourette syndrome and other tic disorders: Summary statement. *European Child and Adolescent Psychiatry*, in press.

Murphy, T. K., Lewin, A. B., Storch, E. A., & Stock, S. (2013). Practice parameter for the assessment and treatment of children and adolescents with tic disorders. *Journal of the American Academy of Child and Adolescent Psychiatry*, *52*, 1341–1359.

Plessen, K. J. (2013). Tic disorders and Tourette's syndrome. *European Child & Adolescent Psychiatry*, *22*(Suppl 1), S55–S60.

Pringsheim, T., Doja, A., Gorman, D., McKinlay, D., Day, L., Billinghurst, L., et al. (2012). Canadian guidelines for the evidence-based treatment of tic disorders: Pharmacotherapy. *Canadian Journal of Psychiatry*, *57*(3), 133–143.

Pringsheim, T., Holler-Managan, Y., Okun, M. S., Jankovic, J., Piacentini, J., Cavanna, A. E., et al. (2019). Comprehensive systematic review summary: Treatment of tics in people with Tourette syndrome and chronic tic disorders. *Neurology*, *92*(19), 907–915.

Pringsheim, T., Okun, M. S., Müller-Vahl, K., Martino, D., Jankovic, J., Cavanna, A. E., et al. (2019). Practice guideline recommendations summary: Treatment of tics in people with Tourette syndrome and chronic tic disorders. *Neurology*, *92*(19), 896–906.

Qi, H., Liu, R., Zheng, W., Zhang, L., Ungvari, G. S., Ng, C. H., et al. (2020). Efficacy and safety of traditional Chinese medicine for Tourette's syndrome: A meta-analysis of randomized controlled trials. *Asian Journal of Psychiatry*, *47*, 101853.

Rickards, H., & Cavanna, A. E. (2009). Gilles de la Tourette: The man behind the syndrome. *Journal of Psychosomatic Research*, *67*(6), 469–474.

Rickards, H., Cavanna, A. E., & Worrall, R. (2012). Treatment practices in Tourette syndrome: The European perspective. *European Journal of Paediatric Neurology*, *16*(4), 361–364.

Robertson, M. M., & Eapen, V. (2014). Tourette's: Syndrome, disorder or spectrum? Classificatory challenges and an appraisal of the DSM criteria. *Asian Journal of Psychiatry*, *11*, 106–113.

Roessner, V., Eichele, H., Stern, J. S., Skov, L., Rizzo, R., Debes, N. M., et al. (2022). European clinical guidelines for Tourette syndrome and other tic disorders-version 2.0. Part III: Pharmacological treatment. *European Child and Adolescent Psychiatry*. in press.

Roessner, V., Ludolph, A. G., Müller-Vahl, K., Neuner, I., Rothenberger, A., Woitecki, K., et al. (2014). DSM-5 Kommentar: Tourette Syndrom und andere tic-Störungen im DSM-5 [Tourette syndrome and other tic disorders in DSM-5: A comment]. *Zeitschrift für Kinder- und Jugendpsychiatrie und Psychotherapie*, *42*(2), 129–134.

Roessner, V., Plessen, K. J., Rothenberger, A., Ludolph, A. G., Rizzo, R., Skov, L., et al. (2011). European clinical guidelines for Tourette syndrome and other tic disorders. Part II: Pharmacological treatment. *European Child and Adolescent Psychiatry*, *20*(4), 173–196.

Rong, P., Ma, R., Han, X., & Wu, H. (2019). Guideline for TCM pediatrics clinical diagnosis and treatment: Tic disorder. *Journal of Pediatric Traditional Chinese Medicine*, *15*(6), 1–6.

Sackett, D. L., Rosenberg, W. M. C., Gray, J. A., Haynes, R. B., & Richardson, W. S. (1996). Evidence based medicine: What it is and what it isn't. *British Medical Journal*, *312*, 71–72.

Steeves, T., McKinlay, D., Gorman, D., Billinghurst, L., Day, L., Carroll, A., et al. (2012). Canadian guidelines for the evidence-based treatment of tic disorders: Behavioural therapy, deep brain stimulation and transcranial magnetic stimulation. *Canadian Journal of Psychiatry*, *57*(3), 144–151.

Szejko, N., Robinson, S., Hartmann, A., Ganos, C., Debes, N. M., Skov, L., et al. (2022). European clinical guidelines for Tourette syndrome and other tic disorders-version 2.0. Part I: Assessment. *European Child and Adolescent Psychiatry*, in press.

Szejko, N., Worbe, Y., Hartmann, A., Visser-Vandewalle, V., Ackermans, L., Ganos, C., et al. (2022). European clinical guidelines for Tourette syndrome and other tic disorders-version 2.0. Part IV: Deep brain stimulation. *European Child and Adolescent Psychiatry*, in press.

Termine, C., Grossi, E., Anelli, V., Derhemi, L., & Cavanna, A. E. (2021). Possible tics diagnosed as stereotypies in patients with severe autism spectrum disorder: A video-based evaluation. *Neurological Sciences*, *42*(4), 1559–1561.

Tourette Syndrome Classification Study Group. (1993). Definitions and classification of tic disorders. *Archives of Neurology*, *50*, 1013–1016.

Verdellen, C., van de Griendt, J., Hartmann, A., Murphy, T., & ESSTS Guidelines Group. (2011). European clinical guidelines for Tourette syndrome and other tic disorders. Part III: Behavioural and psychosocial interventions. *European Child and Adolescent Psychiatry*, *20*(4), 197–207.
Whittington, C., Pennant, M., Kendall, T., Glazebrook, C., Trayner, P., Groom, M., et al. (2016). Treatments for Tourette syndrome in children and young people: A systematic review. *Journal of Child Psychology and Psychiatry*, *57*(9), 988–1004.
Woods, D. W., & Thomsen, P. H. (2014). Tourette and tic disorders in ICD-11: Standing at the diagnostic crossroads. *Revista Brasileira de Psiquiatria*, *36*(1), 51–58.
World Health Organization (WHO). (1993). *International statistical classification of diseases and related health problems, tenth edition (ICD-10)*. Geneva: World Health Organization.
World Health Organization (WHO). (2019). *International statistical classification of diseases and related health problems, eleventh edition (ICD-11)*. Geneva: World Health Organization.
Yang, C., Hao, Z., Zhang, L. L., Zhu, C. R., Zhu, P., & Guo, Q. (2019). Comparative efficacy and safety of antipsychotic drugs for tic disorders: A systematic review and Bayesian network meta-analysis. *Pharmacopsychiatry*, *52*(1), 7–15.
Yang, C., Hao, Z., Zhu, C., Guo, Q., Mu, D., & Zhang, L. (2016). Interventions for tic disorders: An overview of systematic reviews and meta analyses. *Neuroscience and Biobehavioral Reviews*, *63*, 239–255.
Yang, C., Zhang, Z., Zhang, L., Tian, J., Yu, D., Wang, J., et al. (2018). Quality assessment of clinical practice guidelines on tic disorders with AGREE II instrument. *Psychiatry Research*, *259*, 385–391.
Zhao, R. Z., Xin, Y., Wang, W. X., & Sun, K. X. (2020). Meta-analysis of clinical effect of acupuncture on tic disorder. *Shanghai Journal of Acupuncture and Moxibustion*, *39*, 244–252.

CHAPTER FOUR

Update and recent progress in the neurobiology of Tourette syndrome

Natalia Szejko[a,b,*]
[a]Department of Neurology, Medical University of Warsaw, Poland
[b]Department of Bioethics, Medical University of Warsaw, Poland
*Corresponding author: e-mail address: natalia.szejko@wum.edu.pl

Contents

1. Introduction

Pathophysiology of Tourette syndrome (TS) is complex, which also explains clinical variability of symptoms on the verge between neurology and psychiatry. Previous studies have demonstrated that the main pathway implicated in occurrence of tics is cortical–striatal–thalamic–cortical (CSTC) circuit (Levy, 2011; Ramkiran, Heidemeyer, Gaebler, Shah, & Neuner, 2019; Scholl, Baladron, Vitay, & Hamker, 2021; Wang et al., 2011; Worbe et al., 2015). Nevertheless, many other studies have demonstrated abnormal neurotransmission in other brain regions (Felling & Singer, 2011), such as basal

International Review of Movement Disorders, Volume 3
ISSN 2666-7878
https://doi.org/10.1016/bs.irmvd.2021.11.003

ganglia (Greene et al., 2017; Mink, 2001; Peterson et al., 2003), prefrontal cortex (Ganos et al., 2014; Greene et al., 2017; Yamamuro et al., 2015), corpus callosum (Cavanna et al., 2010; Draganski et al., 2010; Mostofsky, Wendlandt, Cutting, Denckla, & Singer, 1999; Wolff et al., 2016), midbrain (Devinsky, 1983; Garraux et al., 2006), hypothalamus (Sandyk, Iacono, & Bamford, 1987), thalamus (Makki, Behen, Bhatt, Wilson, & Chugani, 2008; Miller et al., 2010) or cerebellum (Sigurdsson, Jackson, Jolley, Mitchell, & Jackson, 2020; Tobe et al., 2010). Similarly, diverse neurotransmission systems were found to be involved in pathophysiology of tics: dopaminergic (Hienert, Gryglewski, Stamenkovic, Kasper, & Lanzenberger, 2018; Liu et al., 2010; Maia & Conceição, 2017, 2018; Müller-Vahl, Loeber, Kotsiari, Müller-Engling, & Frieling, 2017), glutaminergic (Kanaan et al., 2017; Singer, Morris, & Grados, 2010), serotoninergic (Müller-Vahl et al., 2019; Wong et al., 2008), noradrenergic (Stöber et al., 1999), histaminergic (Hartmann, Worbe, & Arnulf, 2012; Rapanelli & Pittenger, 2016), cholinergic (Albin, Minderovic, & Koeppe, 2017; Kataoka et al., 2010), GABA-ergic (Draper et al., 2014; Tübing & Münchau, 2015) and endocannabinoid (Müller-Vahl et al., 2003, 2020, 2021; Szejko et al., 2020). Therefore, up to this date it is not certain where the primary injury is located. Probably TS occur as a consequence of changes in diverse neurocircuits involving different brain regions and different neurotransmitters.

The aim of this chapter is to familiarize the reader with the recent updates regarding neurobiology of TS and indicate clinical implications of these findings. In order to organize the presentation of the findings, the information was gathered in categories related to particular neurotransmission: dopaminergic, glutaminergic, serotoninergic, histaminergic, GABA-ergic, endocannabinoid and others. For the purpose of this book, only studies in humans were analyzed. The summary of the most important studies exploring the topic of involvement of diverse neurotransmission in pathophysiology of TS is presented in Table 1.

Table 1 Summary of selected studies exploring neurobiology of Tourette syndrome

Study	Methodology	No. of subjects	No. of controls	Main findings
Neuroimaging MRS/MRI studies				
Mostofsky et al. (1999)	MRI	10 with TS, 9 with TS+ADHD	22	No difference in corpus callosum between groups
Garraux et al. (2006)	VBM	31	31	Increased gray matter mainly in the left mesencephalon in 31 TS patients
Makki et al. (2008)	MRI-DTI	23	35	Microstructural dysfunction measured by MRI-DTI in component regions of the FST circuit contribute to the pathophysiology in TS
Cavanna et al. (2010)	MRI-DTI	2	–	Mean corpus callosum FA values were significantly lower in the TS twin than in the unaffected twin
Miller et al. (2010)	MRI	149	134	Enlargement in TS-affected thalami compared to controls
Tobe et al. (2010)	MRI	163	147	The TS group demonstrated reduced volumes of the cerebellar hemispheres bilaterally
Draper et al. (2014)	MRS	15	15	GABA concentrations within the SMA are paradoxically elevated in individuals with TS
Ganos et al. (2014)	VBM	14	15	Grey matter volumes in the right inferior frontal gyrus and the left frontal pole were reduced in patients relative to healthy controls
Puts et al. (2015)	MRS	23	67	Reduced GABAergic inhibition and abnormal sensory symptoms in children with Tourette syndrome

Continued

Table 1 Summary of selected studies exploring neurobiology of Tourette syndrome—cont'd

Study	Methodology	No. of subjects	No. of controls	Main findings
Worbe et al. (2015)	MRI	49	28	Altered structural connectivity of cortico-striato-pallido-thalamic networks in Gilles de la Tourette syndrome
Yamamuro et al. (2015)	NIRS	10	10	Oxy-Hb changes in the prefrontal cortex were significantly smaller in the TS group compared with the control group, especially in the left dorsolateral prefrontal cortex
Freed et al. (2016)	MRS	15	36	Anterior cingulate cortex GABA levels were lower in participants with TS compared with control subjects
Wolff et al. (2016)	DTI	26	24	Reduced AD of the CC in treatment-naïve boys with pure TS in comparison to healthy controls
Fan et al. (2017)	MRS	23 TS, 20 OCD	22	The 3-group comparison showed a significant difference in choline concentration in the thalamus. Thalamic choline was highest in OCD patients, showing a significant difference with TS, and a trend compared to controls. Glutamine in dACC correlated negatively with tic severity scores in TS patients, while glutamate in thalamus correlated positively with anxiety severity scores in OCD patients
Greene et al. (2017)	VBM	103	103	The TS group demonstrated lower WM volume bilaterally in orbital and medial prefrontal cortex, and greater GM volume in posterior thalamus, hypothalamus and midbrain
Kanaan et al. (2017)	MRS	37	36	A significant reduction in glutamate concentration in striatum and thalamus in TS patients compared to controls. No differences between the group of patients treated with aripiprazole and those without any treatment when it comes to metabolite concentrations. In multiple regression analysis tic severity was associated with striatal glutamate and premonitory urge with thalamic glutamate

Naaijen et al. (2017)	MRS	TS = 15, ADHD 39, TS + ADHD = 29	53	No evidence for glutamatergic neuropathology in TS
Mahone, Puts, Edden, Ryan, and Singer (2018)	MRS	32	43	Glutamate was significantly higher in the TS group, compared to controls, with no significant group differences in GABA observed
Tübing et al. (2018)	TMS	15	19	The dysfunction of the SMA in GTS is not primarily related to altered plasticity in SMA-M1 connections
Sigurdsson et al. (2020)	VBM	28	36	Reduced GM volume in cerebellar lobule involved in higher-order cognitive functions and sensorimotor processing, in patients
Neuroimaging PET and SPECT studies				
George et al. (1994)	SPECT	15	6	Unmediated TS patients showed no differences from controls
Turjanski et al. (1994)	PET	10	34	Striatal D2 receptors are normal in patients with TS in comparison to healthy controls
Malison et al. (1995)	SPECT	5	5	Striatal [123I]beta-CIT binding was a mean of 37% higher in the subjects TS than in the comparison subjects, and each TS patient had a higher level than his or her paired comparison subject
Ernst et al. (1999)	FDOPA PET	11	10	Subjects with TS showed higher FDOPA accumulation than controls in the left caudate nucleus and right midbrain
Singer et al. (2002)	PET	7	5	Greater putamen dopamine release was seen in adults with TS than in comparison subjects after a pharmacologic challenge with amphetamine

Continued

Table 1 Summary of selected studies exploring neurobiology of Tourette syndrome—cont'd

Study	Methodology	No. of subjects	No. of controls	Main findings
Berding et al. (2004)	[123I]AM281 SPECT	6	–	Specific binding of [123I]AM281 to CB1 receptors can be detected in TS patients using SPECT
Yeh et al. (2006)	(99m)Tc-TRODAT-1 SPECT	6	17	No significant difference in DAT activity between patients with TS and healthy controls was found in the striatum and its sub-regions.
Yeh et al. (2007)	(99m)Tc-TRODAT-1 SPECT	8	8	A functional abnormality of the dopaminergic system in the right caudate nucleus of TS patients
Liu et al. (2010)	(99m)Tc-TRODAT-1 SPECT	18	8	TS patients showed significantly higher uptake of (99m)Tc-TRODAT-1 in bilateral striatum in comparison with the controls
Lerner et al. (2012)	PET	11	11	Decreased binding of GABA receptors in TS patients bilaterally in the ventral striatum, globus pallidus, thalamus, amygdala and right insula
Albin et al. (2017)	$[^{18}F]$FEOBV-PET	12	10	No differences between TS patients and controls regarding acetylcholine neurotransmission in $[^{18}F]$FEOBV PET study
Hienert et al. (2018)	Meta-analysis of PET and SPECT studies exploring DAT in TS	111	93	A nonsignificant trend toward lower dopamine 2/3 receptor binding in striatum of TS patients
Müller-Vahl et al. (2019)	SPECT	10 TS, 10 TS − OCD, 8 TS + OCD and 5 OCD	10	Alterations in the serotonergic system in TS are related to comorbid OCD and do not represent the primary cause of the disease

Functional neuroimaging studies				
Peterson et al. (1998)	fMRI	22	–	Significant changes in signal intensity were seen in the basal ganglia and thalamus and in anatomically connected cortical regions believed to subserve attention-demanding tasks
Marsh, Zhu, Wang, Skudlarski, and Peterson (2007)	fMRI	66	70	Reduced default-mode processing in ventral prefrontal and posterior cingulate cortices in adults with TS suggests the presence of greater difficulty in engaging mental processes that become active when freed from the demands of more challenging cognitive task
Kawohl, Brühl, Krowatschek, Ketteler, and Herwig (2009)	fMRI	1	–	The comparisons of "tics" to tics suppressed" and of "tics" to "tics imitated" showed similar activation in the anterior cingulate cortex
Wang et al.	fMRI	13	21	Tics are caused by the combined effects of excessive activity in motor pathways and reduced activation in control portions of cortico-striato-thalamo-cortical circuits
Neuner et al. (2014)	fMRI	10	–	The temporal pattern of tic generation follows the cortico-striato-thalamo-cortical circuit, and that cortical structures precede subcortical activation
Polyanska, Critchley, and Rae (2017)	fMRI, PET meta-analysis	314	337	TS participants showed distributed differences in the activation of prefrontal, anterior cingulate, and motor preparation cortices. Within TS participants, tic severity correlated with engagement of SMA, precentral gyrus, and middle frontal gyrus
van der Salm et al. (2018)	fMRI	15	22	TS patients use a characteristic limbic suppression strategy

Continued

Table 1 Summary of selected studies exploring neurobiology of Tourette syndrome—cont'd

Study	Methodology	No. of subjects	No. of controls	Main findings
Botteron et al. (2019)	fMRI	15	30	The TS group blinked more than twice as often during the blink suppression block, and reported higher baseline discomfort, smaller excursion from baseline to peak discomfort during the blink suppression block, and slower return of discomfort to baseline during the recovery block
Ramkiran et al. (2019)	Resting state fMRI	28	28	TS is characterized by increased connectivity and functional integration of multiple basal ganglia-thalamo-cortical circuits
Bhikram, Crawley, Arnold, Abi-Jaoude, and Sandor (2021)	fMRI	40	20	Compared to controls, patients exhibited greater activity in the superior temporal gyrus and midcingulate during the inhibition of urges
Studies exploring metabolites of neurotransmitters				
Janik, Kalbarczyk, Gutowicz, Barańczyk-Kuźma, and Kwieciński (2010)	High-performance liquid chromatography	67	57	The TS group had higher glutamic acid and lower GABA levels. Patients with concomitant OCD and severe tics had higher glutamate levels. Glutamate concentrations correlated positively with the number of comorbid psychiatric disorders and GABA concentrations correlated negatively with the number of behavioral problems in patients with comorbidities

Müller-Vahl et al. (2020)	CSF levels of using liquid-liquid lipid extraction and simultaneous quantification by liquid chromatography multiple reaction monitoring	19	20	CSF levels of endocannabinoids were significantly increased in TS compared with controls
Capetian et al. (2021)	The urinary concentrations of the TIQ derivatives	17 TS, 12 TS + ADHD	19	NSAL concentrations were elevated significantly in TS and TS + ADHD compared with controls
Postmortem studies				
Singer, Hahn, and Moran (1991)	Postmortem study, high-performance liquid chromatography	3	3	Neuronal dopamine uptake carrier sites were significantly increased in number over control in the putamen
Minzer, Lee, Hong, and Singer (2004)	Postmortem study, high-performance liquid chromatography	3	3	TS subjects had increased densities of prefrontal D2 receptor protein, greater than 140% of their matched control
Kataoka et al. (2010)	Postmortem study	5	5	Decreased number of parvalbumin and cholinergic interneurons in the striatum of individuals with TS

Continued

Study	Methodology	No. of subjects	No. of controls	Main findings
Genetic/epigenetic studies				
Gelernter, Pauls, Leckman, Kidd, and Kurlan (1994)	Study of alleles at the DRD2 A system	17/47[a]	–	No support for an association between alleles at the A system of DRD2 and TS
Nöthen et al. (1994)	HRR for association analysis			
	61	61	*DRD2* is not a modifying gene in TS	
Stöber et al. (1999)	Single-strand conformation analysis for genetic variants of NET	43	46	No mutation of NET was found that distinguished TS patients from healthy controls
Gadzicki et al. (2004)	Systematic mutation scan of *CNR1* gene	40	81	No association between TS and mutations in *CNR1* gene
Lee et al. (2005)	PCR aiming to identify the Taq I DRD2 and DRD2 (H313H) polymorphisms of the *DRD2* gene	151	183	*DRD2* gene polymorphisms are associated in Taiwanese children with TS

Table 1 Summary of selected studies exploring neurobiology of Tourette syndrome—cont'd

Adamczyk et al. (2011)	Genetic and functional characterization of sequence variants in *SLC1A3* gene	256	224	E219D is a functional *SLC1A3* variant that is presented in a small number of individuals with Tourette syndrome
Crane et al. (2011)	Genetic and functional characterization of sequence variants in *DLGAP3* gene	1288	–	Initial association between two polymorphisms of *DLGAP3* gene and TS was not confirmed after correction
Moya et al. (2013)	Genotyping of the refined SERT-linked polymorphic region	151	858	A unique combination of common and rare variants in one gene in TS, all of which are associated with SERT gain of function
Che et al. (2015)	Genotyping by PCR and sequencing	261	–	A link between *GRIN2B* gene polymorphisms and TS
Yuan et al. (2015)	Meta-analysis of all studies investigating the risk of TS with *DRD2/ANKK1* TaqIA	523	564	*DRD2/ANKK1* TaqIA polymorphism might contribute to TS susceptibility, especially in Caucasian population

Continued

Table 1 Summary of selected studies exploring neurobiology of Tourette syndrome—cont'd

Study	Methodology	No. of subjects	No. of controls	Main findings
Müller-Vahl et al. (2017)	Peripheral DNA methylation of *DAT*, *DRD2*, and *COMT* genes	51	51	TS is associated with hypermethylation of the *DRD2* gene
Mufford et al. (2019)	GWAS	4644 for genetic studies; 30,717 of neuroimaging studies	8695 for genetic studies	Concordance for genetic variation involved in disorder risk and subcortical brain volumes in TS
Liu et al. (2020)	WES	300	524	*ASH1L* is a TS risk gene
Szejko et al. (2020)	SNP of *CNR1* analysis	262	279	C allele of rs2023239 polymorphism of the *CNR1* gene is a risk factor of TS in Polish population
Hildonen et al. (2021)	Genotyping, expression analysis, methylation analysis	72	87	*SLC6A4* expression was upregulated in TS individuals compared to controls

[a]Different numbers of patients were included in primary and secondary analyses.

Studies are presented according to thematical and chronological order. No studies with medication are included in this Table.

PET, positron emission tomography; SPECT, single-photon emission computed tomography; TS, Tourette syndrome; CB1, cannabinoid receptor 1; MRS, magnetic resonance spectroscopy; fMRI, functional magnetic resonance imaging; ADHD, attention deficit hyperactivity disorder; TIQ, tetrahydroisoquinolines; MRI-DTI, magnetic resonance imaging diffusion tensor imaging; FA, fractional anisotropy; SMA, supplementary motor area; SMA-M1, supplementary motor area—motor cortex; M1GABA, γ-aminobutyric acid; OCD, obsessive-compulsive disorder; dACC, dorsal anterior cingulate cortex; PCR, polymerase chain reaction; VBM, voxel-based morphometry; GM, grey matter; WM, white matter; DAT, dopamine transporter binding; *DRD2*, dopamine receptor D2 gene; WES, whole exome sequencing; FST, the fronto-striato-thalamic circuit; [123I]beta-CIT, [123I]2 beta-carbomethoxy-3 beta-(4-iodophenyl)tropane ([123I]beta-CIT); ADHD, attention deficit hyperactivity disorder; GWAS, genome wide association study; CSF, cerebrospinal fluid; COMT, catechol-O-methyltransferase; HRR, haplotype relative risk; NET, the norepinephrine transporter gene; SNP, single nucleotide polymorphism; *CNR1*, cannabis receptor 1 gene; TMS, transcranial magnetic stimulation; CSTC, cortico-striato-thalamo-cortical circuits; AD, axial diffusivity; CC, corpus callosum; NIRS, near-infrared spectroscopy; Hb, hemoglobin; *ANKK1,* Ankyrin Repeat And Kinase Domain Containing 1 Gene.

2. Dopaminergic system

By far, the biggest amount of evidence supports the engagement of dopaminergic system in TS pathophysiology (Dwyer, 2018; Maia & Conceição, 2018). First studies regarding this matter date back to 60' and 70' (Messiha, Knopp, Vanecko, O'Brien, & Corson, 1971; Yvonneau & Bezard, 1970) when dopamine-blocking agents were used for the first time in treatment of tics. Therefore, up to this date, the most important argument behind the dopaminergic etiology of tics is the efficacy of antipsychotics for tic reduction that are one of the first-line drugs recommended in guidelines of the European Society for the Study of Tourette Syndrome (Müller-Vahl et al., 2021) and the American Academy of Neurology (Pringsheim et al., 2019). This statement is also based on the established efficacy of neuroleptics confirmed in variety of randomized double blind placebo controlled studies (RCT).

Numerous approaches were used to determine the nature of abnormalities of dopaminergic system in TS. In animal models, altered dopaminergic function in dorsal striatum induced tic-like behaviors in rats (Nespoli, Rizzo, Boeckers, Schulze, & Hengerer, 2018). As for neuroimaging studies, these included mainly positron emission tomography (PET) and single-photon emission computed tomography (SPECT) studies. First studies have been published already in the 90s (Ernst et al., 1999; George et al., 1994; Malison et al., 1995; Turjanski et al., 1994), but results were contradictory, which could be the repercussion of diverse methodological approaches and small sample sizes. In the recent years, these preliminary findings have been widely explored. Liu et al. (Liu et al., 2010) explored the dopamine transporter (DAT) binding in drug-naive TS patients and matched controls using (99m)Tc-TRODAT-1 SPECT/CT imaging. As expected, TS patients had higher levels of DAT compared to controls. Importantly, DAT binding was negatively correlated with TS duration. The authors concluded that there might be certain adaptive changes that happen in brain of TS patients with age and, therefore, excessive dopamine activity mainly occurs at the beginning of TS. Similar study was done by Yeh et al. (Yeh et al., 2006), but the authors did not find any differences between TS subjects and healthy controls when it comes to DAT binding in (99m) Tc-TRODAT-1 SPECT/CT imaging. Again, it was suspected that disturbance in dopamine neurotransmission only happen at the beginning of the disease and, therefore, could not be detected in this study since all subjects were adults with longer disease duration. The same group (Yeh et al., 2007) applied alike strategy, but also challenged the subjects with methylphenidate. Interestingly, the DAT-binding decreased significantly after drug administration in both groups. However, there was no relationship between tic severity and DAT-binding. Hienert et al. (Hienert et al., 2018) conducted meta-analysis of PET and SPECT studies on the topic of TS. They identified 49 PET and SPECT studies. Out of these, 8 studies investigated the dopamine transporter (DAT) with 111 TS patients and 93 healthy controls. As a result, they found significantly increased DAT binding in TS, but this result was not significant after correcting for age.

It is also not clear what is the exact nature behind the dysregulation of dopaminergic neurotransmission in TS. Tentative mechanism suggested so far are supersensitivity of dopamine receptors, (2) overactivity of dopamine transporters, (3) presynaptic dysfunction in dopamine neurons, or (4) dopaminergic hyperinnervation (Maia & Conceição, 2018). The biggest amount of evidence indicate that there is an abnormality in tonic-phasic dopamine release that leads to occurrence of tics (Maia & Conceição, 2017; Singer et al., 2002).

Another group of evidence supporting the dopamine hypothesis of TS comes from the area of biochemistry. Already in 1976 van Woert et al. (Van Woert, Jutkowitz, Rosenbaum, & Bowers, 1976) explored profile of dopamine metabolites in CSF of 5 TS patients and found an elevated probenecid-induced accumulation of homovanillic acid (HVA), the major metabolite of dopamine. Similarly, Singer et al. (Singer, Butler, Tune, Seifert, & Coyle, 1982) tested how treatment with haloperidol influence profile of dopamine metabolites in CSF of 7 TS patients. Before treatment, CSF HVA levels were significantly reduced in all patients, whereas after the therapy with haloperidol these levels significantly increased. Just recently, Capetian et al. (Capetian et al., 2021) published results of their study exploring the urine concentrations of diverse tetrahydroisoquinolines (TIQs), modulators of dopaminergic neurotransmission, in patients with TS, TS +ADHD and healthy controls. They found elevated levels of TIQs in patients with TS and ADHD compared with healthy controls and concluded that these results support dopamine hyperactivity in TS.

Several postmortem studies have also examined the topic of dopaminergic system in TS. Nevertheless, these studies are limited by very small sample sizes, since, due to obvious reasons, tissue specimens are very difficult to obtain. Therefore, these findings are precluded to only few studies completed in the 90s. Singer et al. (Singer et al., 1991) evaluated dopamine metabolites in postmortem striatum of 3 adults with TS. Neuronal dopamine uptake was significantly increased in TS patients over controls in the caudate and the putamen supporting dopaminergic hypothesis behind TS. The same group (Minzer et al., 2004) conducted similar study in 3 TS patients and 3 age- and sex-matched controls. As a result, the authors have demonstrated that prefrontal cortex had the greatest number of changes in the two typical TS cases, including increases for dopamine 2 receptors (D2) or dopamine transporter. All 3 TS subjects had increased densities of prefrontal D2 receptor protein.

Also genetic studies suggest involvement of dopaminergic system in TS pathophysiology, but again results have been somewhat conflicting. Main genes investigated so far are Dopamine Receptor D2 Gene *(DRD2)* and TaqIA polymorphism of the ankyrin repeat and protein kinase domain-containing protein (*ANKK1*) gene. First studies from 90s explored whether the A1 allele of the Taq I polymorphism at the *DRD2* gene had impact on the TS severity, but failed to prove such relationship (Gelernter et al., 1994; Nöthen et al., 1994). On the contrary, Lee et al. (Lee et al., 2005) showed an association between the *DRD2* gene and TS. Yuan et al. (Yuan et al., 2015) performed a meta-analysis of the relationship between *DRD2/ANKK1 TaqIA* polymorphism and TS. This analysis suggested that the *DRD2/ ANKK1 TaqIA* polymorphism might contribute to TS susceptibility, especially in the Caucasian population. Another report by Liu et al. (Liu et al., 2020) showed that mutations in ASH1 Like Histone Lysine Methyltransferase *(ASH1L)* gene, which leads to hyper-activation and elevated dopamine release, are present in TS. Following the same line of evidence, not only genetic, but also epigenetic changes in dopaminergic system have been found in TS (Müller-Vahl et al., 2017).

3. Glutamatergic system

Two main neurotransmitters that regulate neurotransmission in CSTC pathway, the region of the brain that is hypothesized to be responsible for the occurrence of tics, are γ-aminobutyric acid (GABA) and glutamate. As for the role of glutamate in TS pathophysiology, this has been mainly investigated in neuroimaging studies. Naaijen et al. (Naaijen et al., 2017) studied glutamatergic transmission in children with TS and attention deficit hyperactivity disorder (ADHD). In order to tackle this problem the authors applied magnetic resonance spectroscopy (MRS) in children with pure TS, ADHD, TS+ADHD and healthy controls, but failed to demonstrate any evidence for glutamatergic pathology in TS or ADHD. Kanaan et al. (Kanaan et al., 2017) explored neurochemical characteristics of 3 cortico-striato-thalamo-cortical regions of 37 TS patients and 36 matched healthy controls via MRS at 3 T. The authors also investigated the influence of 4-week treatment with aripiprazole on metabolite concentrations. There was a significant reduction in glutamate concentration in striatum and thalamus in TS patients compared to controls. No differences were detected between the group of patients treated with aripiprazole and those without any treatment when it comes to metabolite concentrations. Finally, multiple regression analysis demonstrated that tic severity was associated with striatal glutamate and premonitory urge with thalamic glutamate. Mahone et al. (Mahone et al., 2018) enrolled 75 participants in their study (32 with TS, 43 controls) who underwent ^{1}H MRS at 7T. GABA and glutamate were measured in different regions of the brain, in particular, in dorsolateral prefrontal cortex (DLPFC), ventromedial prefrontal cortex (VMPFC), premotor cortex (PMC), and striatum. In addition, also neuropsychological assessment was conducted. Glutamate, but not GABA was significantly higher in the TS group. Moreover, elevated levels of glutamate were associated with improved neuropsychological performance.

One study analyzed selected neurotransmitter concentrations, also glutamate, in serum of TS patients and compared those with healthy controls (Janik et al., 2010). The TS group had higher glutamic acid and lower GABA levels, while glycine concentrations were comparable. Interestingly enough, there was no difference in drug naive and patients on treatment. Higher glutamate levels were correlated with tic severity and comorbid obsessive-compulsive disorder (OCD). The authors concluded that imbalance between excitatory and inhibitory systems in the brains of TS patients may be reflected by glutamate and GABA serum level changes.

Also genetic studies have explored the importance of glutamatergic abnormalities in TS. One gene that was studied in this context was Disks-large associated protein 3 (*DLGAP3*) that encodes SAP90/PSD95-associated protein 3 (SAPAP3/DLGAP3), that is highly expressed in glutamatergic system (Crane et al., 2011), however, initial association between two polymorphisms of these genes and TS was not confirmed after correction. Adamczyk et. (Adamczyk et al., 2011) examined whether missense variant in a genes related to glutamate transporter (*SLC1A3*) increase the risk of TS. The results indicated that E219D is a functional *SLC1A3* variant that is presented in a small number of individuals with Tourette syndrome. Another gene explored in this context was *GRIN2B*, gene encoding N-methyl-d-aspartate (NMDA) receptor 2B subunit (Che et al., 2015). The study demonstrated that there was a link between *GRIN2B* gene polymorphisms and TS, however these findings should be interpreted with caution since sample was relatively small and restricted to Chinese population.

Finally, one recent RCT was testing efficacy and safety of N-acetylcysteine, the agent that modulates glutamatergic system, in patients with TS (Bloch et al., 2016). The authors failed to show any effect of this drug on tic reduction. Also glutamate agonist (D-serine) and antagonist (riluzole) were studied as potential therapy for TS, but were not effective (Lemmon et al., 2015).

4. Serotoninergic system

Another neurobiological underprint of TS is serotonergic neurotransmission. Already in the 70s and 80s several case reports and theoretical studies suggested that serotonin antagonists could induce tic exacerbation (Crosley, 1979; Sandyk & Bamford, 1988). This topic was also investigated in neuroimaging studies. Just recently, Müller-Vahl et al. (Müller-Vahl et al., 2019) investigated for the first time the serotonin transporter (SERT) binding in TS patients with and without OCD (TS - OCD) compared to both healthy controls and OCD patients as well as the influence of escitalopram. SERT binding was normal in patients with OCD and TS - OCD, but significantly increased in those with TS + OCD, particularly in caudate and midbrain compared to both healthy controls and TS - OCD. Treatment with escitalopram resulted in a significant overall reduction in SERT binding. These results provided further evidence that alterations in the serotonergic system in TS are related to comorbid OCD and not TS itself.

These reports have been supplemented by the genetic studies. Moya et al. (Moya et al., 2013) evaluated the hypothesis that overexpression of the SERT gene (solute carrier family 6, member 4, *SLC6A4*) could be detected in TS. For this purpose, they enrolled an impressively large group of TS patients (n = 151) and controls (n = 858). Higher expression of 5-HTTLPR/rs25531 LA allele was found more frequently in TS group than in controls. Surprisingly enough, this association was significantly more frequent in pure TS than TS + OCD. Just recently, Hildonen et al. (Hildonen et al., 2021) reported about elevated expression of *SLC6A4* gene encoding the SERT in TS compared to healthy controls.

Finally, Silvestri et al. (Silvestri et al., 1994) conducted a preliminary trial with fluoxetine, a 5-HT reuptake inhibitor, but it included only 2 TS patients. This trial gave promising results according to which fluoxetine leaded to reduction of tics and OCD.

5. GABA-ergic system

What is the exact nature of abnormal GABAergic neurotransmission in TS is still a matter of debate. Previously mentioned study by Mahone et al. (Mahone et al., 2018) explored both the changes in GABA and glutamate in TS using 7T MRS. While there was significantly higher glutamate level in TS group, no differences were detected when it comes to GABA. Couple of years later, Draper et al. (Draper et al., 2014) reported about increased GABA in brain of patients with TS that contributed to enhanced control over motor excitability. Again, the authors used the newest modality of MRS, 7T MRS. Paradoxically, GABA concentrations in the supplementary motor area (SMA), the region of the brain that has been previously shown to be implicated in TS (Tübing et al., 2018) was elevated in patients with tics and was inversely related to fMRI blood oxygen level-dependent activation. Importantly, GABA levels in regions of the brain responsible for movement control did not differ from healthy controls. The authors have also shown that GABA concentrations in SMA were inversely correlated with cortical excitability in primary motor cortex. Also Lerner et al. (Lerner et al., 2012) investigated the nature of changes in GABAergic system of TS patients, especially in basal ganglia and cerebellum. In this study 11 TS patients and matched healthy controls were included and all were subjected to PET study. It was established that TS subjects were characterized by decreased binding to GABA receptors in several brain regions (ventral striatum, globus pallidus, thalamus, amygdala and right insula) compared to healthy controls. Moreover, the GABA receptor binding was elevated in the bilateral substantia nigra, left periaqueductal grey, right posterior cingulate cortex and bilateral cerebellum. Another study was targeting GABA changes in adolescents and young adults with TS (Freed et al., 2016). In this study GABA levels were measured using PET. Lower levels of GABA were detected in anterior cingulate of TS participants compared to healthy controls. GABA levels did not correlate with any clinical measures in TS group. Similar findings were reported by Puts et al. (Puts et al., 2015), who showed reduced GABAergic inhibition and abnormal sensory symptoms in children with TS. This was another MRS study complemented by a battery of vibrotactile tasks to investigate the role of GABA and atypical sensory processing in children with TS. The authors demonstrated reduced GABA concentrations in primary sensorimotor cortex in patients with TS compared to healthy controls. Importantly, these findings correlated with diminished performance on tactile detection and adaptation tasks. Finally, brain GABA concentration in children with TS correlated with motor tic severity.

6. Endocannabinoid system

One of the most important systems involved in neuromodulation is endocannabinoid system (ECS) (Lu & Mackie, 2021; Peters, Cheer, & Tonini, 2021). Regarding the involvement of ECS in TS pathophysiology, the most important evidence comes from clinical studies (Artukoglu & Bloch, 2019; Szejko, Saramak, Lombroso, & Müller-Vahl, 2021). On the other hand, in spite of growing evidence supporting the use of cannabis-based medicine (CBM) in TS, the majority of studies are still limited to those with low level of evidence, case reports, case series and open studies (Artukoglu & Bloch, 2019). Till now, only 2 RCTs, that included small sample of TS patients using tetrahydrocannabinol (THC, dronabinol), have shown efficacy and safety of this treatment in TS (Müller-Vahl et al., 2002, 2003). On the other hand, recently published results of the RCT testing efficacy and safety of Lu AG06466, a modulator of endocannabinoid system, was not efficacious in treatment of tics. In line with these findings, both the guidelines of the European Society for the Study of Tourette Syndrome (Müller-Vahl, Szejko, et al., 2021; Roessner et al., 2011) and American Academy of Neurology (Pringsheim et al., 2019) recommend this treatment only in treatment resistant cases in which established therapy did not lead to symptom alleviation.

From the pathophysiological perspective, previous studies have shown that density of cannabinoid receptor 1 (CB1) is especially high in central nervous system (CNS) and basal ganglia (Davis et al., 2018) which further suggests importance of ECS in motor control. Furthermore, it has been suggested that changes in ECS lead, in turn to dopaminergic disturbances and this causes tics (Müller-Vahl et al., 2020). This hypothesis was supported by recent study conducted in Hannover Medical School (Müller-Vahl et al., 2020) in which the authors measured cerebrospinal fluid (CSF) concentrations of diverse endocannabinoids in 20 adults with TS and 19 matched healthy controls. Contrary to the primary hypothesis, CSF concentration of endocannabinoids was significantly higher in TS group. Interestingly, levels of 2-arachidonoylglycerol (2-AG) correlated with the severity of ADHD. As for other neurotransmitters, there has been several studies investigating ECS in TS using diverse modalities of neuroimaging techniques. Berding et al. (Berding et al., 2004) conducted a SPECT study in 6 patients with TS and, in addition, studied whether treatment with THC influenced CB1 receptor binding. As no control group was included in this study, no conclusions could be made regarding the changes in CB1 receptor binding in TS. Similarly, genetic studies exploring genes determining ECS neurotransmission have been inconclusive (Gadzicki et al., 2004; Szejko et al., 2020).

7. Acetylcholine

Some studies suggested that the use of nicotine, either as patch or smoking, reduce tics and therefore suggested that abnormality of acetylcholine neurotransmission can lead to occurrence of tics (McEvoy & Allen, 2002; Stahl & Berger, 1982). Later reports regarding this matter have demonstrated conflicting results. Small postmortem study including 5 TS subjects and healthy controls (Kataoka et al., 2010) showed that TS subjects had a 50%–60% decrease of both parvalbumin (PV) and choline acetyltransferase (ChAT)+ cholinergic interneurons in the caudate and the putamen. For example, Albin et al. (Albin et al., 2017) failed to show differences between TS patients and controls regarding acetylcholine neurotransmission in [^{18}F] FEOBV PET study. Another MRS study targeting to compare choline concentrations in TS, OCD and healthy controls (Fan et al., 2017) has shown that there was a significant inter-group difference in choline concentration in thalamus, with the highest levels in OCD patients, compared to TS patients. As for genetic underprints, a pilot study conducted at University of California investigated RNA expression of genes involved in TS has shown that many genes correlated with tic severity, but GABA and acetylcholine related genes were among the most overrepresented. The authors also tested whether alternative splicing regarding these genes were present in TS. Accordingly, many GABA and acetylcholine related genes were alternatively spliced in TS, in particular, genes encoding GABA receptors GABRA4 and GABRG1, the nicotinic ACh receptor CHRNA4 and cholinergic differentiation factor. Finally, one study tested the efficacy and safety of a nicotinic, acetylcholine antagonist, mecamylamine, in the treatment of TS. Although the results were positive, this was an open, retrospective study and these findings should be interpreter with caution.

8. Functional studies

In recent years there has been increasing number of functional neuroimaging studies exploring neuronal activity change during certain tasks or under specific conditions in patients with TS (Hsu, Wong, Wang, & Lee, 2020; Zapparoli, Porta, & Paulesu, 2015). The most relevant question that these studies tried to investigate was the origin of tics (Neuner et al., 2014; Wang et al., 2011), tic suppression (Kawohl et al., 2009; Peterson et al., 1998), premonitory urge (Bhikram et al., 2021) and the comparison of tics to voluntary movements (Botteron et al., 2019; Marsh et al., 2007; van der Salm et al., 2018). The most recent reports suggest that tics occur as a result of overactivity in motor pathways (Wang et al., 2011), mainly in cortico-striato-thalamo-cortical circuit (Neuner et al., 2014). Recent meta-analysis of fMRI studies testing neuronal determinants of tics included 14 task-based studies (Polyanska et al., 2017) with 25 experiments and 651 participants. Compared to controls, TS subjects demonstrated in activation of the following brain regions: prefrontal (inferior, middle, and superior frontal gyri), anterior cingulate, and motor preparation cortices (lateral premotor cortex and supplementary motor area; SMA), sensory and temporo-parietal association cortices. Importantly, tic severity correlated with activity in the following regions: SMA, precentral gyrus, and middle frontal gyrus.

9. Conclusions: An interplay of diverse pathways and neurotransmitters

It can therefore be concluded that although it is clear that dopaminergic system, especially in the CSTC neurocircuit, is involved in TS pathophysiology, the exact nature of abnormality is still not clear. It also seems that variety of other neurotransmitters could play a role in occurrence of tics and comorbidities that frequently co-exist with TS. Therefore, it could be speculated that clinical variability is the reflexion of the complex genetic and neurobiological architecture of TS. Future endeavors should be aiming to merge genetic and neuroimaging findings, such as the initiative launched by the ENIGMA-TS working group (Mufford et al., 2019).

References

Adamczyk, A., Gause, C. D., Sattler, R., Vidensky, S., Rothstein, J. D., Singer, H., et al. (2011). Genetic and functional studies of a missense variant in a glutamate transporter, SLC1A3, in Tourette syndrome. *Psychiatric Genetics*, *21*(2), 90–97. https://doi.org/10.1097/YPG.0b013e328341a307.

Albin, R. L., Minderovic, C., & Koeppe, R. A. (2017). Normal striatal vesicular acetylcholine transporter expression in Tourette syndrome. *eNeuro*, *4*(4). https://doi.org/10.1523/eneuro.0178-17.2017.

Artukoglu, B. B., & Bloch, M. H. (2019). The potential of cannabinoid-based treatments in Tourette syndrome. *CNS Drugs*, *33*(5), 417–430. https://doi.org/10.1007/s40263-019-00627-1.

Berding, G., Müller-Vahl, K., Schneider, U., Gielow, P., Fitschen, J., Stuhrmann, M., et al. (2004). [123I]AM281 single-photon emission computed tomography imaging of central cannabinoid CB1 receptors before and after Delta9-tetrahydrocannabinol therapy and whole-body scanning for assessment of radiation dose in Tourette patients. *Biological Psychiatry*, *55*(9), 904–915. https://doi.org/10.1016/j.biopsych.2004.01.005.

Bhikram, T., Crawley, A., Arnold, P., Abi-Jaoude, E., & Sandor, P. (2021). Neuroimaging the emotional modulation of urge inhibition in Tourette syndrome. *Cortex*, *135*, 341–351. https://doi.org/10.1016/j.cortex.2020.10.010.

Bloch, M. H., Panza, K. E., Yaffa, A., Alvarenga, P. G., Jakubovski, E., Mulqueen, J. M., et al. (2016). N-Acetylcysteine in the treatment of pediatric Tourette syndrome: Randomized, double-blind, placebo-controlled add-on trial. *Journal of Child and Adolescent Psychopharmacology*, *26*(4), 327–334. https://doi.org/10.1089/cap.2015.0109.

Botteron, H. E., Richards, C. A., Nishino, T., Ueda, K., Acevedo, H. K., Koller, J. M., et al. (2019). The urge to blink in Tourette syndrome. *Cortex*, *120*, 556–566. https://doi.org/10.1016/j.cortex.2019.07.010.

Capetian, P., Roessner, V., Korte, C., Walitza, S., Riederer, F., Taurines, R., et al. (2021). Altered urinary tetrahydroisoquinoline derivatives in patients with Tourette syndrome: Reflection of dopaminergic hyperactivity? *Journal of Neural Transmission (Vienna)*, *128*(1), 115–120. https://doi.org/10.1007/s00702-020-02289-6.

Cavanna, A. E., Stecco, A., Rickards, H., Servo, S., Terazzi, E., Peterson, B., et al. (2010). Corpus callosum abnormalities in Tourette syndrome: An MRI-DTI study of monozygotic twins. *Journal of Neurology, Neurosurgery, and Psychiatry*, *81*(5), 533–535. https://doi.org/10.1136/jnnp.2009.173666.

Che, F., Zhang, Y., Wang, G., Heng, X., Liu, S., & Du, Y. (2015). The role of GRIN2B in Tourette syndrome: Results from a transmission disequilibrium study. *Journal of Affective Disorders*, *187*, 62–65. https://doi.org/10.1016/j.jad.2015.07.036.

Crane, J., Fagerness, J., Osiecki, L., Gunnell, B., Stewart, S. E., Pauls, D. L., et al. (2011). Family-based genetic association study of DLGAP3 in Tourette syndrome. *American Journal of Medical Genetics. Part B, Neuropsychiatric Genetics*, *156b*(1), 108–114. https://doi.org/10.1002/ajmg.b.31134.

Crosley, C. J. (1979). Decreased serotoninergic activity in Tourette syndrome. *Annals of Neurology*, *5*(6), 596–597. https://doi.org/10.1002/ana.410050622.

Davis, M. I., Crittenden, J. R., Feng, A. Y., Kupferschmidt, D. A., Naydenov, A., Stella, N., et al. (2018). The cannabinoid-1 receptor is abundantly expressed in striatal striosomes and striosome-dendron bouquets of the substantia nigra. *PLoS One*, *13*(2), e0191436. https://doi.org/10.1371/journal.pone.0191436.

Devinsky, O. (1983). Neuroanatomy of Gilles de la Tourette's syndrome. Possible midbrain involvement. *Archives of Neurology*, *40*(8), 508–514. https://doi.org/10.1001/archneur.1983.04210070048013.

Draganski, B., Martino, D., Cavanna, A. E., Hutton, C., Orth, M., Robertson, M. M., et al. (2010). Multispectral brain morphometry in Tourette syndrome persisting into adulthood. *Brain*, *133*(Pt. 12), 3661–3675. https://doi.org/10.1093/brain/awq300.

Draper, A., Stephenson, M. C., Jackson, G. M., Pépés, S., Morgan, P. S., Morris, P. G., et al. (2014). Increased GABA contributes to enhanced control over motor excitability in Tourette syndrome. *Current Biology*, *24*(19), 2343–2347. https://doi.org/10.1016/j.cub.2014.08.038.

Dwyer, J. B. (2018). A developmental perspective of dopaminergic dysfunction in Tourette syndrome. *Biological Psychiatry*, *84*(5), e33–e35. https://doi.org/10.1016/j.biopsych.2018.07.008.

Ernst, M., Zametkin, A. J., Jons, P. H., Matochik, J. A., Pascualvaca, D., & Cohen, R. M. (1999). High presynaptic dopaminergic activity in children with Tourette's disorder. *Journal of the American Academy of Child and Adolescent Psychiatry*, *38*(1), 86–94. https://doi.org/10.1097/00004583-199901000-00024.

Fan, S., Cath, D. C., van den Heuvel, O. A., van der Werf, Y. D., Schöls, C., Veltman, D. J., et al. (2017). Abnormalities in metabolite concentrations in tourette's disorder and obsessive-compulsive disorder—A proton magnetic resonance spectroscopy study. *Psychoneuroendocrinology*, 77, 211–217. https://doi.org/10.1016/j.psyneuen.2016.12.007.

Felling, R. J., & Singer, H. S. (2011). Neurobiology of Tourette syndrome: Current status and need for further investigation. *The Journal of Neuroscience*, *31*(35), 12387–12395. https://doi.org/10.1523/jneurosci.0150-11.2011.

Freed, R. D., Coffey, B. J., Mao, X., Weiduschat, N., Kang, G., Shungu, D. C., et al. (2016). Decreased anterior cingulate cortex γ-aminobutyric acid in youth with Tourette's disorder. *Pediatric Neurology*, *65*, 64–70. https://doi.org/10.1016/j.pediatrneurol.2016.08.017.

Gadzicki, D., Müller-Vahl, K. R., Heller, D., Ossege, S., Nöthen, M. M., Hebebrand, J., et al. (2004). Tourette syndrome is not caused by mutations in the central cannabinoid receptor (CNR1) gene. *American Journal of Medical Genetics. Part B, Neuropsychiatric Genetics*, *127b*(1), 97–103. https://doi.org/10.1002/ajmg.b.20159.

Ganos, C., Kühn, S., Kahl, U., Schunke, O., Brandt, V., Bäumer, T., et al. (2014). Prefrontal cortex volume reductions and tic inhibition are unrelated in uncomplicated GTS adults. *Journal of Psychosomatic Research*, *76*(1), 84–87. https://doi.org/10.1016/j.jpsychores.2013.10.014.

Garraux, G., Goldfine, A., Bohlhalter, S., Lerner, A., Hanakawa, T., & Hallett, M. (2006). Increased midbrain gray matter in Tourette's syndrome. *Annals of Neurology*, *59*(2), 381–385. https://doi.org/10.1002/ana.20765.

Gelernter, J., Pauls, D. L., Leckman, J., Kidd, K. K., & Kurlan, R. (1994). D2 dopamine receptor alleles do not influence severity of Tourette's syndrome. Results from four large kindreds. *Archives of Neurology*, *51*(4), 397–400. https://doi.org/10.1001/archneur.1994.00540160099012.

George, M. S., Robertson, M. M., Costa, D. C., Ell, P. J., Trimble, M. R., Pilowsky, L., et al. (1994). Dopamine receptor availability in Tourette's syndrome. *Psychiatry Research*, *55*(4), 193–203. https://doi.org/10.1016/0165-1781(95)91245-9.

Greene, D. J., Williams Iii, A. C., Koller, J. M., Schlaggar, B. L., Black, K. J., & The Tourette Association of America Neuroimaging, C. (2017). Brain structure in pediatric Tourette syndrome. *Molecular Psychiatry*, *22*(7), 972–980. https://doi.org/10.1038/mp.2016.194.

Hartmann, A., Worbe, Y., & Arnulf, I. (2012). Increasing histamine neurotransmission in Gilles de la Tourette syndrome. *Journal of Neurology*, *259*(2), 375–376. https://doi.org/10.1007/s00415-011-6171-1.

Hienert, M., Gryglewski, G., Stamenkovic, M., Kasper, S., & Lanzenberger, R. (2018). Striatal dopaminergic alterations in Tourette's syndrome: A meta-analysis based on 16 PET and SPECT neuroimaging studies. *Translational Psychiatry*, *8*(1), 143. https://doi.org/10.1038/s41398-018-0202-y.

Hildonen, M., Levy, A. M., Dahl, C., Bjerregaard, V. A., Birk Møller, L., Guldberg, P., et al. (2021). Elevated expression of SLC6A4 encoding the serotonin transporter (SERT) in Gilles de la Tourette syndrome. *Genes (Basel)*, *12*(1). https://doi.org/10.3390/genes12010086.

Hsu, C. J., Wong, L. C., Wang, H. P., & Lee, W. T. (2020). The multimodality neuroimage findings in individuals with Tourette syndrome. *Pediatrics and Neonatology*, *61*(5), 467–474. https://doi.org/10.1016/j.pedneo.2020.03.007.

Janik, P., Kalbarczyk, A., Gutowicz, M., Barańczyk-Kuźma, A., & Kwieciński, H. (2010). The analysis of selected neurotransmitter concentrations in serum of patients with Tourette syndrome. *Neurologia i neurochirurgia polska*, *44*(3), 251–259. https://doi.org/10.1016/s0028-3843(14)60039-6.

Kanaan, A. S., Gerasch, S., García-García, I., Lampe, L., Pampel, A., Anwander, A., et al. (2017). Pathological glutamatergic neurotransmission in Gilles de la Tourette syndrome. *Brain*, *140*(1), 218–234. https://doi.org/10.1093/brain/aww285.

Kataoka, Y., Kalanithi, P. S., Grantz, H., Schwartz, M. L., Saper, C., Leckman, J. F., et al. (2010). Decreased number of parvalbumin and cholinergic interneurons in the striatum of individuals with Tourette syndrome. *The Journal of Comparative Neurology*, *518*(3), 277–291. https://doi.org/10.1002/cne.22206.

Kawohl, W., Brühl, A., Krowatschek, G., Ketteler, D., & Herwig, U. (2009). Functional magnetic resonance imaging of tics and tic suppression in Gilles de la Tourette syndrome. *The World Journal of Biological Psychiatry*, *10*(4 Pt 2), 567–570. https://doi.org/10.1080/15622970802118356.

Lee, C. C., Chou, I. C., Tsai, C. H., Wang, T. R., Li, T. C., & Tsai, F. J. (2005). Dopamine receptor D2 gene polymorphisms are associated in Taiwanese children with Tourette syndrome. *Pediatric Neurology*, *33*(4), 272–276. https://doi.org/10.1016/j.pediatrneurol.2005.05.005.

Lemmon, M. E., Grados, M., Kline, T., Thompson, C. B., Ali, S. F., & Singer, H. S. (2015). Efficacy of glutamate modulators in tic suppression: A double-blind, randomized control trial of D-serine and riluzole in Tourette syndrome. *Pediatric Neurology*, *52*(6), 629–634. https://doi.org/10.1016/j.pediatrneurol.2015.02.002.

Lerner, A., Bagic, A., Simmons, J. M., Mari, Z., Bonne, O., Xu, B., et al. (2012). Widespread abnormality of the γ-aminobutyric acid-ergic system in Tourette syndrome. *Brain*, *135*(Pt. 6), 1926–1936. https://doi.org/10.1093/brain/aws104.

Levy, F. (2011). Cortico-striatal-thalamic-cortical (CSTC) circuits, Tourette's disorder, and OCD. *The ADHD Report*, *19*(2). 8-9,15 https://doi.org/10.1521/adhd.2011.19.2.8.

Liu, H., Dong, F., Meng, Z., Zhang, B., Tan, J., & Wang, Y. (2010). Evaluation of Tourette's syndrome by (99m)Tc-TRODAT-1 SPECT/CT imaging. *Annals of Nuclear Medicine*, *24*(7), 515–521. https://doi.org/10.1007/s12149-010-0389-3.

Liu, S., Tian, M., He, F., Li, J., Xie, H., Liu, W., et al. (2020). Mutations in ASH1L confer susceptibility to Tourette syndrome. *Molecular Psychiatry*, *25*(2), 476–490. https://doi.org/10.1038/s41380-019-0560-8.

Lu, H. C., & Mackie, K. (2021). Review of the Endocannabinoid system. *Biological Psychiatry. Cognitive Neuroscience and Neuroimaging*, *6*(6), 607–615. https://doi.org/10.1016/j.bpsc.2020.07.016.

Mahone, E. M., Puts, N. A., Edden, R. A. E., Ryan, M., & Singer, H. S. (2018). GABA and glutamate in children with Tourette syndrome: A (1)H MR spectroscopy study at 7T. *Psychiatry Research. Neuroimaging*, *273*, 46–53. https://doi.org/10.1016/j.pscychresns.2017.12.005.

Maia, T. V., & Conceição, V. A. (2017). The roles of phasic and tonic dopamine in tic learning and expression. *Biological Psychiatry*, *82*(6), 401–412. https://doi.org/10.1016/j.biopsych.2017.05.025.

Maia, T. V., & Conceição, V. A. (2018). Dopaminergic disturbances in Tourette syndrome: An integrative account. *Biological Psychiatry*, *84*(5), 332–344. https://doi.org/10.1016/j.biopsych.2018.02.1172.

Makki, M. I., Behen, M., Bhatt, A., Wilson, B., & Chugani, H. T. (2008). Microstructural abnormalities of striatum and thalamus in children with Tourette syndrome. *Movement Disorders*, *23*(16), 2349–2356. https://doi.org/10.1002/mds.22264.

Malison, R. T., McDougle, C. J., van Dyck, C. H., Scahill, L., Baldwin, R. M., Seibyl, J. P., et al. (1995). [123I]beta-CIT SPECT imaging of striatal dopamine transporter binding in Tourette's disorder. *American Journal of Psychiatry*, *152*(9), 1359–1361. https://doi.org/10.1176/ajp.152.9.1359.

Marsh, R., Zhu, H., Wang, Z., Skudlarski, P., & Peterson, B. S. (2007). A developmental fMRI study of self-regulatory control in Tourette's syndrome. *American Journal of Psychiatry*, *164*(6), 955–966. https://doi.org/10.1176/ajp.2007.164.6.955.

McEvoy, J. P., & Allen, T. B. (2002). The importance of nicotinic acetylcholine receptors in schizophrenia, bipolar disorder and Tourette's syndrome. *Current Drug Targets. CNS and Neurological Disorders*, *1*(4), 433–442. https://doi.org/10.2174/1568007023339210.

Messiha, F. S., Knopp, W., Vanecko, S., O'Brien, V., & Corson, S. A. (1971). Haloperidol therapy in Tourette's syndrome: Neurophysiological, biochemical and behavioral correlates. *Life Sciences. Pt. I: Physiology & Pharmacology*, *10*(8), 449–457. https://doi.org/10.1016/0024-3205(71)90110-x.

Miller, A. M., Bansal, R., Hao, X., Sanchez-Pena, J. P., Sobel, L. J., Liu, J., et al. (2010). Enlargement of thalamic nuclei in Tourette syndrome. *Archives of General Psychiatry*, *67*(9), 955–964. https://doi.org/10.1001/archgenpsychiatry.2010.102.

Mink, J. W. (2001). Basal ganglia dysfunction in Tourette's syndrome: A new hypothesis. *Pediatric Neurology*, *25*(3), 190–198. https://doi.org/10.1016/s0887-8994(01)00262-4.

Minzer, K., Lee, O., Hong, J. J., & Singer, H. S. (2004). Increased prefrontal D2 protein in Tourette syndrome: a postmortem analysis of frontal cortex and striatum. *Journal of Neurological Science*, *219*(1-2), 55–61. https://doi.org/10.1016/j.jns.2003.12.006.

Mostofsky, S. H., Wendlandt, J., Cutting, L., Denckla, M. B., & Singer, H. S. (1999). Corpus callosum measurements in girls with Tourette syndrome. *Neurology*, *53*(6), 1345–1347. https://doi.org/10.1212/wnl.53.6.1345.

Moya, P. R., Wendland, J. R., Rubenstein, L. M., Timpano, K. R., Heiman, G. A., Tischfield, J. A., et al. (2013). Common and rare alleles of the serotonin transporter gene, SLC6A4, associated with Tourette's disorder. *Movement Disorders*, *28*(9), 1263–1270. https://doi.org/10.1002/mds.25460.

Mufford, M., Cheung, J., Jahanshad, N., van der Merwe, C., Ding, L., Groenewold, N., et al. (2019). Concordance of genetic variation that increases risk for tourette syndrome and that influences its underlying neurocircuitry. *Translational Psychiatry*, *9*(1), 120. https://doi.org/10.1038/s41398-019-0452-3.

Müller-Vahl, K. R., Bindila, L., Lutz, B., Musshoff, F., Skripuletz, T., Baumgaertel, C., et al. (2020). Cerebrospinal fluid endocannabinoid levels in Gilles de la Tourette syndrome. *Neuropsychopharmacology: Official Publication of the American College of Neuropsychopharmacology*, *45*(8), 1323–1329. https://doi.org/10.1038/s41386-020-0671-6.

Müller-Vahl, K. R., Fremer, C., Beals, C., Ivkovic, J., Loft, H., & Schindler, C. (2021). Monoacylglycerol Lipase Inhibition in Tourette syndrome: A 12-week, randomized, controlled study. *Movement Disorders*, *36*(10), 2413–2418. https://doi.org/10.1002/mds.28681.

Müller-Vahl, K. R., Loeber, G., Kotsiari, A., Müller-Engling, L., & Frieling, H. (2017). Gilles de la Tourette syndrome is associated with hypermethylation of the dopamine D2 receptor gene. *Journal of Psychiatric Research*, *86*, 1–8. https://doi.org/10.1016/j.jpsychires.2016.11.004.

Müller-Vahl, K. R., Schneider, U., Koblenz, A., Jöbges, M., Kolbe, H., Daldrup, T., et al. (2002). Treatment of Tourette's syndrome with Delta 9-tetrahydrocannabinol (THC): A randomized crossover trial. *Pharmacopsychiatry*, *35*(2), 57–61. https://doi.org/10.1055/s-2002-25028.

Müller-Vahl, K. R., Schneider, U., Prevedel, H., Theloe, K., Kolbe, H., Daldrup, T., et al. (2003). Delta 9-tetrahydrocannabinol (THC) is effective in the treatment of tics in Tourette syndrome: A 6-week randomized trial. *Journal of Clinical Psychiatry*, *64*(4), 459–465. https://doi.org/10.4088/jcp.v64n0417.

Müller-Vahl, K. R., Szejko, N., Verdellen, C., Roessner, V., Hoekstra, P. J., Hartmann, A., et al. (2021). European clinical guidelines for Tourette syndrome and other tic disorders: Summary statement. *European Child & Adolescent Psychiatry*. https://doi.org/10.1007/s00787-021-01832-4.

Müller-Vahl, K. R., Szejko, N., Wilke, F., Jakubovski, E., Geworski, L., Bengel, F., et al. (2019). Serotonin transporter binding is increased in Tourette syndrome with obsessive compulsive disorder. *Scientific Reports*, *9*(1), 972. https://doi.org/10.1038/s41598-018-37710-4.

Naaijen, J., Forde, N. J., Lythgoe, D. J., Akkermans, S. E., Openneer, T. J., Dietrich, A., et al. (2017). Fronto-striatal glutamate in children with Tourette's disorder and attention-deficit/hyperactivity disorder. *Neuroimage. Clinical*, *13*, 16–23. https://doi.org/10.1016/j.nicl.2016.11.013.

Nespoli, E., Rizzo, F., Boeckers, T., Schulze, U., & Hengerer, B. (2018). Altered dopaminergic regulation of the dorsal striatum is able to induce tic-like movements in juvenile rats. *PLoS One*, *13*(4), e0196515. https://doi.org/10.1371/journal.pone.0196515.

Neuner, I., Werner, C. J., Arrubla, J., Stöcker, T., Ehlen, C., Wegener, H. P., et al. (2014). Imaging the where and when of tic generation and resting state networks in adult Tourette patients. *Frontiers in Human Neuroscience*, *8*, 362. https://doi.org/10.3389/fnhum.2014.00362.

Nöthen, M. M., Hebebrand, J., Knapp, M., Hebebrand, K., Camps, A., von Gontard, A., et al. (1994). Association analysis of the dopamine D2 receptor gene in Tourette's syndrome using the haplotype relative risk method. *American Journal of Medical Genetics*, *54*(3), 249–252. https://doi.org/10.1002/ajmg.1320540311.

Peters, K. Z., Cheer, J. F., & Tonini, R. (2021). Modulating the neuromodulators: Dopamine, serotonin, and the endocannabinoid system. *Trends in Neurosciences*, *44*(6), 464–477. https://doi.org/10.1016/j.tins.2021.02.001.

Peterson, B. S., Skudlarski, P., Anderson, A. W., Zhang, H., Gatenby, J. C., Lacadie, C. M., et al. (1998). A functional magnetic resonance imaging study of tic suppression in Tourette syndrome. *Archives of General Psychiatry*, *55*(4), 326–333. https://doi.org/10.1001/archpsyc.55.4.326.

Peterson, B. S., Thomas, P., Kane, M. J., Scahill, L., Zhang, H., Bronen, R., et al. (2003). Basal ganglia volumes in patients with Gilles de la Tourette syndrome. *Archives of General Psychiatry*, *60*(4), 415–424. https://doi.org/10.1001/archpsyc.60.4.415.

Polyanska, L., Critchley, H. D., & Rae, C. L. (2017). Centrality of prefrontal and motor preparation cortices to Tourette Syndrome revealed by meta-analysis of task-based neuroimaging studies. *Neuroimage. Clinical*, *16*, 257–267. https://doi.org/10.1016/j.nicl.2017.08.004.

Pringsheim, T., Okun, M. S., Müller-Vahl, K., Martino, D., Jankovic, J., Cavanna, A. E., et al. (2019). Practice guideline recommendations summary: Treatment of tics in people with Tourette syndrome and chronic tic disorders. *Neurology*, *92*(19), 896–906. https://doi.org/10.1212/wnl.0000000000007466.

Puts, N. A., Harris, A. D., Crocetti, D., Nettles, C., Singer, H. S., Tommerdahl, M., et al. (2015). Reduced GABAergic inhibition and abnormal sensory symptoms in children with Tourette syndrome. *Journal of Neurophysiology*, *114*(2), 808–817. https://doi.org/10.1152/jn.00060.2015.

Ramkiran, S., Heidemeyer, L., Gaebler, A., Shah, N. J., & Neuner, I. (2019). Alterations in basal ganglia-cerebello-thalamo-cortical connectivity and whole brain functional network topology in Tourette's syndrome. *NeuroImage: Clinical*, *24*, 101998. https://doi.org/10.1016/j.nicl.2019.101998.

Rapanelli, M., & Pittenger, C. (2016). Histamine and histamine receptors in Tourette syndrome and other neuropsychiatric conditions. *Neuropharmacology*, *106*, 85–90. https://doi.org/10.1016/j.neuropharm.2015.08.019.

Roessner, V., Plessen, K. J., Rothenberger, A., Ludolph, A. G., Rizzo, R., Skov, L., et al. (2011). European clinical guidelines for Tourette syndrome and other tic disorders. Part II: pharmacological treatment. *European Child & Adolescent Psychiatry*, *20*(4), 173–196. https://doi.org/10.1007/s00787-011-0163-7.

Sandyk, R., & Bamford, C. R. (1988). Opioid-serotoninergic dysregulation in the pathophysiology of Tourette's syndrome. *Functional Neurology*, *3*(2), 225–235.

Sandyk, R., Iacono, R. P., & Bamford, C. R. (1987). The hypothalamus in Tourette's syndrome. *International Journal of Neuroscience*, *35*(1-2), 99–103. https://doi.org/10.3109/00207458708987117.

Scholl, C., Baladron, J., Vitay, J., & Hamker, F. H. (2021). Enhanced habit formation in Tourette syndrome: Dopamine release and striatal disinhibition modulate shortcut connections in a hierarchical model of cortico-basal ganglia loops. *bioRxiv*. 2021.2002.2008.430235 https://doi.org/10.1101/2021.02.08.430235.

Sigurdsson, H. P., Jackson, S. R., Jolley, L., Mitchell, E., & Jackson, G. M. (2020). Alterations in cerebellar grey matter structure and covariance networks in young people with Tourette syndrome. *Cortex*, *126*, 1–15. https://doi.org/10.1016/j.cortex.2019.12.022.

Silvestri, R., Raffaele, M., De Domenico, P., Tisano, A., Laganà, A., & Di Perri, R. (1994). Serotoninergic agents in the treatment of Gilles de la Tourette's syndrome. *Acta Neurol (Napoli)*, *16*(1-2), 58–63.

Singer, H. S., Butler, I. J., Tune, L. E., Seifert, W. E., Jr., & Coyle, J. T. (1982). Dopaminergic dsyfunction in Tourette syndrome. *Annals of Neurology*, *12*(4), 361–366. https://doi.org/10.1002/ana.410120408.

Singer, H. S., Hahn, I. H., & Moran, T. H. (1991). Abnormal dopamine uptake sites in postmortem striatum from patients with Tourette's syndrome. *Annals of Neurology*, *30*(4), 558–562. https://doi.org/10.1002/ana.410300408.

Singer, H. S., Morris, C., & Grados, M. (2010). Glutamatergic modulatory therapy for Tourette syndrome. *Medical Hypotheses*, *74*(5), 862–867. https://doi.org/10.1016/j.mehy.2009.11.028.

Singer, H. S., Szymanski, S., Giuliano, J., Yokoi, F., Dogan, A. S., Brasic, J. R., et al. (2002). Elevated intrasynaptic dopamine release in Tourette's syndrome measured by PET. *American Journal of Psychiatry*, *159*(8), 1329–1336. https://doi.org/10.1176/appi.ajp.159.8.1329.

Stahl, S. M., & Berger, P. A. (1982). Neuroleptic effects in Tourette syndrome predict dopamine excess and acetylcholine deficiency. *Biological Psychiatry*, *17*(9), 1047–1053.

Stöber, G., Hebebrand, J., Cichon, S., Brüss, M., Bönisch, H., Lehmkuhl, G., et al. (1999). Tourette syndrome and the norepinephrine transporter gene: Results of a systematic mutation screening. *American Journal of Medical Genetics*, *88*(2), 158–163. https://doi.org/10.1002/(sici)1096-8628(19990416)88:2<158::aid-ajmg12>3.0.co;2-w.

Szejko, N., Fichna, J. P., Safranow, K., Dziuba, T., Żekanowski, C., & Janik, P. (2020). Association of a variant of CNR1 gene encoding cannabinoid receptor 1 with Gilles de la Tourette syndrome. *Frontiers in Genetics*, *11*, 125. https://doi.org/10.3389/fgene.2020.00125.

Szejko, N., Saramak, K., Lombroso, A., & Müller-Vahl, K. (2021). Cannabis-based medicine in treatment of patients with Gilles de la Tourette syndrome. *Neurologia i neurochirurgia polska*. https://doi.org/10.5603/PJNNS.a2021.0081.

Tobe, R. H., Bansal, R., Xu, D., Hao, X., Liu, J., Sanchez, J., et al. (2010). Cerebellar morphology in Tourette syndrome and obsessive-compulsive disorder. *Annals of Neurology*, *67*(4), 479–487. https://doi.org/10.1002/ana.21918.

Tübing, J., Gigla, B., Brandt, V. C., Verrel, J., Weissbach, A., Beste, C., et al. (2018). Associative plasticity in supplementary motor area—motor cortex pathways in Tourette syndrome. *Science Reports*, *8*(1), 11984. https://doi.org/10.1038/s41598-018-30504-8.

Tübing, J., & Münchau, A. (2015). Cortical GABAergic activity: A mediator of tic control? *Movement Disorders, 30*(3), 339. https://doi.org/10.1002/mds.26194.

Turjanski, N., Sawle, G. V., Playford, E. D., Weeks, R., Lammerstma, A. A., Lees, A. J., et al. (1994). PET studies of the presynaptic and postsynaptic dopaminergic system in Tourette's syndrome. *Journal of Neurology, Neurosurgery, and Psychiatry, 57*(6), 688–692. https://doi.org/10.1136/jnnp.57.6.688.

van der Salm, S. M. A., van der Meer, J. N., Cath, D. C., Groot, P. F. C., van der Werf, Y. D., Brouwers, E., et al. (2018). Distinctive tics suppression network in Gilles de la Tourette syndrome distinguished from suppression of natural urges using multimodal imaging. *NeuroImage. Clinical, 20*, 783–792. https://doi.org/10.1016/j.nicl.2018.09.014.

Van Woert, M. H., Jutkowitz, R., Rosenbaum, D., & Bowers, M. B., Jr. (1976). Gilles de la Tourette's syndrome: biochemical approaches. *Research publications - Association for Research in Nervous and Mental Disease, 55*, 459–465.

Wang, Z., Maia, T. V., Marsh, R., Colibazzi, T., Gerber, A., & Peterson, B. S. (2011). The neural circuits that generate tics in Tourette's syndrome. *American Journal of Psychiatry, 168*(12), 1326–1337. https://doi.org/10.1176/appi.ajp.2011.09111692.

Wolff, N., Luehr, I., Sender, J., Ehrlich, S., Schmidt-Samoa, C., Dechent, P., et al. (2016). A DTI study on the corpus callosum of treatment-naïve boys with 'pure' Tourette syndrome. *Psychiatry Research. Neuroimaging, 247*, 1–8. https://doi.org/10.1016/j.pscychresns.2015.12.003.

Wong, D. F., Brasić, J. R., Singer, H. S., Schretlen, D. J., Kuwabara, H., Zhou, Y., et al. (2008). Mechanisms of dopaminergic and serotonergic neurotransmission in Tourette syndrome: Clues from an in vivo neurochemistry study with PET. *Neuropsychopharmacology: Official Publication of the American College of Neuropsychopharmacology, 33*(6), 1239–1251. https://doi.org/10.1038/sj.npp.1301528.

Worbe, Y., Marrakchi-Kacem, L., Lecomte, S., Valabregue, R., Poupon, F., Guevara, P., et al. (2015). Altered structural connectivity of cortico-striato-pallido-thalamic networks in Gilles de la Tourette syndrome. *Brain, 138*(2), 472–482. https://doi.org/10.1093/brain/awu311.

Yamamuro, K., Ota, T., Iida, J., Nakanishi, Y., Uratani, M., Matsuura, H., et al. (2015). Prefrontal dysfunction in pediatric Tourette's disorder as measured by near-infrared spectroscopy. *BMC Psychiatry, 15*, 102. https://doi.org/10.1186/s12888-015-0472-3.

Yeh, C. B., Lee, C. H., Chou, Y. H., Chang, C. J., Ma, K. H., & Huang, W. S. (2006). Evaluating dopamine transporter activity with 99mTc-TRODAT-1 SPECT in drug-naive Tourette's adults. *Nuclear Medicine Communications, 27*(10), 779–784. https://doi.org/10.1097/01.mnm.0000239485.53780.16.

Yeh, C. B., Lee, C. S., Ma, K. H., Lee, M. S., Chang, C. J., & Huang, W. S. (2007). Phasic dysfunction of dopamine transmission in Tourette's syndrome evaluated with 99mTc TRODAT-1 imaging. *Psychiatry Research, 156*(1), 75–82. https://doi.org/10.1016/j.pscychresns.2007.01.003.

Yuan, A., Su, L., Yu, S., Li, C., Yu, T., & Sun, J. (2015). Association between DRD2/ANKK1 TaqIA polymorphism and susceptibility with Tourette syndrome: A meta-analysis. *PLoS One, 10*(6), e0131060. https://doi.org/10.1371/journal.pone.0131060.

Yvonneau, M., & Bezard, P. (1970). Apropos of a case of Gilles de la Tourette's disease blocked by sulpiride. Psycho-biological study. *Encephale, 59*(5), 439–459. Sur un cas de maladie des tics bloquée par le sulpiride. Etude psycho-biologique.

Zapparoli, L., Porta, M., & Paulesu, E. (2015). The anarchic brain in action: The contribution of task-based fMRI studies to the understanding of Gilles de la Tourette syndrome. *Current Opinion in Neurology, 28*(6), 604–611. https://doi.org/10.1097/wco.0000000000000261.

SECTION II

Neuroimaging and pathophysiology

CHAPTER FIVE

Recent advances in neuroimaging of Tourette syndrome

Nicholas Cothros[a] and Davide Martino[b,c,*]
[a]Department of Medicine, Queen's University School of Medicine, Kingston, ON, Canada
[b]Department of Clinical Neurosciences, University of Calgary, Calgary, AB, Canada
[c]Hotchkiss Brain Institute, Calgary, AB, Canada
*Corresponding author: e-mail address: davide.martino@ucalgary.ca

Contents

1. Introduction

Tics are brief, repetitive, stereotyped, intermittent movements or sounds/utterances that are most associated with Tourette syndrome (TS), a neurodevelopmental disorder of childhood origin (Sanger, Chen,

International Review of Movement Disorders, Volume 3
ISSN 2666-7878
https://doi.org/10.1016/bs.irmvd.2022.02.001

Fehlings, et al., 2010). Tics can be voluntarily suppressed and are commonly preceded by sensory phenomena, often characterized as a premonitory urge (Kurvits, Martino, & Ganos, 2020; Kwak, Dat Vuong, & Jankovic, 2003; Leckman, Walker, & Cohen, 1993; Martino, Ganos, & Pringsheim, 2017; Patel, Jankovic, & Hallett, 2014). Postmortem studies (Hong, Loiselle, Yoon, et al., 2004; Minzer, Lee, Hong, & Singer, 2004; Singer, Hahn, & Moran, 1991) and some animal models of TS (Bronfeld & Bar-Gad, 2013; Bronfeld, Yael, Belelovsky, & Bar-Gad, 2013; Godar, Mosher, Giovanni, & Bortolato, 2014; McCairn & Isoda, 2013) suggest that tics arise from dysfunction in the basal ganglia and excessive dopamine release, permitting excessive motor output. These studies partially reinforce the basal ganglia as part of a wider dysfunctional circuit in TS, the cortico-striato-thalamo-cortical circuit (Ganos, Roessner, & Münchau, 2013).

The pathophysiology of TS is the subject of ongoing research. Imaging studies have generated several insights over the last two decades, corresponding to a large body of literature. Structural, functional, and spectroscopic imaging studies to date suggest the pathophysiology extends beyond the basal ganglia and dopaminergic dysfunction, indicating that different neurotransmitter systems are likely involved, along with brain networks outside the above circuit.

Structural imaging in TS has identified gray matter volume and sulcal morphology changes in various regions associated with sensorimotor processing, such as parietal, temporal, and occipital cortical regions, plus abnormalities in frontal regions possibly related to deficiencies in inhibitory function. Spectroscopic imaging indicates dysfunction in GABAergic and glutamatergic neurotransmitter systems. Interestingly, functional magnetic resonance imaging (fMRI) prior to the appearance of tics and at tic onset suggest the neural correlates of the premonitory urge and tic release are physiologically distinct and encompass brain areas related to motor planning and interoception, such as the supplementary motor area, the insula, and cingulate cortices. The cerebellum has also emerged as an important node in tic-suppression and tic-generation networks.

Resting-state fMRI represents a growing area of research in this field and substantiates ideas of network dysfunction in tic disorders, suggesting hyperconnectivity between subcortical areas and cortical regions related to motor planning, sensory processing, and interoception; and hypoconnectivity between frontal areas and subcortical regions, and between the cerebellum and cortical areas. Connectivity studies perhaps reinforce ideas of increased sensorimotor noise, conveyed by input from cortical to subcortical areas, where reduced inhibition permits excessive motor output. They may also support hypotheses related to deficiencies in inhibitory capacity and that TS results from anomalies in brain development or maturation, though it should be noted that the literature is generally cross-sectional rather than longitudinal.

Notably, imaging studies have also highlighted areas of overlap with psychiatric comorbidities, thus affording a deeper characterization of tic disorders along certain clinical dimensions, suggesting the neural bases of behavioral endophenotypes. Finally, task-based and behavioral studies of individuals with tics, using functional imaging, help reveal not only the correlates of tic phenomenology, but possibly additional neural correlates that clarify the biological trait of having tics.

2. Brain structural changes associated with the diagnosis of TS and persistent tic disorders

Important morphological abnormalities have been documented in patients with TS and other persistent tic disorders (Table 1), despite study heterogeneity for patients' age group, exposure to medications, comorbid behavioral disorders, and imaging technique. A meta-analysis of voxel-based morphometry (VBM) studies which applied the seed-based d mapping method corroborated the notion that fronto-subcortical networks connecting the prefrontal cortex to striatum and thalami are highly involved in these disorders (Wan et al., 2021). Smaller gray matter volume (GMV) in the bilateral inferior frontal gyri and greater GMV in the right putamen and bilateral pulvinar nucleus of the thalamus were key findings of this meta-analysis, confirmed also by another voxel quantitative meta-analysis that applied activation likelihood estimation (Wen et al., 2021). A conspicuous body of work had already shown decreased prefrontal cortical thickness in TS, which correlated with greater tic severity in some studies (Draganski, Martino, Cavanna, et al., 2010; Fahim, Yoon, Das, et al., 2010; Greene et al., 2017; Müller-Vahl et al., 2009; Sowell, Kan, Yoshii, et al., 2008; Worbe, Gerardin, Hartmann, et al., 2010). This finding has generally been interpreted as supporting an association between inferior frontal gyri or other prefrontal regions and defective inhibition of prepotent motor programs, hence related to tic generation and suppression.

Abnormalities identified by the whole body of structural imaging studies extend, however, to most brain regions associated with sensory-motor processing, including parietal, temporal, insula, anterior cingulate, and occipital cortical regions (Ganos et al., 2013; Greene et al., 2017; Hashemiyoon, Kuhn, & Visser-Vandewalle, 2017; Kong et al., 2020; Liu

Table 1 Main findings of the relevant anatomical MRI studies ($n = 12$) in patients with Tourette syndrome (TS) published in the last 5 years.

References	Age and number of participants	Medication-free period (or proportion of medicated patients)	Imaging methods	Statistically significant findings in patients
Greene, Williams Iii, Koller, Schlaggar, and Black (2017)	103 TS 11.9 ± 2.1 years 103 HC 11.9 ± 2.1 years	67/103 on medications	VBM	Lower white matter volume bilaterally in orbital and medial prefrontal cortex. Greater gray matter volume in posterior thalamus, hypothalamus, and midbrain.
Forde, Zwiers, Naaijen, et al. (2017)	47 TS 10.5 ± 1.4 years 39 ADHD 10.7 ± 1.3 years 55 HC 11 ± 1 years	Proportion of medicated patients unclear (at least 11/47)	Basal ganglia volumetry and morphology	None
Wen, Yan, Yu, et al. (2021)	44 TS 9 ± 3.1 years 41 HC 10.3 ± 3.2 years	All drug-naïve	Probabilistic tractography with graph theoretical analysis	Decreased global and local efficiency, increased shortest path length and small worldness. Decreased nodal efficiency, mainly in the default mode, language, visual and sensorimotor systems.
Martino et al. (2017)	13 TS 33 ± 10.3 years 13 HC 32.8 ± 6.8 years	5/13 on medications	DTI and probabilistic tractography of interhemispheric corpus callosum	Higher FA in both M1-M1 and SMA-SMA callosal fiber tracts.

Schlemm et al. (2017)	13 TS 34.5 ± 8.9 years 13 HC 34.6 ± 9.1 years	>3 weeks	DTI and probabilistic tractography with graph theoretical analysis	Reduced connectivity in right hemispheric networks and reduced local graph parameters (local clustering, efficiency and strength). Increased normalized global efficiency of whole brain and right hemispheric networks (positively correlating with tic severity).
Sigurdsson et al. (2018)	35 TS 14 ± 3.3 years 35 HC 13.9 ± 3.3 years	8/35 on medications	Diffusion MRI analyzed using TBSS and probabilistic tractography	Decreased axial diffusivity and altered white matter connectivity. Positive correlation between M1-caudate connectivity and tic severity. Positive correlation between M1-insula connectivity and premonitory urge severity.
Atkinson-Clement et al. (2020*b*)	29 TS with intermittent explosive outbursts 31.45 ± 11.17 years 29 TS without intermittent explosive outbursts 31.03 ± 11.55 years	38% of TS with intermittent explosive outbursts and 31% of TS without intermittent explosive outbursts	DTI	Structural changes in the right SMA and in the right hippocampus (increased FA), and in the left orbitofrontal cortex (decreased mean diffusivity) in TS with intermittent explosive outbursts.

Continued

Table 1 Main findings of the relevant anatomical MRI studies ($n = 12$) in patients with Tourette syndrome (TS) published in the last 5 years.—cont'd

References	Age and number of participants	Medication-free period (or proportion of medicated patients)	Imaging methods	Statistically significant findings in patients
Kong, Lv, Wu, et al. (2020)	60 TS 8.79 ± 3.07 years 52 HC 10.89 ± 3.33 years	Medication-naïve	Cortical thickness, surface area, curvature, and sulcus, and local gyrification index measures on images processed with FreeSurfer software	Reduced cortical thickness in TS in left postcentral cortex, left superior parietal cortex, left rostral anterior cingulate gyrus, right paracentral gyrus, right precuneus, and right lateral occipital gyrus. Increased cortical thickness in TS in right frontal pole, right medial orbitofrontal gyrus, right precentral gyrus, and right postcentral gyrus. Lower cortical sulcus in TS in left superior temporal gyrus, left lingual, right medial orbitofrontal gyrus, right superior parietal gyrus, right lateral occipital gyrus and deeper in left middle temporal gyrus, left supramarginal gyrus, and right inferior parietal gyrus. Decreased cortical curvature in TS in left rostral middle frontal gyrus, left caudal anterior cingulate cortex and right inferior parietal gyrus, and right lateral occipital gyrus. Increased cortical curvature in left supramarginal gyrus. Different local gyrification index in superior frontal gyrus, left inferior temporal gyrus and left insula.
Sigurdsson, Jackson, Jolley, Mitchell, and Jackson (2020)	28 TS 14.6 ± 3.4 years 36 HC 14.4 ± 3.2 years	10/28 on medications	VBM and "seed-to-voxel" structural covariance network mapping	Reduced gray matter volume in cerebellar lobule Crus I and lobule VIIIa in TS. Increased gray matter structural covariance between Crus I seed region and sensorimotor and frontal cortex areas, mid-cingulate cortex, and cerebellar lobule VI. Decreased covariance with area encompassing the calcarine gyrus and precuneus.

Tikoo, Suppa, Tommasin, et al. (2021)	16 "pure TS" 9.7 ± 2.1 years 14 TS + OCD 10.2 ± 2.1 years 11 OCD 10.7 ± 2.5 years 12 HC 10 ± 1.2 years	Medication-naïve	Cerebellar lobular volumes using the SUIT toolbox; TBSS and DTI of cerebellar peduncles	Lower gray matter in bilateral Crus I and lobule VIIIb in OCD patients compared to "pure TS." Higher FA in cerebellar peduncles in "pure TS" and TS + OCD.
Wan, Zhang, Wang, et al. (2021)	331 TS and 327 HC from 10 studies selected for metaanalysis	–	VBM	Decreased gray matter volume in the bilateral inferior frontal gyrus. Increased gray matter volume in the cerebellum, right striatum, and bilateral thalami.
Wen et al. (2021)	247 TS and 236 HC from 6 studies selected for metaanalysis	–	VBM	Decreased gray matter volume in the bilateral postcentral gyrus, bilateral anterior cingulate, left sub-lobar insula, left medial frontal gyrus, and left posterior cingulate. Increased gray matter volume in sub-regions of the following brain structures: bilateral thalamus (pulvinar), left red nucleus, right precentral gyrus, right hypothalamus, left postcentral gyrus, left inferior parietal lobule, right putamen, and left insula.

To review findings from anatomical studies published prior to 2017, please consult Martino, Ganos & Worbe, 2018; 143:65–108, HC: healthy control subjects. VBM: voxel-based morphometry. DTI: diffusion tensor imaging. TBSS: tract-based spatial statistics. SMA: supplementary motor area. FA: fractional anisotropy. M1: primary motor cortex. OCD: obsessive-compulsive disorder.

et al., 2013; Neuner, Schneider, & Shah, 2013; Worbe, Lehericy, & Hartmann, 2015). Importantly, TS patients exhibit enhanced cortico-basal ganglia connectivity within circuits connecting the primary motor and somatosensory cortical regions with striatum and thalamus, which also correlated with tic severity (Worbe, Marrakchi-Kacem, Lecomte, et al., 2015). A greater representation of complex tics was also associated with more extended cortical thinning (Worbe et al., 2010). Cortical sulcal morphology in the central and superior temporal sulcus was also found to correlate with tic severity in adults (Muellner, Delmaire, Valabregue, et al., 2015). In line with this, the most prominent alteration in the local gyrification index of TS children was identified in the inferior temporal gyrus and insula (Kong et al., 2020). Abnormal connectivity of cerebellar structures has been recently documented by a voxel-based morphometry that used a "seed-to-voxel" structural covariance network mapping approach and employed a cerebellar specific template in adolescents and young adults with TS (Sigurdsson et al., 2020). Reduced GMV was identified in the cerebellar lobules involved in higher-order cognitive functions (Crus I) and sensorimotor processing (lobule VIIIa). Not all studies, however, have replicated this finding (Tikoo et al., 2021). With respect to inter-hemispheric connectivity, axial diffusivity of all five corpus callosum segments was found to be reduced in 26 treatment-naïve boys with pure TS compared to 24 healthy control subjects, and correlated negatively with tic severity (Wolff, Luehr, Sender, et al., 2016).

There is a dearth of longitudinal data evaluating the association between the natural course of tics and structural connectivity. Debes et al. (2015) followed up 22 TS youths and 21 age-matched control subjects, observing a decrease in parallel and perpendicular diffusivity within caudate, thalamus and frontal lobe in patients with persistent tics at the end of adolescence. Conversely, patients undergoing remission at the beginning of adulthood were more similar to control subjects.

3. Brain functional and neurochemical changes associated with the diagnosis of TS and other chronic tic disorders

3.1 Functional MRI

The functional neuroanatomy associated with an idle "ticcing" state has been reviewed by several articles is summarized in Table 2. Functional connectivity assessed using fMRI studies consistently show a global reduction in

Table 2 Main findings of the relevant functional (functional MRI [fMRI], magnetic resonance spectroscopy [MRS], magnetoencephalography [MEG] and electroencephalography [EEG]) studies ($n = 27$) in patients with Tourette syndrome (TS) published in the last 5 years.

References	Age and number of participants	Medication-free period (or proportion of medicated patients)	Imaging methods	Statistically significant findings in TS patients
Liao et al. (2017)	24 TS age range 7–15 years 32 HC age distribution not provided	19/24 on medications	Resting state fMRI	Abnormal intrinsic functional connectivity in the bilateral prefronto-striatum-midbrain networks and bilateral sensorimotor and temporal cortices. Negative correlation between functional connectivity of bilateral aCC and tic severity.
Eddy, Cavanna, Rickards, & Hansen, 2017	25 TS 31.5 ± 11.5 years 25 HC 29.9 ± 10.1 years	10/25 on medications	Theory of Mind task-related fMRI	Increased activity of the R TPJ, R amygdala and posterior cingulate cortex. Increased TPJ activity correlated with tic, premonitory urges, impulse control problems and echophenomena severity. Amygdala activity correlated with premonitory urges severity and left TPJ activity with ratings of non-obscene socially inappropriate behaviors.
Naaijen et al. (2016)	15 TS 10.4 ± 1.2 years 39 ADHD 10.7 ± 1.2 years 29 TS + ADHD 10.7 ± 1.6 years 53 HC 10 ± 1 years	2/15 TS at least 9/29 TS + ADHD	Point resolved MRS	Obsessive-compulsive symptoms positively correlated with aCC glutamate concentration.

Continued

Table 2 Main findings of the relevant functional (functional MRI [fMRI], magnetic resonance spectroscopy [MRS], magnetoencephalography [MEG] and electroencephalography [EEG]) studies ($n = 27$) in patients with Tourette syndrome (TS) published in the last 5 years.—cont'd

References	Age and number of participants	Medication-free period (or proportion of medicated patients)	Imaging methods	Statistically significant findings in TS patients
Kanaan, Gerasch, Garcia-Garcia, et al. (2017)	37 TS at baseline 38.3 ± 11.1 years 15 TS after treatment with aripiprazole 40.1 ± 13.1 years 36 HC 38.4 ± 11.1 years	Aripiprazole medication part of the study design	Point resolved MRS	Decreased glutamate and glutamine concentrations and glutamine/glutamate ratio in L striatum. Glutamate concentration in the thalamus was partially normalized by aripiprazole.
Fan, Cath, van den Heuvel, et al. (2017)	18 TS 35.1 ± 12.6 years 19 OCD 46.1 ± 9.7 years 22 HC 43.3 ± 14.2 years	9/18 TS on medications	MRS	Glutamine concentrations in dorsal aCC correlated negatively with tic severity.
Liu, Wang, Zhang, et al. (2017)	21 TS 8.7 ± 3 years 29 HC 10.1 ± 3.1 years	Not specified	Resting state fMRI computing ALFF and fALFF	Abnormal ALFF or fALFF in vision-related structures including the calcarine sulcus, cuneus, fusiform gyrus, and L insula. Decreased regional homogeneity in R cerebellum (positively correlated with TS duration).

Zapparoli, Seghezzi, Devoto, et al. (2020)	28 TS 28.7 ± 12 years 24 HC 28.8 ± 14 years	17/28 on medications	fMRI related to finger opposition task analyzed through a dynamic causal modeling	Increased intrinsic connectivity in the premotor network, with stronger connections from the SMA, the dorsolateral premotor cortex and the putamen to the R superior frontal gyrus. Positive correlation between connectivity strength from the R striatum to the R M1 and tic severity. Negative correlation between connectivity strength from the R SMA to the R M1 and tic severity.
Fan, van den Heuvel, Cath, et al. (2018)	23 TS 34.6 ± 12.3 years 22 HC 43.3 ± 14.2 years	9/23 TS on medications (antipsychotics)	rs-fMRI and independent component analysis of networks	Greater involvement of the dorsal medial prefrontal cortex in the connectivity of the DMN and smaller involvement of the inferior parietal cortex in the connectivity of the FPN. Stronger coupling between DMN and left FPN. Negative correlation between DMN connectivity and tic severity.
Fan, Cath, van der Werf, et al. (2018)	18 TS without OCD 35.1 ± 12.6 years 18 OCD 46.9 ± 9.4 years 20 HC 41.4 ± 13.5 years	9/18 TS and 6/18 OCD on medications	fMRI related to a Stop-Signal task using region-of-interest analyses	Higher right inferior parietal cortex activation during inhibition in TS; higher left cerebellum, right midbrain, and right insular activation during error processing in TS; higher error-related supplementary motor area activity in TS, and positive correlation of error-related left cerebellar activity with tic severity.

Continued

Table 2 Main findings of the relevant functional (functional MRI [fMRI], magnetic resonance spectroscopy [MRS], magnetoencephalography [MEG] and electroencephalography [EEG]) studies ($n = 27$) in patients with Tourette syndrome (TS) published in the last 5 years.—cont'd

References	Age and number of participants	Medication-free period (or proportion of medicated patients)	Imaging methods	Statistically significant findings in TS patients
Mahone, Puts, Edden, Ryan, and Singer (2018)	32 TS 9.9 ± 1.9 years 43 HC 8.1 ± 1.9 years	>3 months	MRS at 7T	Increased glutamate concentrations in the premotor cortex, associated with selective motor inhibition.
Rae, Polyanska, Gould van Praag, et al. (2018)	21 TS mean age 33 years (range 18–51) 21 HC mean age 34 years (range 19–55)	2/21 TS on antipsychotics 6/21 TS on serotonin reuptake inhibitors	fMRI during a face perception task (neutral and angry expressions)	In TS increased functional connectivity of insula with presupplementary motor area, premotor cortex, primary motor cortex, and the putamen. Insula functional connectivity with the globus pallidus and thalamus varied in proportion to tic severity, while supplementary motor area connectivity varied in proportion to premonitory sensations, with insula connectivity to these regions increasing to a greater extent in patients with worse symptom severity. Occipital face area increased functional connectivity with posterior cortical regions, including primary somatosensory cortex.

Akkermans, van Rooij, Naaijen, et al. (2019)	14 TS without ADHD 10.6 ± 1.2 years 33 TS + ADHD 10.8 ± 1.5 years 29 ADHD 10.8 ± 1.4 years 48 HC 11 ± 1 years	3/14 TS without ADHD lifetime exposure to antipsychotics and 1/14 to clonidine 15/33 TS with ADHD lifetime exposure to stimulants, 1/33 to atomoxetine, 7/33 to antipsychotics and 6/33 to clonidine 21/29 ADHD lifetime exposure to stimulants, 2/29 to atomoxetine, 1/29 to antipsychotics	fMRI related to a monetary incentive delay task	ADHD severity across all groups marginally related to hypoactivation of the right nucleus accumbens during reward anticipation; this effect was not moderated by TS diagnosis. No associations of neural activation with TS.
Niccolai, Korczok, Finis, et al. (2019)	12 TS 37 ± 9 years	>6 months	MEG to explore oscillatory cortical correlates of motor preparation during a Go-NoGo task	During the second before tic onset, initial increase of beta power over the left-hemispheric channels overlying the motor cortex, subsequently replaced by a decrease in beta power. Initial increase in beta power positively correlated with intensity of premonitory urges.
Ramkiran, Heidemeyer, Gaebler, Shah, and Neuner (2019)	28 TS 32 ± 11 years 28 HC 31 ± 9 years	14/28 on medications	rs-fMRI and graph theoretical measures of centrality, functional segregation, and functional integration	Increased basal ganglia-cortical and thalamo-cortical connectivity and decreased cortico-cerebellar connectivity in TS. Increased global efficiency through the basal ganglia, thalamus, and cerebellum. Decreased average path length within the default mode and the salience networks.

Continued

Table 2 Main findings of the relevant functional (functional MRI [fMRI], magnetic resonance spectroscopy [MRS], magnetoencephalography [MEG] and electroencephalography [EEG]) studies ($n=27$) in patients with Tourette syndrome (TS) published in the last 5 years.—cont'd

References	Age and number of participants	Medication-free period (or proportion of medicated patients)	Imaging methods	Statistically significant findings in TS patients
Atkinson-Clement, Porte, de Liege, et al. (2020)	31 TS unmedicated 30.4 ± 11.4 years 19 TS medicated 31 ± 9.4 years 26 HC 32 ± 10.7 years	Separate TS groups according to medication exposure (atypical antipsychotics)	rs-fMRI, plus regions-of-interest approach on gray matter signal and functional connectivity analysis – correlation analyses with performances on a stop signal task	Increased gray matter signal from the right SMA and inferior frontal gyrus correlated with reactive motor impulsivity in medicated TS. In medicated TS, decreased resting-state spontaneous activity of the left putamen, higher functional connectivity between the inferior frontal gyrus and the superior temporal gyri (bilaterally), lower functional connectivity between the cerebellum and the right subthalamic nucleus.
Atkinson-Clement, Sofia, Fernandez-Egea, et al. (2020)	29 TS with intermittent explosive outbursts 31.45 ± 11.17 years 29 TS without intermittent explosive outbursts 31.03 ± 11.55 years	38% of TS with intermittent explosive outbursts and 31% of TS without intermittent explosive outbursts	rs-fMRI, plus regions-of-interest approach on gray matter signal and functional connectivity analysis	Lower connectivity within the sensori-motor cortico-basal ganglia network and altered connectivity pattern among the orbito-frontal cortex, amygdala and hippocampus in TS with intermittent explosive outbursts.
Tikoo, Cardona, Tommasin, et al. (2020)	16 TS without OCD 9.7 ± 2.1 years 14 TS + OCD 10.2 ± 2.1 years 10 OCD 10.9 ± 2.5 years 11 HC 9.9 ± 1.3 years	Medication-naïve	rs-fMRI and independent component analysis of networks	Higher FC in the basal ganglia, sensorimotor, cerebellum and default mode networks in TS with and without OCD. Lower FC in the fronto-parietal and salience networks in TS with and without OCD. Tic severity negatively correlated with FC in cerebellum and fronto-parietal networks.

Openneer, Marsman, van der Meer, et al. (2020)	27 TS without ADHD 10.7 ± 1.3 years 19 TS + ADHD 10.7 ± 1.4 years 23 ADHD 11 ± 1.3 years 40 HC 10.9 ± 1 years	8/27 TS without ADHD 10/19 TS + ADHD 16/23 ADHD on medications	rs-fMRI and measures of global efficiency, local efficiency, and clustering coefficient per acquired network	Local efficiency and clustering coefficient lower in children TS without ADHD in the default mode network compared with HC, and in the frontoparietal network compared with ADHD. Negative association of tic severity with local efficiency and clustering coefficient in the default mode network and in the frontoparietal network.
Lou, Li, Wang, et al. (2020)	79 TS 9.56 ± 2.5 years 63 HC 9.35 ± 2.19 years	25/79 on medications	rs-fMRI and measures of frequency-specific ReHo and independent component analysis	Increased ReHo in the left precentral gyrus and decreased ReHo in the right operculum in TS. Increased connectivity of the right superior frontal gyrus within the left executive control network in TS. Negative correlation of tic severity with ReHo values of the right operculum in the highest frequency bands (0.198–0.25 Hz); positive correlation of tic severity with connectivity of the right superior frontal gyrus.
Naro, Billeri, Colucci, et al. (2020)	13 TS 13 ± 1 years 11 persistent tic disorders 11 ± 3 years 15 HC 14 ± 2 years	Not reported	Scalp EEG during resting state to analyze the eLORETA current source densities and the lagged phase synchronization to estimate nonlinear functional connectivity between cortical areas	FPN functional connectivity in delta band more deteriorated in more severe TS. Sensorimotor functional connectivity in beta-2 band stronger in more severe TS and persistent tic disorders.

Continued

Table 2 Main findings of the relevant functional (functional MRI [fMRI], magnetic resonance spectroscopy [MRS], magnetoencephalography [MEG] and electroencephalography [EEG]) studies ($n=27$) in patients with Tourette syndrome (TS) published in the last 5 years.—cont'd

References	Age and number of participants	Medication-free period (or proportion of medicated patients)	Imaging methods	Statistically significant findings in TS patients
Nielsen, Gratton, Church, et al. (2020)	39 TS children 10.9 ± 1.6 years 23 TS adolescents 14.5 ± 0.86 years 39 TS adults 25.9 ± 5.3 years 39 HC children 10.7 ± 1.5 years 23 adolescent HC 14.6 ± 1.1 years 39 adult HC 25.9 ± 4.6 years	19/39 TS children, 14/23 TS adolescents, 18/39 TS adults on medications	rs-fMRI and multivariate classification methods	Patterns of functional connections that best distinguished TS from HC were age-specific. Developmental patterns among these functional connections used for diagnostic classification deviated from typical development. Brain networks in childhood TS appeared "older" and brain networks in adulthood TS appeared "younger" compared to typically developing individuals.
Rae, Parkinson, Betka, et al. (2020)	23 TS mean age 34 years (range 18–51) 21 HC mean age 35 years (range 19–55)	2/23 TS on antipsychotics 6/23 TS on serotonin reuptake inhibitors	fMRI during an intentional inhibition task	In TS, hyperactivation of the inferior frontal gyrus, insula and basal ganglia during voluntary action and inhibition, and reactive inhibition; hyperactivation of the caudate nucleus and primary motor cortex during intentional and reactive inhibition. Severity of PU correlated with connectivity of the pre-SMA to caudate, globus pallidus and thalamus during voluntary action.

Zapparoli et al. (2020)	25 TS 25.7 ± 3.8 years 25 HC 26.3 ± 9.4 years	19/25 TS on antipsychotics	fMRI during a temporal judgment task measuring the intentional binding phenomenon	In TS, loss of the association between magnitude of the intentional binding and activity of a premotor–parietal–cerebellar network.
Bhikram, Arnold, Crawley, Abi-Jaoude, and Sandor (2020)	40 TS 37.1 ± 13.1 years 20 HC 36.9 ± 17.2 years	25/40 on medications	rs-fMRI and seed-based functional connectivity of the supplementary motor area, orbitofrontal cortex, insula, caudate, and putamen	Greater connectivity between the temporal gyri, insula and putamen, and between the OFC and cingulate cortex in TS. Tic severity associated with greater connectivity between the putamen and the sensorimotor cortex.
Bhikram, Crawley, Arnold, Abi-Jaoude, and Sandor (2020)	40 TS 37.1 ± 13.1 years 20 HC 36.9 ± 17.2 years	25/40 on medications	Viewing blocks of OCS-provoking pictures relating to washing, checking and symmetry symptoms, and to disgusting and neutral scenes during fMRI	Negative associations between OCS severity and activity in the supramarginal gyrus, inferior frontal gyrus, sensorimotor cortex, precuneus and visual processing regions among the provocation conditions. Tic severity associated with activity in the anterior cingulate cortex for the symmetry condition.

Continued

Table 2 Main findings of the relevant functional (functional MRI [fMRI], magnetic resonance spectroscopy [MRS], magnetoencephalography [MEG] and electroencephalography [EEG]) studies ($n=27$) in patients with Tourette syndrome (TS) published in the last 5 years.—cont'd

References	Age and number of participants	Medication-free period (or proportion of medicated patients)	Imaging methods	Statistically significant findings in TS patients
Bhikram, Crawley, Arnold, Abi-Jaoude, and Sandor (2021)	40 TS 37.1 ± 13.1 years 20 HC 36.9 ± 17.2 years	25/40 on medications	Emotional blink suppression paradigm during fMRI	Greater activity in the superior temporal gyrus and midcingulate during inhibition of PU. Greater tic severity associated with activity in the superior frontal gyrus during the angry inhibition contrast as compared to neutral. Greater PU severity associated with activity in hippocampus, middle temporal gyrus, subcortex. Blink inhibition negatively associated with activity in thalamus and insula.
Openneer, van der Meer, Marsman, et al. (2021)	28 TS without ADHD 10.32 ± 1.29 years 23 TS + ADHD 10.14 ± 1.54 years 52 HC 10.53 ± 0.1 years	4/28 TS without ADHD 2/23 TS + ADHD on medications	fMRI-based stop-signal task	Stronger neural activation during failed inhibition in the inferior frontal gyrus and temporal and parietal areas in TS + ADHD compared to HC

To review findings from functional studies published prior to 2017, please consult Martino, Ganos & Worbe, 2018; 143:65–108, aCC: anterior cingulate cortex. ADHD: attention-deficit hyperactivity disorder. ALFF: amplitude of low frequency fluctuations. fALFF: fractional amplitude of low frequency fluctuations. DMN: default mode network. FC: functional connectivity. FPN: fronto-parietal network. HC: healthy controls. M1: primary motor cortex. ReHo: regional homogeneity. OCD: obsessive-compulsive disorder. OCS: obsessive-compulsive symptoms. OFC: orbitofrontal cortex. PU: premonitory urges. SMA: supplementary motor area. TPJ: temporo-parietal junction.

long range connectivity (involving parietal, cingulate, insular, temporal motor cortical regions and cerebellum), and an increase in short distance connectivity associated with motor processing in TS (Church, Fair, Dosenbach, et al., 2009; Worbe, Malherbe, Hartmann, et al., 2012), and during resting state (Bhikram, Arnold, et al., 2020; Cui et al., 2014; Fan, Cath, et al., 2018; Fan, van den Heuvel, et al., 2018; Liu et al., 2017; Ramkiran et al., 2019; see also Fig. 1). There is also a certain degree of inconsistency across studies, however, which may be attributed to differences in methodological approach; for example, a resting-state fMRI study that used independent component analysis found increased resting state functional connectivity in children with TS across different networks (basal ganglia, sensorimotor, cerebellar, the latter associated with Crus I, an area involved in higher cognitive functions; Tikoo et al., 2020). Analyses of the connectivity of the default mode network have confirmed lower integration between short-range connections within the default mode network and the frontoparietal network (Fan, Cath, et al., 2018; Fan, van den Heuvel, et al., 2018; Openneer et al., 2020; Tikoo et al., 2020). These findings could indicate "immature" brain development in people with TS. Using frequency-specific regional homogeneity and independent component analysis, Lou et al. (2020) revealed increased connectivity of the right superior frontal gyrus and an association between higher tic severity and decreased regional homogeneity values of the right operculum in higher frequency bands (0.198–0.25 Hz), supporting important connectivity alterations within the executive control network. Support vector machine classification indicated discriminatory capacity of resting state fMRI between TS and controls (Greene, Church, Dosenbach, et al., 2016), which could be age specific (Nielsen et al., 2020). Abnormal interhemispheric communication could also correlate to tic severity, particularly across homotopic voxels of the anterior cingulate cortices (Liao et al., 2017).

3.2 Molecular imaging: Nuclear medicine and magnetic resonance spectroscopy

Despite a postulated hyperdopaminergic state associated with enhanced phasic release of dopamine in the striatum in TS, more recent positron emission tomography (PET) studies of dopamine receptor availability yielded some inconsistencies (Abi-Jaoude, Segura, Obeso, et al., 2015; Denys, de Vries, Cath, et al., 2013). There seems to be a diffuse decrease in $GABA_A$ receptor binding potential, as indicated by a study that used the [^{11}C]-flumazenil in 11 adults with TS Lerner, Bagic, Simmons, et al., 2012). A reduced GABA

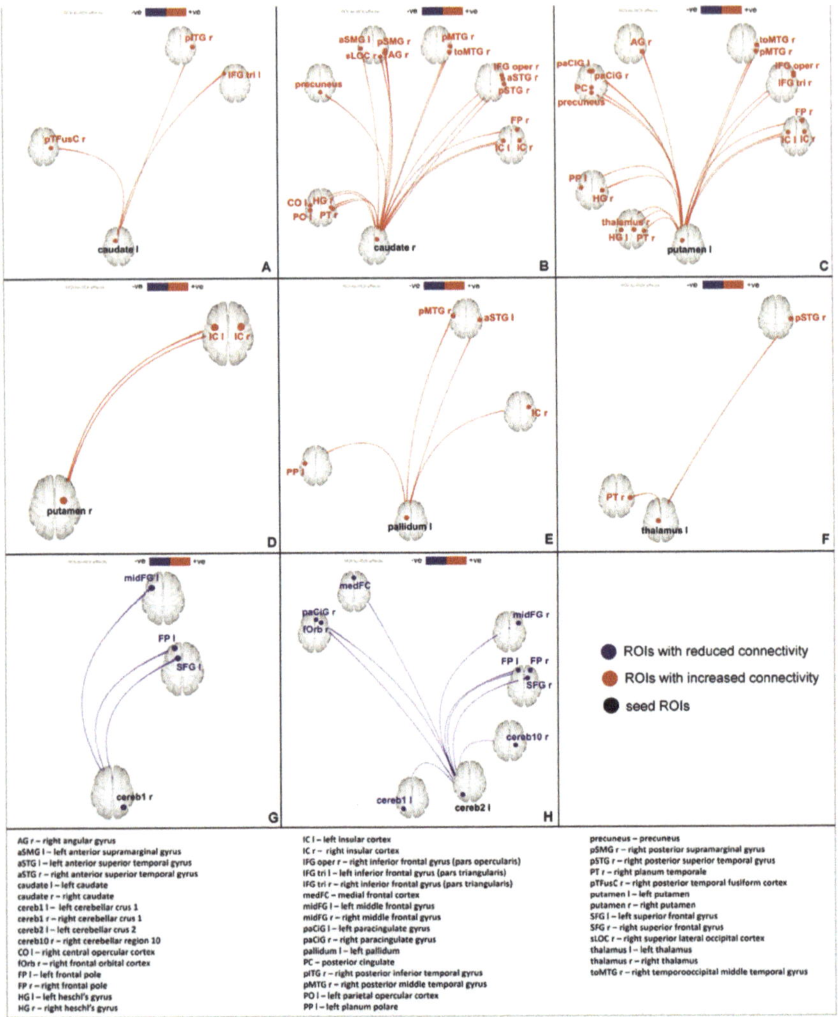

Fig. 1 Functional connectivity alterations in patients with Tourette syndrome, compared to age-matched healthy volunteers increased connectivity of left caudate (A), right caudate (B), left putamen (C), right putamen (D), left pallidum (E) and left thalamus (F) with several other regions; reduced connectivity of right cerebellar crus 1 (G) and left cerebellar crus 2 (H) with other brain regions. Red—increased connectivity, Blue—reduced connectivity. *From Ramkiran, S., Heidemeyer, L., Gaebler, A., Shah, N.J., & Neuner, I. (2019). Alterations in basal ganglia-cerebello-thalamo-cortical connectivity and whole brain functional network topology in Tourette's syndrome.* Neuroimage Clinic, 24, *101998.*

peak in the primary somatosensory cortex was demonstrated with magnetic resonance spectroscopy (MRS) in TS children (Puts, Harris, Crocetti, et al., 2015), whereas a larger, subsequent study that performed 7-Tesla ^{1}H-MRS did not confirm this finding in the dorsolateral and ventromedial prefrontal cortex, premotor cortex, and striatum (Mahone et al., 2018). The same study and another (Kanaan et al., 2017) found increased concentrations of glutamate in the premotor cortex and in the striatum, respectively. Premotor cortex glutamate concentrations correlated with motor inhibition performances (Mahone et al., 2018), whereas higher tic severity was associated with lower striatal glutamate (Kanaan et al., 2017). MRS studies also show discrepancies, with one study failing to show significant differences between TS adults and control participants (Fan et al., 2017). Despite the substantive heterogeneity of results, probably reflecting patient group heterogeneity, nuclear medicine imaging and MRS indicate a complex interplay between several neurotransmitter systems, which may be acting in parallel, or contributing to the abnormal regulation of striatal dopamine levels in TS.

4. Neural correlates of premonitory urges

A characteristic of tics is that of premonitory urges (PU), defined as regional or generalized sensory phenomena or urges that precede the appearance of a tic (Kurvits et al., 2020; Martino et al., 2017). Patients frequently report that performance of a tic temporarily mollifies or eliminates the PU (Patel et al., 2014). Premonitory phenomena are common in tic disorders (Leckman et al., 1993), reported by 92% in one sample (Kwak et al., 2003). In most of those patients, the anatomical distribution of PU involved the face/head, neck, and shoulders (Kwak et al., 2003), roughly corresponding to the most frequently affected areas in which tics appear (Ganos & Martino, 2015). The description of PU varies across patients and thus its measurement presents a challenge. The Premonitory Urge for Tics Scale (PUTS) has been used in both clinical assessment of PU and behavioral studies (Brandt, Beck, Sajin, et al., 2016; McGuire, McBride, Piacentini, et al., 2016; Woods, Piacentini, Himle, & Chang, 2005).

The idea that PU give rise to tics has received considerable attention. An early account of tics by Bliss (1980) characterized them as voluntary motor responses to involuntary somatosensory phenomena (i.e., a response to the PU). Sensory abnormalities in tic disorders may relate to the perception of external and internal stimuli (Ganos & Martino, 2015; Rae, Larsson,

Garfinkel, & Critchley, 2019). The mechanisms through which sensory phenomena may give rise to tics are not well understood.

Functional imaging patterns prior to tic onset may reflect neural correlates of PU. Bohlhalter, Goldfine, Matteson, et al. (2006) used fMRI to compare free ticcing and imitation of tics, in patients with TS. Two seconds before tic expression, the study authors found increased activation in the bilateral posterior middle cingulate cortex, insula, supplementary motor area, and parietal opercular cortex. These regions have been identified as part of an interoceptive network (Jackson, Parkinson, Kim, Schuermann, & Eickhoff, 2011; O'Neill, Piacentini, & Peterson, 2019) and may be interpreted in this study as a neural correlate of the PU, though it should be noted that PU were not measured directly and so these regions may play a different role in tic generation.

An examination of natural urges, to provide comparison with PU in patients with TS, was provided by Jackson et al. (2011), whose meta-analysis encompassed PET, single photon emission computed tomography, and fMRI studies of normal urges (e.g., swallowing, micturition, yawning) and results from Bohlhalter et al. (2006). They found the greatest overlap in the bilateral posterior middle cingulate cortex and right insula (Jackson et al., 2011; O'Neill et al., 2019), suggesting a PU network. This is also consistent with research in humans and non-human primates suggesting the insula receives several types of sensory input plus limbic and cingulate input, which are relayed back to limbic and cingulate areas, striatum, somatosensory cortices, and motor cortices (Benarroch, 2019). Draper, Jackson, Morgan, and Jackson (2016) provided further evidence of the relationship between brain structural abnormalities and PU, as the severity of PU was found to be inversely correlated with gray matter thickness of sensorimotor cortex and insula.

Resting-state fMRI (rs-fMRI) outlines the spatial organization of neural networks and their connectivity (Soares, Magalhães, Moreira, et al., 2016). These include the default mode (DMN), frontoparietal (FPN), and salience networks (Fan, van den Heuvel, et al., 2018). Posterior cingulate cortex, precuneus, medial frontal, inferior parietal, and temporal regions comprise the DMN, which is thought to play a role in cognitive-emotional integration and environmental monitoring (Fan, van den Heuvel, et al., 2018). The FPN encompasses the anterior prefrontal cortex, anterior cingulate cortex, anterior insula, anterior inferior parietal lobule, and the caudate nucleus; and is thought to process attention and underlie higher-order cognition (Fan, van den Heuvel, et al., 2018). The salience network includes the bilateral anterior insula, anterior cingulate cortex, pre-supplementary motor area, amygdala, ventral striatum, and substantia nigra; and is thought to control allocation of attention, response to salient events (e.g., error detection), and goal-directed action (Fan, van den Heuvel, et al., 2018).

Some researchers have suggested PU and tic release are reflected by disordered connectivity, which may relate to interoception, or excessive cortical input to subcortical areas, or vice versa. Using rs-fMRI to examine whole-brain functional connectivity in adults with TS, Tinaz, Malone, Hallett, and Horovitz (2015) used networks as defined by Bohlhalter et al. (2006), adding the dorsal anterior insula as a node for PU, thereby defining a composite urge-tic network. In TS, they found increased functional connectivity between the right anterior insula and bilateral supplementary motor area, and this correlated with PUTS scores. Tics were not captured. Connectivity parameters did not correlate significantly with tic severity (Tinaz et al., 2015), suggesting PU and tics are physiologically distinct. The authors suggested that PU arise from increased input from interoceptive cortical areas to motor planning/output areas, though inferring directional effects is limited by the statistical methodology determining functional connectivity.

Sigurdsson et al. (2018) appraised PU and brain connectivity in their study of 28 children and adolescents with TS and 30 controls, using diffusion tensor imaging. They found PUTS scores were significantly associated with local efficiency on three nodes. The left caudate was negatively associated with PUTS scores; and the right posterior insula and left anterior insula were positively associated with PUTS scores. The authors supported the role of the insula in PU. These findings may be in keeping with the notion of a dorsal-posterior insula network through which the insula receives and sends inputs between sensorimotor, premotor, and supplementary motor areas, temporal cortices, and visual cortices, combining sensory/environmental monitoring and goal-directed motor behavior (Benarroch, 2019).

Bhikram, Arnold, et al. (2020) sought to characterize functional connectivity patterns associated not only with tics, but also with obsessive-compulsive symptoms and sensory phenomena (including PU) using rs-fMRI in their study of 39 adults with TS and 20 controls. Patients with TS showed stronger connectivity between temporal gyri and both insula and putamen, as well as stronger connectivity between the orbitofrontal cortex and the anterior cingulate cortex. Bhikram, Arnold, et al. (2020) argued for the insula's contribution to PU. The authors also highlighted the role of the anterior cingulate cortex in error detection and therefore the possibility of sensory phenomena arising from this region, which is subject to inhibition by the orbitofrontal cortex, suggested by hyperconnectivity between orbitofrontal and anterior cingulate cortical regions. Higher PUTS scores were associated with reduced connectivity between orbitofrontal cortex and primary motor and somatosensory regions, as well as with reduced connectivity between the inferior frontal gyrus, putamen, and insula. As the inferior frontal gyrus is purported to play a role in response inhibition, reduced connectivity between inferior frontal gyrus, insula and putamen was argued to reflect impairment in suppressing PU arising from the insula (Bhikram, Arnold, et al., 2020).

Observing that the intrinsic cortical networks revealed by rs-fMRI are mirrored by structural covariance networks of cortical thickness or gray matter volume, Jackson et al. (2020) used voxel-based morphometry techniques to investigate structural covariance networks associated with insular cortex in 28 children and adolescents with TS compared with 36 controls. PUTS scores were positively correlated with volume in the anterodorsal region of the right insula. The authors also found that PU were not correlated with tic severity, supporting the idea of PU as physiologically distinct from tics. Relatedly, PU severity and tic severity were found to be associated with different regions of the right insula—PU positively correlated with gray matter volume in the anterodorsal region, whereas motor tic severity negatively correlated with gray matter volume in the posterior region (see also Fig. 2).

The issue of lower temporal resolution of fMRI was indirectly addressed by Niccolai et al. (2019). In 12 adults with TS, Niccolai et al. (2019) used magnetoencephalography to investigate two time periods preceding tic

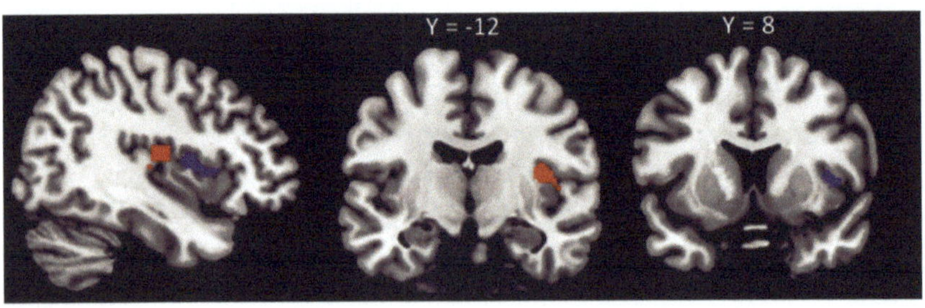

Fig. 2 This figure shows the insula voxels that are significantly negatively correlated with the motor tic severity score (in red) and significantly positively correlated with the premonitory urge score (in blue). *From Jackson, S.R., Loayza, J., Crighton, M., Sigurdsson, H.P., Dyke, K. & Jackson, G.M. (2020). The role of the insula in the generation of motor tics and the experience of the premonitory urge-to-tic in Tourette syndrome.* Cortex, 126, *119–133.*

onset, from 1-to-0.5 s prior to tic onset (Tw1) and 0.5 s before tic onset to the time of tic onset (Tw2). The authors relied on the high temporal resolution of magnetoencephalography and known correlates of motor preparation (beta oscillations of 15 to 30 Hz) in motor cortical regions. The authors used a Go/NoGo task, analyzing beta power modulation alongside PUTS scores, and comparing tic-free trials to those with tics. Beta power increased during Tw1 in cortical motor areas of the left hemisphere, and this correlated with a selected motor item of the PUTS score, which the authors interpreted as PU. During Tw2, beta power decreased significantly, marking voluntary movement. This evolution was interpreted as transitioning from motor inhibition (i.e., tic suppression) to voluntary motor preparation, which was construed as PU build-up giving way to tic release. This highlights the difficulty of differentiating between voluntary inhibition of tics and the physiological underpinnings of the PU sensory experience.

Bhikram et al. (2021) used a blink suppression task as an analogue of PU, comparing 39 adults with TS and 20 controls, instructing participants to alternate between free blinking and inhibiting blinks. The researchers used fMRI to establish differences between patients and controls in the neural correlates of urge suppression and how these were affected by tic severity and obsessive-compulsive symptom severity. They also examined how the neural correlates of urge suppression varied with emotional stimuli (angry versus neutral faces). There was no real-time measurement of PU or tics. Much like YGTSS scores, PUTS scores were not associated with brain activity in the general, inhibition-versus-free-blinking contrast. However, with exposure to angry faces, higher PUTS scores were associated with greater activity in the thalamus, hippocampus/parahippocampus, tail of the caudate, middle temporal gyrus, precuneus, and mid-cingulate cortex. Overall, the findings reinforce the difficulty in interpreting neural correlates of urge suppression, as these may reflect compensatory activity to control urges, or the neural mechanisms that give rise to urges. Additional correlates to PU were recently identified in other task-based fMRI studies that used social perception and Go/No-Go tasks (Rae et al., 2018, 2020), which are summarized in the chapter section on neural correlates of cognitive functions in TS.

Combining MRI with electroencephalography (EEG) may provide novel insights into the origins of PU and tics. Naro et al. (2020) combined rs-fMRI and EEG to test their hypotheses of hypo-connectivity of the frontoparietal network in children with TS or chronic tic disorder (cTD) versus controls; and that frontoparietal network dysfunction accounts for tic severity. The rs-fMRI analyses were combined with frequency band analysis of EEG data, accounting for periods in which tics occurred. In 13 children with TS and 11 with persistent tic disorders, their findings showed decreased connectivity within the frontoparietal network, with weaker connectivity in an anterior-to-posterior gradient between dorsolateral prefrontal and supplementary motor cortical regions to primary sensorimotor and posterior cingulate cortices. This decreased connectivity correlated with YGTSS scores. Brain oscillatory activity at tic onset showed bilateral alpha and beta deactivation in sensorimotor areas, which correlated with YGTSS scores. Just prior to tic onset, those with TS showed gamma and beta frequency deactivation in posterior cingulate cortex and supplementary motor area. There was also a negative correlation between functional connectivity within fronto-parieto-temporal areas in the gamma band and PU. Overall, YGTSS score, brain oscillatory activity during tics, PU, and psychiatric comorbidity all correlated with frontoparietal network disconnection. The researchers argued that beta deactivation represents a reduction in sensory gating, contributing to pathologically elevated cortical excitability. While the study was limited by small sample size and low spatial resolution afforded by EEG, the findings of fast oscillations in dorsolateral prefrontal cortex and supplementary motor area preceding tic onset and occurrence of PU, and of slower frequencies in posterior cingulate and sensorimotor cortical regions at tic onset, help delineate between PU and tic generation, demonstrating a relationship with symptom severity.

5. Neural correlates of tics

As tics phenomenologically resemble voluntary movement and are amenable to voluntary inhibition (Ganos, Rothwell, & Haggard, 2018), researchers have posited that tics arise from dysfunction in the neural control of voluntary movement (Neuner et al., 2013; Rae et al., 2019). The long-held, presumed pathophysiology of tics is centred around the cortico-striato-thalamo-cortical (CSTC) circuit (Ganos et al., 2013). Different explanatory frameworks have been put forward, including the idea that tics may be related to excessive cortical input into basal ganglia, through which insufficient inhibition permits excessive direct pathway motor output (Lerner et al., 2012; Neuner et al., 2013). However, varying concepts of disinhibition, from cellular to behavioral, have introduced challenges in understanding tic pathophysiology (Kurvits et al., 2020). As noted above, the role of PU in generating tics has also been considered, although timely measurement of PU while acquiring functional images proves challenging.

Functional imaging patterns around tic onset have provided some insights into the neural mechanisms of tic generation. At tic onset, Bohlhalter et al. (2006) found increased activation in the bilateral posterior middle cingulate cortex -though less so than prior to tic onset- plus increased activation in the bilateral superior parietal cortex and cerebellum.

Another exploration of pre-tic and tic activity by Hampson, Tokoglu, King, Constable, and Leckman (2009) compared 16 controls and 16 adults with TS during spontaneous tics versus simulated tics, testing the hypothesis that the supplementary motor area would show more fMRI activity prior to tics. They found a nearly identical degree of cross-correlation between primary motor cortex and supplementary motor area between imitated tics and genuine tics, though in TS the supplementary motor area showed a broader profile of cross-correlation with primary motor cortex, with abnormally elevated activity in the seconds preceding and following tics. Their secondary analysis yielded a significant interaction effect implicating supplementary motor area activation in tics, for those with TS and a history of PU. Noting the cortico-striato-thalamo-cortical circuit (Ganos et al., 2013), these findings may support the idea that tics and PU stem from overactivity in brain areas associated with movement planning.

Wang et al. (2011) compared 21 controls and 13 adults with TS during spontaneous tics versus simulated tics. Granger causality analysis of effective connectivity found stronger activation during tics (versus simulated tics) in somatosensory and posterior parietal cortices, putamen, and amygdalohippocampal areas, as well as weaker activation in caudate and anterior cingulate cortex; and these findings were correlated with tic severity. Linking their findings to the cortico-striato-thalamo-cortical circuit, Wang et al. (2011) argued the stronger activity patterns in TS represent the PU, while weaker activation represents underactivity in regions that inhibit motor pathways (anterior cingulate cortex, caudate). Their data suggested causal influences were stronger during spontaneous tics (vs. controls) in connections from premotor to supplementary motor area and vice-versa, from primary somatosensory to primary motor cortical regions, from substantia nigra to caudate and putamen, and from pallidum to supplementary motor area via thalamus. These findings suggest hyperactivity in the cortico-striato-thalamo-cortical circuit and partially support the idea of excessive cortical input into basal ganglia.

Neuner, Werner, Arrubla, et al. (2014) used rs-fMRI to identify patterns of brain activation prior to tics and at tic onset in 16 adult patients with TS. Two seconds before tic onset, they found increased activation of bilateral posterior middle cingulate cortex, supplementary motor area, ventral primary motor cortex, primary somatosensory cortex and parietal operculum. One second prior to tic onset, there was increased activation of the anterior cingulate cortex, putamen, insula, amygdala, cerebellum, and extra-striatal visual cortex. The results suggest stages of cortical pre-tic activity, beginning in areas related to motor planning (supplementary motor area), motor output (primary motor cortex), interoception and cognitive-emotional integration (posterior middle cingulate cortex), and somatosensory processing (parietal operculum), and subsequently followed by activity in areas related to attentional control and responding to salient events (anterior cingulate cortex), and interoception (insula). The authors suggested the possibility of PU arising from supplementary motor area activity, and from activity in sensory areas related to interoception and unpleasant stimuli, noting the insula and parietal operculum, respectively. At tic onset, there was increased fMRI activation in the thalamus, central operculum, and primary motor and somatosensory cortical regions. Thalamic involvement implies motor output via relay nuclei, and this was deemed consistent with scintigraphic evidence of thalamic disinhibition as a tic-generating mechanism (Lerner et al., 2012).

Using rs-fMRI and focusing on the basal ganglia-cerebellar-thalamo-cortical system, Ramkiran et al. (2019) applied graph theory metrics and, in participants with TS, found increased basal ganglia-cortical and thalamo-cortical functional connectivity, plus increased parallel communication through the basal ganglia, thalamus, and cerebellum. They also found a reduced average path length within the default mode and salience networks. The findings in TS included increased functional and effective connectivity between the basal ganglia and several cortical regions but decreased functional and effective connectivity between the cerebellum and several cortical regions. The authors thus speculated that the cerebellum plays a role in tic suppression. They alluded to increased connectivity between the basal ganglia and the insula as evidence of a possible neural correlate of PU and similarly postulated that increased connectivity between the basal ganglia and default mode network may be correlated with self-referential thoughts and with the PU.

The cerebellum has emerged as an important node in the tic-generating network, with a recent computational model asserting that subthalamic

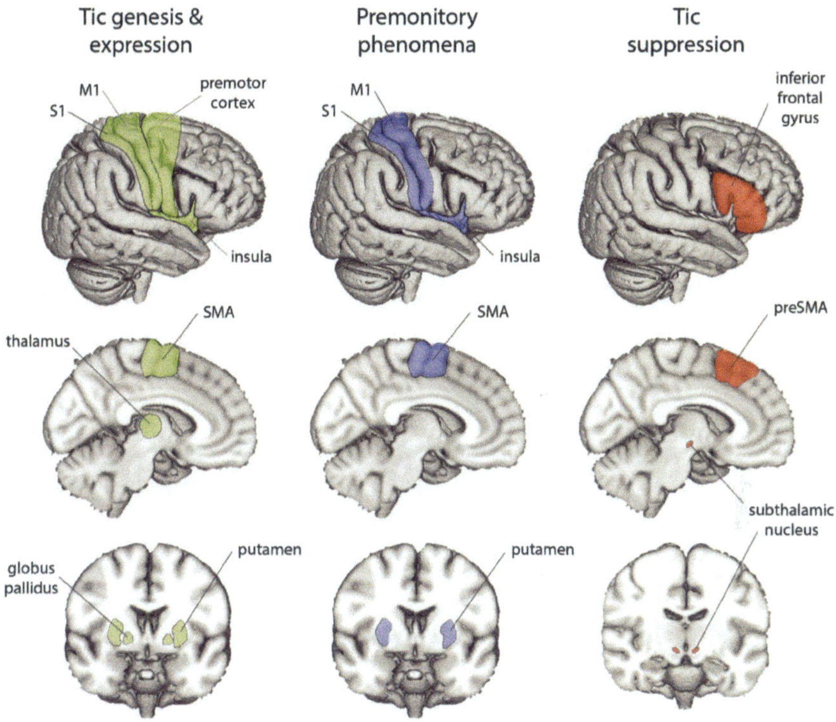

Fig. 3 This figure illustrates the cortical and subcortical regions associated with the phenomenological features of Tourette syndrome. Tic genesis and expression (green) is associated with premotor cortex, SMA, insula, M1, S1, the putamen, globus pallidus, and thalamus; premonitory sensations (blue) with the SMA, insula, M1, S1, and the putamen; tic suppression (red) with the inferior frontal gyrus, preSMA, and the subthalamic nucleus. *From Rae, C.L., Critchley, H.D., Seth, A.K. (2019). A Bayesian Account of the Sensory-Motor Interactions Underlying Symptoms of Tourette Syndrome.* Front Psychiatry, 10, *29.*

nucleus output to the cerebellum underlies tic release (Caligiore, Mannella, Arbib, & Baldassarre, 2017; Fig. 3). Along with fMRI evidence implicating the cerebellum in tic generation (Bohlhalter et al., 2006; Neuner et al., 2014), there is direct physiological evidence of cerebellar involvement from a macaque study of bicuculline microinjections into putamen, resulting in tic-like orofacial movements, reliably preceded by local field potential responses in M1 and the cerebellum, showing a basal ganglia-driven cerebellar tic generator (McCairn & Isoda, 2013). Using FDG-PET, Pourfar, Feigin, Tang, et al. (2011) showed a TS pattern of hypermetabolism in the bilateral premotor cortex and cerebellum. The precise nature of cerebellar involvement in TS remains unclear, as a PET study of $GABA_A$ receptor activity showed increased flumazenil binding in the cerebellum (Lerner et al., 2012), which may be in line with neurodevelopmental, maturational abnormalities involving GABAergic systems (Kalanithi, Zheng, Kataoka, et al., 2005; Puts et al., 2015), but this may be confounded with the development of compensatory mechanisms to inhibit tics (Jackson, Draper, Dyke, Pépés, & Jackson, 2015).

6. Imaging correlates of comorbid behavioral symptoms in TS

6.1 Attention deficit hyperactivity disorder (ADHD) in the context of TS

ADHD is the most common neurodevelopmental comorbidity in TS patients and may share common neurobiological underpinnings with the persistent tic disorder. Behavioral and neuroimaging studies, in some cases combining the two methodologies, indicate that partially overlapping fronto-striatal circuits may be specifically involved in TS and ADHD. The performance on response inhibition and other inhibitory control tasks has been linked to volume decreases in fronto-striatal circuits in the presence of ADHD diagnosis, with changes in motor cortico-basal networks being more directly associated with tic severity (Draganski et al., 2010; Plessen, Royal, & Peterson, 2007; Worbe et al., 2010). Imaging studies comparing patients with TS without ADHD, TS + ADHD, and ADHD without tics are, however, not numerous. Forde et al. (2017) did not detect differences in basal ganglia volume and shape between children with TS and/or ADHD within an 8 to 12 years age range. Other studies demonstrated morphological changes in the amygdala, hippocampal complex and left inferior frontal gyrus in TS + ADHD, with particular emphasis on the left inferior frontal gyrus as a structure strongly correlated to this comorbid phenotype (Ludolph, Pinkhardt, Tebartz van Elst, et al., 2008; Peterson, Choi, Hao, et al., 2007; Wittfoth, Bornmann, Peschel, et al., 2012).

In line with these morphological studies, the application of a whole-brain approach to investigate neural activation linked to performance on a stop-signal task has led to the observation of specific differences in youth with TS + ADHD compared to TS without ADHD. The latter group exhibited an intermediate pattern between TS + ADHD and typically developing, albeit closer to TS + ADHD subjects. Openneer et al. (2021) studied 103 children in the 8–12 years age range (23 TS + ADHD, 28 TS without ADHD, 52 typically developing) to demonstrate slower reaction time during response inhibition (stop trials) in a stop-signal task in TS + ADHD children. They also found that TS + ADHD children manifested stronger activation in the right inferior frontal gyrus, left insula, middle and temporal gyri particularly during failed inhibition, which was at odds with previous studies' results in youth with ADHD, in whom frontal gyri activity was decreased during failed inhibition. This pattern of hyperactivation in brain regions known to be involved in inhibitory performance might result from the co-existence of urges and tics, since prefrontal and insular hyperactivation are associated with active suppression of tics and urges, respectively. More studies are needed to better clarify the influence of OCD comorbidity in this group of cognitive tasks. Another recent study in medicated adult TS patients (Atkinson-Clement, Porte, et al., 2020) confirmed that reactive motor impulsivity -measured using the stop signal task- is associated with increased gray matter signal from the right SMA and inferior frontal gyrus, with higher functional connectivity between the inferior frontal gyrus and the superior temporal gyri bilaterally, as well as decreased functional connectivity between cerebellum and right subthalamic nucleus and lower resting activity of the left putamen. Interestingly, this finding was not influenced by the presence of a diagnosis of comorbid ADHD.

Other, specific functional activation changes may be a correlate of ADHD comorbidity in TS, being these shared by patients with TS + ADHD and patients with ADHD without tics. Using a task testing reward anticipation and receipt during fMRI in youth aged 8–12 years, Akkermans et al. (2019) showed a weak but significant association between ADHD severity and hypoactivation of the right nucleus accumbens (ventral striatum) regardless of co-existing TS diagnosis. Despite the small number of TS patients without ADHD and the high number of participants on stimulant medications, this result supports that nucleus accumbens hypoactivation could be another dimensional marker of ADHD severity, irrespective of primary diagnosis.

Finally, basal ganglia morphology may differ substantially between ADHD and TS. By means of imaging-based measures and semi-supervised machine learning algorithms, Bansal, Staib, Laine, et al. (2012) reported that patterns of variations in the surface morphology of discrete sub-regions of the globus pallidus, caudate nucleus, putamen, thalamus, and amygdalo-hippocampal complex accurately discriminate patients with ADHD from patients with TS.

6.2 Obsessive-compulsive disorder (OCD) in the context of TS

In a similar fashion to the relationship between TS and ADHD, the frequent co-occurrence of TS and OCD (Martino et al., 2017), likely associated with a specific phenomenological subtype, prompted the exploration of network connectivity to evaluate whether TS + OCD shows more similarities with either the "pure" TS or OCD phenotypes, or has features that are intermediate between the two conditions. A critical appraisal of the imaging studies that included TS patients with comorbid OCD reveals volumetric and connectivity changes indicating that TS + OCD is indeed an intermediate disorder, with some characteristics that are closer to "pure" OCD and others more resembling "pure" TS.

The presence of comorbid OCD in TS was observed to be associated with reduced caudate nucleus volume (Bloch, Leckman, Zhu, & Peterson, 2005), as well as with reduced thickness of the ventromedial prefrontal cortex and hippocampus (Worbe et al., 2010). These structural changes are aligned to changes observed also in OCD. The severity of compulsive behaviors was shown to be inversely correlated with the activity of the ventromedial prefrontal cortex (Worbe et al., 2012). Moreover, these structural changes were associated on fMRI with reduced functional activation of the ventromedial prefrontal cortex and of the ventral striatum (limbic loop of the cortico-striato-thalamo-cortical circuitry), leading to insufficient learning of a simple probabilistic task due to impaired sensitivity to reward. This cognitive phenotype related to the incentive motivation neuronal network is in common between OCD without tics and TS + OCD, as OCD patients without tics also display abnormal activation in the same region of the prefrontal cortex (Jung, Kang, Gu, et al., 2011; Kaufmann, Beucke, Preuβe, et al., 2013). Other cognitive correlates may, on the other hand, differentiate more evidently between TS (with or without OCD) and "pure" OCD. A trans-diagnostic comparison of task-related fMRI using a Stop-Signal task showed higher error-related supplementary motor area activity in TS compared to OCD and healthy control subjects, with error-related left cerebellar activity correlating positively with tic severity (Fan, Cath, et al., 2018). At a specific symptom level, symmetry and ordering behaviors related to "just-right" phenomena were, on the other hand, associated with the over-activation of a limbic network involving anterior cingulate, supplementary motor and inferior frontal cortices (de Vries, van den Heuvel, Cath, et al., 2013). Other areas of overlap between TS + OCD and OCD without tics can be observed at a large-scale brain network level, consisting of abnormal functional and structural connectivity within associative and limbic cortico-basal ganglia loops. Applying small-world theory to resting state fMRI and diffusion tensor imaging, Worbe et al. (2012); Worbe, Lehericy, & Hartmann (2015); Worbe, Marrakchi-Kacem, et al. (2015) observed a correlation between OCD severity in TS and the strength of connections of the medial orbito-frontal and dorsolateral-prefrontal cortices with other cortical areas and with the basal ganglia. OCD patients who do not exhibit tics display similar structural (Chamberlain, Menzies, Hampshire, et al., 2008) and functional connectivity (Sakai, Narumoto, Nishida, et al., 2011) features, linked to lack of a cognitive and attention flexibility (Menzies et al., 2008). OCD patients without tics also display reduced connectivity between putamen and dorsolateral prefrontal cortex, which is associated with greater obsessive-compulsive symptom severity (Vaghi, Hampshire, Fineberg, et al., 2017). Patients with TS + OCD manifest a relative enlargement of several cerebellar regions, including crus and lobules VI, VIIB, and VIIIA, in proportion to the increasing severity of obsessive-compulsive symptoms, and to the decreasing severity of tics (Tobe, Bansal, Xu, et al., 2010). These findings led to the hypothesis that both "tic-related" and "pure" OCD rely on partially overlapping neural circuits and share underlying cognitive mechanisms.

More recent evidence, on the other hand, has identified some important commonalities between TS + OCD and TS without OCD. Exploring resting-state functional connectivity, Tikoo et al. (2020) applied independent component analysis to show that TS + OCD does not differ in functional connectivity of six major resting state networks from "pure" TS, but shows lower functional connectivity in the cerebellar and fronto-parietal networks compared to "pure" OCD, with the functional connectivity of these two networks correlating positively with obsessive-compulsive symptom severity in "pure" OCD and negatively with tic severity in TS (with or without OCD). In a follow-up multimodal study (Tikoo et al., 2021), the same group of authors identified that cerebellar-prefrontal connectivity (highly involved in executive functioning) correlates negatively with tic severity in youth with TS (with or without co-morbid OCD), and positively with obsessive-compulsive symptom severity in youth with "pure" OCD. These findings suggested that changes in fronto-parietal and cerebellar connectivity could be exploited as a neural marker to distinguish TS from OCD. Moreover, Tikoo et al. (2021) showed decreased functional connectivity between dentate nucleus and precentral gyri or cerebellar crus II in "pure" TS compared to TS + OCD. A closer neurobiological resemblance of TS + OCS/OCD to "pure" TS than to "pure" OCD is supported also by the report of a negative correlation between OCS severity in adults with TS and connectivity within the "tic-associated" sensorimotor loop of the cortico-striato-thalamo-cortical circuitry using seed-based fMRI (Bhikram, Arnold, et al., 2020), as well as by the lack of significant difference in the activation level of caudate and orbitofrontal cortex (associated with the associative and limbic loops of the CSTC circuitry) during obsessive-compulsive symptoms provocation tasks in the same group of adults with TS + OCD (Bhikram, Crawley, et al., 2020). This closer neurobiological resemblance seems reflected by the phenomenology of obsessive-compulsive symptoms in TS, which consists typically of symmetry and counting-related compulsions that may be associated with achieving a visceral sense of completeness or the relief of unpleasant sensory experiences, rather than dampening anxiety by averting a feared consequence as in "pure" OCD. Overall, the available neuroimaging findings favor the TS + OCD comorbidity as an intermediate condition between the "pure" versions of these two disorders. At the same time, there is initial evidence that the specific obsessive-compulsive phenomenology may influence whether neuroanatomical changes are closer to the tic disorder as opposed to the obsessive-compulsive one.

6.3 Intermittent-explosive outbursts in the context of TS

Intermittent-explosive outbursts (IEO), often referred to as "rage attacks", are abrupt episodes of verbal or physical aggression, not preceded by any warning sign and disproportionate to the context. Most patients with TS who exhibit IEO have a comorbid diagnosis of ADHD (Budman, Bruun, Park, Lesser, & Olson, 2000; Chen, Budman, Diego Herrera, et al., 2013). Atkinson-Clement, Sofia, et al. (2020) investigated 55 TS patients to explore the structural and functional correlates of IEO in the absence of comorbid ADHD. These authors reported two main findings associated with the presence of IEO. The first was a decreased connectivity within the sensorimotor loop of the cortico-basal ganglia circuitry, particularly involving the SMA, known to be associated with action selection and top-down behavioral control, and its connections to the putamen, subthalamic nucleus, as well as temporal and dorsal cingulate cortices, the latter linked to anger control and threat appraisal. Their second main finding was structural and lower functional connectivity within limbic circuits, particularly between the orbitofrontal cortex, the amygdala, the hippocampus, and the cerebellum, suggestive of inefficient processing of socio-emotional information, regulation of emotions, and control/decision-making related to the initiation and suppression of aggressive impulses. In addition, they detected increased functional connectivity between the orbitofrontal cortex and the temporal pole, possibly suggestive of higher sensitivity to social cues in TS patients with IEO.

7. Neural correlates of cognitive control in TS: Insights from social cognition, habit formation and cognitive control of motor behavior

7.1 Neural correlates of social cognition in TS

Social cognition is the ability to understand mental states (e.g., beliefs and emotions of others) in order to successfully interact within society, and comprises the domains of social perception abilities (e.g., facial recognition), Theory of Mind, empathy, and social behavior (including imitation abilities). About 25%–30% of TS patients experience urges to perform socially inappropriate behaviors (obscene or non-obscene; Eddy & Cavanna, 2013). Brain networks implicated in social decision-making has been proposed as relevant to TS pathophysiology (Albin, 2018; Vicario & Martino, 2018). On fMRI, adults with TS exhibit overactivation of the orbitofrontal cortex, posterior cingulate, right amygdala, and right temporo-parietal junction, as well as decreased activity in the left inferior parietal cortex, while performing a social perception task involving the recognition of emotionally salient expressions of human eyes (Eddy, Cavanna, & Hansen, 2017). The same group reported lower activation of the temporo-parietal junction, right amygdala, and posterior cingulate cortex while performing a Theory of Mind task comprising reasoning about incorrect beliefs and outdated physical representations (Eddy, Cavanna, Rickards, & Hansen, 2017). Moreover, urge severity positively correlated with amygdala activation, and non-obscene socially inappropriate behaviors positively correlated with activity of the left temporo-parietal junction.

The role of the insula has been elegantly demonstrated in relation also to the impact of social cues on the generation of PU and tics. Rae et al. (2018) recorded the functional activity of the insula in TS patients and healthy volunteers while processing facial expression stimuli. First, they observed that TS patients manifested additional responses, compared to healthy, to neutral and angry faces within the anterior insula. Moreover, tic severity directly correlated with the functional connectivity of the insula with the globus pallidus and thalamus, while insula connectivity correlated positively with the supplementary motor area in proportion to the intensity of premonitory sensations. Based on their findings, these authors suggested that an increased representation within the insula of embodied affective social information probably influences the reactivity of subcortical motor pathways, supporting motor programs involved in the genesis of tics. As a corollary to their findings, the insula can be considered a promising potential target of neuromodulation in the treatment of tics.

7.2 Neural correlates of habitual control in TS

Tics may be perceived as intentional actions performed in a habitual way in response to sensory stimuli (Leckman & Riddle, 2000). This has led to explore neuronal networks related to goal-directed and habitual systems in TS. TS patients tend to rely on habitual control at expense of goal-directed behavioral control, which is predicted by the structural connectivity of the white matter within the premotor cortex and putamen (Delorme, Salvador, Valabregue, et al., 2016). Habit reversal training in the context of behavioral intervention for tics is characterized by a process that enhances awareness of premonitory urges and involves the selection, initiation and learning of competing motor responses that counteract the execution of tics. Deckersbach, Chou, Britton, et al. (2014) reported a significant decrease in the activity of the putamen comparing pre-treatment and post-treatment time points in TS patients undergoing this intervention, thus hinting to a potential neuroanatomical correlate of the mechanism of habit reversal training in this clinical population.

7.3 Neural correlates of cognitive control of motor behavior and sense of agency in TS

"Self-regulation" refers to the ability to suppress competing attentional and behavioral responses. The development of this ability continues throughout childhood and is facilitated by the maturation of a component of the frontal cortex and the basal ganglia that contribute to it. Several models of cognitive control suggest that different regions of the frontal cortex are engaged in programming and executing context-appropriate motor behaviors; the basal ganglia are involved in inhibiting competing, undesired motor behaviors (Casey, Tottenham, & Fossella, 2002). TS patients may exhibit dysfunctional cognitive control of motor behaviors associated with abnormal maturation of these fronto-striatal circuits.

The earliest studies investigating cognitive control applied a version of the Stroop task, which requires subjects to inhibit an automatized response (e.g., reading) to activate another response (e.g., color naming). In healthy subjects, performance on this task improves from childhood to adulthood, in associated with an increased activation of fronto-striatal networks, and greater deactivation of the ventral prefrontal and posterior cingulate cortical regions (Marsh, Maia, & Peterson, 2009). When TS patients perform this version of the Stroop task, they exhibit reduced deactivation of the ventral prefrontal and posterior cingulate cortical regions with the advancing age, as well compensatory increased activation of the dorsolateral prefrontal cortex, striatum, globus pallidus, and thalamus (Marsh, Zhu, Wang, Skudlarski, & Peterson, 2007). Children with TS without comorbidities exhibited an association between slower performance on this task and greater tic severity. Brain imaging demonstrated an association between greater tic severity and greater activity in the substantia nigra/ventral tegmental area, striatum, thalami, and subthalamic regions (Baym, Corbett, Wright, & Bunge, 2008), and hyperactivation of the left prefrontal cortex compared to healthy volunteers during task performance. Thomalla, Jonas, Baumer, et al. (2014) used the Go/NoGo reaction time task to assess action inhibition, demonstrating slower performance and lower brain activitation in primary and secondary motor cortical regions exclusively during the Go trials in TS patients.

Also using a Go/NoGo task and fMRI, Rae et al. (2020) compared 23 adults with TS and 21 controls, in a paradigm that included free-choice trials in which participants could choose a Go or NoGo response. Imaging data suggested the neural processes for choosing to act or to withhold action are anatomically similar between TS and controls. When choosing Go, patients with greater PU showed increased functional connectivity between pre-supplementary motor area and thalamus, globus pallidus, and caudate. Compared to controls, those with TS showed increased activity in primary motor cortex even when no actions were made. Rae et al. (2020) argued that TS is associated with elevated primary motor cortex reactivity and permissive, dysfunctional connectivity between motor planning and motor output areas, including basal ganglia, allowing emergence of tics. The authors further argued that the pathways underlying voluntary movement are affected by hyperconnectivity in those with TS and more severe PU. These findings may also be in line with the idea that tic generation arises from increased sensorimotor noise, leading to abnormal release of motor signals, appearing as tics (Ganos et al., 2018; Misirlisoy, Brandt, Ganos, et al., 2015; Rawji, Modi, Latorre, et al., 2020).

Models of motor control based on neurocognitive research state that accurate action attribution is crucial to the experience of agency, i.e., the ability to attribute the responsibility for our actions and their consequences to ourselves. There is increasing evidence that TS patients exhibit abnormalities of the sense of agency. In a behavioral and fMRI study, Zapparoli et al. (2020) assessed the phenomenon of intentional binding, consisting in the perceived temporal compression between volitional behavior and its external consequences. This phenomenon is considered an implicit index of the sense of agency. In their TS patient sample, these Authors showed that tic severity increased with the reduction of individual intentional binding. Interestingly, whereas healthy volunteers exhibited an association between intentional binding and the activity of a premotor-parietal-cerebellar network, TS patients did not exhibit this association, which was interpreted as an altered activation of the brain network associated with sense of agency. A possible interpretation of this finding is that TS patients are less accurate in discriminating the external consequences produced by their own voluntary behaviors from those produced by other people's behaviors.

To summarize these findings, TS patients exhibit greater prefrontal cortical activity and greater activity of cortico-basal ganglia networks during tasks involving cognitive control systems, which may be interpreted as compensatory activation changes for the relative inability to inhibit unwanted motor actions (tics and socially inappropriate behaviors). Moreover, TS patients exhibit a dissociation between intentional binding and premotor-parietal-cerebellar connectivity, which may represent a neuroimaging correlate of an alteration of the sense of agency in these patients.

8. Final commentary and areas for future research

The functional neuroanatomy of the urge-tic complex has been further clarified in the past 5 years, thanks to new single-modality functional imaging studies exploring brain activation or oscillatory activity in conjunction to tasks involving motor inhibition and motor preparation, or combined functional imaging methodologies, e.g., rs-fMRI and EEG. These studies have remarked and elucidated the importance to urge generation/control of alterations in the connectivity between *cortical regions controlling interoceptive processing* (both anterior and posterior regions of the insula) and *cortico-subcortical networks modulating motor planning and output*, linking supplementary motor and primary sensorimotor cortical areas to the striatum. At the same time, *connections regulating attentional control and stimulus salience* (between orbitofrontal and anterior cingulate cortices) and the *integration between cognitive and emotional experiences* (including, among other areas, the mid-cingulate cortex, precuneus and temporal gyri) have been shown to be associated with the processing of urges.

Magneto- and electroencephalographic studies have provided important insight in the oscillatory activities linked to the time periods preceding or corresponding to the onset of tics. *Fast oscillations in the dorsolateral prefrontal cortex and supplementary motor area* preceding the onset of tics, and therefore potentially time-locked to the occurrence of urges, as well as *beta deactivation in the posterior cingulate cortex, supplementary motor, and sensorimotor cortical regions* around onset of tics, have been revealed as interesting, non-invasively detectable physiomarkers of the urge-tic complex. More recent neuro-computational models highlighted a potential role of *cerebellum-basal ganglia connectivity* in the generation of tics, possibly linked to maturational abnormalities involving GABAergic neurotransmission. This concept is, to date, mostly supported by toxic hyperexcitability primate models of tic-like movements and needs to be further substantiated by studies in humans.

Recent structural and functional imaging studies have provided additional support to the association between alterations in the maturation trajectory of different neural networks and diagnosis of TS, as well as specific phenomenological and natural history features of this condition. The *increased connectivity of shorter range connections*, including those forming the *sensorimotor loop of the cortico-basal ganglia circuitry*, has been confirmed by newer reports and meta-analyses, alongside the potential association between persistence of tics over time and volume and thickness of prefrontal cortical areas. Of all phenomenological domains of tics, complexity seems the most strongly associated with increased connectivity along the cortico-basal ganglia sensorimotor loop. Global tic severity, on the other hand, was found to correlate with connectivity along several other networks and brain regions, including temporal gyri, networks associated with executive control and encompassing right superior frontal gyrus and right operculum, as well as bi-hemispheric connectivity of the anterior cingulate cortices through the corpus callosum. Immaturity of brain development in TS could be observed at the level of different resting state networks, particularly in the form of a *disconnection of the fronto-parietal network (long-range connection)*.

The comorbidity profile of TS has also been associated with differences in functional brain connectivity. A comorbid diagnosis of ADHD was associated with *decreased connectivity along fronto-striatal circuits*, whereas a behavioral pattern of increased reactive motor impulsivity correlated with *increased connectivity between the inferior frontal and superior temporal gyri*, and *decreased connectivity between cerebellum and basal ganglia* (subthalamic nucleus, putamen). OCD comorbidity, on the other hand, was confirmed to be associated with *decreased connectivity in the limbic loop of the cortico-basal ganglia circuitry*, linking the ventral striatum to the ventromedial prefrontal cortex and hippocampus. The TS + OCD comorbidity is reflected by a condition of functional connectivity abnormalities that is intermediate between TS and OCD. Interestingly, the existence of intermittent explosive outbursts in TS patients was found to be associated with changes combining alterations observed in ADHD (decreased connectivity between supplementary motor area and basal ganglia) and in OCD (decreased connectivity within limbic circuits involving the orbitofrontal cortex, amygdala, hippocampus, and cerebellum).

Finally, TS patients exhibit activation changes associated with specific cognitive performances, e.g. cognitive control of motor output, that appear to be compensatory in nature, e.g. *enhanced functional connectivity between prefrontal cortical regions and basal ganglia*. Similarly, recent advances have suggested a functional anatomical correlate of changes of sense of agency in TS, particularly involving the *connectivity between premotor and parietal regions and cerebellum*.

Despite recent advances in the elucidation of the functional neuroanatomy of tics and tic disorders, several crucial areas remain underinvestigated. In particular, a closer understanding of age-related developmental changes in neurobiology in TS compared to typically developing individuals is needed. Moreover, it remains undefined whether early developmental course patterns explored by neuroimaging might help predicting the persistence of tics beyond adolescence. The sexually dimorphic course of severity of tic disorders (Garcia-Delgar, Servera, Coffey, et al., 2021; Girgis, Martino, & Pringsheim, 2021) also needs to be explored with respect to developmental trajectories and related neuroanatomical changes. Finally, the role of medications and behavioral therapies in the shaping of these developmental trajectories requires more investigation. Large, multisite, collaborative prospective studies represent the design approach that would be best equipped to provide answers to these critical questions.

References

Abi-Jaoude, E., Segura, B., Obeso, I., et al. (2015). Similar striatal D2/D3 dopamine receptor availability in adults with Tourette syndrome compared with healthy controls: A [^{11}C]-(+)-PHNO and [^{11}C]raclopride positron emission tomography imaging study. *Human Brain Mapping, 36*, 2592–2601.

Akkermans, S. E. A., van Rooij, D., Naaijen, J., et al. (2019). Neural reward processing in paediatric Tourette syndrome and attention-deficit/hyperactivity disorder. *Psychiatry Research: Neuroimaging, 292*, 13–22.

Albin, R. L. (2018). Tourette syndrome: A disorder of the social decision-making network. *Brain, 141*, 332–347.

Atkinson-Clement, C., Porte, C. A., de Liege, A., et al. (2020). Neural correlates and role of medication in reactive motor impulsivity in Tourette disorder. *Cortex, 125*, 60–72.

Atkinson-Clement, C., Sofia, F., Fernandez-Egea, E., et al. (2020). Structural and functional abnormalities within sensori-motor and limbic networks underpin intermittent explosive symptoms in Tourette disorder. *Journal of Psychiatric Research*, *125*, 1–6.

Bansal, R., Staib, L. H., Laine, A. F., et al. (2012). Anatomical brain images alone can accurately diagnose chronic neuropsychiatric illnesses. *Proceedings of the Library of Sciences One*, *7*, e50698.

Baym, C. L., Corbett, B. A., Wright, S. B., & Bunge, S. A. (2008). Neural correlates of tic severity and cogntiive control in children with Tourette syndrome. *Brain*, *131*, 165–179.

Benarroch, E. E. (2019). Insular cortex: Functional complexity and clinical correlations. *Neurology*, *93*, 932–938.

Bhikram, T., Arnold, P., Crawley, A., Abi-Jaoude, E., & Sandor, P. (2020). The functional connectivity profile of tics and obsessive-compulsive symptoms in Tourette syndrome. *Journal of Psychiatric Research*, *123*, 128–135.

Bhikram, T., Crawley, A., Arnold, P., Abi-Jaoude, E., & Sandor, P. (2020). Examining the functional activity of different obsessive-compulsive symptom dimensions in Tourette syndrome. *Neuroimage Clinics*, *26*, 102198.

Bhikram, T., Crawley, A., Arnold, P., Abi-Jaoude, E., & Sandor, P. (2021). Neuroimaging the emotional modulation of urge inhibition in Tourette syndrome. *Cortex*, *135*, 341–351.

Bliss, J. (1980). Sensory experiences of Gilles de la Tourette syndrome. *Archives of General Psychiatry*, *37*, 1343–1347.

Bloch, M. H., Leckman, J. F., Zhu, H., & Peterson, B. S. (2005). Caudate volumes in childhood predict symptom severity in adults with Tourette syndrome. *Neurology*, *65*, 1253–1258.

Bohlhalter, S., Goldfine, A., Matteson, S., et al. (2006). Neural correlates of tic generation in Tourette syndrome: An event-related functional MRI study. *Brain*, *129*, 2029–2037.

Brandt, V. C., Beck, C., Sajin, V., et al. (2016). Temporal relationship between premonitory urges and tics in Gilles de la Tourette syndrome. *Cortex*, *77*, 24–37.

Bronfeld, M., & Bar-Gad, I. (2013). Tic disorders: What happens in the basal ganglia? *Neuroscientist*, *19*(1), 101–108. https://doi.org/10.1177/1073858412444466. Epub 2012 May 16. PMID: 22596263.

Bronfeld, M., Yael, D., Belelovsky, K., & Bar-Gad, I. (2013). Motor tics evoked by striatal disinhibition in the rat. *Frontiers in Systems Neuroscience*, 7, 50. https://doi.org/10.3389/fnsys.2013.00050. PMID: 24065893. PMCID: PMC3776161.

Budman, C. L., Bruun, R. D., Park, K. S., Lesser, M., & Olson, M. (2000). Explosive outbursts in children with Tourette's disorder. *Journal of the American Academy of Child and Adolescent Psychiatry*, *39*, 1270–1276.

Caligiore, D., Mannella, F., Arbib, M. A., & Baldassarre, G. (2017). Dysfunctions of the basal ganglia-cerebellar-thalamo-cortical system produce motor tics in Tourette syndrome. *Proceedings of the Library of Ssciences Computational Biology*, *13*, e1005395.

Casey, B. J., Tottenham, N., & Fossella, J. (2002). Clinical, imaging, lesion, and genetic approaches toward a model of cognitive control. *Developmental Psychobiology*, *40*, 237–254.

Chamberlain, S. R., Menzies, L., Hampshire, A., et al. (2008). Orbitofrontal dysfunction in patients with obsessive-compulsive disorder and their unaffected relatives. *Science*, *321*, 421–422.

Chen, K., Budman, C. L., Diego Herrera, L., et al. (2013). Prevalence and clinical correlates of explosive outbursts in Tourette syndrome. *Psychiatry Research*, *205*, 269–275.

Church, J. A., Fair, D. A., Dosenbach, N. U., et al. (2009). Control networks in paediatric Tourette syndrome show immature and anomalous patterns of functional connectivity. *Brain*, *132*, 225–238.

Cui, Y., Jin, Z., Chen, X., He, Y., Liang, X., & Zheng, Y. (2014). Abnormal baseline brain activity in drug-naïve patients with Tourette syndrome: A resting-state fMRI study. *Frontiers in Human Neuroscience*, *7*, 913.

de Vries, F. E., van den Heuvel, O. A., Cath, D. C., et al. (2013). Limbic and motor circuits involved in symmetry behavior in Tourette's syndrome. *CNS Spectrums*, *18*, 34–42.

Debes, N. M., Jeppesen, S., Raghava, J. M., Groth, C., Rostrup, E., & Skov, L. (2015). Longitudinal magnetic resonance imaging (MRI) analysis of the developmental changes of Tourette syndrome reveal reduced diffusion in the cortico-striato-thalamo-cortical pathways. *Journal of Child Neurology*, *30*, 1315–1326.

Deckersbach, T., Chou, T., Britton, J. C., et al. (2014). Neural correlates of behavior therapy for Tourette's disorder. *Psychiatry Research*, *224*, 269–274.

Delorme, C., Salvador, A., Valabregue, R., et al. (2016). Enhanced habit formation in Gilles de la Tourette syndrome. *Brain*, *139*, 605–615.

Denys, D., de Vries, F. E., Cath, D., et al. (2013). Dopaminergic activity in Tourette syndrome and obsessive-compulsive disorder. *European Neuropsychopharmacology*, *23*, 1423–1431.

Draganski, B., Martino, D., Cavanna, A. E., et al. (2010). Multispectral brain morphometry in Tourette syndrome persisting into adulthood. *Brain*, *133*, 3661–3675.

Draper, A., Jackson, G. M., Morgan, P. S., & Jackson, S. R. (2016). Premonitory urges are associated with decreased grey matter thickness within the insula and sensorimotor cortex in young people with Tourette syndrome. *Journal of Neuropsychology*, *10*, 143–153.

Eddy, C. M., & Cavanna, A. E. (2013). Altered social cognition in Tourette syndrome: Nature and implications. *Behavioural Neurology*, *27*, 15–22.

Eddy, C. M., Cavanna, A. E., & Hansen, P. C. (2017). Empathy and aversion: The neural signature of mentalizing in Tourette syndrome. *Psychological Medicine*, *47*, 507–517.

Eddy, C. M., Cavanna, A. E., Rickards, H. E., & Hansen, P. C. (2017). Temporo-parietal dysfunction in Tourette syndrome: Insights from an fMRI study of theory of mind. *Journal of Psychiatric Research*, *81*, 102–111.

Fahim, C., Yoon, U., Das, S., et al. (2010). Somatosensory-motor bodily representation cortical thinning in Tourette: Effects of tic severity, age and gender. *Cortex*, *46*, 750–760.

Fan, S., Cath, D. C., van den Heuvel, O. A., et al. (2017). Abnormalities in metabolite concentrations in Tourette's disorder and obsessive-compulsive disorder—A proton magnetic resonance spectroscopy study. *Psychoneuroendocrinology*, *77*, 211–217.

Fan, S., Cath, D. C., van der Werf, Y. D., et al. (2018). Trans-diagnostic comparison of response inhibition in Tourette's disorder and obsessive-compulsive disorder. *World Journal of Biological Psychiatry*, *19*, 527–537.

Fan, S., van den Heuvel, O. A., Cath, D. C., et al. (2018). Altered functional connectivity in resting state networks in Tourette's disorder. *Frontiers in Human Neuroscience*, *12*, 363.

Forde, N. J., Zwiers, M. P., Naaijen, J., et al. (2017). Basal ganglia structure in Tourette's disorder and/or attention-deficit/hyperactivity disorder. *Movement Disorders*, *32*, 601–604.

Ganos, C., & Martino, D. (2015). Tics and Tourette syndrome. *Neurologic Clinics*, *33*, 115–136.

Ganos, C., Roessner, V., & Münchau, A. (2013). The functional anatomy of Gilles de la Tourette syndrome. *Neuroscience and Biobehavioral Reviews*, *37*, 1050–1062.

Ganos, C., Rothwell, J., & Haggard, P. (2018). Voluntary inhibitory motor control over involuntary tic movements. *Movement Disorders*, *33*, 937–946.

Garcia-Delgar, B., Servera, M., Coffey, B. J., et al. (2021). Tic disorders in children and adolescents: Does the clinical presentation differ in males and females? A report by the EMTICS group. *European Child & Adolescent Psychiatry*. https://doi.org/10.1007/s00787-021-01751-4. Epub ahead of print. PMID: 33944988.

Girgis, J., Martino, D., & Pringsheim, T. (2021). Influence of sex on tic severity and psychiatric comorbidity profile in patients with pediatric tic disorder. *Developmental Medicine and Child Neurology*. https://doi.org/10.1111/dmcn.15088 (Epub ahead of print).

Godar, S. C., Mosher, L. J., Giovanni, G. D., & Bortolato, M. (2014). Animal models of tic disorders: A translational perspective. *Journal of Neuroscience Methods*, *236*, 54–69.

Greene, D. J., Church, J. A., Dosenbach, N. U., et al. (2016). Multivariate pattern classification of pediatric Tourette syndrome using functional connectivity MRI. *Developmental Science*, *2016*(19), 581–598.

Greene, D. J., Williams Iii, A. C., Koller, J. M., Schlaggar, B. L., & Black, K. J. (2017). Brain structure in pediatric Tourette syndrome. *Molecular Psychiatry*, *22*, 972–980.

Hampson, M., Tokoglu, F., King, R. A., Constable, R. T., & Leckman, J. F. (2009). Brain areas coactivating with motor cortex during chronic motor tics and intentional movements. *Biological Psychiatry*, *65*, 594–599.

Hashemiyoon, R., Kuhn, J., & Visser-Vandewalle, V. (2017). Putting the pieces together in Gilles de la Tourette syndrome: Exploring the link between clinical observations and the biological basis of dysfunction. *Brain Topography*, *30*, 3–29.

Hong, J. J., Loiselle, C. R., Yoon, D. Y., et al. (2004). Microarray analysis in Tourette syndrome postmortem putamen. *Journal of the Neurological Sciences*, *225*, 57–64.

Jackson, S. R., Parkinson, A., Kim, S. Y., Schuermann, M., & Eickhoff, S. B. (2011). On the functional anatomy of the urge-for-action. *Cognitive Neuroscience*, *2*, 227–257.

Jackson, G. M., Draper, A., Dyke, K., Pépés, S. E., & Jackson, S. R. (2015). Inhibition, disinhibition, and the control of action in Tourette syndrome. *Trends in Cognitive Sciences*, *19*(11), 655–665. https://doi.org/10.1016/j.tics.2015.08.006. Epub 2015 Oct 1. PMID: 26440120.

Jackson, S. R., Loayza, J., Crighton, M., Sigurdsson, H. P., Dyke, K., & Jackson, G. M. (2020). The role of the insula in the generation of motor tics and the experience of the premonitory urge-to-tic in Tourette syndrome. *Cortex*, *126*, 119–133.

Jung, W. H., Kang, D. H., Gu, B. M., et al. (2011). Aberrant ventral striatal responses during incentive processing in unmedicated patients with obsessive-compulsive disorder. *Acta Psychiatrica Scandinavica*, *123*, 376–386.

Kalanithi, P. S., Zheng, W., Kataoka, Y., et al. (2005). Altered parvalbumin-positive neuron distribution in basal ganglia of individuals with Tourette syndrome. *Proceedings of the National Academy of Sciences of the USA*, *102*, 13307–13312.

Kanaan, A. S., Gerasch, S., Garcia-Garcia, I., et al. (2017). Pathological glutamatergic neurotransmission in Gilles de la Tourette syndrome. *Brain*, *140*, 218–234.

Kaufmann, C., Beucke, J. C., Preuβe, F., et al. (2013). Medial prefrontal brain activation to anticipated reward and loss in obsessive-compulsive disorder. *Neuroimage Clinics*, *2*, 212–220.

Kong, L., Lv, B., Wu, T., et al. (2020). Altered structural cerebral cortex in children with Tourette syndrome. *European Journal of Radiology*, *129*, 109119.

Kurvits, L., Martino, D., & Ganos, C. (2020). Clinical features that evoke the concept of disinhibition in Tourette syndrome. *Frontiers in Psychiatry*, *11*, 21.

Kwak, C., Dat Vuong, K., & Jankovic, J. (2003). Premonitory sensory phenomenon in Tourette's syndrome. *Movement Disorders*, *18*, 1530–1533.

Leckman, J. F., & Riddle, M. A. (2000). Tourette's syndrome: When habit-forming systems form habits of their own? *Neuron*, *28*, 349–354.

Leckman, J. F., Walker, D. R., & Cohen, D. J. (1993). Premonitory urges in Tourette's syndrome. *American Journal of Psychiatry*, *150*, 98–102.

Lerner, A., Bagic, A., Simmons, J. M., et al. (2012). Widespread abnormality of the gamma-aminobutyric acid-ergic system in Tourette syndrome. *Brain*, *135*, 1926–1936.

Liao, W., Yu, Y., Miao, H. H., Feng, Y. X., Ji, G. J., & Feng, J. H. (2017). Inter-hemispheric intrinsic connectivity as a neuromarker for the diagnosis of boys with Tourette syndrome. *Molecular Neurobiology*, *54*, 2781–2789.

Liu, Y., Miao, W., Wang, J., Gao, P., Yin, G., Zhang, L., et al. (2013). Structural abnormalities in early Tourette syndrome children: A combined voxel-based morphometry and tract-based spatial statistics study. *PLoS One*, *8*(9), e76105. https://doi.org/10.1371/journal.pone.0076105. PMID: 24098769. PMCID: PMC3786886.

Liu, Y., Wang, J., Zhang, J., et al. (2017). Altered spontaneous brain activity in children with early Tourette syndrome: A resting-state fMRI study. *Scientific Reports*, *7*, 4808.

Lou, Y. T., Li, X. L., Wang, Y., et al. (2020). Frequency-specific regional homogeneity alterations in Tourette syndrome. *Frontiers in Psychiatry*, *11*, 543049.

Ludolph, A. G., Pinkhardt, E. H., Tebartz van Elst, L., et al. (2008). Are amygdalar volume alterations in children with Tourette syndrome due to ADHD comorbidity? *Developmental Medicine and Child Neurology*, *50*, 524–529.

Mahone, E. M., Puts, N. A., Edden, R. A. E., Ryan, M., & Singer, H. S. (2018). GABA and glutamate in children with Tourette syndrome: A (1)H MR spectroscopy study at 7T. *Psychiatry Research: Neuroimaging*, *273*, 46–53.

Marsh, R., Zhu, H., Wang, Z., Skudlarski, P., & Peterson, B. S. (2007). A developmental fMRI study of self-regulatory control in Tourette's syndrome. *American Journal of Psychiatry*, *164*, 955–966.

Marsh, R., Maia, T. V., & Peterson, B. S. (2009). Functional disturbances within frontostriatal circuits across multiple childhood psychopathologies. *American Journal of Psychiatry*, *166*, 664–674.

Martino, D., Ganos, C., & Pringsheim, T. M. (2017). Tourette syndrome and chronic tic disorders: The clinical spectrum beyond tics. In *vol. 134. International Review of Neurobiology* (pp. 1461–1490). San Diego: Elsevier Science Publishing Co Inc.

McCairn, K. W., & Isoda, M. (2013). Pharmacological animal models of tic disorders. In *vol. 112. International review of neurobiology* (pp. 178–209). San Diego: Elsevier Science Publishing Co Inc.

McGuire, J. F., McBride, N., Piacentini, J., et al. (2016). The premonitory urge revisited: An individualized premonitory urge for tics scale. *Journal of Psychiatric Research*, *83*, 176–183.

Menzies, L., Chamberlain, S. R., Laird, A. R., Thelen, S. M., Sahakian, B. J., & Bullmore, E. T. (2008). Integrating evidence from neuroimaging and neuropsychological studies of obsessive-compulsive disorder: The orbitofronto-striatal model revisited. *Neuroscience and Biobehavioral Reviews*, *32*, 525–549.

Minzer, K., Lee, O., Hong, J. J., & Singer, H. S. (2004). Increased prefrontal D2 protein in Tourette syndrome: A postmortem analysis of frontal cortex and striatum. *Journal of the Neurological Sciences*, *219*, 55–61.

Misirlisoy, E., Brandt, V., Ganos, C., et al. (2015). The relation between attention and tic generation in Tourette syndrome. *Neuropsychology*, *29*, 658–665.

Muellner, J., Delmaire, C., Valabregue, R., et al. (2015). Altered structure of cortical sulci in gilles de la Tourette syndrome: Further support for abnormal brain development. *Movement Disorders*, *30*, 655–661.

Müller-Vahl, K. R., Kaufmann, J., Grosskreutz, J., Dengler, R., Emrich, H. M., & Peschel, T. (2009). Prefrontal and anterior cingulate cortex abnormalities in Tourette syndrome: Evidence from voxel-based morphometry and magnetization transfer imaging. *BMC Neuroscience*, *10*, 47.

Naaijen, J., Forde, N. J., Lythgoe, D. J., Akkermans, S. E., Openneer, T. J., Dietrich, A., Zwiers, M. P., et al. (2016). Fronto-striatal glutamate in children with Tourette's disorder and attention-deficit/hyperactivity disorder. *NeuroImage: Clinical*, *13*, 16–23. https://doi.org/10.1016/j.nicl.2016.11.013. PMID: 27909683. PMCID: PMC5124361.

Naro, A., Billeri, L., Colucci, V. P., et al. (2020). Brain functional connectivity in chronic tic disorders and Gilles de la Tourette syndrome. *Progress in Neurobiology*, *194*, 101884.

Neuner, I., Schneider, F., & Shah, N. J. (2013). Functional neuroanatomy of tics. In *vol. 112*. *International review of neurobiology* (pp. 35–71). San Diego: Elsevier Science Publishing Co Inc.

Neuner, I., Werner, C. J., Arrubla, J., et al. (2014). Imaging the where and when of tic generation and resting state networks in adult Tourette patients. *Frontiers in Human Neuroscience*, *8*, 362.

Niccolai, V., Korczok, S., Finis, J., et al. (2019). A peek into premonitory urges in Tourette syndrome: Temporal evolution of neurophysiological oscillatory signatures. *Parkinsonism & Related Disorders*, *65*, 153–158.

Nielsen, A. N., Gratton, C., Church, J. A., et al. (2020). Atypical functional connectivity in Tourette syndrome differs between children and adults. *Biological Psychiatry*, *87*, 164–173.

O'Neill, J., Piacentini, J. C., & Peterson, B. S. (2019). Cingulate role in Tourette syndrome. *Handbook of Clinical Neurology*, *166*, 165–221.

Openneer, T. J. C., Marsman, J. C., van der Meer, D., et al. (2020). A graph theory study of resting-state functional connectivity in children with Tourette syndrome. *Cortex*, *126*, 63–72.

Openneer, T. J. C., van der Meer, D., Marsman, J. C., et al. (2021). Impaired response inhibition during a stop-signal task in children with Tourette syndrome is related to ADHD symptoms: A functional magnetic resonance imaging study. *World Journal of Biological Psychiatry*, *22*, 350–361.

Patel, N., Jankovic, J., & Hallett, M. (2014). Sensory aspects of movement disorders. *Lancet Neurology*, *13*, 100–112.

Peterson, B. S., Choi, H. A., Hao, X., et al. (2007). Morphologic features of the amygdala and hippocampus in children and adults with Tourette syndrome. *Archives of General Psychiatry*, *64*, 1281–1291.

Plessen, K. J., Royal, J. M., & Peterson, B. S. (2007). Neuroimaging of tic disorders with co-existing attention-deficit/hyperactivity disorder. *European Child and Adolescent Psychiatry*, *16*(Suppl 1), 60–70.

Pourfar, M., Feigin, A., Tang, C. C., et al. (2011). Abnormal metabolic brain networks in Tourette syndrome. *Neurology*, *76*, 944–952.

Puts, N. A., Harris, A. D., Crocetti, D., et al. (2015). Reduced GABAergic inhibition and abnormal sensory symptoms in children with Tourette syndrome. *Journal of Neurophysiology*, *114*, 808–817.

Rae, C. L., Polyanska, L., Gould van Praag, C. D., et al. (2018). Face perception enhances insula and motor network reactivity in Tourette syndrome. *Brain*, *141*, 3249–3261.

Rae, C. L., Larsson, D. E. O., Garfinkel, S. N., & Critchley, H. D. (2019). Dimensions of interoception predict premonitory urges and tic severity in Tourette syndrome. *Psychiatry Research*, *271*, 469–475.

Rae, C. L., Parkinson, J., Betka, S., et al. (2020). Amplified engagement of prefrontal cortex during control of voluntary action in Tourette syndrome. *Brain Communications*, 2 (eCollection 2020).

Ramkiran, S., Heidemeyer, L., Gaebler, A., Shah, N. J., & Neuner, I. (2019). Alterations in basal ganglia-cerebello-thalamo-cortical connectivity and whole brain functional network topology in Tourette's syndrome. *Neuroimage Clinic*, *24*, 101998.

Rawji, V., Modi, S., Latorre, A., et al. (2020). Impaired automatic but intact volitional inhibition in primary tic disorders. *Brain*, *143*, 906–919.

Sakai, Y., Narumoto, J., Nishida, S., et al. (2011). Corticostriatal functional connectivity in non-medicated patients with obsessive-compulsive disorder. *European Psychiatry*, *26*, 463–469.

Sanger, T. D., Chen, D., Fehlings, D. L., et al. (2010). Definition and classification of hyperkinetic movements in childhood. *Movement Disorders, 25*, 1538–1549.

Sigurdsson, H. P., Pépés, S. E., Jackson, G. M., Draper, A., Morgan, P. S., & Jackson, S. R. (2018). Alterations in the microstructure of white matter in children and adolescents with Tourette syndrome measured using tract-based spatial statistics and probabilistic tractography. *Cortex, 104*, 75–89.

Sigurdsson, H. P., Jackson, S. R., Jolley, L., Mitchell, E., & Jackson, G. M. (2020). Alterations in cerebellar grey matter structure and covariance networks in young people with Tourette syndrome. *Cortex, 126*, 1–15.

Singer, H. S., Hahn, I. H., & Moran, T. H. (1991). Abnormal dopamine uptake sites in postmortem striatum from patients with Tourette's syndrome. *Annals of Neurology, 30*, 558–562.

Soares, J. M., Magalhães, R., Moreira, P. S., et al. (2016). A hitchhiker's guide to functional magnetic resonance imaging. *Frontiers in Neuroscience, 10*, 515.

Sowell, E. R., Kan, E., Yoshii, J., et al. (2008). Thinning of sensorimotor cortices in children with Tourette syndrome. *Nature Neuroscience, 11*, 637–639.

Thomalla, G., Jonas, M., Baumer, T., et al. (2014). Costs of control: Decreased motor cortex engagement during a go/NoGo task in Tourette's syndrome. *Brain, 137*, 122–136.

Tikoo, S., Cardona, F., Tommasin, S., et al. (2020). Resting-state functional connectivity in drug-naive pediatric patients with Tourette syndrome and obsessive-compulsive disorder. *Journal of Psychiatric Research, 129*, 129–140.

Tikoo, S., Suppa, A., Tommasin, S., et al. (2021). The cerebellum in drug-naive children with Tourette syndrome and obsessive-compulsive disorder. *Cerebellum*. https://doi.org/10.1007/s12311-021-01327-7 (Epub ahead of print).

Tinaz, S., Malone, P., Hallett, M., & Horovitz, S. G. (2015). Role of the right dorsal anterior insula in the urge to tic in Tourette syndrome. *Movement Disorders, 30*, 1190–1197.

Tobe, R. H., Bansal, R., Xu, D., et al. (2010). Cerebellar morphology in Tourette syndrome and obsessive-compulsive disorder. *Annals of Neurology, 67*, 479–487.

Vaghi, M. M., Hampshire, A., Fineberg, N. A., et al. (2017). Hypoactivation and dysconnectivity of a frontostriatal circuit during goal-directed planning as an endophenotype for obsessive-compulsive disorder. *Biological Psychiatry Cognitive Neuroscience Neuroimaging, 2*, 655–663.

Vicario, C. M., & Martino, D. (2018). Social communication in Tourette syndrome: A glimpse at the contribution of the insula and the prefrontal cortex. *Brain, 141*, e58.

Wan, X., Zhang, S., Wang, W., et al. (2021). Gray matter abnormalities in Tourette syndrome: A meta-analysis of voxel-based morphometry studies. *Translational Psychiatry, 11*, 287.

Wang, Z., Maia, T. V., Marsh, R., Colibazzi, T., Gerber, A., & Peterson, B. S. (2011). The neural circuits that generate tics in Tourette's syndrome. *American Journal of Psychiatry, 168*, 1326–1337.

Wen, F., Yan, J., Yu, L., et al. (2021). Grey matter abnormalities in Tourette syndrome: An activation likelihood estimation meta-analysis. *BMC Psychiatry, 21*, 184.

Wittfoth, M., Bornmann, S., Peschel, T., et al. (2012). Lateral frontal cortex volume reduction in Tourette syndrome revealed by VBM. *BMC Neuroscience, 13*, 17.

Wolff, N., Luehr, I., Sender, J., et al. (2016). A DTI study on the corpus callosum of treatment-naïve boys with 'pure' Tourette syndrome. *Psychiatry Research: Neuroimaging, 247*, 1–8.

Woods, D. W., Piacentini, J., Himle, M. B., & Chang, S. (2005). Premonitory urge for tics scale (PUTS): Initial psychometric results and examination of the premonitory urge phenomenon in youths with tic disorders. *Journal of Developmental and Behavioral Pediatrics, 26*, 397–403.

Worbe, Y., Gerardin, E., Hartmann, A., et al. (2010). Distinct structural changes underpin clinical phenotypes in patients with Gilles de la Tourette syndrome. *Brain, 133*, 3649–3660.

Worbe, Y., Malherbe, C., Hartmann, A., et al. (2012). Functional immaturity of cortico-basal ganglia networks in Gilles de la Tourette syndrome. *Brain, 135*, 1937–1946.

Worbe, Y., Lehericy, S., & Hartmann, A. (2015). Neuroimaging of tic genesis: Present status and future perspectives. *Movement Disorders, 30*, 1179–1183.

Worbe, Y., Marrakchi-Kacem, L., Lecomte, S., et al. (2015). Altered structural connectivity of cortico-striato-pallido-thalamic networks in Gilles de la Tourette syndrome. *Brain, 138*, 472–482.

Zapparoli, L., Seghezzi, S., Devoto, F., et al. (2020). Altered sense of agency in Gilles de la Tourette syndrome: Behavioural, clinical and functional magnetic resonance imaging findings. *Brain Communications, 2*, fcaa204.

CHAPTER SIX

Mechanistic insight into the pathophysiological basis of Tourette syndrome

Charlotte L. Rae[a,*] and Hugo D. Critchley[b,c]
[a]School of Psychology, University of Sussex, United Kingdom
[b]Department of Neuroscience, Brighton & Sussex Medical School, United Kingdom
[c]Sussex Partnership NHS Foundation Trust, United Kingdom
*Corresponding author: e-mail address: c.rae@sussex.ac.uk

Contents

International Review of Movement Disorders, Volume 3
ISSN 2666-7878
https://doi.org/10.1016/bs.irmvd.2021.11.001

1. Introduction

Tourette syndrome is a developmental tic disorder characterized primarily by the chronic expression of multiple motor and phonic (often vocal) tics. Motor tics are typically brief (clonic) involuntary or "unvoluntary" actions, which range from the contraction of single or proximally-related muscle groups to more complex sequences of actions. Phonic tics again represent a spectrum from simple brief sniffs and coughs to more complex clicks and vocalizations onto complex sequences of words and other modulations of speech (Jankovic, 1997). Ultimately, there are a variety of ways in which tics are expressed: Tonic (or dystonic) tics persist for longer durations than the (more common) clonic muscle twitches, jerks, blinks and noises. Many tics occur in the absence of forewarning sensations, while others are preceded by localized, consciously perceived feelings of tension and urge. Such premonitory sensations (Leckman, Walker, Goodman, Pauls, & Cohen, 1994) are often relieved by expression of the tic and thereby can enable some voluntary control of the tic (i.e. suppression/withholding) in response to the sensory compulsion. Tics often occur in bouts, suggesting the existence of tic-prone brain states, often related to suggestibility, social environment, emotional context, and/or level of psychological arousal. Longer term fluctuation in tic expression (over weeks, months and years) is also characteristic of TS. A distinction is made between simple TS and "full-blown TS" in which complex tics are prominent, including the expression of copro- (obscene), echo- (imitative) and pali- (repetitive) phenomena. In addition, TS+ has been used to describe the common co-occurrence of TS with common comorbidities, including attention deficit hyperactivity disorder (ADHD), obsessive compulsive disorder (OCD) and autism (Robertson, 2000). As Robertson (2000) said, "TS is not a unitary condition but a cluster of multiple phenotypes" (Cavanna, Servo, Monaco, & Robertson, 2009; Robertson, 2000). Given this complexity, a comprehensive understanding of the pathophysiology of TS needs to draw together multiple lines of evidence to account tor developmental trajectories, phenomenological variability, and the distinct constellations of symptoms that might present in the context of fluctuating, yet enduring, expression of multiple motor and phonic tics.

2. Tic generation

Clinical inference and neuroscientific evidence indicate that tics occur through aberrant interactions between frontal cortices and the basal ganglia, expressed via cortico-striato-thalamo-cortical (CSTC) loops. In these circuits, signals from prefrontal and premotor cortex are processed through the basal ganglia and thalamus, which then releases signals for action to the primary motor cortex. Our understanding of how these processes become dysfunctional in TS is informed by classical models of CSTC function and anatomy, post-mortem histology data, and modern brain imaging (fMRI and PET) studies. Together, these data suggest that tics are generated through a combination of supplementary motor area (SMA) hyperactivity, a dominance of direct over indirect pathway activation in the basal ganglia, and tonic hyperexcitability of the primary motor cortex.

2.1 The direct and indirect pathways

In classical models of CSTC function, there are two primary routes through the basal ganglia: the "direct" pathway, which evokes the production of action, and the "indirect" pathway, which evokes instead the opposite: a withholding or prevention of movement (Alexander & Crutcher, 1990). These two pathways are so-named in reference to the anatomical routes by which neural signals pass through the basal ganglia. The direct pathway has fewer synaptic stage-posts than the indirect pathway, and so neural signals travel from the striatum to the thalamus more "directly."

In these CSTC loops, signals from premotor cortex and SMA representing action plans enter the basal ganglia via the dorsolateral portions of the striatum (putamen and caudate nucleus) (Fig. 1). Here, two populations of GABAergic medium spiny neurons carry signals through the direct and indirect pathways respectively.

In the direct pathway, which promotes movement, excitatory glutamatergic inputs from cortex synapse on to a population of medium spiny neurons in the striatum that project to the internal globus pallidus (Fig. 2). Since these striatal projections to the internal globus pallidus are GABAergic, the effect of direct pathway activation is to decrease internal globus pallidus activity. The internal globus pallidus then synapses on to the thalamus. This is another GABAergic projection; the effect being that when activity within the internal globus pallidus is low, having been inhibited by the striatum, it no longer exerts an inhibitory influence over the thalamus. As a result, the thalamus increases in neural firing rate, and via glutamatergic outputs to primary motor cortex, releases an instruction to move to the spinal cord. It is via the double inhibitory link through the basal ganglia that the direct pathway ultimately evokes the production of movement.

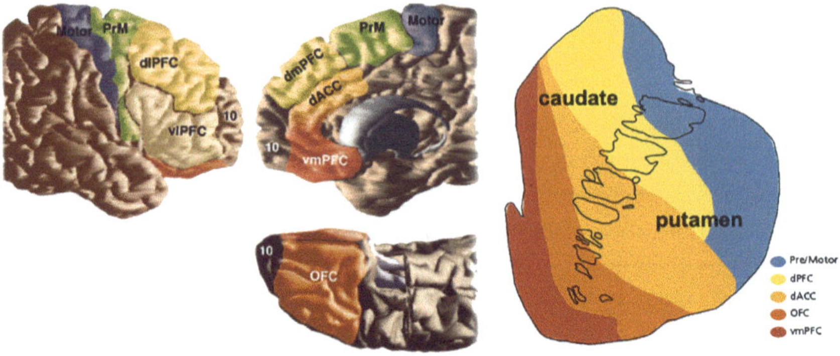

Fig. 1 The topography of frontal inputs to the rostral striatum. dACC, dorsal anterior cingulate cortex; dlPFC, dorsolateral prefrontal cortex; dmPFC, dorsomedial prefrontal cortex; dPFC, dorsal prefrontal cortex; OFC, orbitofrontal cortex; PrM, premotor cortex; vlPFC, ventrolateral prefrontal cortex; vmPFC, ventromedial prefrontal cortex. The location of inputs from premotor cortex in the dorsolateral striatum (caudate and putamen) are shown in blue. *Adapted from Haber, S. (2016). Corticostriatal circuitry.* Dialogues in Clinical Neuroscience, 18*(1), 7–21. https://doi.org/10.31887/DCNS.2016.18.1/SHABER.*

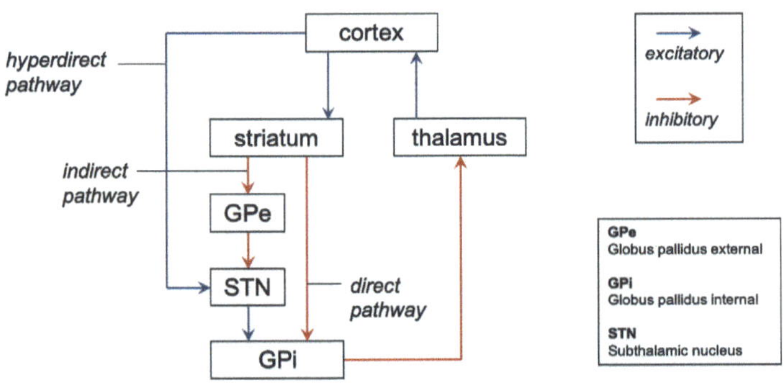

Fig. 2 The direct, indirect, and hyperdirect pathways through the basal ganglia. The hyperdirect pathway by-passes the usual striatal entry point to the basal ganglia for fast, abortive, stopping of action. Blue arrows indicate excitatory (glutamatergic) inputs; red arrows indicate inhibitory (GABAergic) inputs.

In contrast, the indirect pathway evokes the withholding or prevention of movement via an alternative route. Here, excitatory glutamatergic inputs from cortex synapse onto a population of medium spiny neurons in the striatum that project to the external, rather than internal, globus pallidus (Fig. 2). As in the direct pathway, the striatal outputs to the external globus pallidus are GABAergic, and thus the effect of indirect pathway activation is to decrease neural firing within the external globus pallidus. From the external globus pallidus, there is an onward, inhibitory, synapse to the subthalamic nucleus. With the external globus pallidus inhibited by the striatum, it no longer exerts this inhibitory influence on the subthalamic nucleus. As a result, subthalamic nucleus activity can increase, and via excitatory projections to the internal globus pallidus, ramp up internal globus pallidus activity. The internal globus pallidus can then exert its inhibitory influence over the thalamus, thereby pausing motor outflow to primary motor cortex, and thus, preventing movement.

Although these sequelae of events through the direct and indirect pathways are complex, the essence is that the chain of activity leading to either the production, or prevention, of movement, depends upon whether the thalamic outflow to primary motor cortex is facilitated (direct pathway) or paused (indirect pathway). In other words, in order to act, the brake upon the motor cortex exerted by the thalamus must be released.

These classical models of basal ganglia anatomy and function tell us that tics likely arise either through over-amplification of the direct pathway, or through attenuation of the indirect pathway—or both. The net effect is that the thalamic brake upon the motor cortex is too easily released, producing tics.

2.2 Basal ganglia function in TS

A dominance of direct over indirect pathway activity in TS implies that there will be differences in how the basal ganglia regulate cortical inputs in people with tics. One key way that cortical inputs to the striatum are regulated is via inhibitory interneurons. These interneurons synapse on to the medium spiny neurons that form the pallidal projections out of the striatum, and help determine the balance of activity flow through the direct and indirect pathways (Gittis, Nelson, Thwin, Palop, & Kreitzer, 2010). This enables the motor system to exist in a state of stability with a relative lack of movement, until the direct pathway is activated to produce an action.

Although inhibitory interneurons are present in both pathways, they preferentially target direct pathway medium spiny neurons (Gittis et al., 2010). Post-mortem histological studies of the basal ganglia of people with TS show that the presence of these inhibitory interneurons in the striatum is decreased by up to 50% compared to controls (Kalanithi et al., 2005; Kataoka et al., 2010; Lennington et al., 2016). However, these sample sizes are small, and the findings need replicating in larger studies. Nevertheless, such data is compelling evidence that the balance of activity between the two pathways is likely disrupted in TS. This leads to a tendency for the direct pathway to become active, even if there are fewer cortical inputs signaling the action (although there is likely a combined effect of reduced striatal inhibitory interneurons, and increased cortical signals from the SMA to act) (Fig. 3).

Animal models also inform our perspective on what is happening in the basal ganglia in TS (Nespoli, Rizzo, Boeckers, Hengerer, & Ludolph, 2016). Injection of the GABA antagonist bicuculline into the striatum of macaques or rodents leads to repetitive limb and orofacial movements that bear a striking resemblance to human tics (Bronfeld, Israelashvili, & Bar-Gad, 2013; McCairn, Bronfeld, Belelovsky, & Bar-Gad, 2009). This generation of tic-like movements through a bicuculline-mediated reduction in GABAergic transmission in the striatum aligns with the post-mortem evidence that excitatory cortical inputs to the basal ganglia are not sufficiently regulated by direct pathway inhibitory interneurons in TS.

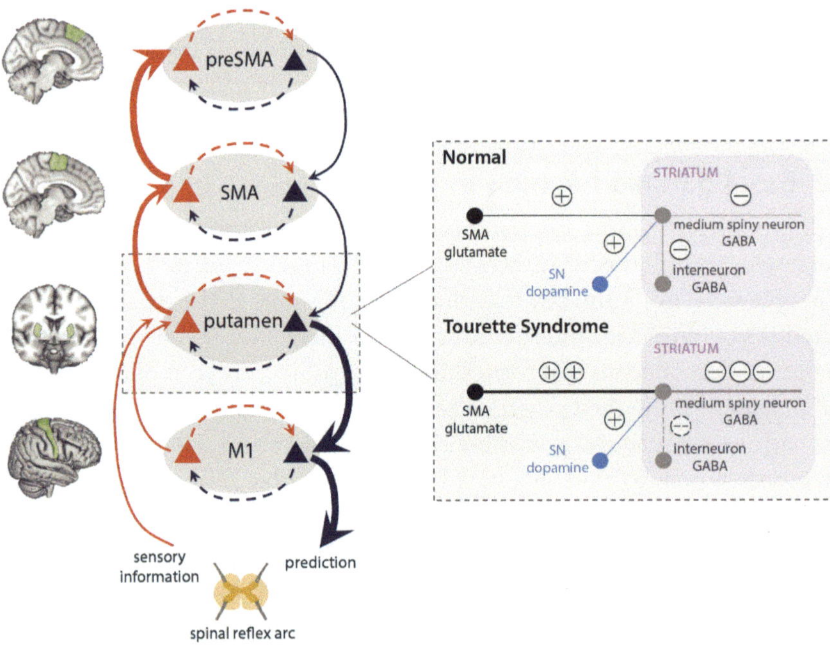

Fig. 3 Increased glutamatergic inputs from the SMA to the striatum, combined with reduced number of inhibitory interneurons, lead to an amplification of activity in the direct pathway in TS. See Rae, Critchley, and Seth (2019) for further details on how active inference (which combines prior predictions with sensory likelihoods) within cortico-striatal hierarchies in TS may generate symptoms. *Reprinted from Rae, C. L., Critchley, H. D., & Seth, A. K. (2019). A Bayesian account of the sensory-motor interactions underlying symptoms of Tourette syndrome Frontiers in Psychiatry, 10(March). https://doi.org/10.3389/fpsyt.2019.00029.*

At a more global level, voxel-based morphometry meta-analysis of T1 structural MRI scans suggests that overall striatal volumes may be altered in TS, with some evidence for increased putamen volume (Wen et al., 2021). However, it is still unclear how any such global increases or decreases in basal ganglia nuclei volume relate to microstructural neuronal alterations in the basal ganglia, and how this drives the generation of tics in TS.

2.3 The role of dopamine

The two populations of medium spiny neuron populations within the striatum that broadly correspond to the direct and indirect pathways carry dopaminergic, as well as glutamatergic, receptors. Thus, in addition to increasing their firing rate in response to excitatory signals from cortex, their activity can be further modulated by dopamine release from neurons projecting from the substantia nigra. The medium spiny neurons of the direct pathway have a predominance of excitatory D1 receptors (Calabresi, Picconi, Tozzi, Ghiglieri, & Di Filippo, 2014). In contrast, neurons in the indirect pathway predominately express inhibitory D2 receptors (Calabresi et al., 2014). Thus, the combined effect of the release of dopamine is to amplify activity within the direct pathway, while attenuating activity within the indirect pathway, and thus, promoting the production of movement.

The hypokinenetic neurodegenerative condition of Parkinson's disease demonstrates the crucial importance of dopamine transmission within the basal ganglia, and what happens when there is *insufficient* dopamine, due to alphasynuclienopathy in the substantia nigra: an attenuation of activity in the direct pathway, leading to bradykinesia (Albin, Young, & Penney, 1989; Redgrave et al., 2010). The natural inference from our knowledge of Parkinson's disease pathophysiology is that, in contrast, there may be *too much* dopamine in TS, leading to an amplification of activity in the direct pathway, and/or an attenuation of activity in the indirect pathway, resulting in hyperkinesia.

The fact that dopamine blockers can successfully reduce expression of tics in some patients with TS also suggests that a dopamine-driven dominance of activity in the direct versus indirect pathway could be at play in the generation of tics (Hartmann & Worbe, 2013; Jankovic, 2020; Robertson, 2000). However, dopamine antagonists do not always completely reduce tic expression in patients with TS, and, in some individuals, can be largely ineffective in treating tics (Hartmann & Worbe, 2013). The implication is that dysfunctional dopaminergic processes are only one causal factor in the generation of tics in TS.

Different medications can have different profiles of blocking action on D2 versus D1 versus receptors. This means that disentangling the precise mechanism of their therapeutic action can be challenging. However, D2 receptor antagonism is a primary action of the most commonly prescribed medications with efficacy in reducing tics: Aripiprazole (in fact a partial D2 agonist), Risperidone, Sulpiride and Haloperidol (Buse, Schoenefeld, Münchau, & Roessner, 2013; Hartmann & Worbe, 2013). Thus, it may be that the core therapeutic mechanism of dopamine-blockers is to amplify activity in the indirect pathway, with the result that the thalamic brake upon primary motor cortex is exerted, and tics are less likely to be released. However, selective D1 blocking agents, notably Ecopipam, are showing promise (Gilbert et al., 2018).

Although the direct and indirect pathways, and the differential expression of D1 and D2 receptors in their two populations of medium spiny neurons, is a commonly accepted model of basal ganglia function, the complexity of functional anatomy in these nuclei is still an area of active exploration. For example, direct pathway medium spiny neurons also often contain D2 receptors, and indirect neurons D1 receptors (Calabresi et al., 2014). This complexity highlights that even subtle structural alterations can potentially lead to differential expression of neurochemistry, and an altered balance of activity between the promotion and prevention of action.

Meta-analysis of PET and SPECT imaging studies, however, does not suggest strong evidence for a structural alteration in number of D2 receptors in TS (Abi-Jaoude et al., 2015; Albin et al., 2009; Hienert, Gryglewski, Stamenkovic, Kasper, & Lanzenberger, 2018). It may be that in fact there is no dopaminergic dysfunction in TS—and the medications work simply by ramping up indirect pathway activity to counteract the impoverished inhibitory interneuron regulation of cortical inputs to the direct pathway. In addition to a reduction in striatal inhibitory interneurons in TS, there are also strong contributions to the generation of tics from alterations in cortical function.

2.4 Cortical tic generation sites

Corroborating evidence for altered functional anatomy of the basal ganglia and connected cortical regions in people with TS comes from non-invasive human brain imaging studies, mostly using functional magnetic resonance imaging (fMRI). Three key studies measured brain activity in people with TS during "free ticcing" states, in which they freely express tics during fMRI scanning. These studies reveal a core network associated with tic generation, comprising the SMA, premotor cortex, insula, striatum, globus pallidus, thalamus, primary motor cortex (M1), and primary somatosensory cortex (S1) (Bohlhalter et al., 2006; Neuner et al., 2014; Wang, Maia, Marsh, Gerber, & Peterson, 2011) (Fig. 4). Interestingly, the thalamus and primary motor cortex are engaged at the moment of tic expression, while the SMA, insula, and striatum become active just prior to the release of the tic (Neuner et al., 2014). As with the post-mortem histology data, more studies on larger samples are needed to replicate these data. However, these findings provide compelling insights into how cortex interacts with the basal ganglia and thalamus in the generation of tics.

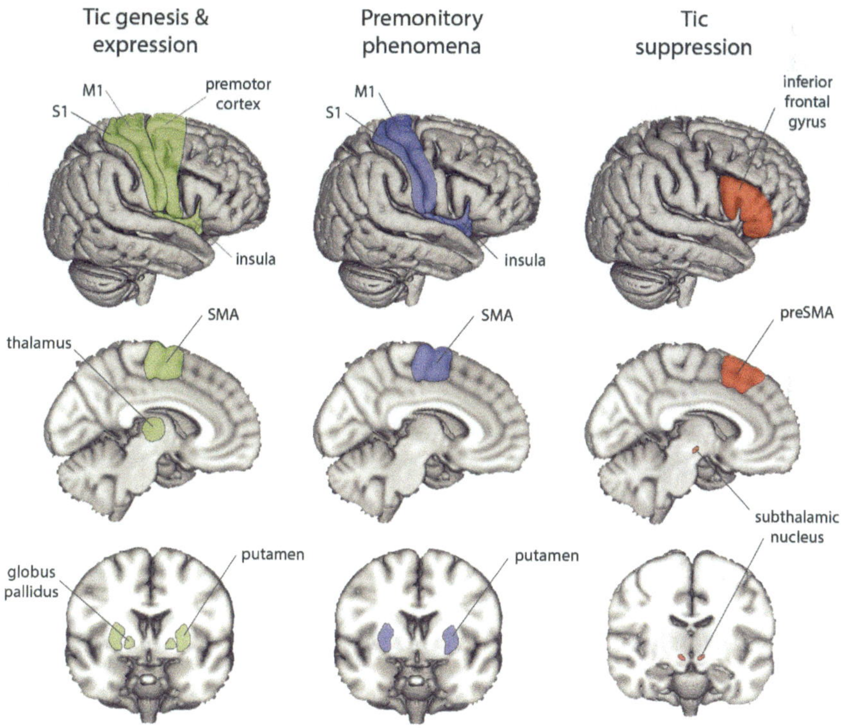

Fig. 4 Cortical and subcortical regions associated with the symptoms and features of TS. Tic genesis (green) is associated with premotor cortex, SMA, insula, M1, S1, putamen, globus pallidus, and thalamus; premonitory sensations (blue) with SMA, insula, M1, S1, and putamen; tic suppression (red) with inferior frontal gyrus, preSMA, and STN. *Reprinted from Rae, C. L., Critchley, H. D., & Seth, A. K. (2019). A Bayesian account of the sensory-motor interactions underlying symptoms of Tourette syndrome.* Frontiers in Psychiatry, 10*(March). https://doi.org/10.3389/fpsyt.2019.00029.*

As noted by other authors, during tic generation SMA activity appears to precede activity in the basal ganglia (Ganos, Roessner, & Münchau, 2013; Neuner, Schneider, & Shah, 2013; Zapparoli, Porta, & Paulesu, 2015). Notably, our meta-analysis of task-based fMRI studies specifically identified greater SMA activity as a correlate of tic severity (Polyanska, Critchley, & Rae, 2017). In addition, among brain regions targeted in brain stimulation studies using repetitive transcranial magnetic stimulation (rTMS) or transcranial direct current stimulation (tDCS) as a potential therapy for TS, the SMA in particular shows the best evidence for eliciting tic reduction (Dyke, Jackson, Nixon, & Jackson, 2019; Grados, Huselid, & Duque-Serrano, 2018). Thus altering activity levels in the SMA seems to reduce tics, and putatively the functional plasticity of SMA may be an underpinning mechanism at play in adolescents whose tics reduce with age (Draper et al., 2014; Jackson, Draper, Dyke, Pépés, & Jackson, 2015). Altogether, SMA hyperactivity likely initiates a cascade of activity through CSTC circuits, which combined with insufficient interneuron regulation of the direct pathway, leads to tics (Fig. 3).

The final cortical stage before tic-generating motor signals are released to the spinal cord is the primary motor cortex (M1). Evidence for aberrant M1 function in TS comes from TMS studies of corticospinal excitability, which indicate that thresholds for action are tonically altered (Draper, Jude, Jackson, & Jackson, 2015; Orth, 2009). Furthermore, during volitional tic inhibition, the greater the reduction in corticospinal excitability, the more successful a patient is at suppressing their tics (Ganos, Rocchi, et al., 2018). Complementing TMS data, in an fMRI study of intentional inhibition and reactive inhibition on a Go/NoGo task, we observed that when controls inhibit actions (either volitionally or when instructed to do so), M1 activity is suppressed (Rae et al., 2020). In contrast, there was no suppression of M1 activity below baseline in the TS participants, despite them producing no overt motor output. Together, these data suggest that there are lower thresholds for the release of action from M1 in TS. Combined with SMA hyperactivity and striatal dysregulation, the passage of motor signals can run unhindered through CSTC circuits, generating and releasing tics.

Diffusion MRI data from patients with TS suggest that there are structural differences in white matter tracts connecting these key CSTC areas. Tractography analyses highlight increased connectivity amongst the SMA, basal ganglia nuclei, thalamus, and M1, with increased M1 connectivity also correlating with tic severity (Worbe et al., 2015). Differences in diffusion tensor imaging (DTI) parameters include higher fractional anisotropy

(FA) and lower radial diffusivity (RD) in cortico-striatal pathways, which also indicates stronger white matter connectivity. However, two subsequent studies in children and adolescents found the opposite (i.e. lower FA and higher axial diffusivity (AD)), both of which indicate poorer white matter structure in TS (Sigurdsson et al., 2018; Wen et al., 2016). This may imply that neurodevelopmental change takes place in the white matter of CSTC circuits. Longitudinal studies and a larger literature for meta-analyses would be very valuable in helping resolve such questions.

The pre-supplementary motor area (preSMA) is a cardinal site of voluntary action, and given this strong association, it is perhaps surprising that the preSMA does not more often appear in functional imaging studies as a site of overt dysfunction in TS. Previously, we hypothesized that the preSMA is not explicitly involved in the generation of tics, but does play a role in ascribing an "unvoluntary" feeling to tics that have been produced by dysfunctional circuitry lower in the motor hierarchy (Rae, Critchley, & Seth, 2019). Interestingly, overall levels of activity in the preSMA are no different to controls during volitional decisions to move, or to not move (Rae et al., 2020), suggesting that the preSMA may function similarly to controls in terms of voluntary action decisions. In addition, the capacity of many people with TS to voluntarily withhold their tics is a striking indicator of an intact capacity for "freely willed" action decisions, likely mediated by the preSMA (Rae, Critchley, & Seth, 2019).

In summary, tics are generated via a sequence of events including SMA hyperactivity, amplified direct pathway engagement through the basal ganglia, and altered thresholds for the release of action from M1. However, there is also a role in the generation of tics for the insular cortex and the creation of premonitory urge sensations: feeling states that appear to cue the production of tics.

3. Feeling states: Premonitory urge sensations and the compulsive cueing of tics

Premonitory urge sensations were first comprehensively described by Leckman, Walker, and Cohen (1993), who reported that the vast majority of adults with TS identified experiencing a sensation of "an urge," "a feeling," "an impulse," or "a need" to tic prior to tic release. Historically, tics had been conceptualized as involuntary, unwilled, and unchosen. However, Leckman's seminal paper highlighted the fact that many patients report their tics to be a semi-voluntary response to premonitory urge sensations.

We now recognize that for many individuals, for many (although not all) of their tics, it is the experience of an (involuntary) feeling or urge to move that generates a (semi-voluntary) response to release the tic in order to relieve the urge (Kwak, Dat Vuong, & Jankovic, 2003). More recently, therefore, tics have been better described as "unvoluntary" actions (Cavanna & Nani, 2013), indicating they occupy a space part-way along a spectrum of "voluntariness."

Premonitory urge sensations show a number of curious features. They are typically described as uncomfortable feelings of itch or pressure that are relieved once the tic is released (Kwak et al., 2003). In this regard, they share many similarities to non-tic bodily feelings that are strong motivators for action, and are accompanied by a feeling of relief upon execution, such as scratching an itchy skin, yawning, and micturition (Jackson, Parkinson, Kim, Schüermann, & Eickhoff, 2011). Just as an itchy skin compels an immediate action to scratch the affected area, premonitory urge sensations compel an immediate tic. The compulsive nature of tics (and related obsessive-compulsive behaviors and complex tics such as forced touching) is likely strongly driven, therefore, by uncomfortable premonitory urge sensations, at least for some tics.

Many premonitory urge sensations are linked to the precise bodily location at which a tic subsequently emerges, although some are more diffuse general feelings of unresolved tension (Leckman et al., 1993). The fact that premonitory urge sensations are such strong harbingers of tics is capitalized on by our primary behavioral therapies for tics, such as habit reversal therapy (HRT), exposure and response prevention (ERP), and cognitive behavioral intervention for tics (CBIT) (Andrén et al., 2021). These utilize the awareness of premonitory feelings experienced by patients to either alter the expression of a given tic to another action, or to habituate to the feelings of discomfort and ultimately attempt to suppress a tic entirely (Woods et al., 2015). The fact that some premonitory urge sensations can be localized to a specific location or muscle group implies that regions of the brain that support the perception of physiological sensations and bodily feelings are involved in their genesis.

3.1 The functional anatomy of premonitory urge sensations

The insular cortex is a territory approximately 5–7 cm long, underlying the frontal and temporal lobes. Signals from the visceral organs carrying information representing physiological state are relayed to the brainstem via the vagus nerve. From here, physiological state signals are conveyed first to the thalamus, and then the posterior insular cortex, which contains a viscerotopic representation of the heart, lungs, and gut (Craig, 2002). The posterior insular cortex thereby rapidly receives information from the body on its autonomic state.

These viscerotopic representations are then re-represented and integrated in the mid-insula. The anterior insular cortex, projecting into orbitofrontal cortex, supports conscious awareness and interrogation of these bodily feelings (Critchley, 2005; Critchley, Wiens, Rotshtein, Öhman, & Dolan, 2004), enabling us to answer the question, "How do I feel?" (Craig, 2009, 2014). This extends from perception of specific sensations arising in particular organs (such as a stomach pain) to the underpinning of emotional states, founded on autonomic signals (Critchley & Garfinkel, 2017).

Our understanding of the role the insular cortex plays in bodily perception and affective states tells us that this region likely plays a part in the generation of premonitory urge sensations in TS. In addition, structural and functional neuroimaging evidence points towards the insula as a critical site for premonitory urge sensations (Fig. 4).

Two of the three "free ticcing" fMRI studies described above identify insula activation prior to tic release (Bohlhalter et al., 2006; Neuner et al., 2014). Positron emission tomography (PET) studies also report reduced $GABA_A$ receptor density along the anterior-posterior extent of the insula of adults with TS (Lerner et al., 2012), highlighting a potential causal basis of heightened insular activity. In addition, two studies of children and adolescents with TS report associations between insular grey matter volume and worse premonitory urge sensations, as measured by the premonitory urge for tics scale (PUTS) (Woods, Piacentini, Himle, & Chang, 2005). However, one of these found a negative correlation (Draper, Jackson, Morgan, & Jackson, 2016), while the other found a positive association between grey matter volume and PUTS score (Jackson et al., 2020). Nevertheless, these data link insula integrity to premonitory urge sensations. Furthermore, fMRI data from non-patient participants undertaking a blink suppression task suggests that the insula is associated with physiological feelings of an uncomfortable urge (Berman, Horovitz, Morel, & Hallett, 2012).

How does the insular representation of uncomfortable bodily sensations motivate the production of tics? A key piece of evidence comes from resting-state fMRI data: The strength of functional connectivity between the insula and SMA correlates with severity of patients' premonitory urge sensations (Tinaz, Malone, Hallett, & Horovitz, 2015). We replicated this relationship in a task-based connectivity analysis of the insular cortex, observing greater connectivity with the SMA in patients who had worse premonitory sensations (Rae et al., 2018). Together, this suggests the insula may function as a tic trigger-site, cueing the production of actions (tics) to relieve the uncomfortable premonitory urge sensations, plausibly via the SMA (Rae, Critchley, & Seth, 2019). With this increase in SMA activity, an onward cascade of CSTC activity then drives the production of tics.

We also previously proposed that there may be a role for somato- and viscero-topic regions of the striatum in the generation of premonitory urge sensations (Rae, Critchley, & Seth, 2019). Interestingly, in the macaque, there are projections from the mid and posterior insula to a region of the striatum that largely overlaps with the areas subserving motor function (Chikama, McFarland, Amaral, & Haber, 1997; Haber, 2016). In addition to the SMA route to tic generation, the insula (and its representation of premonitory sensations) also directly interfaces with the striatum. Here, inputs from insula may be insufficiently regulated by the lack of inhibitory interneurons, and perhaps combined with heightened SMA inputs, may inappropriately drive the direct pathway. Notably, the insula and putamen are also highlighted as likely sites underpinning compulsive actions motivated by feeling states in other neuropsychiatric conditions: trichotillomania, skin picking disorder, and restless legs syndrome (Badenoch, Searle, Watson, & Cavanna, 2020).These obsessive-compulsive conditions also share with TS the subjective experience of relief reported by patients upon executing the compulsive action. The typical timeline of an urge intensity increasing prior to tics, and decreasing after tic release, is approximately 10 seconds (Brandt et al., 2016), indicating a tight temporal coupling between perception of the urge, performance of an associated action, and sensation of relief. Asking patients with TS to suppress their tics typically increases the urge (Brandt et al., 2016), creating mounting pressure to tic in order to relieve the uncomfortable sensation. A failure to act and release the tic may generate increasingly uncomfortable bodily sensations via ascending prediction errors to the anterior insula, for premonitory urges that were generated more posteriorly in the insular cortex (Rae, Critchley, & Seth, 2019). Thus, the increasing strength of inputs from anterior insula may accumulate within SMA, until it releases signals via the basal ganglia to trigger a tic.

3.2 Psychological representation of interoceptive information in premonitory urge sensations

The insular cortex underpins our ability to gain conscious access to interoceptive information: The sensing and perceiving of physiological state and inner bodily signals (e.g. Critchley et al., 2004). Thus, structural differences in insular cortex may shape the degree to which an individual may be more or less sensitive to internal arousal states and bodily sensations. A study that used a heartbeat detection task to measure sensitivity to interoceptive sensations (aiming to quantify how accurately a person can perceive their heart beating at rest) found that interoceptive performance was lower in people with TS than in a non-TS comparison group (Ganos et al., 2015). In addition, and somewhat paradoxically, better interoceptive sensitivity was associated with greater severity of premonitory sensations. Subsequently, a similar study of children and adolescents with TS also observed lower accuracy on a heartbeat detection task, suggesting that reduced interoceptive accuracy may be a feature of TS throughout the developmental time-course (Pile, Lau, Topor, Hedderly, & Robinson, 2018).

In addition to objective performance on heartbeat perception tasks, patients may differ on other aspects of interoceptive sensitivity, notably more subjective self-report measures and the "metacognitive" correspondence between objective vs subjective measures of sensing inner bodily signals. These three dimensions of interoception were formulated into a dimensional model as (1) interoceptive accuracy, (2) interoceptive sensibility and (3) interoceptive awareness (Garfinkel, Seth, Barrett, Suzuki, & Critchley, 2015). Following on from Ganos et al. (2015) and Pile et al. (2018), we examined these three dimensions of interoception in adults with TS (Rae, Larsson, Garfinkel, & Critchley, 2019). First, we replicated a numerically lower performance on the heartbeat counting task across the TS group, although the result was not a statistically significant different from a closely-matched non-TS comparison group. We also observed a tendency for patients with TS to report higher interoceptive sensibility on a questionnaire measure of bodily sensitivity. However, on calculating trait interoceptive prediction error, which represents the magnitude of discrepancy between 'objective' interoceptive accuracy and self-reported sensibility (Garfinkel et al., 2016), we found a significantly higher mismatch in TS. This indicates a greater tendency for patients with TS to experience "interoceptive surprise" through heightened sensitivity to poorly predicted bodily sensations, an effect observed in other neurodevelopmental conditions and found to predict symptoms of anxiety in autistic individuals (Garfinkel et al., 2016).

Of all the dimensions examined, it was interoceptive sensibility specifically that was found to correlate with tic severity, tic impairment, and premonitory sensation severity (Rae, Larsson, et al., 2019). These data highlight that it may be the subjective experience of bodily sensations that is most relevant for understanding symptom expression in TS. It is also important to note that the heartbeat counting task used in these studies of interoception has well-established limitations, since its performance may be influenced by demand characteristics and top-down knowledge including an individual's knowledge of own heart rate (Brener & Ring, 2016).

While such methodological issues constrain inferences, the evident relationship between interoception and premonitory urge sensations has unclear causality: We do not know if altered interoception causes premonitory urge sensations, if the experience of premonitory urge sensations causes altered interoception, or if structural insular changes in TS cause both, and altered interoception is simply a corollary of premonitory urge sensations. Nevertheless, it may be that training interoceptive ability so as to reduce the mismatch between objective and subjective representations of bodily signals may be a useful approach to alleviate symptoms of TS. Aligning dimensions of interoception in this way reduces symptoms of anxiety in autism (Quadt et al., 2021), and may also hold therapeutic potential for TS.

4. Tic triggering

A notable feature of tics is that they appear to be remarkably responsive to both internal bodily state, and external environment. Many patients report a subjective experience of tics worsening during moments of stress, anxiety, anger, or excitement; and during times of upheaval or change in home life (Robertson, 2015). This indicates an important role of autonomic reactivity in the generation of tics.

In addition, many patients experience echophenomena: Tics that copy, or echo, the actions or vocalizations of others—or even of themselves (Ganos, Ogrzal, Schnitzler, & Münchau, 2012). This highlights that tics can be evoked by what happens in the world around them, although echoed tics are most commonly already part of their repertoire (Finis et al., 2012). The evoking of tics by stimuli in the external environment also extends to the impact of social stimuli, including the stressful scrutiny of others, which further interacts with autonomic processes to invoke a worsening of tics, often at moments when patients feel the most "visible" publicly, with significant impacts on quality of life (Eapen, Cavanna, & Robertson, 2016).

4.1 Stress and autonomic reactivity

In addition to anecdotal patient reports, some experimental evidence suggests that tics worsen during states of sympathetic nervous system arousal

(Hawksley, Cavanna, & Nagai, 2015). For example, tics increase during arousal biofeedback training in which patients learn to amplify sympathetic arousal via higher electrodermal activity, while tics lessen during relaxation biofeedback in which patients learn instead to attenuate sympathetic arousal via lower electrodermal activity (Nagai, Cavanna, & Critchley, 2009). Although not yet formally tested, it is likely that these effects are mediated by changing activity in an interoceptive network, including interactions between brainstem nuclei, the thalamus, and insular cortex. Neuroimaging studies indicate engagement of ventromedial prefrontal cortex, an insular projection region, during effective biofeedback relaxation (Nagai, Critchley, Featherstone, Trimble, & Dolan, 2004). It is also possible that such biofeedback training reduces tic expression via the direct connections from the insula to the striatum mentioned previously (Chikama et al., 1997; Haber, 2016).

Interestingly, two small scale studies that applied vagus nerve stimulation (VNS) in TS found a therapeutic reduction in tics (Diamond, Kenney, & Jankovic, 2006; Sperling, Reulbach, Maihöfner, Kornhuber, & Bleich, 2008). In VNS, an implanted device directly stimulates the vagus nerve to modulate interoceptive function. Increasingly, studies using a non-invasive approach, such as by stimulating the skin of the ear in a region innervated by a branch of the vagus nerve, has also shown modulatory effects similar to VNS (Paciorek & Skora, 2020). Larger clinical trials are needed to delineate more precisely the therapeutic potential of such approaches in TS (Hawksley et al., 2015). However, our knowledge of VNS as an experimental tool indicates that it appears to amplify the strength of afferent signals from visceral organs, and increases insular activity as seen by fMRI. Furthermore, there is evidence of therapeutic efficacy in other neuropsychiatric conditions beyond TS, such as PTSD (Paciorek & Skora, 2020). In TS, VNS and its transauricular equivalent may offer a means to alter insular function and ultimately reduce the severity and impact of premonitory urges.

4.2 Echophenomena

Even in the earliest description of the condition, echophenomena were noted as a feature of TS (Ganos et al., 2012). The generation of echophenomena at a neural level is proposed to arise through one of two mechanisms—an over-active mirror neuron system, or, exhaustion of the prefrontal control of action that many patients regularly engage to suppress tics (Ganos et al., 2012). However, strong evidence for either of these neural mechanisms driving echophenomena in TS is sparse.

On the basis of data from healthy control participants, SMA is likely to play a key role: In non-TS individuals, stimulating the SMA with high frequency TMS is sufficient to induce echophenomena (Finis et al., 2013). In addition, the temporo-parietal junction (TPJ) may also contribute; this region is implicated in the detection and interpretation of other people's movements and in empathic processes, such as experiencing pain when seeing another in pain (known as "mirror pain synaesthesia"). In such mirror pain synaesthetes, there is greater connectivity of the TPJ with the insula when viewing others in pain (Grice-Jackson, Critchley, Banissy, & Ward, 2017b). This may reflect a blurring of boundaries between the self and other (Grice-Jackson, Critchley, Banissy, & Ward, 2017a). Complementing these data, TPJ activity in TS during a theory-of-mind task correlates with echophenomena severity (Eddy, Cavanna, Rickards, & Hansen, 2016). Consequently, altered TPJ function may "underline problems with self-other distinctions in TS" (Eddy, 2018). It would be useful to know, via connectivity analyses of fMRI data, if TPJ activity drives the SMA and subsequent basal ganglia direct pathway activation during echophenomena in TS.

4.3 Social triggers

Many echoed tics arise through seeing others' actions or hearing their vocalizations. Beyond echophenomena, social stimuli may cue the production of tics in a less specific manner by worsening their frequency or severity. This likely occurs via autonomic reactivity processes as described above, centred on the insula as an interoceptive nexus.

We found evidence for a critical role of the anterior insula in cueing motor system activity when adults with TS see social stimuli, namely, faces (Rae et al., 2018). Upon viewing faces which wore either an angry or neutral expression, patients with TS showed activation in the insula, when controls had no response. Interestingly, this insular response in TS was present regardless of whether the face bore a neutral or angry expression, suggesting that simply the presence of a social cue (or possibly the lack of a positive social cue) was sufficient to evoke insular activation. Connectivity analyses linked this higher insula reactivity to increased fiunctional coupling with preSMA, premotor cortex, putamen, and M1. Furthermore, the greater the insula connectivity with the globus pallidus and thalamus, the worse the tics; and the greater the insula connectivity with the SMA, the worse the premonitory sensations.

These data shed light on the neural processes through which stressful social stimuli may evoke a worsening of tics: Insula hyper-reactivity leads to a cascade of onward activity through CSTC circuits. In addition, the observation that insula-SMA connectivity correlates with premonitory sensation severity mirrors the earlier finding of an association between resting-state insula-SMA connectivity and PUTS scores (Tinaz et al., 2015). This lends further credence to the hypothesis that the insula represents uncomfortable premonitory urge sensations, which when passed to the SMA, drives the production of tics via signals to the basal ganglia.

5. Complex tics

Echophenomena are one expression of a complex tic, noteworthy as inducible in social contexts. In contrast to simple tics, which are restricted to one localized group of muscles or small brief sniffs and noises, complex tics describe those that involve several muscle groups, sequences of movement (or vocalizations) and often purposeful actions including touching, sniffing objects, and use of words and phrases (Eapen & Robertson, 2015). Complex tics are common; only around 10% of people with TS have tic only "pure-TS" with simple tics (Eapen & Robertson, 2015). Echo- and palilalia and praxia (copying or repeating vocalizations or actions), coprophenomena (see below), self-injurious behaviors and other compulsive actions are subsumed in this category of tics, though the boundary with obsessive compulsive behaviors (intrinsic to TS) becomes increasingly blurred. This picture is further confounded by the presence in many people with TS of comorbid OCD and/or ADHD.

Pathophysiologically, this relationship suggests that the expression of TS varies as a consequence of a spectrum of developmental pathoetiological insults centred on the core involvement of corticostriatal interactions (through CSTC circuitry), yet with differential involvement of paralimbic, prefrontal, cingulate, and insular cortices. The "multistrike model" of neurodevelopmental conditions (Spessot, Plessen, & Peterson, 2004) highlights the interaction of genetic, epigenetic and environmental factors that would shape the profile of TS and related neurodevelopmental disorders, determining symptom expression (see also, Eapen & Robertson, 2015; Nespoli et al., 2016). By way of illustration, frontal involvement may bias symptoms toward expression of socially inappropriate coprophenomena and non-obscene socially inappropriate (NOSI) behaviors.

Obsessive compulsive experiences are both core to TS and, when severe and associated with anxiety/distress, may represent a comorbid OCD (Robertson, 2000). While across populations there are general differences in the overall clusters of obsessions and compulsions experienced by people with and without tics (TS obsessionality often features numbers, symmetry and evening up, in contrast to more classic ego-dystonic "contamination fears"), there is considerable overlap, and historically TS has been classified alongside OCD (a common genetic basis has previously been supposed) (Browne et al., 2015; Robertson, 2000). The arbitrariness of cleaving complex tics from obsessive compulsive actions suggest that a comprehensive understating of TS pathophysiology must draw upon the relevant literature relating to OCD (Neal & Cavanna, 2013; Worbe et al., 2010).

In fact, the known neuropathological evidence pertaining to OCD highlights dysfunctional cortico-striatal-thalamic interactions with emphasis on orbitofrontal and cingulate cortex. Correspondingly, neurosurgical approaches to managing intractable OCD have sought to target fiber tracts between (orbital) prefrontal cortex, basal ganglia and thalamus (limbic leucotomy, subcaudate tractotomy anterior capsulotomy, cingulotomy and deep brain stimulation of anterior limb of internal capsule) (e.g. Haber, Yendiki, & Jbabdi, 2021). Neuroimaging studies of intrusive perseverative, ruminatory thoughts implicate dysfunction of orbital and ventromedial prefrontal cortices (alongside insula and anterior cingulate (Makovac, Fagioli, Rae, Critchley, & Ottaviani, 2020), consistent with neuropsychological studies of perseveration and dysexecutive inflexibility (Insel, Donnelly, Lalakea, Alterman, & Murphy, 1983). Similarly, early neuroimaging studies highlighted orbitofrontal hyperactivity, usually also with caudate hyperactivity, in patients with OCD (Baxter et al., 1988; Swedo et al., 1989). Subsequent imaging studies reinforce the evidence implicating wider cortico-striatal circuitry and CSTC loops in the expression of OCD and the associated different symptom profiles (Calzà et al., 2019; Rauch et al., 2007; Saxena & Rauch, 2000). A synthesis of different interpretive models of such findings brings together the crucial role of CSTC circuitry with interaction of ascending dopaminergic and serotonin neuromodulatory systems to account for deficits in the effective control of repetitive behaviors in OCD (Aouizerate et al., 2004). Of interest is the implicit association of orbitofrontal regions (rather than insular cortex) with the valenced experiential feelings of urge compulsion in the context of OCD.

5.1 Coprophenomena

Coprolalia was first named by Gilles de la Tourette. Among complex tics, coprophenomena (coprolalia, i.e. obscene vocal tics, and copropraxia, obscene gestures) arguably carry the most social burden. While infrequent among the majority of people with TS (affecting a maximum of 20% of patients over their lives; Freeman et al., 2009), coprophenomena are commonly associated with public understanding of TS, in part by the exemplification of a broader range of difficulties associated with tic expression. Obscene utterances are part of a general language repertoire in the absence of tics, and may be used or evoked for a number of reasons e.g. in response to acute pain. In the wider neurological context, seizures, frontal lobe damage and behavioral variant frontotemporal dementia can be associated with coprolalia. Nevertheless, the expression of coprolalia (and copropraxia) has more compulsive and tic-like characteristics, not least in the premonitory sensations that typically precede and drive the tic. Moreover, coprophenomena share features of the wider range of non-obscene socially inappropriate (NOSI) compulsive behaviors (e.g. shouting 'bomb' in an airport) and negativism that reflect the spectrum between tic and obsessive-compulsive behaviors typically present in TS (Eddy & Cavanna, 2013). Pathophysiologically, therefore, the basis to coprophenomena is at the interface between the mechanisms described above relating to simple (typically motor) tic expression, and the broader pathological mechanisms underpinning and maintaining obsessive compulsive experiences. Coprophenomena and NOSI behaviors can also be formulated in terms of function within a psychosocial context: The target, reasons, and effect of the socially-inappropriate utterance or act (including social mimicry) provide a contextual explanation for the triggering of such tics, not least through amplification of a premonitory urge that requires (stress) release (Senberg, Münchau, Münte, Beste, & Roessner, 2021).

Coprophenomena are seen as part of the clinical syndrome of "full-blown TS" and linked to severity of tics and their impact. The general symptom burden of patients predicts the presence of coprophenomena, not least the co-occurrence of comorbid ADHD and OCD that respectively compromise an individual's (attention-dependent) capacity for impulse control and exacerbate tensions between compulsive urges and control of negativistic action. Neurobiological studies suggest the basis for coprolalia extends beyond left hemispheric regions supporting language comprehension and speech production. For example, the ability to swear may persist after dense aphasia following left hemispheric stroke. Neuroimaging studies of the processing of taboo words also described engagement of right prefrontal and dorsomedial prefrontal cortices, likely in an inhibitory capacity (Wagner-Altendorf et al., 2020). As is the case with complex tics in general, fewer studies however have examined neural correlates of coprophenomena in TS, or are small scaled such that inferences about differences in activity across multiple frontal and temporo-parietal regions are anecdotal (Stern et al., 2000). The role of dysfunctional corticostriatal interactions in coprolalia and copropraxia is assumed from extrapolating our general neurobiological understanding of tic generation and obsessive-compulsive disorder.

6. Motor control: Suppressing tics

A notable feature of TS is the remarkable capacity of many patients to voluntarily suppress their tics (Ganos, Rothwell, & Haggard, 2018). Indeed, this ability is such a defining feature of TS that it is often used diagnostically to differentiate the condition from other hyperkinetic movement disorders such as myoclonous (Ganos, Rothwell, & Haggard, 2018). It also speaks to the "unvoluntary" nature of tics, which feel compulsive, yet can be suppressed by conscious effort (Cavanna & Nani, 2013). As Ganos, Rothwell, and Haggard (2018) say, "Voluntary tic inhibition involves an intentional, goal-directed, and effortful cognitive activity, decided and initiated by the patient themselves…[it] thus has many of the cognitive features of willed action, although its motor consequences are to reduce movement, not promote it." Tic suppression also underpins behavioral therapies, notably CBIT, which utilizes patients' abilities to temporarily withhold tic expression together with their capacity to attend to premonitory urge sensations in order to habituate to urges without tic release (Woods et al., 2015).

Our understanding of the neural mechanisms by which patients with TS suppress tics comes from analogous studies on motor inhibition in control samples; literature explicitly examining inhibition of tics, or non-tic actions, in TS patients; and investigation of which areas generate a beneficial reduction in tics when stimulated with deep brain stimulation (DBS). These indicate that broadly speaking, the prefrontal cortex leverages control over downstream areas in the CSTC motor hierarchy, particularly the basal ganglia, to prevent motor programs being released by the thalamus to primary motor cortex (Fig. 4).

6.1 The hyperdirect pathway

Revisiting the direct and indirect pathway model of the basal ganglia sheds light on the neural processes by which patients with TS might be suppressing tics. We know that the indirect pathway supports the withholding of action, and that dopaminergic medications likely exert a beneficial reduction in tics by amplifying indirect pathway activity. Thus, it is plausible that when patients elect to suppress tics, they do so by activating the indirect pathway.

However, patients report a subjective experience during tic suppression that the tic often does not "go away." This suggests that the tic motor program may still be active in the direct pathway. An alternative route to blocking motor outflow to thalamus is via the hyperdirect pathway (Fig. 2). This fast, rapid route to action stopping occurs via a monosynaptic link from cortex to the subthalamic nucleus, by-passing the usual entry point to the basal ganglia of the striatum (Nambu, Tokuno, & Takada, 2002). This means that motor outflow can very quickly be paused by activating the subthalamic nucleus, which in turn activates the internal globus pallidus, and thus exerts an inhibitory brake over the thalamic output to primary motor cortex.

This hyperdirect pathway is known to support the fast, rapid motor inhibition demanded by the stop signal task, in which participants are required to stop actions currently in preparation (Aron, Herz, Brown, Forstmann, & Zaghloul, 2016). The hyperdirect pathway enables a quicker reaction to the stop signal than the indirect pathway, since the indirect pathway involves a larger number of synaptic stageposts. It is thus plausible that when patients with TS suppress tics, they invoke the hyperdirect pathway, in order to beat the tic traveling through the direct pathway before it is released to primary motor cortex. When healthy controls undertake the stop signal task during fMRI, the right inferior frontal gyrus is observed as a key cortical site that invokes the stop process, possibly by augmenting inputs from the preSMA to the STN (Aron, Behrens, Smith, Frank, & Poldrack, 2007; Rae, Hughes, Anderson, & Rowe, 2015). This implies that the right inferior frontal gyrus is a crucial prefrontal locus for tic suppression.

6.2 Prefrontal control of action

Indeed, the inferior frontal gyrus was identified in one crucial study explicitly examining tic suppression during fMRI (Fig. 5A). When patients with TS suppressed tics in the scanner, there was greater activity in the inferior frontal gyrus compared to a free ticcing state (Ganos et al., 2014). In addition,

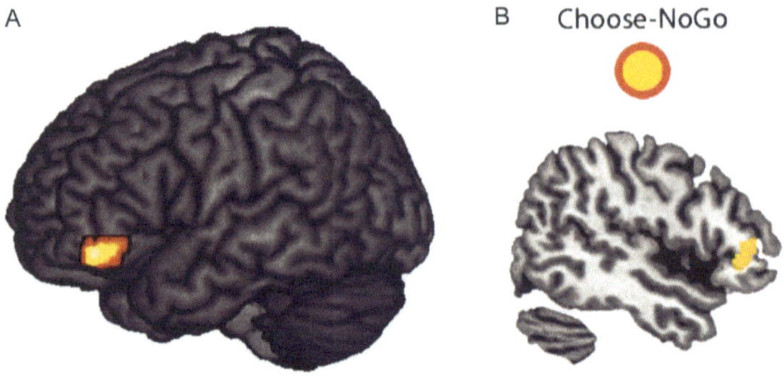

Fig. 5 The inferior frontal gyrus is associated with both (A) voluntary tic suppression, and (B) intentional inhibition of non-tic actions in TS. *Adapted from (A) Ganos, C., Kahl, U., Brandt, V., Schunke, O., Bäumer, T., Thomalla, G., et al. (2014). The neural correlates of tic inhibition in Gilles de la Tourette syndrome.* Neuropsychologia, 65, *297–301. https://doi.org/10.1016/j.neuropsychologia.2014.08.007 and (B) Rae, C. L., Parkinson, J., Betka, S., Gouldvan Praag, C. D., Bouyagoub, S., Polyanska, L., et al. (2020). Amplified engagement of prefrontal cortex during control of voluntary action in Tourette syndrome.* Brain Communications, 2*(2). https://doi.org/10.1093/BRAINCOMMS/FCAA199.*

the greater the inferior frontal gyrus activity, the greater the patient's success at tic suppression—both during the scanning session, and outside of the scanner. This is compelling evidence that prefrontal cortex—specifically, the inferior frontal gyrus—initiates a tic suppression process.

This is complemented by data from fMRI studies of non-tic motor inhibition in patients with TS. We studied intentional inhibition, in which participants choose whether to act or not, by electing to either make a simple button press, or to withhold the act (Rae et al., 2020). Compared to controls, patients with TS showed greater activity in the inferior frontal gyrus—in the exact same locus previously identified for tic suppression (Ganos et al., 2014) (Fig. 5B). This suggests that the same prefrontal mechanisms may be at play for the volitional withholding of both tic and non-tic actions in TS. In addition, we saw that activity in the downstream area of M1 was reduced below baseline by controls when they chose to inhibit, but was not in TS, despite no overt motor behavior. This suggests that people with TS require greater inferior frontal gyrus leverage over CSTC circuits in order to successfully withhold actions, so as to overcome the strong downstream drive for action. However, connectivity analyses did not identify the hyperdirect pathway during intentional inhibition. This may be due to insufficient MRI field strength.

Interestingly, there is also some meta-analysis evidence for inferior frontal gyrus hyperactivation in TS in studies of externally-cued response inhibition, such as those using the stop signal task (Zapparoli et al., 2015), although behaviorally, patients with TS do not consistently appear to have a general response inhibition deficit, unless there is comorbid ADHD or OCD (Mancini et al., 2018; Openneer et al., 2021). The inferior frontal gyrus activations that we see can therefore be interpreted less as evidence of dysfunction, and more as evidence for a greater requirement of prefrontal resources in order to overcome competing downstream motor drive in the basal ganglia. This also accords with the subjective report by patients that suppressing tics is an effortful and tiring process (Ganos, Rothwell, & Haggard, 2018).

As a cardinal site of voluntary action, the preSMA is strongly associated with "whether" type decisions to execute or withhold actions in healthy controls (Zapparoli, Seghezzi, & Paulesu, 2017). In our study of intentional inhibition, we did not see any differences in preSMA activity between TS and control participants (Rae et al., 2020). This could suggest that volitional "free will" processes, driven by the preSMA, are not fundamentally altered in TS—that is, they do not require "more volition" in their decision to act, or to withhold—but they do require greater prefrontal braking, via the inferior frontal gyrus and hyperdirect pathway.

6.3 Deep brain stimulation

DBS can have therapeutic benefits for the most refractive cases of TS (Martinez-Ramirez et al., 2018). Electrodes are typically implanted in the internal globus pallidus or thalamus (Martinez-Ramirez et al., 2018). Notably, these are the two key basal ganglia stageposts that are influenced by activation of the subthalamic nucleus, in the hyperdirect pathway (Fig. 2). The fact that many patients see a reduction in tics during pallidal or thalamic DBS suggests that the stimulation works by pausing thalamic outflow to primary motor cortex. It is plausible that volitional tic suppression happens in a similar way but requires the cortical input from inferior frontal gyrus to achieve the same outcome of reduced motor output. In this respect, DBS might be akin to a permanent state of tic suppression—without the cognitive effort.

In summary, tic suppression appears to proceed via the hyperdirect pathway, with the inferior frontal gyrus driving subthalamic activity in order to pause motor outflow. However, direct evidence, in the form of fMRI connectivity analyses during active tic suppression, are needed to confirm this.

7. Further insights on pathophysiology from therapeutic approaches

In the preceding sections, a model for understanding the pathogenesis of TS centres on the evidence implicating corticostriatal interaction and the putative dysfunction of CSTC loops in the generation of tics. Here, emphasis is given to the direct and indirect pathways. In addition, the role of insular cortex as a substrate for affective and premonitory feeling states is relevant for translating psychophysiological or interoceptive arousal states (including socio-emotional challenges) into activity driving or predisposing to tics in patients with TS. The phenomenological overlap and co-occurrence of tics (especially complex tics) with obsessive compulsive behaviors and clinical OCD further suggest a role for prefrontal cortices (notably orbitofrontal cortex) in TS pathophysiology.

Such understanding is informed also by which treatments work and which do not: The role of dopamine D1 and D2 receptors in the governance of motor facilitation is dealt with above. The impact of deep brain stimulation as a means of modulating tic expression (and OCD) is described with corticostriatal pathways targeted in accordance with the suggested model of tic generation and control of compulsive and repetitive actions. The targeting of interoceptive insular pathways is compatible with the wider model of social and affective stress triggering bouts of tics. Interestingly, however, oscillatory exteroceptive stimulation (e.g. through electrocutaneous stimulation in the median nerve dermatome) can also inhibit tic expression, plausibly through entrainment of incompatible brain rhythms (Maiquez et al., 2020). Here, we consider further what the different treatment approaches may say about the pathophysiology of TS. There are multiple reviews on treatments of TS, including evidence-based evaluations and clinical guidelines (Andrén et al., 2021; Eddy, Rickards, & Cavanna, 2011; Jankovic, 2020; Müller-Vahl et al., 2021; Pringsheim et al., 2019; Robertson, 2000; Roessner et al., 2011; Szejko et al., 2021; Verdellen et al., 2011). With the probable exception of DBS, these focus on what treatments work, based on clinical evidence rather than inferred neuropathological knowledge. In this context, it is fair to say that beyond dopamine receptor inhibition (with commonly used D2 receptor antagonists putatively enhancing indirect pathway inhibition, and emerging D1 antagonists acting on the direct pathway), pharmacotherapies appear less proximate to the motor pathways. Dopamine depleting agents (e.g. Tetrabenzine) do still retain a role in tic management for similar reasons. Mostly in children, alpha-2 agonists (Clonidine and Guanfacine) highlight the involvement of noradrenaline in tics. As with the treatment of comorbid ADHD with other noradrenergic agents (e.g. Methylphenidate or Atomoxetine), some of the benefit on tic expression may arise through the improved deployment of cognitive attentional resources toward tic suppression and avoidance of tic-triggering situations. These medicines also diminish anxiety and arousal in body (hence decrease interoceptive signaling of stress states), and the sedative effect is sleep promoting, which may have a positive impact on attention-dependent cognitive (frontal) control. However, alpha-2 receptors are heavily located within the basal ganglia (MacDonald & Scheinin, 1995) and are implicated in dopamine regulation (Yamamoto & Novotney, 1998). Thus, therapeutic effects of alpha-2 agonists likely arise due to a combination of impacts on cortical function and CSTC circuitry.

In many guidelines for tic management, the role of serotonergic agents is underplayed, yet selective serotonin reuptake inhibitors (SSRIs) remain at the forefront of management options for TS-relevant conditions including severe OCD and social anxiety. The targeting of premonitory sensations (and cravings and urges) with SSRIs suggest interoceptive mechanisms are relevant to their usefulness in tic management. However, SSRIs normalize abnormal orbitofrontal-caudate activity in OCD patients, suggesting again a more direct modulatory role in CSTC circuit function, in part through regulation of dopamine release. The emerging story with cannabinoids as potential therapy for TS suggest again the combination of indirect action in obviating tic-prone brain states in conjunction with basal ganglia function (Jankovic, 2020; Müller-Vahl et al., 2021).

Behavioral therapies, notably HRT and ERP, share similar mechanisms in promoting enhanced awareness of premonitory sensations (akin perhaps to our speculations above with respect to training interoception) to allow for prolongation of tolerance of the urge, and enable for one or more competing response to be actioned as an alternative to the expression of the emergent tic. Perhaps relatedly, the treatment of tics with botulinum toxin injection into target muscle(s), including even vocal cords, not only aborts the expression of target tics, but typically diminishes premonitory sensations and may elicit enduring effects, despite the return from paralysis of the muscle (Porta, Maggioni, Ottaviani, & Schindler, 2004). These observations, along with the typically sustained impact of effective psychological therapies, attest to plasticity of the neurobiological substrates underpinning tic expression. This plasticity may also underpin the often frustrating fluctuating course and changing pattern of tics characteristic of TS. However, it is also a basis for therapeutic optimism and natural improvement in symptoms that occurs in the majority of patients with TS.

8. Conclusions

There is increasing knowledge regarding the core neurobiological substrates for tic expression and the influence at multiple levels of understanding how contextual, genetic, environmental, and social factors underpin symptom expression of TS. In parallel, animal models of TS are able to tackle the complexity of TS as a neurodevelopmental condition more robustly. Such reverse translation can only grow as more research tools become available for fine-grained dissection of functional neural circuits. These advances, in parallel with improving resolution and replicability of human neuroimaging studies, will continue to enhance our understanding of the pathophysiology of TS and inform the optimization and availability of converging treatments.

References

Abi-Jaoude, E., Segura, B., Obeso, I., Cho, S. S., Houle, S., Lang, A. E., et al. (2015). Similar striatal D2/D3 dopamine receptor availability in adults with Tourette syndrome compared with healthy controls: A [11C]-(+)-PHNO and [11C]raclopride positron emission tomography imaging study. *Human Brain Mapping*, *36*(7), 2592–2601. https://doi.org/10.1002/hbm.22793.

Albin, R. L., Koeppe, R. A., Wernette, K., Zhuang, W., Nichols, T., Kilbourn, M. R., et al. (2009). Striatal [11C]dihydrotetrabenazine and [11C] methylphenidate binding in Tourette syndrome. *Neurology*, *72*(16), 1390–1396. https://doi.org/10.1212/WNL.0b013e3181a187dd.

Albin, R. L., Young, A. B., & Penney, J. B. (1989). The functional anatomy of basal ganglia disorders. *Trends in Neurosciences*, *12*(10), 366–375. https://doi.org/10.1016/0166-2236(89)90074-X.

Alexander, G. E., & Crutcher, M. D. (1990). Functional architecture of basal ganglia circuits: Neural substrates of parallel processing. *Trends in Neurosciences*, *13*(7), 266–271. https://doi.org/10.1016/0166-2236(90)90107-L.

Andrén, P., Jakubovski, E., Murphy, T. L., Woitecki, K., Tarnok, Z., Zimmerman-Brenner, S., et al. (2021). European clinical guidelines for Tourette syndrome and other tic disorders—version 2.0. Part II: Psychological interventions. *European Child and Adolescent Psychiatry*, 1–21. https://doi.org/10.1007/s00787-021-01845-z.

Aouizerate, B., Guehl, D., Cuny, E., Rougier, A., Bioulac, B., Tignol, J., et al. (2004). Pathophysiology of obsessive-compulsive disorder: A necessary link between phenomenology, neuropsychology, imagery and physiology. *Progress in Neurobiology*, *72*(3), 195–221. https://doi.org/10.1016/j.pneurobio.2004.02.004.

Aron, A. R., Behrens, T. E., Smith, S., Frank, M. J., & Poldrack, R. A. (2007). Triangulating a cognitive control network using diffusion-weighted magnetic resonance imaging (MRI) and functional MRI. *Journal of Neuroscience*, *27*(14), 3743–3752. https://doi.org/10.1523/JNEUROSCI.0519-07.2007.

Aron, A. R., Herz, D. M., Brown, P., Forstmann, B. U., & Zaghloul, K. (2016). Frontosubthalamic circuits for control of action and cognition. *Journal of Neuroscience*, *36*(45), 11489–11495. https://doi.org/10.1523/JNEUROSCI.2348-16.2016.

Badenoch, J., Searle, T., Watson, I., & Cavanna, A. E. (2020). Sensory symptoms in body-focused repetitive behaviors, restless legs syndrome, and Tourette syndrome: An overlap? *Neuroscience and Biobehavioral Reviews*, *119*, 320–332. https://doi.org/10.1016/j.neubiorev.2020.10.008.

Baxter, L. R., Schwartz, J. M., Mazziotta, J. C., Phelps, M. E., Pahl, J. J., Guze, B. H., et al. (1988). Cerebral glucose metabolic rates in nondepressed patients with obsessive-compulsive disorder. *American Journal of Psychiatry, 145*(12), 1560–1563. https://doi.org/10.1176/ajp.145.12.1560.

Berman, B. D., Horovitz, S. G., Morel, B., & Hallett, M. (2012). Neural correlates of blink suppression and the buildup of a natural bodily urge. *NeuroImage, 59*(2), 1441–1450. https://doi.org/10.1016/J.NEUROIMAGE.2011.08.050.

Bohlhalter, S., Goldfine, A., Matteson, S., Garraux, G., Hanakawa, T., Kansaku, K., et al. (2006). Neural correlates of tic generation in Tourette syndrome: An event-related functional MRI study. *Brain, 129*(8), 2029–2037. https://doi.org/10.1093/brain/awl050.

Brandt, V. C., Beck, C., Sajin, V., Baaske, M. K., Bäumer, T., Beste, C., et al. (2016). Temporal relationship between premonitory urges and tics in Gilles de la Tourette syndrome. *Cortex, 77*, 24–37. https://doi.org/10.1016/j.cortex.2016.01.008.

Brener, J., & Ring, C. (2016). Towards a psychophysics of interoceptive processes: The measurement of heartbeat detection. *Philosophical Transactions of the Royal Society B: Biological Sciences, 371*(1708). https://doi.org/10.1098/rstb.2016.0015.

Bronfeld, M., Israelashvili, M., & Bar-Gad, I. (2013). Pharmacological animal models of Tourette syndrome. *Neuroscience and Biobehavioral Reviews, 37*(6), 1101–1119. https://doi.org/10.1016/j.neubiorev.2012.09.010.

Browne, H. A., Hansen, S. N., Buxbaum, J. D., Gair, S. L., Nissen, J. B., Nikolajsen, K. H., et al. (2015). Familial clustering of tic disorders and obsessive-compulsive disorder. *JAMA Psychiatry, 72*(4), 359–366. https://doi.org/10.1001/jamapsychiatry.2014.2656.

Buse, J., Schoenefeld, K., Münchau, A., & Roessner, V. (2013). Neuromodulation in Tourette syndrome: Dopamine and beyond. *Neuroscience and Biobehavioral Reviews, 37*(6), 1069–1084. https://doi.org/10.1016/j.neubiorev.2012.10.004.

Calabresi, P., Picconi, B., Tozzi, A., Ghiglieri, V., & Di Filippo, M. (2014). Direct and indirect pathways of basal ganglia: A critical reappraisal. *Nature Neuroscience, 17*(8), 1022–1030. https://doi.org/10.1038/nn.3743.

Calzà, J., Gürsel, D. A., Schmitz-Koep, B., Bremer, B., Reinholz, L., Berberich, G., et al. (2019). Altered cortico-striatal functional connectivity during resting state in obsessive-compulsive disorder. *Frontiers in Psychiatry, 10*(MAY), 319. https://doi.org/10.3389/fpsyt.2019.00319.

Cavanna, A. E., & Nani, A. (2013). Tourette syndrome and consciousness of action. *Tremor and Other Hyperkinetic Movements, 3*, 3. https://doi.org/10.5334/tohm.169.

Cavanna, A. E., Servo, S., Monaco, F., & Robertson, M. M. (2009). The behavioral spectrum of gilles de la Tourette syndrome. *Journal of Neuropsychiatry and Clinical Neurosciences, 21*(1), 13–23. https://doi.org/10.1176/jnp.2009.21.1.13.

Chikama, M., McFarland, N. R., Amaral, D. G., & Haber, S. N. (1997). Insular cortical projections to functional regions of the striatum correlate with cortical cytoarchitectonic organization in the primate. *Journal of Neuroscience, 17*(24), 9686–9705. https://doi.org/10.1523/jneurosci.17-24-09686.1997.

Craig, A. D. (2002). How do you feel? Interoception: The sense of the physiological condition of the body. *Nature Reviews Neuroscience, 3*(8), 655–666. https://doi.org/10.1038/nrn894.

Craig, A. D. (2009). How do you feel—Now? The anterior insula and human awareness. *Nature Reviews Neuroscience, 10*(1), 59–70. https://doi.org/10.1038/nrn2555.

Craig, A. D. (2014). How do you feel?: An interoceptive moment with your neurobiological self. In *How Do You Feel?: An Interoceptive Moment with Your Neurobiological Self* Princeton University Press. https://doi.org/10.3399/bjgp15x687985.

Critchley, H. D. (2005). Neural mechanisms of autonomic, affective, and cognitive integration. *Journal of Comparative Neurology, 493*(1), 154–166. https://doi.org/10.1002/cne.20749.

Critchley, H. D., & Garfinkel, S. N. (2017). Interoception and emotion. *Current Opinion in Psychology, 17*, 7–14. https://doi.org/10.1016/j.copsyc.2017.04.020.

Critchley, H. D., Wiens, S., Rotshtein, P., Öhman, A., & Dolan, R. J. (2004). Neural systems supporting interoceptive awareness. *Nature Neuroscience*, 7(2), 189–195. https://doi.org/10.1038/nn1176.

Diamond, A., Kenney, C., & Jankovic, J. (2006). Effect of vagal nerve stimulation in a case of Tourette's syndrome and complex partial epilepsy. *Movement Disorders*, *21*(8), 1273–1275. https://doi.org/10.1002/mds.20949.

Draper, A., Jackson, G. M., Morgan, P. S., & Jackson, S. R. (2016). Premonitory urges are associated with decreased grey matter thickness within the insula and sensorimotor cortex in young people with Tourette syndrome. *Journal of Neuropsychology*, *10*(1), 143–153. https://doi.org/10.1111/jnp.12089.

Draper, A., Jude, L., Jackson, G. M., & Jackson, S. R. (2015). Motor excitability during movement preparation in Tourette syndrome. *Journal of Neuropsychology*, *9*(1), 33–44. https://doi.org/10.1111/jnp.12033.

Draper, A., Stephenson, M. C., Jackson, G. M., Pépés, S., Morgan, P. S., Morris, P. G., et al. (2014). Increased GABA contributes to enhanced control over motor excitability in tourette syndrome. *Current Biology*, *24*(19), 2343–2347. https://doi.org/10.1016/j.cub.2014.08.038.

Dyke, K., Jackson, G. M., Nixon, E., & Jackson, S. R. (2019). Effects of single-session cathodal transcranial direct current stimulation on tic symptoms in Tourette's syndrome. *Experimental Brain Research*, *237*(11), 2853–2863. https://doi.org/10.1007/s00221-019-05637-5.

Eapen, V., Cavanna, A. E., & Robertson, M. M. (2016). Comorbidities, social impact, and quality of life in tourette syndrome. *Frontiers in Psychiatry*, 7(JUN), 5–10. https://doi.org/10.3389/fpsyt.2016.00097.

Eapen, V., & Robertson, M. M. (2015). Are there distinct subtypes in Tourette syndrome? Pure-Tourette syndrome versus Tourette syndrome-plus, and simple versus complex tics. *Neuropsychiatric Disease and Treatment*, *11*, 1431–1436. https://doi.org/10.2147/NDT.S72284.

Eddy, C. M. (2018). Social cognition and self-other distinctions in neuropsychiatry: Insights from schizophrenia and Tourette syndrome. In *Vol. 82. Progress in neuropsychopharmacology and biological psychiatry* (pp. 69–85). Prog Neuropsychopharmacol Biol Psychiatry. https://doi.org/10.1016/j.pnpbp.2017.11.026.

Eddy, C. M., & Cavanna, A. E. (2013). On being your own worst enemy: An investigation of socially inappropriate symptoms in Tourette syndrome. *Journal of Psychiatric Research*, *47*(9), 1259–1263. https://doi.org/10.1016/J.JPSYCHIRES.2013.05.019.

Eddy, C. M., Cavanna, A. E., Rickards, H. E., & Hansen, P. C. (2016). Temporo-parietal dysfunction in Tourette syndrome: Insights from an fMRI study of Theory of Mind. *Journal of Psychiatric Research*, *81*, 102–111. https://doi.org/10.1016/j.jpsychires.2016.07.002.

Eddy, C. M., Rickards, H. E., & Cavanna, A. E. (2011). Treatment strategies for tics in Tourette syndrome. *Therapeutic Advances in Neurological Disorders*, *4*(1), 25–45. https://doi.org/10.1177/1756285610390261.

Finis, J., Enticott, P. G., Pollok, B., Münchau, A., Schnitzler, A., & Fitzgerald, P. B. (2013). Repetitive transcranial magnetic stimulation of the supplementary motor area induces echophenomena. *Cortex*, *49*(7), 1978–1982. https://doi.org/10.1016/j.cortex.2012.08.019.

Finis, J., Moczydlowski, A., Pollok, B., Biermann-Ruben, K., Thomalla, G., Heil, M., et al. (2012). Echoes from childhood—imitation in Gilles de la Tourette syndrome. *Movement Disorders*, *27*(4), 562–565. https://doi.org/10.1002/MDS.24913.

Freeman, R. D., Zinner, S. H., Müller-Vahl, K. R., Fast, D. K., Burd, L. J., Kano, Y., et al. (2009). Coprophenomena in Tourette syndrome. *Developmental Medicine and Child Neurology*, *51*(3), 218–227. https://doi.org/10.1111/j.1469-8749.2008.03135.x.

Ganos, C., Garrido, A., Navalpotro-Gómez, I., Ricciardi, L., Martino, D., Edwards, M. J., et al. (2015). Premonitory urge to tic in tourette's is associated with interoceptive awareness. *Movement Disorders*, *30*(9), 1198–1202. https://doi.org/10.1002/mds.26228.

Ganos, C., Kahl, U., Brandt, V., Schunke, O., Bäumer, T., Thomalla, G., et al. (2014). The neural correlates of tic inhibition in Gilles de la Tourette syndrome. *Neuropsychologia*, *65*, 297–301. https://doi.org/10.1016/j.neuropsychologia.2014.08.007.

Ganos, C., Ogrzal, T., Schnitzler, A., & Münchau, A. (2012). The pathophysiology of echopraxia/echolalia: Relevance to Gilles De La Tourette syndrome. *Movement Disorders*, *27*(10), 1222–1229. https://doi.org/10.1002/mds.25103.

Ganos, C., Rocchi, L., Latorre, A., Hockey, L., Palmer, C., Joyce, E. M., et al. (2018). Motor cortical excitability during voluntary inhibition of involuntary tic movements. *Movement Disorders*, *33*(11), 1804–1809. https://doi.org/10.1002/mds.27479.

Ganos, C., Roessner, V., & Münchau, A. (2013). The functional anatomy of Gilles de la Tourette syndrome. *Neuroscience and Biobehavioral Reviews*, *37*(6), 1050–1062. https://doi.org/10.1016/j.neubiorev.2012.11.004.

Ganos, C., Rothwell, J., & Haggard, P. (2018). Voluntary inhibitory motor control over involuntary tic movements. *Movement Disorders*, *33*(6), 937–946. https://doi.org/10.1002/mds.27346.

Garfinkel, S. N., Seth, A. K., Barrett, A. B., Suzuki, K., & Critchley, H. D. (2015). Knowing your own heart: Distinguishing interoceptive accuracy from interoceptive awareness. *Biological Psychology*, *104*, 65–74. https://doi.org/10.1016/j.biopsycho.2014.11.004.

Garfinkel, S. N., Tiley, C., O'Keeffe, S., Harrison, N. A., Seth, A. K., & Critchley, H. D. (2016). Discrepancies between dimensions of interoception in autism: Implications for emotion and anxiety. *Biological Psychology*, *114*, 117–126. https://doi.org/10.1016/j.biopsycho.2015.12.003.

Gilbert, D. L., Murphy, T. K., Jankovic, J., Budman, C. L., Black, K. J., Kurlan, R. M., et al. (2018). Ecopipam, a D1 receptor antagonist, for treatment of tourette syndrome in children: A randomized, placebo-controlled crossover study. *Movement Disorders*, *33*(8), 1272–1280. https://doi.org/10.1002/MDS.27457.

Gittis, A. H., Nelson, A. B., Thwin, M. T., Palop, J. J., & Kreitzer, A. C. (2010). Distinct roles of GABAergic interneurons in the regulation of striatal output pathways. *Journal of Neuroscience*, *30*(6), 2223–2234. https://doi.org/10.1523/JNEUROSCI.4870-09.2010.

Grados, M., Huselid, R., & Duque-Serrano, L. (2018). Transcranial magnetic stimulation in tourette syndrome: A historical perspective, its current use and the influence of comorbidities in treatment response. *Brain Sciences*, *8*(7). https://doi.org/10.3390/brainsci8070129.

Grice-Jackson, T., Critchley, H. D., Banissy, M. J., & Ward, J. (2017a). Common and distinct neural mechanisms associated with the conscious experience of vicarious pain. *Cortex*, *94*, 152–163. https://doi.org/10.1016/j.cortex.2017.06.015.

Grice-Jackson, T., Critchley, H. D., Banissy, M. J., & Ward, J. (2017b). Consciously feeling the pain of others reflects atypical functional connectivity between the pain matrix and frontal-parietal regions. *Frontiers in Human Neuroscience*, *11*. https://doi.org/10.3389/fnhum.2017.00507.

Haber, S. (2016). Corticostriatal circuitry. *Dialogues in Clinical Neuroscience*, *18*(1), 7–21. https://doi.org/10.31887/DCNS.2016.18.1/SHABER.

Haber, S. N., Yendiki, A., & Jbabdi, S. (2021). Four deep brain stimulation targets for obsessive-compulsive disorder: Are they different? *Biological Psychiatry*, *10*, 667–677. https://doi.org/10.1016/j.biopsych.2020.06.031.

Hartmann, A., & Worbe, Y. (2013). Pharmacological treatment of Gilles de la Tourette syndrome. *Neuroscience and Biobehavioral Reviews*, *37*(6), 1157–1161. https://doi.org/10.1016/j.neubiorev.2012.10.014.

Hawksley, J., Cavanna, A., & Nagai, Y. (2015). The role of the autonomic nervous system in Tourette syndrome. *Frontiers in Neuroscience*, *9*(Mar). https://doi.org/10.3389/fnins.2015.00117.

Hienert, M., Gryglewski, G., Stamenkovic, M., Kasper, S., & Lanzenberger, R. (2018). Striatal dopaminergic alterations in Tourette's syndrome: A meta-analysis based on 16 PET and SPECT neuroimaging studies. *Translational Psychiatry*, *8*(1). https://doi.org/10.1038/s41398-018-0202-y.

Insel, T. R., Donnelly, E. F., Lalakea, M. L., Alterman, I. S., & Murphy, D. L. (1983). Neurological and neuropsychological studies of patients with obsessive-compulsive disorder. *Biological Psychiatry*, *18*(7), 741–751. https://pubmed.ncbi.nlm.nih.gov/6615935/.

Jackson, G. M., Draper, A., Dyke, K., Pépés, S. E., & Jackson, S. R. (2015). Inhibition, disinhibition, and the control of action in Tourette syndrome. *Trends in Cognitive Sciences*, *19*(11), 655–665. https://doi.org/10.1016/j.tics.2015.08.006.

Jackson, S. R., Loayza, J., Crighton, M., Sigurdsson, H. P., Dyke, K., & Jackson, G. M. (2020). The role of the insula in the generation of motor tics and the experience of the premonitory urge-to-tic in Tourette syndrome. *Cortex*, *126*, 119–133. https://doi.org/10.1016/j.cortex.2019.12.021.

Jackson, S. R., Parkinson, A., Kim, S. Y., Schüermann, M., & Eickhoff, S. B. (2011). On the functional anatomy of the urge-for-action. *Cognitive Neuroscience*, *2*(3–4), 227–243. https://doi.org/10.1080/17588928.2011.604717.

Jankovic, J. (1997). Phenomenology and classification of tics. *Neurologic Clinics*, *15*(2), 267–275. https://doi.org/10.1016/S0733-8619(05)70311-X.

Jankovic, J. (2020). Treatment of tics associated with Tourette syndrome. *Journal of Neural Transmission*, *127*(5), 843–850. https://doi.org/10.1007/s00702-019-02105-w.

Kalanithi, P. S. A., Zheng, W., Kataoka, Y., DiFiglia, M., Grantz, H., Saper, C. B., et al. (2005). Altered parvalbumin-positive neuron distribution in basal ganglia of individuals with Tourette syndrome. *Proceedings of the National Academy of Sciences of the United States of America*, *102*(37), 13307–13312. https://doi.org/10.1073/pnas.0502624102.

Kataoka, Y., Kalanithi, P. S. A., Grantz, H., Schwartz, M. L., Saper, C., Leckman, J. F., et al. (2010). Decreased number of parvalbumin and cholinergic interneurons in the striatum of individuals with tourette syndrome. *Journal of Comparative Neurology*, *518*(3), 277–291. https://doi.org/10.1002/cne.22206.

Kwak, C., Dat Vuong, K., & Jankovic, J. (2003). Premonitory sensory phenomenon in Tourette's syndrome. *Movement Disorders*, *18*(12), 1530–1533. https://doi.org/10.1002/mds.10618.

Leckman, J. F., Walker, D. E., & Cohen, D. J. (1993). Premonitory urges in Tourette's syndrome. *The American Journal of Psychiatry*, *150*(1), 98–102. https://doi.org/10.1176/AJP.150.1.98.

Leckman, J. F., Walker, D. E., Goodman, W. K., Pauls, D. L., & Cohen, D. J. (1994). "Just right" perceptions associated with compulsive behavior in Tourette's syndrome. *American Journal of Psychiatry*, *151*(5), 675–680. https://doi.org/10.1176/ajp.151.5.675.

Lennington, J. B., Coppola, G., Kataoka-Sasaki, Y., Fernandez, T. V., Palejev, D., Li, Y., et al. (2016). Transcriptome analysis of the human striatum in Tourette syndrome. *Biological Psychiatry*, *79*(5), 372–382. https://doi.org/10.1016/j.biopsych.2014.07.018.

Lerner, A., Bagic, A., Simmons, J. M., Mari, Z., Bonne, O., Xu, B., et al. (2012). Widespread abnormality of the γ-aminobutyric acid-ergic system in Tourette syndrome. *Brain*, *135*(6), 1926–1936. https://doi.org/10.1093/brain/aws104.

MacDonald, E., & Scheinin, M. (1995). Distribution and pharmacology of alpha 2-adrenoceptors in the central nervous system—PubMed. *Journal of Physiology and Pharmacology*, *46*(3), 241–258. https://pubmed.ncbi.nlm.nih.gov/8527807/.

Maiquez, B. M., Sigurdsson, H. P., Dyke, K., Clarke, E., McGrath, P., Pasche, M., et al. (2020). Entraining movement-related brain oscillations to suppress tics in Tourette syndrome. *Current Biology*, *30*(12), 2334–2342.e3. https://doi.org/10.1016/J.CUB.2020.04.044.

Makovac, E., Fagioli, S., Rae, C. L., Critchley, H. D., & Ottaviani, C. (2020). Can't get it off my brain: Meta-analysis of neuroimaging studies on perseverative cognition. *Psychiatry Research - Neuroimaging*, *295*(August 2019), 111020. https://doi.org/10.1016/j.pscychresns.2019.111020.

Mancini, C., Cardona, F., Baglioni, V., Panunzi, S., Pantano, P., Suppa, A., et al. (2018). Inhibition is impaired in children with obsessive-compulsive symptoms but not in those with tics. *Movement Disorders*, *33*(6), 950–959. https://doi.org/10.1002/mds.27406.

Martinez-Ramirez, D., Jimenez-Shahed, J., Leckman, J. F., Porta, M., Servello, D., Meng, F. G., et al. (2018). Efficacy and safety of deep brain stimulation in tourette syndrome the international tourette syndrome deep brain stimulation public database and registry. *JAMA Neurology*, *75*(3), 353–359. https://doi.org/10.1001/jamaneurol.2017.4317.

McCairn, K. W., Bronfeld, M., Belelovsky, K., & Bar-Gad, I. (2009). The neurophysiological correlates of motor tics following focal striatal disinhibition. *Brain*, *132*(8), 2125–2138. https://doi.org/10.1093/brain/awp142.

Müller-Vahl, K. R., Szejko, N., Verdellen, C., Roessner, V., Hoekstra, P. J., Hartmann, A., et al. (2021). European clinical guidelines for Tourette syndrome and other tic disorders: Summary statement. *European Child and Adolescent Psychiatry*, *1*, 1–6. https://doi.org/10.1007/s00787-021-01832-4.

Nagai, Y., Cavanna, A., & Critchley, H. D. (2009). Influence of sympathetic autonomic arousal on tics: Implications for a therapeutic behavioral intervention for Tourette syndrome. *Journal of Psychosomatic Research*, *67*(6), 599–605. https://doi.org/10.1016/J.JPSYCHORES.2009.06.004.

Nagai, Y., Critchley, H. D., Featherstone, E., Trimble, M. R., & Dolan, R. J. (2004). Activity in ventromedial prefrontal cortex covaries with sympathetic skin conductance level: A physiological account of a "default mode" of brain function. *NeuroImage*, *22*(1), 243–251. https://doi.org/10.1016/j.neuroimage.2004.01.019.

Nambu, A., Tokuno, H., & Takada, M. (2002). Functional significance of the cortico-subthalamo-pallidal "hyperdirect" pathway. *Neuroscience Research*, *43*(2), 111–117. https://doi.org/10.1016/S0168-0102(02)00027-5.

Neal, M., & Cavanna, A. E. (2013). Not just right experiences in patients with Tourette syndrome: Complex motor tics or compulsions? *Psychiatry Research*, *210*(2), 559–563. https://doi.org/10.1016/j.psychres.2013.06.033.

Nespoli, E., Rizzo, F., Boeckers, T. M., Hengerer, B., & Ludolph, A. G. (2016). Addressing the complexity of Tourette's syndrome through the use of animal models. *Frontiers in Neuroscience*, *10*(APR), 133. https://doi.org/10.3389/fnins.2016.00133.

Neuner, I., Schneider, F., & Shah, N. J. (2013). Functional neuroanatomy of tics. In *Vol. 112. International Review of Neurobiology* (1st ed.). Elsevier Inc. https://doi.org/10.1016/B978-0-12-411546-0.00002-0.

Neuner, I., Werner, C. J., Arrubla, J., Stöcker, T., Ehlen, C., Wegener, H. P., et al. (2014). Imaging the where and when of tic generation and resting state networks in adult Tourette patients. *Frontiers in Human Neuroscience*, *8*(May), 1–16. https://doi.org/10.3389/fnhum.2014.00362.

Openneer, T. J. C., van der Meer, D., Marsman, J. B. C., Forde, N. J., Akkermans, S. E. A., Naaijen, J., et al. (2021). Impaired response inhibition during a stop-signal task in children with Tourette syndrome is related to ADHD symptoms: A functional magnetic resonance imaging study. *World Journal of Biological Psychiatry*, *22*(5), 350–361. https://doi.org/10.1080/15622975.2020.1813329.

Orth, M. (2009). Transcranial magnetic stimulation in Gilles de la Tourette syndrome. *Journal of Psychosomatic Research*, *67*(6), 591–598. https://doi.org/10.1016/j.jpsychores.2009.07.014.

Paciorek, A., & Skora, L. (2020). Vagus nerve stimulation as a gateway to interoception. *Frontiers in Psychology*, *11*, 1659. https://doi.org/10.3389/fpsyg.2020.01659.

Pile, V., Lau, J. Y. F., Topor, M., Hedderly, T., & Robinson, S. (2018). Interoceptive accuracy in youth with tic disorders: Exploring links with premonitory urge, anxiety and quality of life. *Journal of Autism and Developmental Disorders*, *48*, 3474–3482. https://doi.org/10.1007/s10803-018-3608-8.

Polyanska, L., Critchley, H. D., & Rae, C. L. (2017). Centrality of prefrontal and motor preparation cortices to Tourette syndrome revealed by meta-analysis of task-based neuroimaging studies. *NeuroImage: Clinical*, *16*(July), 257–267. https://doi.org/10.1016/j.nicl.2017.08.004.

Porta, M., Maggioni, G., Ottaviani, F., & Schindler, A. (2004). Treatment of phonic tics in patients with Tourette's syndrome using botulinum toxin type A. *Neurological Sciences*, *24*(6), 420–423. https://doi.org/10.1007/s10072-003-0201-4.

Pringsheim, T., Okun, M. S., Müller-Vahl, K., Martino, D., Jankovic, J., Cavanna, A. E., et al. (2019). Practice guideline recommendations summary: Treatment of tics in people with Tourette syndrome and chronic tic disorders. *Neurology*, *92*(19), 896–906. https://doi.org/10.1212/WNL.0000000000007466.

Quadt, L., Garfinkel, S. N., Mulcahy, J. S., Larsson, D. E., Silva, M., Jones, A. M., et al. (2021). Interoceptive training to target anxiety in autistic adults (ADIE): A single-center, superiority randomized controlled trial. *EClinicalMedicine*, *39*, 101042. https://doi.org/10.1016/j.eclinm.2021.101042.

Rae, C. L., Critchley, H. D., & Seth, A. K. (2019). A Bayesian account of the sensory-motor interactions underlying symptoms of Tourette syndrome. *Frontiers in Psychiatry*, *10*(March). https://doi.org/10.3389/fpsyt.2019.00029.

Rae, C. L., Hughes, L. E., Anderson, M. C., & Rowe, J. B. (2015). The prefrontal cortex achieves inhibitory control by facilitating subcortical motor pathway connectivity. *Journal of Neuroscience*, *35*(2), 786–794. https://doi.org/10.1523/JNEUROSCI.3093-13.2015.

Rae, C. L., Larsson, D. E. O., Garfinkel, S. N., & Critchley, H. D. (2019). Dimensions of interoception predict premonitory urges and tic severity in Tourette syndrome. *Psychiatry Research*, *271*(September 2018), 469–475. https://doi.org/10.1016/j.psychres.2018.12.036.

Rae, C. L., Parkinson, J., Betka, S., Gouldvan Praag, C. D., Bouyagoub, S., Polyanska, L., et al. (2020). Amplified engagement of prefrontal cortex during control of voluntary action in Tourette syndrome. *Brain Communications*, *2*(2). https://doi.org/10.1093/BRAINCOMMS/FCAA199.

Rae, C. L., Polyanska, L., Gould Van Praag, C. D., Parkinson, J., Bouyagoub, S., Nagai, Y., et al. (2018). Face perception enhances insula and motor network reactivity in Tourette syndrome. *Brain*, *141*(11), 3249–3261. https://doi.org/10.1093/brain/awy254.

Rauch, S. L., Wedig, M. M., Wright, C. I., Martis, B., McMullin, K. G., Shin, L. M., et al. (2007). Functional magnetic resonance imaging study of regional brain activation during implicit sequence learning in obsessive-compulsive disorder. *Biological Psychiatry*, *61*(3), 330–336. https://doi.org/10.1016/j.biopsych.2005.12.012.

Redgrave, P., Rodriguez, M., Smith, Y., Rodriguez-Oroz, M. C., Lehericy, S., Bergman, H., et al. (2010). Goal-directed and habitual control in the basal ganglia: Implications for Parkinson's disease. *Nature Reviews Neuroscience*, *11*(11), 760–772. https://doi.org/10.1038/nrn2915.

Robertson, M. M. (2000). Tourette syndrome, associated conditions and the complexities of treatment. *Brain*, *123*(3), 425–462. https://doi.org/10.1093/brain/123.3.425.

Robertson, M. M. (2015). A personal 35 year perspective on Gilles de la Tourette syndrome: Prevalence, phenomenology, comorbidities, and coexistent psychopathologies. *The Lancet Psychiatry*, *2*(1), 68–87. https://doi.org/10.1016/S2215-0366(14)00132-1.

Roessner, V., Plessen, K. J., Rothenberger, A., Ludolph, A. G., Rizzo, R., Skov, L., et al. (2011). European clinical guidelines for Tourette syndrome and other tic disorders. Part II: Pharmacological treatment. *European Child and Adolescent Psychiatry*, *20*(4), 173–196. https://doi.org/10.1007/s00787-011-0163-7.

Saxena, S., & Rauch, S. L. (2000). Functional neuroimaging and the neuroanatomy of obsessive-compulsive disorder. *Psychiatric Clinics of North America*, *23*(3), 563–586. https://doi.org/10.1016/S0193-953X(05)70181-7.

Senberg, A., Münchau, A., Münte, T., Beste, C., & Roessner, V. (2021). Swearing and coprophenomena—A multidimensional approach. *Neuroscience & Biobehavioral Reviews*, *126*, 12–22. https://doi.org/10.1016/J.NEUBIOREV.2021.03.016.

Sigurdsson, H. P., Pépés, S. E., Jackson, G. M., Draper, A., Morgan, P. S., & Jackson, S. R. (2018). Alterations in the microstructure of white matter in children and adolescents with Tourette syndrome measured using tract-based spatial statistics and probabilistic tractography. *Cortex*, *104*, 75–89. https://doi.org/10.1016/j.cortex.2018.04.004.

Sperling, W., Reulbach, U., Maihöfner, C., Kornhuber, J., & Bleich, S. (2008). Vagus nerve stimulation in a patient with Gilles de la Tourette Syndrome and major depression. *Pharmacopsychiatry*, *41*(3), 117–118. https://doi.org/10.1055/s-2008-1062698.

Spessot, A. L., Plessen, K. J., & Peterson, B. S. (2004). Neuroimaging of developmental psychopathologies: The importance of self-regulatory and neuroplastic processes in adolescence. *Annals of the New York Academy of Sciences*, *1021*(1), 86–104. https://doi.org/10.1196/ANNALS.1308.010.

Stern, E., Silbersweig, D. A., Chee, K. Y., Holmes, A., Robertson, M. M., Trimble, M., et al. (2000). Functional neuroanatomy of tics in Tourette syndrome. *Archives of General Psychiatry*, *57*(8), 741–748. https://doi.org/10.1001/archpsyc.57.8.741.

Swedo, S. E., Schapiro, M. B., Grady, C. L., Cheslow, D. L., Leonard, H. L., Kumar, A., et al. (1989). Cerebral glucose metabolism in childhood-onset obsessive-compulsive disorder. *Archives of General Psychiatry*, *46*(6), 518–523. https://doi.org/10.1001/archpsyc.1989.01810060038007.

Szejko, N., Worbe, Y., Hartmann, A., Visser-Vandewalle, V., Ackermans, L., Ganos, C., et al. (2021). European clinical guidelines for Tourette syndrome and other tic disorders—Version 2.0. Part IV: Deep brain stimulation. *European Child & Adolescent Psychiatry*, *2021*, 1–19. https://doi.org/10.1007/S00787-021-01881-9.

Tinaz, S., Malone, P., Hallett, M., & Horovitz, S. G. (2015). Role of the right dorsal anterior insula in the urge to tic in tourette syndrome. *Movement Disorders*, *30*(9), 1190–1197. https://doi.org/10.1002/mds.26230.

Verdellen, C., Van De Griendt, J., Hartmann, A., Murphy, T., Androutsos, C., Aschauer, H., et al. (2011). European clinical guidelines for Tourette syndrome and other tic disorders. Part III: Behavioural and psychosocial interventions. *European Child and Adolescent Psychiatry*, *20*(4), 197–207. https://doi.org/10.1007/s00787-011-0167-3.

Wagner-Altendorf, T. A., Gottschlich, C., Robert, C., Cirkel, A., Heldmann, M., & Münte, T. F. (2020). The suppression of Taboo word spoonerisms Is associated with altered medial frontal negativity: An ERP study. *Frontiers in Human Neuroscience*, *14*, 368. https://doi.org/10.3389/fnhum.2020.00368.

Wang, Z., Maia, T. V., Marsh, R., Gerber, A., & Peterson, B. S. (2011). The neural circuits that generate tics in Tourette's syndrome. *American Journal of Psychiatry*, *168*, 1326–1337.

Wen, H., Liu, Y., Wang, J., Rekik, I., Zhang, J., Zhang, Y., et al. (2016). Combining tract- and atlas-based analysis reveals microstructural abnormalities in early Tourette syndrome children. *Human Brain Mapping*, *37*(5), 1903–1919. https://doi.org/10.1002/hbm.23146.

Wen, F., Yan, J., Yu, L., Wang, F., Liu, J., Li, Y., et al. (2021). Grey matter abnormalities in Tourette syndrome: An activation likelihood estimation meta-analysis. *BMC Psychiatry*, *21*(1). https://doi.org/10.1186/s12888-021-03187-1.

Woods, D. W., Piacentini, J., Chang, S., Deckersbach, T., Ginsburg, G., Peterson, A., et al. (2015). Managing Tourette syndrome. In *Managing Tourette Syndrome*, Oxford University Press. https://doi.org/10.1093/MED:PSYCH/9780195341287.001.0001.

Woods, D. W., Piacentini, J., Himle, M. B., & Chang, S. (2005). Premonitory urge for tics scale (PUTS): Initial psychometric results and examination of the premonitory urge phenomenon in youths with tic disorders. *Journal of Developmental and Behavioral Pediatrics*, *26*(6), 397–403. https://doi.org/10.1097/00004703-200512000-00001.

Worbe, Y., Mallet, L., Golmard, J. L., Béhar, C., Durif, F., Jalenques, I., et al. (2010). Repetitive behaviours in patients with Gilles de la Tourette syndrome: Tics, compulsions, or both? *PLoS One*, *5*(9), e12959. https://doi.org/10.1371/journal.pone.0012959.

Worbe, Y., Marrakchi-Kacem, L., Lecomte, S., Valabregue, R., Poupon, F., Guevara, P., et al. (2015). Altered structural connectivity of cortico-striato-pallido-thalamic networks in Gilles de la Tourette syndrome. *Brain*, *138*(2), 472–482. https://doi.org/10.1093/brain/awu311.

Yamamoto, B. K., & Novotney, S. (1998). Regulation of extracellular dopamine by the norepinephrine transporter. *Journal of Neurochemistry*, *71*(1), 274–280. https://doi.org/10.1046/j.1471-4159.1998.71010274.x.

Zapparoli, L., Porta, M., & Paulesu, E. (2015). The anarchic brain in action: The contribution of task-based fMRI studies to the understanding of Gilles de la Tourette syndrome. *Current Opinion in Neurology*, *28*(6), 604–611. https://doi.org/10.1097/WCO.0000000000000261.

Zapparoli, L., Seghezzi, S., & Paulesu, E. (2017). The what, the when, and the whether of intentional action in the brain: A meta-analytical review. *Frontiers in Human Neuroscience*, *11*(May), 1–11. https://doi.org/10.3389/fnhum.2017.00238.

CHAPTER SEVEN

Course of TS from a clinical and neuroimaging perspective

Laura Bogut Andersen*, Camilla Groth*, Liselotte Skov, and Nanette Mol Debes
Pediatric Department, Herlev University Hospital, Herlev, Denmark
*Corresponding author: e-mail address: nanette.marinette.monique.debes@regionh.dk

Contents

1. Introduction

The clinical symptoms in Tourette syndrome (TS) change with age. Some clinical longitudinal studies have examined the development of symptoms over time. Cortico-striato-thalamo-cortical pathways are believed to be involved in the pathophysiology of TS, but knowledge on the influence of the development of the brain during childhood on the development of clinical symptoms is scarce.

In this chapter, we focus on longitudinal aspects of the development of TS from clinical and neuroradiological perspectives. Firstly, we will describe the clinical development and course of tics and comorbidities and secondly the findings of neurodevelopmental brain imaging studies in TS.

International Review of Movement Disorders, Volume 3
ISSN 2666-7878
https://doi.org/10.1016/bs.irmvd.2021.09.003

2. Clinical development and course of tics and comorbidities

Onset of tics commonly first appears between the ages of 4 and 6 years, with simple motor tics developing into more complex tics and vocal tics (Bloch & Leckman, 2009; Freeman et al., 2000; Groth, Mol Debes, Rask, Lange, & Skov, 2017; Leckman et al., 1998). Retrospective studies have found symptoms of TS associated comorbidities as OCD, ADHD, behavioral problems or anxiety symptoms presenting before the onset of tic (Hirschtritt et al., 2015; Leckman, 2002) and only 40% presenting with tics (Mol Debes, Hjalgrim, & Skov, 2008). Onset of vocal tic or behavioral problems often lead to referrals to other professionals such as ophthalmologist, ear/nose/throat physician or psychologist before TS is diagnosed and can cause an diagnostic delay (Freeman et al., 2000; Mol Debes et al., 2008).

Tics typically occur in bouts over the course of a day, wax and wane over weeks and months.

Recent reviews have estimated the prevalence of TS in childhood to be approximately 0.8% (range 0.3%–5.7%) affecting predominantly boys with a male: female ratio of risk 3–4:1(Khalifa & Knorring, 2003; Robertson, 2015; Scharf et al., 2015). Estimates of adulthood prevalence of TS are more sparse and seem more affected in differences in case identification. A meta-analysis from 2019 estimates an adulthood prevalence of TS of 118 per million adults (CI 95%, 19–751 cases per million adult) based on 3 prevalence studies of adults and with a wide range. However all studies find a lower male: female ratio of risk of 2.33. (Levine, Szejko, & Bloch, 2019).

The course of tic severity has primarily been based on large cross-sectional epidemiological studies (Hirschtritt et al., 2015; Peterson, Pine, Cohen, & Brook, 2001; Peterson, Staib, et al., 2001) and has only been examined longitudinally in a few small clinical studies (Bloch & Leckman, 2009; Leckman et al., 1998). Clinical guidance for patients has been based primarily on clinical experience, prevalence studies and on a retrospective clinical study by Leckman et al. (1998).

2.1 Development of tic severity

The retrospective study by Leckman et al. illustrated expected tic severity from childhood and throughout adolescence. The study was based on tic severity in a cohort of 36 participants aged 2 to 18 years. Parents have

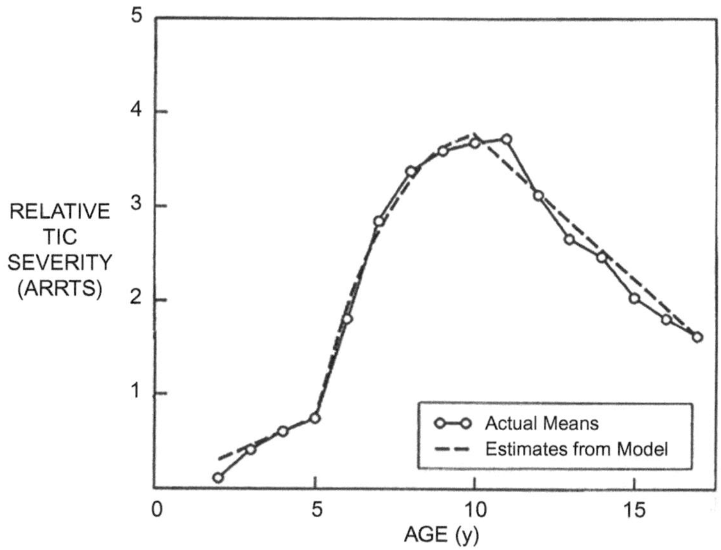

Fig. 1 Clinical course of tic severity in childhood. Plot of average tic severity in a cohort of 36 children aged 2 to 18 years. Parents have retrospectively rated their children's tic severity on a six-point ordinal scale; absent [0], least severe, mild, moderate, severe and most severe [6]. Annual rating of relative tic severity (ARRTS). *Reprinted with permission from Elsevier.*

retrospectively rated their children's tic severity on a six-point ordinal scale and age at their worst ever tics (Leckman et al., 1998) (see Fig. 1).

This study illustrated tic-onset in the age of 4 and 6 years and peak in severity between the age of 10 and 12 years, followed by a decline in tic severity during adolescence. The worst ever tics age was between 6 and 15 years (mean 10.0 years). Half of the cohort had marked or severe worst ever tic with a Yale Global Tic Severity Scale Score (YGTSS) score above 30. The level of tic severity during the worst ever tics period was positively associated with the patient's age. The average YGTTS score at follow up, aged 18 years was 7.92 (SD 9.53). 47.2% were entirely tic-free during the week before, 11.1% had minimal tic symptom (YGTSS $<$ 10), 27.7% had mild symptoms (YGTSS between 10 and 20) and only 11.1% had a moderate or marked level of tic severity (YGTSS between 20 and 40) (Leckman et al., 1998).

In a prospective epidemiological longitudinal study, decline in severity of tics during adolescence was confirmed (Peterson, Pine, et al., 2001; Peterson, Staib, et al., 2001).

In 2006, Bloch et al. published a study following 46 participants aged 7–14 years at baseline and 16–23 years at follow-up examining the course of tics and OCD. Also in this study, a decline in tic severity was seen (Bloch et al., 2006). At baseline, the participants aged 11.4 years (range 7.5–13.9 years) had tics of at least moderate severity (YGTSS score >= 20). At follow-up, participants were aged 19.0 years (range 16.0–22.8) and had a total tic score of the YGTSS of 10.0 (median 7,5, range 0–39). 33% reported being tic-free during the last week and 22% had moderate or marked level of tic severity (YGTTS between 20 and 40) (Bloch, Peterson, et al., 2006).

In 2011–13, a large clinical follow up study was performed examining the clinical course of tics and comorbidities by Groth et al. 314 participants were clinically examined at baseline (T1) in 2005–07, aged mean 12.4 (range 5–20 years) and at follow-up (T2) 4 to 8 years later (median 5.6 years) aged mean 18.5 (range 11–26 years) where 72.3% of the participants were reexamined. The clinical course of tics showed an age-related decline in mean total tic score on YGTSS of 0.80 points annually. This reflected declines in both motor (0.45) and vocal tics (0.35). The decline in total tic score is indicated by the red line in Fig. 2, in which all tic assessments

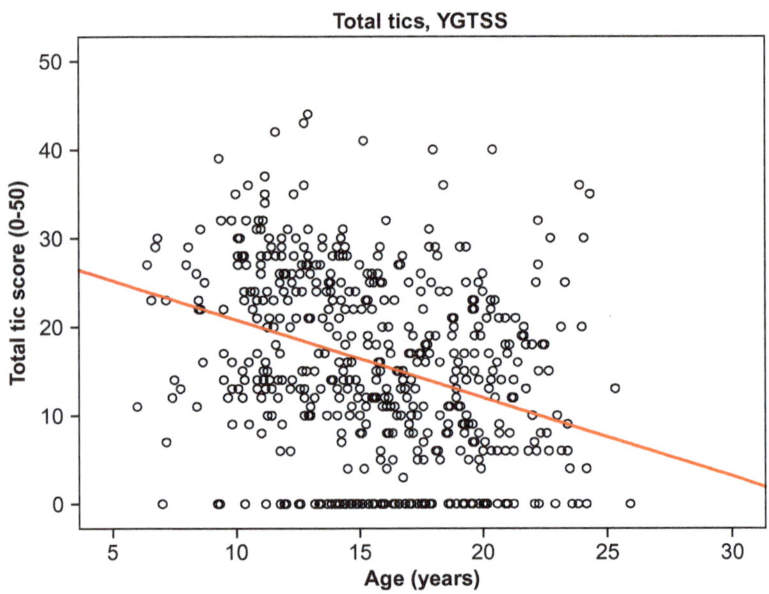

Fig. 2 Clinical course of tics in the age range 5–26 years. Total tic score on the Yale Global Tic Severity Scale (YGTSS) (range 0–44, yearly decline = 0.80). All tics assessments performed with YGTTS (n = 518) are plotted. The red line indicates mean age-related score. *Reprinted with permission from Elsevier.*

Table 1 The severity of total tic scores divided into age subgroups in a clinical cohort with 518 assessments.

	5–10 years	**11–15 years**	**16–20 years**	**21–26 years**	**All assessments**
Tic score	$n=63$	$n=218$	$n=182$	$n=55$	n=518
Absence (0)	4 (6.3%)	31 (14.2%)	32 (17.6%)	10 (18.2%)	77 (14.9%)
Minimal (1–9)	3 (4.8%)	14 (6.4%)	39 (21.4%)	14 (25.5%)	57 (11.0%)
Mild (10–19)	19 (30.2%)	70 (32.1%)	70 (38.5%)	18 (32.7%)	134 (25.9%)
Moderate (20–39)	37 (58.7%)	99 (45.4%)	39 (21.4%)	13 (23.6%)	244 (47.1%)

Reprinted with permission from Danish Medical Bulletin.

performed with YGTSS ($n=518$) are plotted in a scatterplot (Fig. 2). Among participants above 16 years of age, 18% were tics free, 60% had minimal or mild tics, and 22% had moderate to severe tics The severity of tics between the ages of 5 and 26 years is categorized in Table 1. Age at onset of tics did not significantly affect the clinical course (Groth et al., 2017).

Another long-term follow-up study of participants with TS was published by Lowe et al. in 2019. 45 participants with TS were retrospectively followed for 25–32 years. The vast majority (69%) of participants reported significant decrease in tic severity over time, and they described only subtle tic-related difficulties at follow-up. 31% reported that their TS was "very much improved," 38% "much improved," 13% "somewhat improved," 13% "no change" and 4% as "somewhat worse."

In summary, all studies of the clinical course of tic severity point in direction of decline in tic severity during adolescent into early adulthood. However, as shown in Fig. 2 the variability is significant and at above age 16 still 22% of the participants had moderate or severe tics (Bloch, Peterson, et al., 2006; Groth et al., 2017).

2.2 Development of comorbidities

Cross-sectional studies have shown high prevalences of comorbidities in patients with TS. Robertson reported that 90% of a TS population had comorbidities with coexisting emotional and behavioral psychopathologies (Robertson, 2012). Hirschtritt et al. demonstrated a lifetime prevalence of any psychiatric comorbidity in patients with TS of 85.7%, of whom 57.7% of patients had two or more psychiatric disorders (Hirschtritt et al., 2015). A study from Sweden found that 92% of a Swedish school population with a tic disorder (aged 7–15 years) had at least one additional psychiatric diagnosis (Khalifa & Von Knorring, 2006). In a large international clinical study that covered early life to adulthood, Freeman et al. reported TS without comorbidities in 11–12% of the TS population, although there was substantial variation between sites (2–36%) (Freeman et al., 2000).

Some longitudinal studies have examined the development of comorbidities. In the longitudinal study by Groth et al., only 9.7% of the patients were without symptoms of comorbidities at T1. The participants were examined for OCD and ADHD and at T2 screened with Development and Well-Being Assessment (DAWBA) for further coexisting psychopathology. At T2, 37% were without symptoms of comorbidities. Of the 92 participants aged 11–26 years with a comorbidity, 30% had one comorbidity, and 33% had two or more comorbidities. Most frequent comorbidities were ADHD, OCD, emotional disorders like anxiety and depression, behavioral disorders including intermittent explosive disorder, developmental disorders, and psychotic disorders (Groth et al., 2017).

2.2.1 OCD

Regarding the clinical course of TS-associated OCD, findings have not been uniform. The prevalence of TS-associated OCD is approximately 36–50% with a predominance of females (Bloch & Leckman, 2009; Hirschtritt et al., 2015; Robertson, 2012; Robertson, 2015). Onset of comorbid OCD is reported to be in the period of worst-ever tics (10–12 years) (Bloch, Peterson, et al., 2006; Peterson, Pine, et al., 2001; Peterson, Staib, et al., 2001), but can also be earlier (5–7 years) (Hirschtritt et al., 2015; Mol Debes et al., 2008) or appear de novo in early adulthood (Bloch & Leckman, 2009; Leckman, 2002).

Peterson et al. found a clear association of tics and OCD in a prospective longitudinal epidemiological study. They showed that the modal age of onset was age 12, after which the prevalence of OCD declined in late adolescence before increasing in early adulthood (Peterson, Pine, et al., 2001; Peterson, Staib, et al., 2001).

Bloch et al. showed in a clinical follow-up study that TS-associated OCD increased in severity with 6.7 points on CY-BOCS from age 11.4 (7.5–13.9 years) to age 19.0 (16.0–22.8 years). Inclusion criteria for this cohort of 46 patients were IQ above 80 and minimum score of worst-ever symptoms on CY-BOCS of 10. Worst-ever symptoms of

Table 2 The severity of Obsessive Compulsive Disorder (OCD), assessed by Yale-Brown Obsessive-Compulsive Scale for children and adults (CY-BOCS/Y-BOCS), divided into age subgroups in a clinical cohort of children and adolescents with 541 assessments.

	5–10 years	11–15 years	16–20 years	21–26 years	All assessments
OCD score	$n=95$	$n=233$	$n=173$	$n=40$	$n=541$
Normal (0–7)	45 (47.4%)	147 (63.1%)	113 (65.3%)	27 (67.5%)	333 (61.6%)
Subclinical (8–9)	5 (5.3%)	17 (7.3%)	8 (4.6%)	1 (2.5%)	30 (5.5%)
Mild (10–18)	29 (30.5%)	45 (19.3%)	30 (17.3%)	10 (25.0%)	114 (21.1%)
Moderate (19–29)	13 (13.7%)	24 (10.3%)	22 (12.7%)	2 (5.0%)	61 (11.3%)
Severe ($\geq$ 30)	3 (3.2%)	0 (0%)	0 (0%)	0 (0%)	3 (0.6%)

Reprinted with permission from Danish Medical Bulletin.

OCD were described approximately 2 years later than worst-ever tic symptoms (Bloch, Peterson, et al., 2006). Increased childhood IQ was strongly associated with the future development of OCD symptoms, which is consistent with previous studies (Peterson, Pine, et al., 2001; Peterson, Staib, et al., 2001).

On the contrary, in a prospective follow-up study by Groth et al., a decline of OCD symptoms in adolescence was seen in a clinical cohort of children and adolescence aged 5–26 years (Table *2* and Fig. 3). Although symptoms might persist in adulthood, the severity of combined compulsions and obsessions was highly significantly related to age with a modest yearly decline of 0.24 in mean total OCD score on the Yale-Brown Obsessive-Compulsive Scale for children and adults (CY-BOCS/YBOCS) as indicated by the red line in Fig. 3. This reflected a highly significant decline in compulsions (mean score = 0.17), whereas a smaller decline in obsessions (mean score = 0.06) was not significant (Groth et al., 2017). This pattern in the development of OCD symptoms, with compulsions being more dominant in childhood and obsessions becoming more prevalent in adolescence was also found in another study (Selles, Storch, & Lewin, 2014). In the study by Groth et al., all participants with TS were included and there were not used exclusion criteria on OCD score or IQ, which might explain some of the discrepancy between the study by Bloch et al. and Groth et al. (Bloch, Peterson, et al., 2006; Groth et al., 2017).

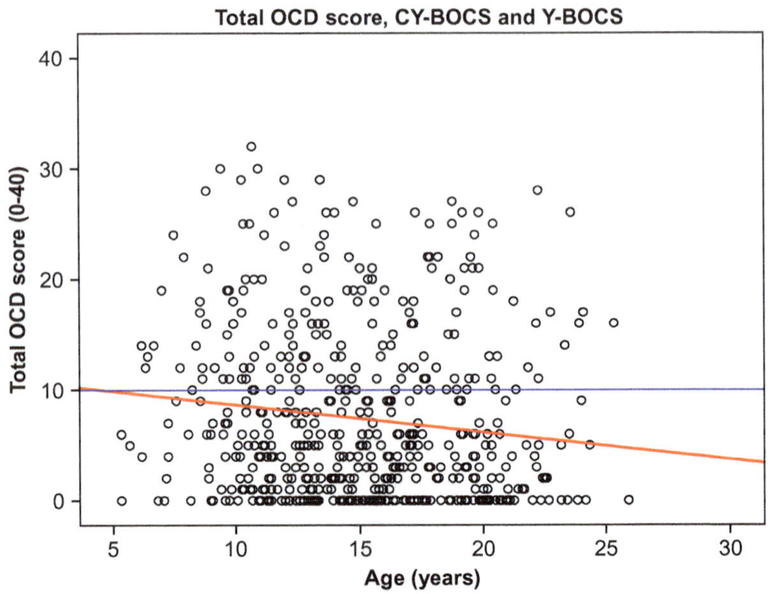

Fig. 3 Clinical course of Obsessive Compulsive Disorder (OCD) in the age range 5–26 years based on 541 assessments. The red line indicates mean age-related scores. The blue line indicates the defined diagnostic cut-off score on Yale-Brown Obsessive-Compulsive Scale for children and adults. *Reprinted with permission from Elsevier.*

2.2.2 ADHD

The prevalence of ADHD in children and adolescence with TS is 50–60% and affects predominantly boys (Freeman et al., 2000; Hirschtritt et al., 2015; Robertson, 2015; Spencer et al., 1998).

The clinical courses of ADHD and TS appear to be independent. ADHD symptoms typically precede tic-onset at approximately 2–6 years of age (Bloch & Leckman, 2009; Hirschtritt et al., 2015; Mol Debes et al., 2008), and there is a greater likelihood of tic remission in adulthood than ADHD remission (Spencer et al., 1998). In childhood, hyperactivity and impulsivity is often dominant, while inattentive difficulties may persist into adulthood (Biederman, Mick, & Faraone, 2000; Faraone, Biederman, & Mick, 2005).

In a longitudinal clinical follow-up study, a highly significantly age-related decline in ADHD symptoms was seen, as indicated by the red line in Fig. 4. Symptoms of inattention, hyperactivity and impulsivity decreased annually although some executive difficulties persisted and many patients

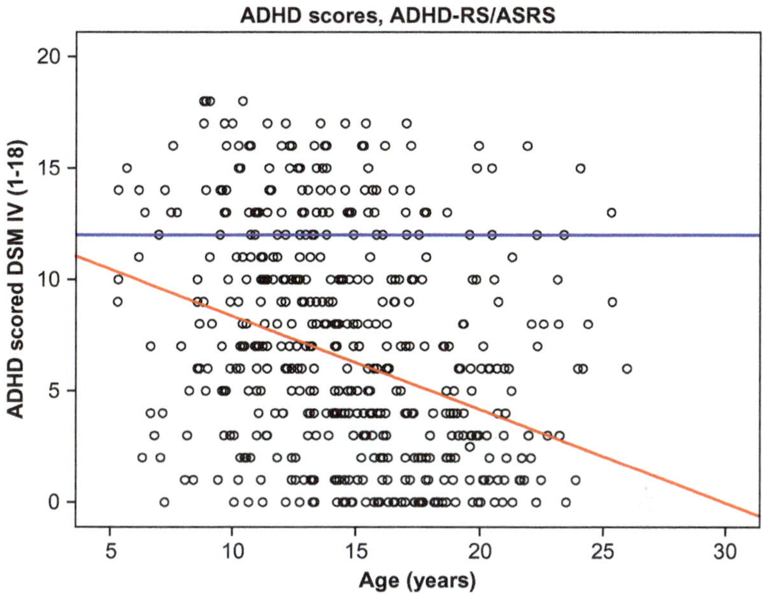

Fig. 4 Development of symptoms of Attention Deficit Hyperactivity Disorder in patients with Tourette Syndrome over time (age range 5–26 years). The red line indicates mean age-related ADHD scores. The blue line indicates the defined diagnostic cut-off score. *Reprinted with permission from Elsevier.*

still needed clinical support to handle difficulties in daily functions including planning, education and relating to their peers in adulthood (Groth et al., 2017).

2.3 Development of phenotypes

Several studies have examined the development of phenotypes over time.

Rizzo et al. longitudinally followed the expression and modification of TS phenotypes in childhood and adolescence in a retrospective study. After 10 years, phenotypes of 100 children aged 3–8 years were assessed retrospectively. Their findings indicated a good long-term clinical course for TS without comorbidities, whereas TS plus comorbidities showed a more severe prognosis (Rizzo, Gulisano, Calì, & Curatolo, 2012).

Groth et al. also examined the developmental trajectory of the TS phenotype expression in a longitudinal follow-up study with median 5.6 years between baseline (T1) and follow-up (T2). TS phenotype expression changed towards fewer comorbidities and approached the TS-only phenotype (TS without OCD and/or ADHD). The TS-only group increased by

15%. A total of 42% of those with comorbidities at T1 moved into the TS-only group at T2, and this trend was most pronounced for the youngest group (aged 5–10 years at T1). In the oldest subgroup (aged 16–20 years at T1), a tendency for those with OCD or ADHD to remain in the same comorbidity group at follow-up was observed. 27% of the TS-only group at T1 developed comorbidities at T2. In general, the TS phenotype development was shown to be dynamic with 53% of individuals changing the phenotype expressed.

2.4 Predictors of the clinical course of tics and comorbidities

Few prospective clinical studies have examined potential predictors of symptom severity for tics, comorbidities, and coexisting psychopathologies (Bloch, Peterson, et al., 2006; Bloch, Sukhodolsky, Leckman, & Schultz, 2006; Goetz, Tanner, Stebbins, Leipzig, & Carr, 1992; Groth, Skov, Lange, & Debes, 2019; Lin et al., 2007). Cross-sectional studies have found associations between tics and comorbidities and although no clear trends have been established, clinical factors and family history have demonstrated some predictive potential (de Groot, Janus, & Bornstein, 1995; Hirschtritt et al., 2015).

Lin et al. found that psychosocial stress predicted depressive symptoms and OCD severity and was a modest predictor of tic severity (Lin et al., 2007).

In a cross-sectional clinical study, Hirschtritt et al. demonstrated that clinical factors including tics, OCD and ADHD could predict the presence of more than one coexisting psychopathology, including emotional disorders (Hirschtritt et al., 2015).

In an epidemiological prospective longitudinal study, tics, OCD, and ADHD were found to predict a continuation of symptoms, e.g., tics in childhood predicted tics in late adolescence. Tics in childhood and early adolescence also predicted later OCD and emotional disorders. OCD in late adolescence predicted tics and ADHD in adulthood. ADHD was a weak predictor of future tics but a stronger predictor of future OCD (Peterson, Pine, et al., 2001; Peterson, Staib, et al., 2001).

Lowe et al. performed a long term follow-up study of 45 patients retrospectively for 25–32 years but did not uncover robust clinical predictors for the course of TS symptoms or other psychosocial variables. However, tendencies were seen for patients with older age of tic onset reporting less improvement in TS severity at follow-up and on the other hand, patients with academic impairment and diagnosis of ADHD at baseline, reported greater improvement at follow-up (Lowe, Capriotti, & McBurnett, 2019).

In a clinical longitudinal follow-up study, childhood clinical factors as possible predictors of the clinical course of tics and comorbidities in early adulthood were examined. Strong clinical predictors of TS associated outcomes were identified for tic severity and diagnoses of OCD, ADHD and emotional disorders. Tic severity at baseline, family history of TS + OCD + AHDH and being teased during childhood were predictors of moderate to severe tics in early adulthood with tic severity at baseline as the strongest predictor. For every 10-points increase in the YGTSS score, participants were 2.4 times more likely to have moderate to severe tics in early adulthood (Groth et al., 2019). This is consistent with the findings of Bloch, Peterson, et al. (2006). OCD score at baseline was the strongest predictor of OCD at follow-up consistent with the results of Peterson, Pine, et al. (2001), Peterson, Staib, et al. (2001) but in discrepancy with Bloch, Peterson, et al. (2006) who did not find any association between OCD severity in childhood and future OCD severity. Peterson, Pine, et al. (2001), Peterson, Staib, et al. (2001) demonstrated that tics and ADHD predicted an increase in future OCD severity supported by Groth et al. (2019) who found ADHD symptoms having an effect on future OCD although not as an independent predictor. The strongest predictor of ADHD in early adulthood was ADHD severity in childhood. Having a family history of ADHD without TS or getting special education in childhood also increased the risk of future ADHD significantly (Groth et al., 2019). The predictive value of a positive family history highlights the complex genetic etiology of TS and ADHD (Hirschtritt et al., 2015; Peterson, Pine, et al., 2001; Peterson, Staib, et al., 2001; Robertson, 2015).

ADHD has been confirmed as a predictor of future coexisting psychopathologies and emotional disorders, suggesting these may be closely related and possibly secondary to comorbid ADHD (Groth et al., 2019; Hirschtritt et al., 2015; Peterson, Pine, et al., 2001; Peterson, Staib, et al., 2001; Robertson, 2015; Spencer et al., 1998). Being a female increased the odds of having a future emotional diagnosis (Groth et al., 2019) consistent with Hirschtritt et al. who found significantly more mood and anxiety disorder in female TS patients (Hirschtritt et al., 2015).

Poor fine motor skills measured by the Purdue Pegboard test in childhood were associated with increased severity of tics in adulthood (Bloch, Sukhodolsky, et al., 2006).

Since TS exhibits a broad clinical heterogeneity expressed in different clinical phenotypes that covers a wide spectrum and develops with age, finding strong predictors for the clinical course of the disease is difficult (Peterson, Pine, et al., 2001; Peterson, Staib, et al., 2001). Furthermore, differences in the methods and aims of studies makes direct comparisons difficult.

3. Neuroimaging studies

Numerous brain imaging studies have been performed in children and adults with TS. Most studies are cross-sectional studies and since the clinical course of TS and comorbidities changes significantly over time, as described in the sections above, this aspect should be considered when interpreting brain imaging studies.

Until now, brain imaging studies have provided strong evidence of anatomical and functional disturbances of the corpus callosum, prefrontal regions, different areas of the basal ganglia and thalamus. However, the findings of different imaging studies still provide conflicting results and brain alterations in TS patients differ remarkably between children and adults.

When investigating the plasticity and developmental changes of the brain in relation to TS, volume and diffusivity are commonly used measures. Volumetric distinctions are measured using magnetic resonance imaging (MRI) and voxel-based morphometry to compare morphological differences in brain areas. Diffusivity is often measured using diffusion tensor imaging (DTI). DTI measures the diffusion of water molecules and holds information tissue architecture. Different informative measures can be derived from DTI including fractional anisotropy (FA) and mean diffusivity (MD). FA is suitable for evaluating microstructural changes in white matter while MD is sensitive to changes in cellular membranes and hereby sensitive to decreased or increased spacing between membrane layers.

3.1 Alterations of volume and diffusivity in children and young adults

In children and young adults with TS the size of the corpus callosum is reduced compared to healthy individuals and asymmetry is altered with dominance of the right side. Reduced size of the corpus callosum as well as reduced caudate volumes and reduced volume of sensorimotor cortices are correlated with tic severity. The reduced size of corpus callosum may limit neuronal trafficking and inhibit input to cortical inhibitory interneurons in the prefrontal cortices. The reduced inhibitory input is thought to enhance prefrontal excitation and thereby enhance tic control (Plessen et al., 2004).

Studies of the prefrontal cortex have shown reduced volume of the white matter in orbital and medial prefrontal cortex and increased volume of the posterior prefrontal cortex (Garraux et al., 2006; Peterson, Pine, et al., 2001; Peterson, Staib, et al., 2001). The functions of the orbital regions are yet to be fully understood, however there is evidence of these regions being involved in decision making and reward. It has been suggested that the orbitofrontal cortex could be involved in the sensory aspects of tics as well (Greene et al., 2017).

The volume of the thalamus is increased in children with TS compared to healthy age matched individuals (Lee et al., 2006; Miller et al., 2010). The thalamic nuclei within cortical-subcortical circuits compose an anatomical crossroad in which enlargement of motor nuclei is thought to represent either activity dependent hypertrophy or an adaptive response that modulate the activity in motor circuits.

Previous studies show mixed results regarding the volume of the putamen in children with TS. Given the important role of putamen in movement, further studies are warranted (Greene et al., 2017).

Moreover, different studies demonstrate reduced gray matter in several cerebellar lobules involved in sensorimotor and executive function and increased gray matter volume in the hypothalamus and midbrain (Sigurdsson, Jackson, Jolley, Mitchell, & Jackson, 2020).

Assessing microstructural changes in children and young adults using DTI confirms the hypothesis of changes in the cortico-striato-thalamo-cortical pathway. Studies have shown less directional organization of white matter microstructure in the corpus callosum (Plessen et al., 2006), increased thickness and tract density in prefrontal areas (Xu et al., 2007) as well as changes in diffusivity in thalamus and different areas of the basal ganglia (Makki, Behen, Bhatt, Wilson, & Chugani, 2008; Xia et al., 2020). Furthermore, studies of diffusion have identified two nodes in the right putamen and left SMA in both children and adults with TS. These nodes may play a role in the pathology of TS in both children (Sigurdsson et al., 2018).

3.2 Alterations of volume and diffusivity in adults

Cross-sectional studies in adults with Tourette syndrome have shown increased volumes of corpus callosum (Peterson et al., 1994; Plessen et al., 2004), increased volume of dorsal prefrontal areas (Peterson, Pine, et al., 2001; Peterson, Staib, et al., 2001), and decreased gray matter in prefrontal areas. Additionally, studies have shown increased volume of thalamus in adults with TS (Greene et al., 2017; Miller et al., 2010).

In adults, the volume of the dorsal prefrontal areas is increased while the volume of gray matter in prefrontal areas is decreased. Overall an inverse association of age and prefrontal volumes has been found.

Using DTI to assess diffusivity in adults with TS, microstructural changes in corpus callosum, the prefrontal areas, thalamus and putamen have been detected. Most of these regions with structural abnormalities also showed correlation with tic severity (Müller-Vahl, Riemann, & Bokemeyer, 2014). Thus, findings regarding microstructural changes in the brain of adults with TS are in accordance with previous findings concerning volumetric changes.

3.3 Longitudinal studies

Longitudinal studies are needed to confirm the findings as the change of phenotype over time is not considered in the patients included in cross-sectional studies. Currently only a single longitudinal neuroimaging study examining the development of tics has been performed (Debes et al., 2015).

The study was published in 2015 examining 22 patients and 21 healthy controls. Both groups were examined clinically and with MRI scans at baseline and reexamined after averagely 4.4 years. Mean age of the participants at baseline was 14.2 years and at follow-up 18.6 years. In total, 8 patients with TS had a decrease of their total tic score on the YGTSS of 10 points or more and were thus defined as being in remission. Fourteen patients had an increase or stable total tic score and defined as patients with persistent tics.

Using MRI scans and volumetric measurements detected no decrease in gray matter volume of left putamen in patients as was the case for the healthy control group.

Regarding diffusivity, changes in mean diffusivity in the right caudate nucleus, right thalamus, and right frontal lobe were found in TS patients. This confirms the hypothesis of the cortico-striato-thalamo-cortical pathways being involved in the pathophysiology of Tourette syndrome. Furthermore, decreased parallel and perpendicular diffusivity in most patients with Tourette syndrome was found. The decrease in parallel diffusivity was less pronounced in the patients with remission of tics compared to patients with persisting tics. These findings suggest that the development of the brain in patients with remission of tics resembles the normal development of the brain more than in the group of patients with persistent tics.

Both the lacking decrease in gray matter volume of left putamen in patients with TS and the findings on diffusivity might reflect a basal change in brain structure or a compensatory mechanism in the brain.

4. Course of TS in a clinical and imaging perspective

In order to examine the course of a disease, longitudinal studies are an important tool. Overall, both the severity of tics and comorbidities decline with age in patients with TS. This means that patients who do not have any significant symptoms anymore as adults, likely not will be included in clinical cross-sectional studies with adult TS patients. The patients with remission of tics and the patients with persistent tics might represent two different phenotypical variants of TS and thus represent two different pathophysiological mechanisms in TS.

Brain imaging studies have shown similar areas in the brain involved in TS in both children and adults (see Table 3). Increased volume of posterior prefrontal cortex and thalamus have been found in both children and adults, and microstructural changes in corpus callosum. Differences between pediatric and adult cross-sectional clinical populations have been shown in volume of corpus callosum. In children and young adults, size reduction of the corpus callosum has been seen, whereas the corpus callosum in adults with TS showed increased average size compared to healthy individuals (Plessen et al., 2004). The causation of the differences in corpus callosum size is not yet inferred however, it has been suggested that the causation of the size difference could either represent the contribution of the corpus callosum to the genesis of tics or to a downstream effect of disturbances elsewhere in the brain (Plessen et al., 2004). It might thus be associated with continued tic symptoms into adulthood.

Changes in parallel diffusivity in thalamus, caudate nucleus and frontal lobe have been found to be most pronounced in patients with persisting tics compared with patients with remission of tics and healthy controls. Therefore, these areas might be most interesting to examine in further studies in order to explore the possibility of different phenotypes regarding the course of TS.

There have not yet been performed longitudinal imaging studies in patients with TS with inclusion of development of severity of comorbidities. Some of the same brain circuits might be involved in the pathogenesis of both TS and comorbidities. Limbic and orbitofrontal basal ganglia-thalamocortical circuits might be involved in the pathogenesis of OCD and dorsolateral

Table 3 Findings in cross-sectional and longitudinal neuroimaging studies with children and adults with Tourette syndrome.

		Children	Adults	Longitudinal
Volumetry	Corpus callosum	Reduced	Increased	
	Caudate volume	Reduced		
	Sensorimotor cortices	Reduced		
	Thalamus	Increased	Increased	
	Putamen	Contradictory		
	Posterior prefrontal cortex	Increased	Increased	
	White matter in orbital and medial prefrontal cortex	Reduced		
	Gray matter in several cerebellar lobules	Reduced		
	Gray matter in hypothalamus	Increased		
	Gray matter in midbrain	Increased		
	Gray matter in prefrontal areas		Decreased	
	Gray matter volume left putamen			Decrease in healthy controls, but not in patients
DTI	Corpus callosum	Microstructural changes	Microstructural changes	
	Prefrontal areas	Thickness and tract density increased	Microstructural changes	
	Thalamus	Changes in diffusivity	Microstructural changes	Changes in mean diffusivity in patients. Decrease parallel diffusivity most pronounced in patients with persisting tics.
	Basal ganglia	Changes in diffusivity		
	Two nodes in right putamen and left SMA	Present	Present	

Table 3 Findings in cross-sectional and longitudinal neuroimaging studies with children and adults with Tourette syndrome.—cont'd

	Children	Adults	Longitudinal
Putamen		Microstructural changes	
Caudate nucleus			Changes in mean diffusivity in patients. Decrease parallel diffusivity most pronounced in patients with persisting tics.
Frontal lobe			Changes in mean diffusivity in patients. Decrease parallel diffusivity most pronounced in patients with persisting tics.

prefrontal and orbitofrontal basal ganglia-thalamocortical circuits in the pathogenesis of ADHD (see for review Sheppard, Bradshaw, Purcell, & Pantelis, 1999). Co-morbid ADHD might be associated with more extensive changes in the excitability of motor cortex circuits compared to TS patients without ADHD or OCD (Orth & Rothwell, 2009). TS patients with comorbid ADHD an increased volume of dorsolateral putamen was found (Draganski et al., 2010). A trend for reduced cortical thickness in the anterior cingulate cortex and altered hippocampal morphology was found in TS patients with comorbid OCD (Worbe et al., 2010). Childhood MRI scans have found smaller caudate volume associated with increased adulthood tic severity and symptoms of OCD, which suggests that morphologic disturbances of the caudate nucleus within cortico-striatal-thalamo-cortical circuits, possibly involving inhibitory interneurons within the basal ganglia, might be central within the pathogenesis and persistence of both tics and symptoms of OCD into adulthood (Bloch, Leckman, Zhu, & Peterson, 2005).

Future longitudinal imaging studies with inclusion of development of tics and comorbidities might provide more knowledge on the hypothesis of two different phenotypical variants of TS.

References

Biederman, J., Mick, E., & Faraone, S. V. (2000). Age-dependent decline of symptoms of attention deficit hyperactivity disorder: Impact of remission definition and symptom type. *The American Journal of Psychiatry*, *157*(5), 816–818. https://doi.org/10.1176/appi.ajp.157.5.816.

Bloch, M. H., & Leckman, J. F. (2009). Clinical course of Tourette syndrome. *Journal of Psychosomatic Research, 67*(6), 497–501. https://doi.org/10.1016/j.jpsychores.2009.09.002.

Bloch, M. H., Leckman, J. F., Zhu, H., & Peterson, B. S. (2005). Caudate volumes in childhood predict symptom severity in adults with Tourette syndrome. *Neurology, 65*(8), 1253–1258. https://doi.org/10.1212/01.wnl.0000180957.98702.69.

Bloch, M. H., Peterson, B. S., Scahill, L., Otka, J., Katsovich, L., Zhang, H., et al. (2006). Adulthood outcome of tic and obsessive-compulsive symptom severity in children with Tourette syndrome. *Archives of Pediatrics & Adolescent Medicine, 160*(1), 65. https://doi.org/10.1001/archpedi.160.1.65.

Bloch, M. H., Sukhodolsky, D. G., Leckman, J. F., & Schultz, R. T. (2006). Fine-motor skill deficits in childhood predict adulthood tic severity and global psychosocial functioning in Tourette's syndrome. *Journal of Child Psychology and Psychiatry, 47*(6), 551–559. https://doi.org/10.1111/j.1469-7610.2005.01561.x.

de Groot, C. M., Janus, M. D., & Bornstein, R. A. (1995). Clinical predictors of psychopathology in children and adolescents with Tourette syndrome. *Journal of Psychiatric Research, 29*(1), 59–70.

Debes, N., Jeppesen, S., Raghava, J. M., Groth, C., Rostrup, E., & Skov, L. (2015). Longitudinal magnetic resonance imaging (MRI) analysis of the developmental changes of Tourette syndrome reveal reduced diffusion in the Cortico-Striato-Thalamo-cortical pathways. *Journal of Child Neurology, 30*(10), 1315–1326. https://doi.org/10.1177/0883073814560629.

Draganski, B., Martino, D., Cavanna, A. E., Hutton, C., Orth, M., Robertson, M. M., et al. (2010). Multispectral brain morphometry in Tourette syndrome persisting into adulthood. *Brain, 133*, 3661–3675.

Faraone, S. V., Biederman, J., & Mick, E. (2005). The age-dependent decline of attention deficit hyperactivity disorder: A meta-analysis of follow-up studies. *Psychological Medicine, 36*(02), 159. https://doi.org/10.1017/S003329170500471X.

Freeman, R. D., Fast, D. K., Burd, L., Kerbeshian, J., Robertson, M. M., & Sandor, P. (2000). An international perspective on Tourette syndrome: Selected findings from 3500 individuals in 22 countries. *Developmental Medicine & Child Neurology, 42*(7), 436–447.

Garraux, G., Goldfine, A., Bohlhalter, S., Lerner, A., Hanakawa, T., & Hallett, M. (2006). Increased midbrain gray matter in Tourette's syndrome. *Annals of Neurology, 59*(2), 381–385. https://doi.org/10.1002/ana.20765.

Goetz, C. G., Tanner, C. M., Stebbins, G. T., Leipzig, G., & Carr, W. C. (1992). Adult tics in Gilles de la Tourette's syndrome: Description and risk factors. *Neurology, 42*(4), 784–788.

Greene, D. J., Williams, A. C., 3rd, Koller, J. M., Schlaggar, B. L., Black, K. J., & The Tourette Association of America Neuroimaging Consortium. (2017). Brain structure in pediatric Tourette syndrome. *Molecular Psychiatry, 22*(7), 972–980. https://doi.org/10.1038/mp.2016.194.

Groth, C., Mol Debes, N., Rask, C. U., Lange, T., & Skov, L. (2017). Course of Tourette syndrome and comorbidities in a large prospective clinical study. *Journal of the American Academy of Child and Adolescent Psychiatry, 56*(4), 304–312. https://doi.org/10.1016/j.jaac.2017.01.010.

Groth, C., Skov, L., Lange, T., & Debes, N. M. (2019). Predictors of the clinical course of Tourette syndrome: A longitudinal study. *Journal of Child Neurology, 34*(14), 913–921. https://doi.org/10.1177/0883073819867245.

Hirschtritt, M. E., Lee, P. C., Pauls, D. L., Dion, Y., Grados, M. A., Illmann, C., et al. (2015). Lifetime prevalence, age of risk, and genetic relationships of comorbid psychiatric disorders in Tourette syndrome. *JAMA Psychiatry, 72*(4), 325. https://doi.org/10.1001/jamapsychiatry.2014.2650.

Khalifa, N., & Knorring, A.-L. (2003). Prevalence of tic disorders and Tourette syndrome in a Swedish school population. *Developmental Medicine & Child Neurology, 45*(5), 315–319.
Khalifa, N., & Von Knorring, A.-L. (2006). Psychopathology in a Swedish population of school children with tic disorders. *Journal of the American Academy of Child & Adolescent Psychiatry, 45*(11), 1346–1353. https://doi.org/10.1097/01.chi.0000251210.98749.83.
Leckman, J. F. (2002). Tourette's syndrome. *The Lancet, 360*(9345), 1577–1586.
Leckman, J. F., Zhang, H., Vitale, A., Lahnin, F., Lynch, K., Bondi, C., et al. (1998). Course of tic severity in Tourette syndrome: The first two decades. *Pediatrics, 102*(1 Pt 1), 14–19.
Lee, J. S., Yoo, S. S., Cho, S. Y., Ock, S. M., Lim, M. K., & Panych, L. P. (2006). Abnormal thalamic volume in treatment-naïve boys with Tourette syndrome. *Acta Psychiatrica Scandinavica, 113*(1), 64–67. https://doi.org/10.1111/j.1600-0447.2005.00666.x.
Levine, J. L. S., Szejko, N., & Bloch, M. H. (2019). Meta-analysis: Adulthood prevalence of Tourette syndrome. *Progress in Neuro-Psychopharmacology and Biological Psychiatry, 95*, 109675. https://doi.org/10.1016/j.pnpbp.2019.109675.
Lin, H., Katsovich, L., Ghebremichael, M., Findley, D. B., Grantz, H., Lombroso, P. J., et al. (2007). Psychosocial stress predicts future symptom severities in children and adolescents with Tourette syndrome and/or obsessive-compulsive disorder. *Journal of Child Psychology and Psychiatry, 48*(2), 157–166. https://doi.org/10.1111/j.1469-7610.2006.01687.x.
Lowe, T. L., Capriotti, M. R., & McBurnett, K. (2019). Long-term follow-up of patients with Tourette's syndrome: Tourette follow-up. *Movement Disorders Clinical Practice, 6*(1), 40–45. https://doi.org/10.1002/mdc3.12696.
Makki, M. I., Behen, M., Bhatt, A., Wilson, B., & Chugani, H. T. (2008). Microstructural abnormalities of striatum and thalamus in children with Tourette syndrome. *Movement Disorders, 23*(16), 2349–2356. https://doi.org/10.1002/mds.22264.
Miller, A. M., Bansal, R., Hao, X., Sanchez-Pena, J. P., Sobel, L. J., Liu, J., et al. (2010). Enlargement of thalamic nuclei in Tourette syndrome. *Archives of General Psychiatry, 67*(9), 955–964. https://doi.org/10.1001/archgenpsychiatry.2010.102.
Mol Debes, N. M. M., Hjalgrim, H., & Skov, L. (2008). Limited knowledge of Tourette syndrome causes delay in diagnosis. *Neuropediatrics, 39*(2), 101–105. https://doi.org/10.1055/s-2008-1081457.
Müller-Vahl, K. R., Riemann, L., & Bokemeyer, S. (2014). Tourette patients' misbelief of a tic rebound is due to overall difficulties in reliable tic rating. *Journal of Psychosomatic Research, 76*(6), 472–476. https://doi.org/10.1016/j.jpsychores.2014.03.003. (Epub 2014 Mar 24).
Orth, M., & Rothwell, J. C. (2009). Motor cortex excitability and comorbidity in Gilles de la Tourette syndrome. *Journal of Neurology, Neurosurgery and Psychiatry, 80*, 29–34.
Peterson, B. S., Leckman, J. F., Duncan, J. S., Wetzles, R., Riddle, M. A., Hardin, M. T., et al. (1994). Corpus callosum morphology from magnetic resonance images in Tourette's syndrome. *Psychiatry Research, 55*(2), 85–99. https://doi.org/10.1016/0925-4927(94)90003-5.
Peterson, B. S., Pine, D. S., Cohen, P., & Brook, J. S. (2001). Prospective, longitudinal study of tic, obsessive-compulsive, and attention-deficit/hyperactivity disorders in an epidemiological sample. *Journal of the American Academy of Child & Adolescent Psychiatry, 40*(6), 685–695.
Peterson, B. S., Staib, L., Scahill, L., Zhang, H., Anderson, C., Leckman, J. F., et al. (2001). Regional brain and ventricular volumes in Tourette syndrome. *Archives of General Psychiatry, 58*(5), 427–440. https://doi.org/10.1001/archpsyc.58.5.427.
Plessen, K. J., Grüner, R., Lundervold, A., Hirsch, J. G., Xu, D., Bansal, R., et al. (2006). Reduced white matter connectivity in the corpus callosum of children with Tourette syndrome. *Journal of Child Psychology and Psychiatry, 47*(10), 1013–1022. https://doi.org/10.1111/j.1469-7610.2006.01639.x.

Plessen, K. J., Wentzel-Larsen, T., Hugdahl, K., Feineigle, P., Klein, J., Staib, L. H., et al. (2004). Altered interhemispheric connectivity in individuals with Tourette's disorder. *American Journal of Psychiatry, 161*(11), 2028–2037. https://doi.org/10.1176/appi.ajp.161.11.2028.
Rizzo, R., Gulisano, M., Calì, P. V., & Curatolo, P. (2012). Long term clinical course of Tourette syndrome. *Brain & Development, 34*(8), 667–673. https://doi.org/10.1016/j.braindev.2011.11.006.
Robertson, M. M. (2012). The Gilles De La Tourette syndrome: The current status. *Archives of Disease in Childhood - Education and Practice, 97*(5), 166–175. https://doi.org/10.1136/archdischild-2011-300585.
Robertson, M. M. (2015). A personal 35 year perspective on Gilles de la Tourette syndrome: Prevalence, phenomenology, comorbidities, and coexistent psychopathologies. *The Lancet Psychiatry, 2*(1), 68–87.
Scharf, J. M., Miller, L. L., Gauvin, C. A., Alabiso, J., Mathews, C. A., & Ben-Shlomo, Y. (2015). Population prevalence of Tourette syndrome: A systematic review and meta-analysis: Meta-analysis of TS prevalence. *Movement Disorders, 30*(2), 221–228. https://doi.org/10.1002/mds.26089.
Selles, R. R., Storch, E. A., & Lewin, A. B. (2014). Variations in symptom prevalence and clinical correlates in younger versus older youth with obsessive–compulsive disorder. *Child Psychiatry & Human Development, 45*(6), 666–674. https://doi.org/10.1007/s10578-014-0435-9.
Sheppard, D. M., Bradshaw, J. L., Purcell, R., & Pantelis, C. (1999). Tourette's and comorbid syndromes: Obsessive compulsive and attention deficit hyperactivity disorder. A common etiology? *Clinical Psychology Review, 19*, 531–552.
Sigurdsson, H. P., Jackson, S. R., Jolley, L., Mitchell, E., & Jackson, G. M. (2020). Alterations in cerebellar grey matter structure and covariance networks in young people with Tourette syndrome. *Cortex, 126*, 1–15. https://doi.org/10.1016/j.cortex.2019.12.022.
Sigurdsson, H. P., Pépés, S. E., Jackson, G. M., Draper, A., Morgan, P. S., & Jackson, S. R. (2018). Alterations in the microstructure of white matter in children and adolescents with Tourette syndrome measured using tract-based spatial statistics and probabilistic tractography. *Cortex, 104*, 75–89. https://doi.org/10.1016/j.cortex.2018.04.004.
Spencer, T., Biederman, J., Harding, M., O'Donnell, D., Wilens, T., Faraone, S., et al. (1998). Disentangling the overlap between Tourette's disorder and ADHD. *Journal of Child Psychology and Psychiatry, and Allied Disciplines, 39*(7), 1037–1044.
Worbe, Y., Gerardin, E., Hartmann, A., Valabrégue, R., Chupin, M., Tremblay, L., et al. (2010). Distinct structural changes underpin clinical phenotypes in patients with Gilles de la Tourette syndrome. *Brain, 133*, 3649–3660.
Xia, X., Lin, Y., Lang, B., Yuan, J., Sheng, L., Yang, D., et al. (2020). Characteristics of diffusion tensor imaging of central nervous system in children with tourette's disease. *Medicine (Baltimore), 99*(22), e20492. https://doi.org/10.1097/MD.0000000000020492.
Xu, D., Hao, X., Bansal, R., Plessen, K. J., Geng, W., Hugdahl, K., et al. (2007). Unifying the analyses of anatomical and diffusion tensor images using volume-preserved warping. *Journal of Magnetic Resonance Imaging, 25*(3), 612–624. https://doi.org/10.1002/jmri.20858.

CHAPTER EIGHT

Altered structural connectivity in Gilles de la Tourette syndrome

Yulia Worbe[a,b,*]
[a]Sorbonne University and Paris Brain Institute, Salpetriere Hospital, Paris, France
[b]Department of Neurophysiology, Saint Antoine Hospital, Paris, France
*Corresponding author: e-mail address: yulia.worbe@aphp.fr

Contents

1. Structural brain abnormalities in children and adults with Tourette disorder and tics

Numerous studies have reported structural abnormalities of cortico-striato-thalamo-cortical circuits at both cortical and basal ganglia levels and in various brain regions in TS (Table 1 summarizes findings from different recent structural studies and metanalyses in TS). Structural alterations have been documented in many brain areas associated with sensorimotor processing (Martino et al., 2018).

On the cortical level, several studies reported widespread cortical thinning and sulcal abnormalities both in children and adults with TS (Fahim et al., 2010; Muellner et al., 2015; Sowell et al., 2008; Worbe et al., 2010). However, a recent study in children with TS showed both thinning of the cortex in postcentral, superior parietal, anterior cingulate, paracentral and pre-cuneus and increased cortical thickness in the frontal pole and

International Review of Movement Disorders, Volume 3
ISSN 2666-7878
https://doi.org/10.1016/bs.irmvd.2022.02.002

Table 1 Main findings of the original and meta-analysis of structural MRI studies ($n=15$) in patients with Tourette syndrome (TS) and tics published between 2019 and 2021.

Reference	Number of participants and the age	Medication/presence of comorbidities	Statistically significant findings in people with TS
Voxel-based morphometry, volumetry, morphometry and surface analysis			
Jackson, Sigurdsson, Dyke, Condon, and Jackson (2021)	39 TS patients, mean age of 14 yo 36 controls, mean age of 14 yo	On medication ADHD, OCD, ASD	Whole-brain VBM: reduction of gray matter in OFC, cerebellum, ACC, MCC. Structural covariance network analysis from ACC and MCC seed: increased structural covariance with cerebellum, inferior frontal cortex, posterior cingulate cortex.
Mirabella et al. (2020)	11 TS, 9 OCD and 12 HC, mean age for all groups 10 yo	Not medicated No other comorbidities	All subjects performed stop-signal tasks out of the scanner In the TS group, normal behavioral performance and VBM measures In the OCD group, gray matter volume of IFG negatively correlated with Stop signal task performance; mid-cingulate and middle frontal gyri volumes associated with severity of OCD and insula and medial OF volumes—negatively correlated with OCD severity and Stop signal task performance.
Sigurdsson, Jackson, Jolley, Mitchell, and Jackson (2020)	28 TS and 36 HC Mean age 14 yo for both groups	10 TS on medication ADHD, OCD, ASD	VBM and "seed to voxel" structural covariance analysis to investigate the cerebellum. In TS, reduction of gray matter volume in motor and cognitive parts of the cerebellum; alteration of structural covariance of the cerebellum with frontal and cingulate cortex.
Kim et al. (2020)	41 Provisional tic disorder (less than 9 months from onset), 7 yo; 34 TS mean age 8 yo, 40 HC, mean age 8 yo	On medication ADHD, OCD	Evaluated twice with 12 months intervals. The larger volume of hippocampus correlated more severe tics on 12-months follow up evaluation.
Kong et al. (2020)	52 TS, mean age 8 yo; 52 HC, mean age 10 yo	NP/NP	Cortical thickness and sulcal morphology have been evaluated. Decreased cortical thickness: postcentral, superior parietal, anterior cingulate, paracentral and pre-cuneus. Increased cortical thickness: frontal pole and medial OFC. Shallower cortical sulci were found in superior temporal, inferior lingual, medial OFC, superior parietal, latero-occipital, but deeper in the middle temporal regions.

Mufford (2019)	Genetic consortium on TS (4644 cases); ENIGMA imaging consortium (30,717 subjects)	NP/NP	Association of TS with lower thalamic volumes; intracranial and hippocampal volumes as well as striatum (putamen and caudate nucleus).
Diffusion tensor imaging			
Bruce et al. (2021)	12 TS, 14 HC Pediatric population		DTI study associated with transcranial magnetic stimulation for assessing inhibitory connections between pre-SMA and motor cortex. In subjects with tics, lower inhibitory capacity from pre-SMA to motor cortex was found and this was correlated with tics severity. Imaging showed increased FA in TS in right superior longitudinal fasciculus, corpus callosum, corona radiata and posterior limb of internal capsule, correlated with lower capacity to suppress tics.
Xia et al. (2020)	15 TS and 15 HC Pediatric population		Lower FA values in globus pallidus and thalamus in TS; decreased FA in thalamus correlates with YGTSS; ADC higher in caudate and thalamus in TS.
Martino et al. (2019)	14 TS and 19 HC Mean age in both groups 30 yo	On medication OCD	Motor-timing task was performed out of scanner. TS showed lower performance in one-hand supra-second time interval condition of the task. TS patients showed a higher FA in SMA-SMA and SMA-putamen tracts.
Multimodal neuroimaging			
Tikoo et al. (2021)	16 TS, mean age 9 yo 14 TS+OCD, mean age 10 yo 11 OCD, mean age 10 yo 12 HC, mean age 10 yo	No medication No comorbidities	In TS and TS+OCD groups, higher FA in cerebellar pedunculi compared to HC and lower FA in cerebellar pedunculi in OCD compared to other groups; altered functional connectivity in cerebello-thalamo-cortical circuits in all 3 groups.

Continued

Table 1 Main findings of the original and meta-analysis of structural MRI studies ($n=15$) in patients with Tourette syndrome (TS) and tics published between 2019 and 2021.—cont'd

Reference	Number of participants and the age	Medication/presence of comorbidities	Statistically significant findings in people with TS
Atkinson-Clement, Sofia, et al. (2020)	29 TS, mean age 31 yo 26 TS with IEO, mean age 31 yo	On medication ADHD, OCD	Intermittent-explosive symptoms were present in 47% of patients and were associated with (i) increased FA in hippocampus and SMA; (ii) decreased MD in OFC; (iii) lower functional connectivity in sensori-motor cortico-basal ganglia network and (iv) altered connectivity among OFC, amygdala and hippocampus.
Atkinson-Clement, Porte, et al. (2020)	50 TS and 26 HC, mean age for both groups 30 yo	On medication ADHD, OCD	Stop signal reaction time task performed out of scanner to measure motor impulsivity. Grater motor impulsivity was found in medicated TS associated with (i) increased gray matter in SMA, IFG, (ii) decreased spontaneous functional activity of putamen and (iii) higher functional connectivity IFG and temporal cortex, (iv) lower functional connectivity of STN with cerebellum.
Metanalyses of structural neuroimaging studies			
Wen et al. (2021)	247 TS, 236 HC Adults and children 6 neuroimaging studies	ON medication Comorbidities in original studies	Increased gray matter volumes: precentral and post-central gyrus, thalamus, lentiform nucleus. Decreased gray matter volumes: anterior cingulate, medial frontal gyrus.
Wan et al. (2021)	331 TS and 327 HC Adults and children 10 neuroimaging studies	ON medication Comorbidities in original studies	Meta-analysis was performed on 10 VBM studies. Decrease of gray matter volume: IFG. Increase of gray matter volume: cerebellum, striatum, thalamus. There were no correlations with tics severity.
Yang et al. (2021)	168 TS and 163 HC Adults and children 8 neuroimaging studies	ON medication Comorbidities in original studies	Meta-analysis was performed on 8 DTI studies. Decreased FA in the corpus callosum, longitudinal fasciculus; no correlation with symptoms severity was found.

To review findings from anatomical studies published before 2013, please consult Neuner et al. (2010); from 2013 to 2018, please consult Martino, Ganos, and Worbe (2018). Abbreviations: HC: healthy controls; VBM: voxel-based morphometry; DTI: diffusion tensor imaging; SMA: supplementary motor area. FA: fractional anisotropy; MD: mean diffusivity; ADC: apparent diffusion coefficient; ADHD: attention deficit hyperactivity disorder; OCD: obsessive-compulsive disorder, ASD: autistic spectrum disorder; OFC: orbitofrontal cortex; ACC: anterior cingulate cortex; MCC: mid-cingulate cortex; NP: information non provided in publication; IEO: intermitted explosive outbursts; IFG: inferior frontal gyrus, STN: subthalamic nucleus.

medial orbitofrontal cortex (Kong et al., 2020), which might indicate a different pattern of cortical changes in children compared to adults with TS. Indeed, different developmental trajectories and/or networks involved in children compared to adults with TS have already been reported in studies that used the functional connectivity approach and resting-state functional imaging (Nielsen et al., 2020). Moreover, results of some structural and functional studies also support the dichotomy of brain development in children and adults with TS: children with TS tend to show advanced maturation patterns when adults exhibited a delayed pattern of both functional and structural connectivity (Nielsen et al., 2020; Worbe et al., 2012, 2015).

Structural changes were also reported in the basal ganglia and less studied in TS research subcortical structures such as hippocampus, amygdala and cerebellum.

In a large group of TS patients, including both children and adults, Peterson et al. (2003) showed a decrease in basal ganglia volume, with caudate nucleus more affected in children and volume reduction in most of the basal ganglia structures showed in adults, with an exception for thalamus that showed 5% enlargement in both groups (Miller et al., 2010). However, the association of basal ganglia volumes with TS remains unclear, as recent metanalysis of voxel-based morphometric studies showed an increase of gray matter volume of the cerebellum, striatum and thalamus with no correlations with the severity of tics (Wan et al., 2021). In contrast, the results of a large study on more than 4000 cases showed that the genetic risk of TS was associated with overall lower intracranial as well as striatal, thalamic and hippocampal volumes (Mufford et al., 2019).

In contrast to the most previously described group comparison studies, one study reviewed the tic-inducing brain lesions and was able to identify a shared network causally related to the occurrence of tics. This study confirmed that the insular and cingulate cortices, striatum, GPi, thalami and the cerebellum are all the regions of the tic-generation network. Moreover, within this network, connectivity to a region within the anterior dorsal putamen was specific to tic-inducing lesions compared with control lesions (Ganos et al., 2022), putting these regions as a central for the occurrence of tics.

Several studies using larger samples were able to disentangle distinct structural changes related to different clinical phenotypes in TS. On the cortical level, one study on adults with TS suggested that these cortical changes might result from neurodevelopmental abnormalities and differed among clinical sub-groups of patients (Worbe et al., 2010). For instance,

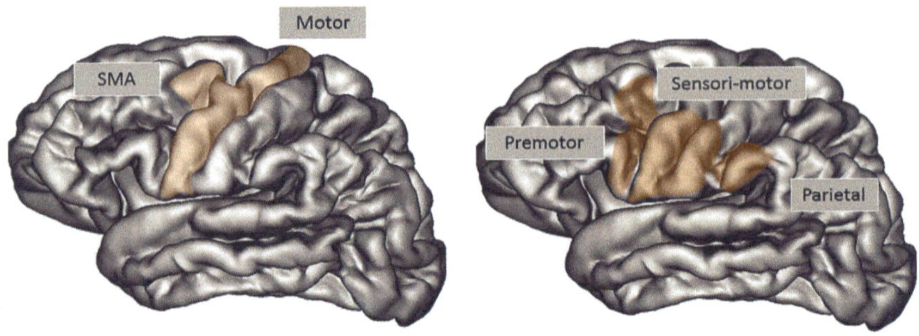

Fig. 1 Structural alterations of cortical regions in relationship with simple (left) and complex (right) tics.

cortical thinning of motor regions was mainly found in patients with simple tics, whereas thinning extended into larger premotor, prefrontal and parietal regions were related to complex tics, and cortical thinning in the ventromedial prefrontal cortex was associated with the co-morbid obsessive-compulsive disorder (Fig. 1).

Similarly to cortical thickness, TS patients had morphological changes of cortical sulci, specifically for patients with tics only and patients with tics associated with OCD. Importantly, some of these features, as the morphology of central and superior temporal sulcus, correlated with the severity of tics, whereas the severity of obsessive-compulsive symptoms was correlated with morphological changes in the left intermediate precentral sulcus and posterior callosum-marginal sulcus (Muellner et al., 2015).

The pattern of cortico-basal ganglia structural connectivity was analyzed using diffusion tension imaging in a large group of TS patients (Worbe et al., 2015). Sensorimotor cortico-striato-thalamic pathway exhibit more important structural connectivity in TS than controls, which also correlated with the severity of tics. Noteworthy, the increase of structural connectivity correlated with the severity of tics was also found in children with TS (Thomalla et al., 2009), albeit inconstantly (Neuner et al., 2010). In contrast, OCD and the severity of obsessive-compulsive symptoms were correlated with orbitofrontal cortex—basal ganglia structural connectivity. Taken together, these studies stressed differences in structural topography in TS and directed attention toward neuroimaging markers of the different clinical presentations.

Aside from cortex and basal ganglia, structural changes were also described in the mid-brain (Garraux et al., 2006) and cerebellum (Tobe et al., 2010) in adult GTS patients, with a reduced gray matter in both motor and cognitive parts of the cerebellum. In children with TS, the same results of reduction of gray matter volume in motor and cognitive parts of the cerebellum were found. In addition, there was an alteration of structural covariance (i.e., structural connectivity) of the cerebellum with the frontal and cingulate cortex (Jackson et al., 2021; Sigurdsson et al., 2020). Overall, these studies support the idea that the cerebellum plays an essential role in tics generation, as was previously suggested by animal (McCairn, Iriki, & Isoda, 2013) and computational modeling studies (Caligiore, Mannella, Arbib, & Baldassarre, 2017).

Widespread white matter connections might also be abnormally developed in patients with TS. The most compelling evidence comes from the studies on interhemispheric connectivity carried out by corpus callosum, in which the motor part modulates finely timed lateralized or bimanual actions. In children with TS, axial diffusivity of all five corpus callosum segments was reduced compared to healthy control and also negatively correlated with the severity of tics (Plessen et al., 2006; Wolff et al., 2016). The widespread alteration of inter-hemispheric connectivity was also found in adults with TS, with a trend of correlation with the severity of tics (Cheng et al., 2014) and in adult monozygotic twins discordant for TS (Cavanna et al., 2010). The tans-callosal right to left supplementary motor cortex connections was also enhanced in adult TS and this alteration correlated with the performance of motor timing task and with a specific deficit of motor timing accuracy with the right (dominant) hand (Martino et al., 2019).

One very large ($n = 103$) voxel-based morphometry study in children with TS showed reduced white matter volume bilaterally in the orbital and medial prefrontal cortex (Greene et al., 2017). In addition, in children, structural axial diffusivity (AD) and mean diffusivity (MD) and volumetric changes were also found in anterior thalamic radiation, right cingulum bundle projecting to the cingulate gyrus and forceps minor (Liu et al., 2013).

2. Structural changes related to the severity of tics and premonitory urges in Tourette syndrome

2.1 Structural changes related to the severity of tics

Correlation analysis of structural changes with tics severity measurements, mostly with Yale Global Tics severity scale—YGTSS, showed relationships with cortical thinning in the sensorimotor cortex and surrounding frontal and parietal and temporal areas (Fahim et al., 2010; Sowell et al., 2008;

Worbe et al., 2010), gray matter increase in the ventral putamen using voxel-based morphometry (Ludolph et al., 2006) and diffusion orientation in the thalamus (Makki, Behen, Bhatt, Wilson, & Chugani, 2008). Tic severity is also negatively correlated with a decrease in white matter volume in the right frontal pole (Liu et al., 2013). Interestingly, in children with TS, alteration of white matter microstructure in the corpus callosum and forceps minor, that contained connections to the prefrontal cortex, also significantly predict tic severity (Jackson et al., 2011).

Several studies tried to identify cerebral structures associated with the severity and persistence of tics during the transition from childhood to adulthood. One study followed up adolescents with TS and controls and showed that people with persistent tics exhibited a decrease in parallel and perpendicular diffusivity within caudate, thalamus and frontal lobe, at difference with patients in remission who shared more similarities with control subjects (Debes et al., 2015). Moreover, in the pediatric population with TS, a smaller volume of the caudate nucleus head also was predictive for the persistence of tics in adulthood (Bloch, Leckman, Zhu, & Peterson, 2005). However, a "New tic study" that followed children 3 months after tics onset for 12 months showed that volume of the striatum was not predictive of the tics' outcome (remission or persistent), but a larger hippocampus at baseline predicted worse severity at follow-up (Black, Kim, Schlaggar, & Greene, 2020; Kim et al., 2020). Taken together, the results stress the structural abnormalities in TS go beyond the sensorimotor regions and cortico-basal ganglia connections, which might reflect compensatory brain reorganization related to the self-regulation mechanisms of tics in youngsters with TS.

2.2 Structural changes correlated with premonitory urges

Only a few imaging studies have investigated premonitory urges (PU) and only 3 of these studies used structural neuroimaging.

In children with TS, cortical thickness analysis demonstrated a negative correlation between PUTS (premonitory urges in TS scale) scores and gray matter thickness within sensorimotor cortical areas as well as within the left insula, with no significant positive correlations of PUTS scores with cortical gray matter thickness (Draper, Jackson, Morgan, & Jackson, 2016). This study demonstrated for the first time a relationship between PU and the structure of the insular cortex as well as conformed the gray matter thickness reduction in these areas in young people with TS compared to an age- and gender-matched control group. Another larger study from the same group confirmed these results demonstrating an association between the frequency of PU and increased structural connectivity between the primary motor cortex and the insula (Sigurdsson et al., 2018). In adults with TS, only one study that performed voxel-based morphometry and diffusion tensor imaging showed a positive correlation between PUTS scores and both gray matter volume and cortical thickness in the left somatosensory and prefrontal cortices and altered structural organization in the parietal portion of the superior longitudinal fascicle in patients with higher PUTS scores (Draganski et al., 2010).

Overall, these findings suggest that, in children and adults with TS, structural abnormalities in the sensorimotor network and insular cortex were associated with greater PU intensity.

3. Structural changes associated with comorbidities in Tourette syndrome

The frequent comorbidity of TS with ADHD and OCD likely results from a shared genetic background, as recently suggested by genetic studies (Brainstorm et al., 2018; Yang et al., 2021) and a common underlying neurobiological substrate, mediated by partly overlapping, albeit still separate, cortico-basal ganglia circuits (Kleimaker et al., 2021). However, neuroimaging studies that directly compared patients with TS, to patients with ADHD or OCD and with the association of TS with ADHD or OCD are very limited in number.

With regards to ADHD, some studies explored the morphology of basal ganglia and deep brain structures in children and adults presenting both ADHD and tics and pointed to abnormalities in the amygdala and hippocampal complex (Ludolph et al., 2008; Plessen, Royal, & Peterson, 2007), whereas the morphology of basal ganglia appeared to be similar to those of the controls (Forde et al., 2017). On a cortical level, the left inferior frontal gyrus might represent a common underlying structural correlate of TS with co-morbid ADHD. However, one study that used semi-supervised machine learning algorithms to identity structural masquer and to evaluate the diagnostic accuracy of various neuropsychiatric disorders (Bansal et al., 2012), showed that surface morphology of deep brain structures was able to distinguish neuropsychiatric disorders. In particular, the morphology of distinct parts of the globus pallidus, caudate nucleus, putamen, thalamus and amygdala-hippocampal complex were able to discriminate between

ADHD and TS, thus suggesting that basal ganglia morphology differs substantially between ADHD and TS, even if these differences are not captured by conventional imaging modalities.

One small study also addressed a question of structural correlates of intermitted-explosive symptoms, manifesting as sudden episodes of verbal or physical aggression, frequently observed in TS with ADHD and sometimes in TS only (Ashurova, Budman, & Coffey, 2021; Atkinson-Clement, Sofia, et al., 2020; Conte, Valente, Fioriello, & Cardona, 2020). In the absence of ADHD, TS patients with intermitted-explosive symptoms compared to those without showed increased fractional anisotropy in the right supplementary motor area and in the right hippocampus and decreased mean diffusivity in the left orbitofrontal cortex. These structural changes were also accompanied by lower functional connectivity within the sensorimotor cortico-basal ganglia network, and an altered connectivity pattern among the orbitofrontal cortex, amygdala and hippocampus (Atkinson-Clement, Sofia, et al., 2020). Overall, the results might suggest intermitted-explosive symptoms resulting from abnormal control to action selection (sensorimotor network) and impairments related to emotional regulation and impulse control (orbitofrontal cortex, amygdala and hippocampus).

Another frequent comorbidity with TS is obsessive-compulsive disorders (OCD) (Martino, Ganos, & Pringsheim, 2017), which are sometimes challenging to detangle as some phenomena lay on the frontier between tics and compulsions/obsessions (Szejko & Muller-Vahl, 2021). In TS, the association with OCD was found to be characterized by volume reduction in the caudate nucleus (Bloch et al., 2005) and with a reduction of cortical thickness in the ventromedial prefrontal cortex and hippocampus (Worbe et al., 2010). Co-morbid OCD in adults with TS was also associated with a relative enlargement of several cerebellar regions, including crus and lobules VI, VIIB, and VIIIA, in proportion to the increasing severity of obsessive-compulsive symptoms, whereas tic severity was correlated to a decrease in volumes of these cerebellar regions (Tobe et al., 2010).

One study specifically evaluated cerebellar involvement in children with TS compared to TS with OCD and OCD and found that TS and TS associated with OCD had higher fractional anisotropy in cerebellar peduncles compared to controls. In contrast, OCD patients without tics showed a lower fractional anisotropy compared to controls, TS and TS + OCD patients (Tikoo et al., 2021). These structural changes were also accompanied by changes in cerebellar functional connectivity, suggesting that cerebellar-thalamo-cortical and cerebello-basal ganglia connection contributes to both tics and OCD.

4. Structural changes and cognitive functions in Tourette syndrome

Among various cognitive functions, inhibitory capacity and cognitive control had received the most attention in TS across behavioral and neuroimaging studies. Indeed, several studies showed that TS patients with a more pronounced decrease of prefrontal cortical thickness or volume also manifested a greater severity of tics, suggesting that tics might be a result of lack of inhibition of prepotent motor programs (Rae et al., 2020). However, the relationship between the inhibitory capacity and tic remains unclear, as both reduced and intact inhibitory capacity has been reported in TS (Ganos et al., 2014, 2014, 2018). In addition, a recent metanalysis on TS performance on different motor inhibition tasks showed that impaired inhibitory capacity was mostly related to the treatment with antipsychotics and the presence of ADHD (Morand-Beaulieu et al., 2017). Some other behavioral and functional imaging studies suggested that specific types of inhibitory action control might be impaired in TS, such as automatic compared to volitional control (Rawji et al., 2020) or capacity to postpone impending action (Atkinson-Clement et al., 2021).

With regards to structural brain abnormalities, in children with TS, worse ability to suppress tics were correlated with less inhibitory capacity and increased fractional anisotropy in the right superior longitudinal fasciculus, corpus callosum, corona radiata and posterior limb of the internal capsule (Bruce et al., 2021). In adults with TS, one study showed a deficit in inhibitory control, measured by stop-signal task, only in medicated with atypical antipsychotics TS patients and that this deficit was related to the structures and by functional connectivity of the fronto-temporal-basal ganglia-cerebellar pathway (Atkinson-Clement, Porte, et al., 2020).

Another cognitive function that has been studied with structural imaging in TS is habit formation, which is the process leading to the automatization of behavioral programs largely supported by the function and structure of the cortico-basal ganglia network (de Wit, Corlett, Aitken, & Dickinson, 2009; de Wit & Dickinson, 2009; Jahanshahi, Obeso, Rothwell, & Obeso, 2015). As tics in TS are sometimes described as intentional actions performed in response to sensory stimuli (such as premonitory urges), they might be considered as voluntary movements performed in the habitual way (Leckman & Riddle, 2000). Habitual behavioral control or habit formation have been explored in several behavioral studies with somewhat discrepant results as both enhanced and reduced habitual control has been reported in TS (Marsh, Alexander, Packard, Zhu, & Peterson, 2005; Shephard, Groom, & Jackson, 2019). These discrepancies might be a result of both the behavioral paradigms used (or how the habit formation has been studied) and of the TS population, as treatment with dopamine blockers (Nelson & Killcross, 2013) and presence of OCD (Gillan et al., 2011) were shown to influence habit formation. Only one study evaluated structural correlates of habit formation in TS. Compared to healthy volunteers, only unmedicated TS patients showed reliance on habitual control, where medicated with antipsychotics patients exhibited intermediate between controls and unmedicated patients' performance. In unmedicated TS, the propensity to rely on habitual behavioral control was predicted by structural connectivity of the white matter pathway connecting the premotor cortex with putamen. Moreover, in unmedicated patients, stronger structural connectivity of the supplementary motor cortex with the sensorimotor putamen also predicted more severe tics (Delorme et al., 2016). Taken together, these results suggested that the sensory-motor system is a common siege for both tics and habits. Indeed, some recent pathophysiological accounts suggested that both abnormalities of inhibitory control within cortico-basal ganglia networks and their dopaminergic up-regulation might serve as tics precursors (Ganos et al., 2021).

5. Structural network in Tourette syndrome: Mapping white matter tracts as predictors for deep brain stimulation response

Deep brain stimulation (DBS) is an effective neuromodulatory therapy that can be applied for the successful treatment of numerous movement disorders. In patients with TS, it was recognized as an effective treatment for disabling tics (Szejko et al., 2021). However, several structures of including an internal segment of globus pallidus intern; centromedial thalamus and nucleus accumbens/anterior limb of the internal capsule, were proposed as possible targets for neuromodulation, and it is still a matter of discussion which of these targets might be the most effective to alleviate tics and comorbidities in TS (Andrade et al., 2020; Servello et al., 2020). Some studies for instance tried to identify the "sweet spots" or the specific regions within the targets that were associated with better clinical outcomes in TS. For instance, one of the largest studies to date on 123 patients were able to identify the "sweet spots" for tics and OCD improvement in thalamic CMT and GPI targets, such as for instance, regions superior or medial to GPi were associated with a greater improvement in OCD than regions inferior to GPi (Johnson et al., 2019).

With the development of diffusion tension imaging and structural connectivity approaches (i.e., identification of white matter connecting various brain regions derived from advanced mathematical algorithms based on the direction of water flow from a voxel-to-voxel allowing to reconstruct of axonal fibers throughout the entire brain) and methods to evaluate the brain tissue activated by specific stimulation parameters, it is now possible to evaluate white matter tracts that are potentially affected by the stimulation. These white matter pathways reconstructions or structural connectivity maps have been used in numerous neurological and psychiatric disorders to identify brain networks related to the best clinical outcomes (Wong et al., 2020).

In TS, several studies focused on the identification of these connectivity maps for different targets and symptoms and showed that GPI target, tics improvement was related to associative pallido-subthalamic tract whereas improvement of obsessive-compulsive behaviors improvement to sensorimotor pallido-subthalamic tract as well as premotor, motor and prefrontal internal capsule pathways (Table 2) (Johnson et al., 2021).

In targets comparison, clinical improvement for GPI included structural connections with caudate, thalamus, cerebellum, as well as limbic and associative network and for CMT with connectivity to the putamen, cerebellum, parieto-temporo-occipital network, to sensori-motor network. For both targets, improvement of obsessive-compulsive behaviors associated with connectivity to the prefrontal cortex, orbitofrontal cortex and cingulate cortex, but only CMT target predicted clinical outcome suggesting that some targets might be more prone to improve associated comorbidities (Fig. 2) (Johnson et al., 2020).

Finally, some studies also tried to identify the structural markers that distinguish good compared to poor responders for DBS. Only two small studies on seven patients each try to address this question. They showed that for the thalamic target the good responders had more connection to the sensorimotor cortex, whereas poor or non-responders to the DBS had a larger amount of the fibers connecting with other frontal cortical regions

Table 2 Structural connectivity and deep brain stimulation in Tourette syndrome.

Reference	Number of participants and the age	Targets	Statistically significant findings
Heiden et al. (2021)	7 TS, adults	GPI, thalamus and NA/ALIC Comparison of responders ($n=5$) vs. non-responder ($n=2$)	Responders had more important connections of the thalamus with S1 (for thalamus target). A specific pattern of connections characterized each target: posterior GPI was more connected to the sensorimotor cortex, anterior GPI—to the frontal cortex, CM—to the amygdala, ventral anterior/ventral lateral thalamic nucleus to the motor cortex.
Andrade et al. (2020)	7 TS, adults	CMT/ventral oralis internus nucleus	Higher projection density between thalamus and sensori-motor cortex in patients with better outcome (YGTSS improvement). Non-responders to the DBS had a larger amount of fibers connecting with other frontal cortical regions.
Jackson et al. (2021)	35 TS, adults	GPI	For tics improvement: associative pallido-subthalamic tract. For OCB improvement: sensori-motor pallido-subthalamic tract; premotor, motor and prefrontal internal capsule pathways.
Johnson et al. (2020)	66 TS, adults	GPI ($n=34$) CMT ($n=32$)	For GPI target: clinical improvement was associated with connectivity to caudate, thalamus, cerebellum as well as a limbic and associative network. For CMT target: clinical improvement was associated with connectivity to putamen, cerebellum, parieto-temporo-occipital network, to sensori-motor network. For OCB and both targets, clinical improvement was associated with connectivity to the prefrontal cortex, orbito-frontal cortex, and cingulate cortex, but only CMT target predicted clinical outcome.
Johnson et al. (2019)	123 TS, adults	GPI ($n=47$) CMT ($n=51$) NA/ALIC ($n=4$) Combined targets ($n=8$)	"Sweet spots" within GPI and CMT improve tics for some patients. Regions within superior or medial GPI were associated with greater improvement of OCB compared to inferior GPI regions.

OCB—obsessive-compulsive behaviors; GPI—globus pallidus intern; CMT—centromedial thalamus; NA/ALIC—nucleus accumbens, anterior limb of the internal capsule.

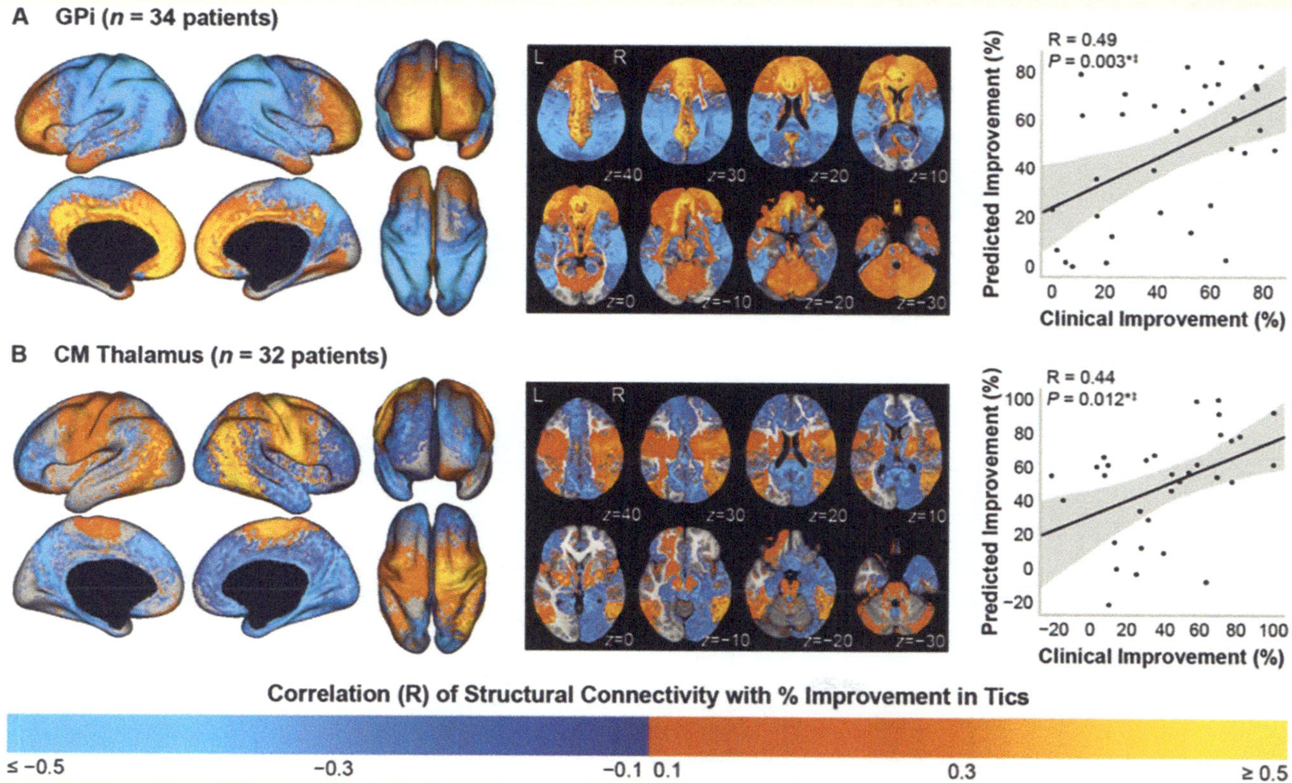

Fig. 2 (A) From GPI target; (B) From CM thalamus target. *From Johnson, K. A., Duffley, G., Anderson, D. N., Ostrem, J. L., Welter, M. L., Baldermann, J. C., … Butson, C. R. (2020). Structural connectivity predicts clinical outcomes of deep brain stimulation for Tourette syndrome.* Brain, 143*(8), 2607–2623. doi:https://doi.org/10.1093/brain/awaa188.*

(Andrade et al., 2020; Heiden et al., 2021). While still preliminary and in need of validation on a larger group of patients, these studies shed light on potential neuroimaging markers of response to DBS treatment and thus adding a pillar to patients' selection (Martino et al., 2021).

References

Andrade, P., Heiden, P., Hoevels, M., Schlamann, M., Baldermann, J. C., Huys, D., et al. (2020). Modulation of fibers to motor cortex during thalamic DBS in Tourette patients correlates with tic reduction. *Brain Sciences, 10*(5). https://doi.org/10.3390/brainsci10050302.

Ashurova, M., Budman, C., & Coffey, B. J. (2021). Ticked off: Anger outbursts and aggressive symptoms in Tourette disorder. *Child and Adolescent Psychiatric Clinics of North America, 30*(2), 361–373. https://doi.org/10.1016/j.chc.2020.10.006.

Atkinson-Clement, C., Porte, C. A., de Liege, A., Klein, Y., Delorme, C., Beranger, B., et al. (2021). Impulsive prepotent actions and tics in Tourette disorder underpinned by a common neural network. *Molecular Psychiatry, 26*(7), 3548–3557. https://doi.org/10.1038/s41380-020-00890-5.

Atkinson-Clement, C., Porte, C. A., de Liege, A., Wattiez, N., Klein, Y., Beranger, B., et al. (2020). Neural correlates and role of medication in reactive motor impulsivity in Tourette disorder. *Cortex, 125*, 60–72. https://doi.org/10.1016/j.cortex.2019.12.007.

Atkinson-Clement, C., Sofia, F., Fernandez-Egea, E., de Liege, A., Beranger, B., Klein, Y., et al. (2020). Structural and functional abnormalities within sensori-motor and limbic networks underpin intermittent explosive symptoms in Tourette disorder. *Journal of Psychiatric Research, 125*, 1–6. https://doi.org/10.1016/j.jpsychires.2020.02.033.

Bansal, R., Staib, L. H., Laine, A. F., Hao, X., Xu, D., Liu, J., et al. (2012). Anatomical brain images alone can accurately diagnose chronic neuropsychiatric illnesses. *PLoS One*, 7(12), e50698. https://doi.org/10.1371/journal.pone.0050698.

Black, K. J., Kim, S., Schlaggar, B. L., & Greene, D. J. (2020). The new tics study: A novel approach to pathophysiology and cause of tic disorders. *Journal of Psychiatry and Brain Science, 5*. https://doi.org/10.20900/jpbs.20200012.

Bloch, M. H., Leckman, J. F., Zhu, H., & Peterson, B. S. (2005). Caudate volumes in childhood predict symptom severity in adults with Tourette syndrome. *Neurology, 65*(8), 1253–1258. 65/8/1253 [pii] https://doi.org/10.1212/01.wnl.0000180957.98702.69.

Brainstorm, C., Anttila, V., Bulik-Sullivan, B., Finucane, H. K., Walters, R. K., Bras, J., et al. (2018). Analysis of shared heritability in common disorders of the brain. *Science, 360*(6395). https://doi.org/10.1126/science.aap8757.

Bruce, A. B., Yuan, W., Gilbert, D. L., Horn, P. S., Jackson, H. S., Huddleston, D. A., et al. (2021). Altered frontal-mediated inhibition and white matter connectivity in pediatric chronic tic disorders. *Experimental Brain Research, 239*(3), 955–965. https://doi.org/10.1007/s00221-020-06017-0.

Caligiore, D., Mannella, F., Arbib, M. A., & Baldassarre, G. (2017). Dysfunctions of the basal ganglia-cerebellar-thalamo-cortical system produce motor tics in Tourette syndrome. *PLoS Computational Biology, 13*(3), e1005395. https://doi.org/10.1371/journal.pcbi.1005395.

Cavanna, A. E., Stecco, A., Rickards, H., Servo, S., Terazzi, E., Peterson, B., et al. (2010). Corpus callosum abnormalities in Tourette syndrome: An MRI-DTI study of monozygotic twins. *Journal of Neurology, Neurosurgery, and Psychiatry, 81*(5), 533–535. https://doi.org/10.1136/jnnp.2009.173666.

Cheng, B., Braass, H., Ganos, C., Treszl, A., Biermann-Ruben, K., Hummel, F. C., et al. (2014). Altered intrahemispheric structural connectivity in Gilles de la Tourette syndrome. *Neuroimage: Clinical, 4*, 174–181. https://doi.org/10.1016/j.nicl.2013.11.011.

Conte, G., Valente, F., Fioriello, F., & Cardona, F. (2020). Rage attacks in Tourette syndrome and chronic tic disorder: A systematic review. *Neuroscience and Biobehavioral Reviews, 119*, 21–36. https://doi.org/10.1016/j.neubiorev.2020.09.019.

de Wit, S., Corlett, P. R., Aitken, M. R. F., Dickinson, A., & Fletcher, P. J. (2009). Differential engagement of the ventromedial prefrontal cortex by goal-directed and habitial behaviour towards food pictures in humans. *Jouranal of Neuroscience, 29*(36), 11330–11338.

de Wit, S., & Dickinson, A. (2009). Associative theories of goal-directed behaviour: A case for animal–human translational models. *Psychological Research, 73*, 463–476.

Debes, N., Jeppesen, S., Raghava, J. M., Groth, C., Rostrup, E., & Skov, L. (2015). Longitudinal magnetic resonance imaging (MRI) analysis of the developmental changes of Tourette syndrome reveal reduced diffusion in the Cortico-Striato-Thalamo-cortical pathways. *Journal of Child Neurology, 30*(10), 1315–1326. https://doi.org/10.1177/0883073814560629.

Delorme, C., Salvador, A., Valabregue, R., Roze, E., Palminteri, S., Vidailhet, M., et al. (2016). Enhanced habit formation in Gilles de la Tourette syndrome. *Brain, 139*, 605–615. https://doi.org/10.1093/brain/awv307.

Draganski, B., Martino, D., Cavanna, A. E., Hutton, C., Orth, M., Robertson, M. M., et al. (2010). Multispectral brain morphometry in Tourette syndrome persisting into adulthood. *Brain, 133*(Pt 12), 3661–3675. awq300 [pii] https://doi.org/10.1093/brain/awq300.

Draper, A., Jackson, G. M., Morgan, P. S., & Jackson, S. R. (2016). Premonitory urges are associated with decreased grey matter thickness within the insula and sensorimotor cortex in young people with Tourette syndrome. *Journal of Neuropsychology, 10*(1), 143–153. https://doi.org/10.1111/jnp.12089.

Fahim, C., Yoon, U., Das, S., Lyttelton, O., Chen, J., Arnaoutelis, R., et al. (2010). Somatosensory-motor bodily representation cortical thinning in Tourette: Effects of tic severity, age and gender. *Cortex, 46*(6), 750–760. S0010-9452(09)00186-5 [pii] https://doi.org/10.1016/j.cortex.2009.06.008.

Forde, N. J., Zwiers, M. P., Naaijen, J., Akkermans, S. E. A., Openneer, T. J. C., Visscher, F., et al. (2017). Basal ganglia structure in Tourette's disorder and/or attention-deficit/hyperactivity disorder. *Movement Disorders, 32*(4), 601–604. https://doi.org/10.1002/mds.26849.

Ganos, C., Al-Fatly, B., Fischer, J.-F., Baldermann, J.-C., Hennen, C., Visser-Vandewalle, V., et al. (2022). A neural network for tics: Insights from causal brain lesions and deep brain stimulation. *Brain*, awac009. https://doi.org/10.1093/brain/awac009.

Ganos, C., Kühn, S., Kahl, U., Schunke, O., Brandt, V., Bäumer, T., et al. (2014). Prefrontal cortex volume reductions and tic inhibition are unrelated in uncomplicated GTS adults. *Journal of Psychosomatic Research, 76*(1), 84–87.

Ganos, C., Kuhn, S., Kahl, U., Schunke, O., Feldheim, J., Gerloff, C., et al. (2014). Action inhibition in Tourette syndrome. *Movement Disorders, 29*(12), 1532–1538.

Ganos, C., Neumann, W. J., Muller-Vahl, K. R., Bhatia, K. P., Hallett, M., Haggard, P., et al. (2021). The phenomenon of exquisite motor control in tic disorders and its pathophysiological implications. *Movement Disorders, 36*(6), 1308–1315. https://doi.org/10.1002/mds.28557.

Ganos, C., Rocchi, L., Latorre, A., Hockey, L., Palmer, C., Joyce, E. M., et al. (2018). Motor cortical excitability during voluntary inhibition of involuntary tic movements. *Movement Disorders, 33*(11), 1804–1809. https://doi.org/10.1002/mds.27479.

Garraux, G., Goldfine, A., Bohlhalter, S., Lerner, A., Hanakawa, T., & Hallett, M. (2006). Increased midbrain gray matter in Tourette's syndrome. *Annals of Neurology, 59*(2), 381–385. https://doi.org/10.1002/ana.20765.

Gillan, C. M., Papmeyer, M., Morein-Zamir, S., Sahakian, B. J., Fineberg, N. A., Robbins, T. W., et al. (2011). Disruption in the balance between goal-directed behavior and habit learning in obsessive-compulsive disorder. *The American Journal of Psychiatry, 168*(7), 718–726. appi.ajp.2011.10071062 [pii] https://doi.org/10.1176/appi.ajp.2011.10071062.

Greene, D. J., Williams Iii, A. C., Koller, J. M., Schlaggar, B. L., Black, K. J., Association, T. T., et al. (2017). Brain structure in pediatric Tourette syndrome. *Molecular Psychiatry, 22*(7), 972–980. https://doi.org/10.1038/mp.2016.194.

Heiden, P., Hoevels, M., Bayram, D., Baldermann, J. C., Schuller, T., Huys, D., et al. (2021). Connectivity patterns of deep brain stimulation targets in patients with Gilles de la Tourette syndrome. *Brain Sciences, 11*(1). https://doi.org/10.3390/brainsci11010087.

Jackson, S. R., Parkinson, A., Jung, J., Ryan, S. E., Morgan, P. S., Hollis, C., et al. (2011). Compensatory neural reorganisation in Tourette syndrome. *Current Biology, 21*(7), 580–585. https://doi.org/10.1016/j.cub.2011.02.047.

Jackson, S. R., Sigurdsson, H. P., Dyke, K., Condon, M., & Jackson, G. M. (2021). The role of the cingulate cortex in the generation of motor tics and the experience of the premonitory urge-to-tic in Tourette syndrome. *Journal of Neuropsychology, 15*(3), 340–362. https://doi.org/10.1111/jnp.12242.

Jahanshahi, M., Obeso, I., Rothwell, J. C., & Obeso, J. A. (2015). A fronto-striato-subthalamic-pallidal network for goal-directed and habitual inhibition. *Nature Reviews. Neuroscience, 16*(12), 719–732. https://doi.org/10.1038/nrn4038.

Johnson, K. A., Duffley, G., Anderson, D. N., Ostrem, J. L., Welter, M. L., Baldermann, J. C., et al. (2020). Structural connectivity predicts clinical outcomes of deep brain stimulation for Tourette syndrome. *Brain, 143*(8), 2607–2623. https://doi.org/10.1093/brain/awaa188.

Johnson, K. A., Duffley, G., Foltynie, T., Hariz, M., Zrinzo, L., Joyce, E. M., et al. (2021). Basal ganglia pathways associated with therapeutic pallidal deep brain stimulation for Tourette syndrome. *Biological Psychiatry: Cognitive Neuroscience and Neuroimaging, 6*(10), 961–972. https://doi.org/10.1016/j.bpsc.2020.11.005.

Johnson, K. A., Fletcher, P. T., Servello, D., Bona, A., Porta, M., Ostrem, J. L., et al. (2019). Image-based analysis and long-term clinical outcomes of deep brain stimulation for Tourette syndrome: A multisite study. *Journal of Neurology, Neurosurgery, and Psychiatry, 90*(10), 1078–1090. https://doi.org/10.1136/jnnp-2019-320379.

Kim, S., Greene, D. J., D'Andrea, C. B., Bihun, E. C., Koller, J. M., O'Reilly, B., et al. (2020). Hippocampal volume in provisional tic disorder predicts tic severity at 12-month follow-up. *Journal of Clinical Medicine, 9*(6). https://doi.org/10.3390/jcm9061715.

Kleimaker, A., Kleimaker, M., Behm, A., Weissbach, A., Baumer, T., Beste, C., et al. (2021). Networks in the field of Tourette syndrome. *Frontiers in Neurology, 12*, 624858. https://doi.org/10.3389/fneur.2021.624858.

Kong, L., Lv, B., Wu, T., Zhang, J., Fan, Y., Ouyang, M., et al. (2020). Altered structural cerebral cortex in children with Tourette syndrome. *European Journal of Radiology, 129*, 109119. https://doi.org/10.1016/j.ejrad.2020.109119.

Leckman, J. F., & Riddle, M. A. (2000). Tourette's syndrome: When habit-forming systems form habits of their own? *Neuron, 28*(2), 349–354.

Liu, Y., Miao, W., Wang, J., Gao, P., Yin, G., Zhang, L., et al. (2013). Structural abnormalities in early Tourette syndrome children: A combined voxel-based morphometry and tract-based spatial statistics study. *PLoS One, 8*(9), e76105. https://doi.org/10.1371/journal.pone.0076105.

Ludolph, A. G., Juengling, F. D., Libal, G., Ludolph, A. C., Fegert, J. M., & Kassubek, J. (2006). Grey-matter abnormalities in boys with Tourette syndrome: Magnetic resonance imaging study using optimised voxel-based morphometry. *The British Journal of Psychiatry*, *188*, 484–485. 188/5/484 [pii] https://doi.org/10.1192/bjp.bp.105.008813.

Ludolph, A. G., Pinkhardt, E. H., Tebartz van Elst, L., Libal, G., Ludolph, A. C., Fegert, J. M., et al. (2008). Are amygdalar volume alterations in children with Tourette syndrome due to ADHD comorbidity? *Developmental Medicine and Child Neurology*, *50*(7), 524–529.

Makki, M. I., Behen, M., Bhatt, A., Wilson, B., & Chugani, H. T. (2008). Microstructural abnormalities of striatum and thalamus in children with Tourette syndrome. *Movement Disorders*, *23*(16), 2349–2356. https://doi.org/10.1002/mds.22264.

Marsh, R., Alexander, G. M., Packard, M. G., Zhu, H., & Peterson, B. S. (2005). Perceptual-motor skill learning in Gilles de la Tourette syndrome. Evidence for multiple procedural learning and memory systems. *Neuropsychologia*, *43*(10), 1456–1465. https://doi.org/10.1016/j.neuropsychologia.2004.12.012.

Martino, D., Deeb, W., Jimenez-Shahed, J., Malaty, I., Pringsheim, T. M., Fasano, A., et al. (2021). The 5 pillars in Tourette syndrome deep brain stimulation patient selection: Present and future. *Neurology*, *96*(14), 664–676. https://doi.org/10.1212/WNL.0000000000011704.

Martino, D., Ganos, C., & Pringsheim, T. M. (2017). Tourette syndrome and chronic tic disorders: The clinical Spectrum beyond tics. *International Review of Neurobiology*, *134*, 1461–1490. https://doi.org/10.1016/bs.irn.2017.05.006.

Martino, D., Ganos, C., & Worbe, Y. (2018). Neuroimaging applications in Tourette's syndrome. *International Review of Neurobiology*, *143*, 65–108. https://doi.org/10.1016/bs.irn.2018.09.008.

Martino, D., Hartmann, A., Pelosin, E., Lagravinese, G., Delorme, C., Worbe, Y., et al. (2019). Motor timing in Tourette syndrome: The effect of movement lateralization and bimanual coordination. *Frontiers in Neurology*, *10*, 385. https://doi.org/10.3389/fneur.2019.00385.

McCairn, K. W., Iriki, A., & Isoda, M. (2013). Global dysrhythmia of cerebro-basal ganglia-cerebellar networks underlies motor tics following striatal disinhibition. *The Journal of Neuroscience*, *33*(2), 697–708. https://doi.org/10.1523/JNEUROSCI.4018-12.2013.

Miller, A. M., Bansal, R., Hao, X., Sanchez-Pena, J. P., Sobel, L. J., Liu, J., et al. (2010). Enlargement of thalamic nuclei in Tourette syndrome. *Archives of General Psychiatry*, *67*(9), 955–964. https://doi.org/10.1001/archgenpsychiatry.2010.102.

Mirabella, G., Upadhyay, N., Mancini, C., Giannì, C., Panunzi, S., Petsas, N., et al. (2020). Loss in grey matter in a small network of brain areas underpins poor reactive inhibition in Obsessive-Compulsive Disorder patients. *Psychiatry Research: Neuroimaging*, *297*, 111044. https://doi.org/10.1016/j.pscychresns.2020.111044.

Morand-Beaulieu, S., Grot, S., Lavoie, J., Leclerc, J. B., Luck, D., & Lavoie, M. E. (2017). The puzzling question of inhibitory control in Tourette syndrome: A meta-analysis. *Neuroscience and Biobehavioral Reviews*, *80*, 240–262. https://doi.org/10.1016/j.neubiorev.2017.05.006.

Muellner, J., Delmaire, C., Valabregue, R., Schupbach, M., Mangin, J. F., Vidailhet, M., et al. (2015). Altered structure of cortical sulci in Gilles de la Tourette syndrome: Further support for abnormal brain development. *Movement Disorders*, *30*(5), 655–661. https://doi.org/10.1002/mds.26207.

Mufford, M., Cheung, J., Jahanshad, N., van der Merwe, C., Ding, L., Groenewold, N., et al. (2019). Concordance of genetic variation that increases risk for Tourette syndrome and that influences its underlying neurocircuitry. *Translational Psychiatry*, *9*(1), 120. https://doi.org/10.1038/s41398-019-0452-3.

Nelson, A. J., & Killcross, S. (2013). Accelerated habit formation following amphetamine exposure is reversed by D1, but enhanced by D2, receptor antagonists. *Frontiers in Neuroscience*, 7, 76. https://doi.org/10.3389/fnins.2013.00076.

Neuner, I., Kupriyanova, Y., Stocker, T., Huang, R., Posnansky, O., Schneider, F., et al. (2010). White-matter abnormalities in Tourette syndrome extend beyond motor pathways. *NeuroImage*, *51*(3), 1184–1193. S1053-8119(10)00217-X [pii] https://doi.org/10.1016/j.neuroimage.2010.02.049.

Nielsen, A. N., Gratton, C., Church, J. A., Dosenbach, N. U. F., Black, K. J., Petersen, S. E., et al. (2020). Atypical functional connectivity in Tourette syndrome differs between children and adults. *Biological Psychiatry*, *87*(2), 164–173. https://doi.org/10.1016/j.biopsych.2019.06.021.

Peterson, B. S., Thomas, P., Kane, M. J., Scahill, L., Zhang, H., Bronen, R., et al. (2003). Basal ganglia volumes in patients with Gilles de la Tourette syndrome. *Archives of General Psychiatry*, *60*(4), 415–424. https://doi.org/10.1001/archpsyc.60.4.415. 60/4/415 [pii].

Plessen, K. J., Gruner, R., Lundervold, A., Hirsch, J. G., Xu, D., Bansal, R., et al. (2006). Reduced white matter connectivity in the corpus callosum of children with Tourette syndrome. *Journal of Child Psychology and Psychiatry*, *47*(10), 1013–1022. https://doi.org/10.1111/j.1469-7610.2006.01639.x.

Plessen, K. J., Royal, J. M., & Peterson, B. S. (2007). Neuroimaging of tic disorders with co-existing attention-deficit/hyperactivity disorder. *European Child & Adolescent Psychiatry*, *16*(Suppl 1), 60–70. https://doi.org/10.1007/s00787-007-1008-2.

Rae, C. L., Parkinson, J., Betka, S., Gouldvan Praag, C. D., Bouyagoub, S., Polyanska, L., et al. (2020). Amplified engagement of prefrontal cortex during control of voluntary action in Tourette syndrome. *Brain Communications*, *2*(2), fcaa 199. https://doi.org/10.1093/braincomms/fcaa199.

Rawji, V., Modi, S., Latorre, A., Rocchi, L., Hockey, L., Bhatia, K., et al. (2020). Impaired automatic but intact volitional inhibition in primary tic disorders. *Brain*, *143*(3), 906–919. https://doi.org/10.1093/brain/awaa024.

Servello, D., Galbiati, T. F., Balestrino, R., Iess, G., Zekaj, E., Michele, S., et al. (2020). Deep brain stimulation for Gilles de la Tourette syndrome: Toward limbic targets. *Brain Sciences*, *10*(5). https://doi.org/10.3390/brainsci10050301.

Shephard, E., Groom, M. J., & Jackson, G. M. (2019). Implicit sequence learning in young people with Tourette syndrome with and without co-occurring attention-deficit/hyperactivity disorder. *Journal of Neuropsychology*, *13*(3), 529–549. https://doi.org/10.1111/jnp.12167.

Sigurdsson, H. P., Jackson, S. R., Jolley, L., Mitchell, E., & Jackson, G. M. (2020). Alterations in cerebellar grey matter structure and covariance networks in young people with Tourette syndrome. *Cortex*, *126*, 1–15. https://doi.org/10.1016/j.cortex.2019.12.022.

Sigurdsson, H. P., Pepes, S. E., Jackson, G. M., Draper, A., Morgan, P. S., & Jackson, S. R. (2018). Alterations in the microstructure of white matter in children and adolescents with Tourette syndrome measured using tract-based spatial statistics and probabilistic tractography. *Cortex*, *104*, 75–89. https://doi.org/10.1016/j.cortex.2018.04.004.

Sowell, E. R., Kan, E., Yoshii, J., Thompson, P. M., Bansal, R., Xu, D., et al. (2008). Thinning of sensorimotor cortices in children with Tourette syndrome. *Nature Neuroscience*, *11*(6), 637–639. nn.2121 [pii] https://doi.org/10.1038/nn.2121.

Szejko, N., & Muller-Vahl, K. R. (2021). Challenges in the diagnosis and assessment in patients with Tourette syndrome and comorbid obsessive-compulsive disorder. *Neuropsychiatric Disease and Treatment*, *17*, 1253–1266. https://doi.org/10.2147/NDT.S251499.

Szejko, N., Worbe, Y., Hartmann, A., Visser-Vandewalle, V., Ackermans, L., Ganos, C., et al. (2021). European clinical guidelines for Tourette syndrome and other tic disorders-version 2.0. Part IV: Deep brain stimulation. *European Child & Adolescent Psychiatry*. https://doi.org/10.1007/s00787-021-01881-9.

Thomalla, G., Siebner, H. R., Jonas, M., Baumer, T., Biermann-Ruben, K., Hummel, F., et al. (2009). Structural changes in the somatosensory system correlate with tic severity in Gilles de la Tourette syndrome. *Brain*, *132*(Pt 3), 765–777. awn339 [pii] https://doi.org/10.1093/brain/awn339.

Tikoo, S., Suppa, A., Tommasin, S., Gianni, C., Conte, G., Mirabella, G., et al. (2021). The cerebellum in drug-naive children with Tourette syndrome and obsessive-compulsive disorder. *Cerebellum*. https://doi.org/10.1007/s12311-021-01327-7.

Tobe, R. H., Bansal, R., Xu, D., Hao, X., Liu, J., Sanchez, J., et al. (2010). Cerebellar morphology in Tourette syndrome and obsessive-compulsive disorder. *Annals of Neurology*, *67*(4), 479–487. https://doi.org/10.1002/ana.21918.

Wan, X., Zhang, S., Wang, W., Su, X., Li, J., Yang, X., et al. (2021). Gray matter abnormalities in Tourette syndrome: A meta-analysis of voxel-based morphometry studies. *Translational Psychiatry*, *11*(1), 287. https://doi.org/10.1038/s41398-021-01394-8.

Wen, F., Yan, J., Yu, L., Wang, F., Liu, J., Li, Y., & Cu, Y. (2021). Grey matter abnormalities in Tourette syndrome: An activation likelihood estimation meta-analysis. *MBC Psychiatry*, *21*(1), 184. https://doi.org/10.1186/s12888-021-03187-1.

Wolff, N., Luehr, I., Sender, J., Ehrlich, S., Schmidt-Samoa, C., Dechent, P., et al. (2016). A DTI study on the corpus callosum of treatment-naive boys with 'pure' Tourette syndrome. *Psychiatry Research: Neuroimaging*, *247*, 1–8. https://doi.org/10.1016/j.pscychresns.2015.12.003.

Wong, J. K., Middlebrooks, E. H., Grewal, S. S., Almeida, L., Hess, C. W., & Okun, M. S. (2020). A comprehensive review of brain Connectomics and imaging to improve deep brain stimulation outcomes. *Movement Disorders*, *35*(5), 741–751. https://doi.org/10.1002/mds.28045.

Worbe, Y., Gerardin, E., Hartmann, A., Valabregue, R., Chupin, M., Tremblay, L., et al. (2010). Distinct structural changes underpin clinical phenotypes in patients with Gilles de la Tourette syndrome. *Brain*, *133*(Pt 12), 3649–3660. https://doi.org/10.1093/brain/awq293.

Worbe, Y., Malherbe, C., Hartmann, A., Pelegrini-Issac, M., Messe, A., Vidailhet, M., et al. (2012). Functional immaturity of cortico-basal ganglia networks in Gilles de la Tourette syndrome. *Brain*, *135*(Pt. 6), 1937–1946. https://doi.org/10.1093/brain/aws056.

Worbe, Y., Marrakchi-Kacem, L., Lecomte, S., Valabregue, R., Poupon, F., Guevara, P., et al. (2015). Altered structural connectivity of cortico-striato-pallido-thalamic networks in Gilles de la Tourette syndrome. *Brain*, *138*(Pt. 2), 472–482. https://doi.org/10.1093/brain/awu311.

Xia, X., Lin, Y., Lang, B., Yuan, J., Sheng, L., Yang, D., et al. (2020). Characteristics of diffusion tensor imaging of central nervous system in children with tourette's disease. *Medicine (Baltimore)*, *99*(22), e20492.

Yang, Z., Wu, H., Lee, P. H., Tsetsos, F., Davis, L. K., Yu, D., et al. (2021). Investigating shared genetic basis across Tourette syndrome and comorbid neurodevelopmental disorders along the impulsivity-compulsivity Spectrum. *Biological Psychiatry*, *90*(5), 317–327. https://doi.org/10.1016/j.biopsych.2020.12.028.

CHAPTER NINE

The neural correlates of tic inhibition

Elia Abi-Jaoude[a,b,*] and Paul Sandor[b,c]

[a]Department of Psychiatry, The Hospital for Sick Children, University of Toronto, Toronto, ON, Canada
[b]Department of Psychiatry, University Health Network, University of Toronto, Toronto, ON, Canada
[c]Youthdale Treatment Centers, Toronto, ON, Canada
*Corresponding author: e-mail address: elia.abi.jaoude@utoronto.ca

Contents

Abbreviations

ACC	anterior cingulate cortex
ADHD	attention-deficit/hyperactivity disorder
ALE	activation likelihood estimation
CBIT	Comprehensive Behavioral Intervention for Tics
CMPf	centromedian and parafascicularis complex
CSTC	cortico-striato-thalamo-cortical
dACC	dorsal ACC
DLPFC	dorsolateral prefrontal cortex
FDG-PET	fluoro-deoxyglucose positron emission tomography

International Review of Movement Disorders, Volume 3
ISSN 2666-7878
https://doi.org/10.1016/bs.irmvd.2021.12.004

fMRI	functional MRI
GPi	globus pallidus internus
IFG	inferior frontal gyrus
NIRS	near-infrared spectroscopy
OCD	obsessive-compulsive disorder
OFC	orbitofrontal cortex
PCC	posterior cingulate cortex
PFC	prefrontal cortex
PPI	prepulse inhibition
rCBF	regional cerebral blood flow
SMA	supplementary motor area
TMS	transcranial magnetic stimulation
TPJ	temporoparietal junction
TS	Tourette syndrome
VBM	voxel-based morphometry
vmPFC	ventromedial PFC

1. Introduction

Tourette syndrome (TS) has been conceptualized as a syndrome of inhibitory deficits involving motor, cognitive, emotional and behavioral domains (Mink, 2001; Stern, Blair, & Peterson, 2008). Phenomenologically, this would be consistent with the manifestations of tics, as well as the high comorbidity with obsessive-compulsive disorder (OCD), attention-deficit/hyperactivity disorder (ADHD), and emotional dysregulation. Tics are considered "semi-compulsory," in that they are temporarily suppressible, until the premonitory urge is such that they can no longer be contained. This is similar to the case of compulsions in OCD. The high comorbidity with ADHD is particularly indicative of problems with inhibitory control (Martino & Mink, 2013; Murphy & Muter, 2012). As well, TS is associated with emotional dysregulation in the form of rage outbursts (Budman, Bruun, Park, Lesser, & Olson, 2000; Budman, Rockmore, Stokes, & Sossin, 2003; Chen et al., 2013; Conte, Valente, Fioriello, & Cardona, 2020; Wright, Rickards, & Cavanna, 2012). Thus, the clinical phenomenology of tics and the common comorbidities of TS suggest inhibitory deficits as being a core feature in TS (Kurvits, Martino, & Ganos, 2020).

On executive functioning tests, patients with TS typically show deficits that are specific to inhibitory function, such as with the Hayling test (Channon, Crawford, Vakili, & Robertson, 2003; Channon, Pratt, & Robertson, 2003; Channon, Sinclair, Waller, Healey, & Robertson, 2004;

Eddy, Rickards, & Cavanna, 2012; Yaniv et al., 2017). A recent study found that children with TS had decreased control of the impulse to mirror environmental stimuli relative to controls (Briguglio et al., 2021). Another interesting study found that while adults with TS performed similarly to controls on the stop signal task, a measure of volitional motor inhibition, this was not the case on the masked prime task, suggesting impaired automatic inhibition in action selection (Rawji et al., 2020). A meta-analysis that included 61 studies of neuropsychological tasks found that overall TS patients exhibited inhibitory deficits compared to control participants (Morand-Beaulieu et al., 2017). While some have argued that most cognitive deficits are attributable to comorbid OCD or ADHD (Greimel et al., 2011; Jung, Jackson, Parkinson, & Jackson, 2013; Lin, Lai, & Gau, 2012; Openneer et al., 2020; Sukhodolsky, Landeros-Weisenberger, Scahill, Leckman, & Schultz, 2010), others have found that executive deficits in TS are independent of comorbidity (Channon et al., 2004; Eddy & Cavanna, 2017; Eddy et al., 2012; Hovik et al., 2017; Jeter et al., 2015; Morand-Beaulieu et al., 2017; Wylie, Claassen, Kanoff, van Wouwe, & van den Wildenberg, 2016; Yaniv et al., 2017). Regardless, given that comorbidity is present in the vast majority of patients with TS, it is clear that the condition is associated with deficits in inhibitory control.

Further evidence for reduced inhibitory control comes from transcranial magnetic stimulation (TMS) studies, which have consistently found TS subjects to have reduced measures of cortical inhibition compared to healthy controls (reviewed by Ganos and colleagues (Ganos, Roessner, & Münchau, 2013)). In a more recent TMS study of 14 adults with TS, voluntary tic suppression was associated with reduced corticospinal excitability, though there were no changes in other cortical excitability measures (Ganos et al., 2018).

Neuroimaging studies have yielded some promising results, suggesting that reduced caudate volume and thinning of motor and sensorimotor cortices reflect the neurological predisposition to tics (Ganos et al., 2013; Jackson, Draper, Dyke, Pépés, & Jackson, 2015; Polyanska, Critchley, & Rae, 2017; Stern et al., 2008; Worbe, Lehericy, & Hartmann, 2015). In addition, hypertrophy of cortical (dorsolateral prefrontal, orbitofrontal, and parietal) and limbic (amygdala and hippocampus) regions and a smaller corpus callosum are thought to reflect compensatory neuroplastic changes. Almost all these studies have been cross-sectional, however, making it difficult to tease out underlying cause from compensatory response. Conflicting findings have also been reported, likely as a result of differences in sample size, demographic characteristics, tic severity, comorbid conditions, medication use, techniques for the acquisition and processing of images, and methods for data analysis. As well, while there has been much interest in the role of striatal dopamine (Buse, Schoenefeld, Münchau, & Roessner, 2013), findings have been inconsistent, and the literature has been characterized by several limitations, including small sample sizes, confounders such as age differences between comparison groups, and medication effects (Abi-Jaoude et al., 2015). Nevertheless, the neuroimaging literature does suggest that tic formation results from a disordered striatum and related dysfunction in cortico-striato-thalamo-cortical (CSTC) circuits, in particular the striatal dopamine system, and cortical control areas (Ganos et al., 2013; Jackson et al., 2015; Polyanska et al., 2017; Stern et al., 2008; Worbe, Lehericy, et al., 2015). This is further supported by neuropathologic studies that show reduced number and altered distribution of inhibitory GABAergic neurons in the sensorimotor areas of the striatum (Kalanithi et al., 2005; Kataoka et al., 2010).

The neuropathologic studies (Kalanithi et al., 2005; Kataoka et al., 2010) are particularly interesting in terms of pointing to a potential account as to the origin of tics (Rae, Critchley, & Seth, 2019). Specifically, these postmortem studies found a reduced number and altered distribution of inhibitory GABAergic parvalbumin-positive and tonically active cholinergic interneurons in the sensorimotor striatum, with an increased number of inhibitory GABAergic parvalbumin-positive interneurons in globus pallidus internus (GPi). These alterations, which are suggestive of an issue with neuronal migration of interneurons during development, may impact the balance between inhibitory, indirect, pathways, and excitatory, direct, pathways within the CSTC loops. GABAergic parvalbumin-positive interneurons are fast-spiking interneurons that regulate medium spiny striatal output neuron by filtering cortical inputs, thereby controlling striatal outflow. Since these GABAergic interneurons preferentially target medium spiny neurons in the direct pathway, the reduction in their number would result in a relative increase in direct pathway over indirect pathway transmission in the CSTC circuits (Rae et al., 2019). The impact on the motor pathway involving the supplementary motor area (SMA) and sensorimotor cortical areas may lead to tics, whereas the impact on the associative/cognitive and emotional/limbic pathways may manifest in the symptoms that are commonly comorbid with TS.

Neuroimaging studies involving a wide variety of approaches frequently implicate prefrontal cortical control areas in TS, though it remains unclear to what extent these reflect mechanisms intrinsic to the pathophysiology of

TS versus compensatory changes. Here we review and summarize this literature, and conclude that these prefrontal areas are likely involved in inhibition of tics and related symptoms.

2. Structural magnetic resonance imaging

The brain areas that have been found to be affected in TS consistently involve regions known to have a role in self-control (Ganos et al., 2013; Jackson et al., 2015; Stern et al., 2008; Worbe, Lehericy, et al., 2015). In a structural magnetic resonance imaging (MRI) study of 155 subjects with TS and 131 healthy controls, Peterson and colleagues found that the dorsolateral prefrontal cortex (DLPFC) region was increased in volume in children but not adults with TS (Peterson et al., 2001). Further, there was an inverse correlation between this increase in DLPFC volumes and tic symptom severity (Peterson et al., 2001). The speculation is that these findings represent compensatory changes associated with tic control in children; in contrast, adults had a relative decrease in DLPFC volumes, which may be related to persistence of tics in adulthood. In the same study, orbitofrontal cortex (OFC) volumes were found to have an inverse correlation with tic symptom severity, particularly in adults (Peterson et al., 2001). An MRI study measuring cortical thickness found sensorimotor cortex and DLPFC thinning in 25 children with TS relative to 35 age- and sex-matched controls, and this correlated to tic symptom severity (Sowell et al., 2008). In another structural MRI study, voxel-based morphometry (VBM) showed decreases in gray matter volumes in various prefrontal areas including the anterior cingulate cortex (ACC), as well as decreases in white matter volumes in the right inferior frontal gyrus (IFG) and left superior frontal gyrus in 19 adults with TS compared to 20 age- and sex-matched controls (Müller-Vahl et al., 2009). In an MRI study of cortical thickness, children and youth with TS showed sensorimotor and right OFC thinning relative to controls (Fahim et al., 2010). In a similar study this time in adults, 60 TS participants had reduced thickness relative to 30 controls in the left motor cortex, left postero-lateral superior frontal and posterior part of the middle frontal gyrus (these latter two overlapping with DLPFC), right IFG and lateral OFC (Worbe et al., 2010). Tic severity correlated negatively with cortical thickness in several regions including OFC and IFG. In yet another cortical thickness MRI study, 40 adults with TS had reductions in thickness relative to 40 age- and gender-matched controls in OFC, ACC, and ventrolateral prefrontal cortex (PFC), and the reductions correlated negatively with tic severity (Draganski et al., 2010). In a subsequent VBM MRI study, 29 adults with TS were found to have decreased left IFG volume relative to 24 controls (Wittfoth et al., 2012). In an MRI probabilistic tractography study, relative to 28 age- and gender-matched controls, 49 adult patients with TS had increased thalamus and striatum structural connectivity to multiple cortical areas including sensorimotor, SMA, OFC and IFG, and decreased connectivity to ACC (Worbe, Marrakchi-Kacem, et al., 2015). A subsequent multicenter MRI VBM study found that 103 children with TS had lower white matter volume in the OFC and medial PFC relative to 103 controls participants matched for age, sex and handedness (Greene, Williams, Koller, Schlaggar, & Black, 2016). Another MRI study found structural differences between 60 children with TS and 52 age- and gender-matched controls in various areas including ACC, medial OFC, and IFG (Kong et al., 2020). A recent whole brain VBM study of 28 youth with TS and 36 controls found decreased gray matter volumes bilaterally in the OFC, ACC, and midcingulate cortex (Jackson, Sigurdsson, Dyke, Condon, & Jackson, 2021). Further, seed-to-voxel structural covariance network mapping found ACC and midcingulate cortex seeds in TS participants to have increased covariance with inferior frontal cortex and posterior cingulate (Jackson et al., 2021).

An interesting multimodal imaging study using the emotional stop signal task found increased reactive motor impulsivity in 19 adults with TS who were on antipsychotic medication compared to 31 unmedicated TS adults, and 26 controls (Atkinson-Clement et al., 2020). Of note, however, 15 of the 19 medicated TS participants were taking the D2 receptor partial agonist aripiprazole, an antipsychotic which has been associated with an increased risk of developing impulse control disorders. This increased reactive motor impulsivity was associated with increased mean gray matter signal from the right SMA and IFG (Atkinson-Clement et al., 2020).

A recent activation likelihood estimation (ALE) meta-analysis of six MRI studies investigating gray matter changes found that in comparison to controls, patients with TS had increased gray matter volumes in multiple cortical and subcortical areas including the left inferior parietal lobule and insula, thalamus bilaterally, and right lentiform nucleus, and decreased gray matter volumes in several overlapping but also different cortical areas, including the insula and ACC bilaterally (Wen et al., 2021). Another, more recent meta-analysis of 10 VBM studies using seed-based mapping found that in comparison to controls, patients with TS patients had smaller gray matter volume in the IFG bilaterally, and greater volume in the right putamen, and pulvinar nucleus of the thalamus bilaterally (Wan et al., 2021).

3. Functional imaging

3.1 Tic suppression and tic generation

Similar brain regions have been implicated in functional MRI (fMRI) studies. In an early study in 22 adults with TS, Peterson and colleagues found that, in comparison to a resting state, tic suppression involved increased activity in right midfrontal cortex and right ventral cortex, both of which correlated positively to each other, and inversely with activity in the putamen, globus pallidus and thalamus (Peterson et al., 1998). This is consistent with the DLPFC being involved in suppressing tics, via excitatory projections to the caudate, which in turn sends inhibitory projections to the putamen (Stern et al., 2008). In an event-related fMRI study, ACC, SMA and premotor areas were found to be activated 2 s prior to and at tic onset in 10 adult TS participants (Bohlhalter et al., 2006). A subsequent study in 16 adults with TS investigated correlations of brain areas to tic-related primary motor areas and found that SMA had broader cross-correlation during tics in comparison to intentional imitation of tics by 16 matched controls (Hampson, Tokoglu, King, Constable, & Leckman, 2009). In another study comparing spontaneous tics by 13 adult participants with TS to voluntary imitation of tics by 21 controls, activity during the former was relatively increased in somatosensory and posterior parietal cortices and putamen (Wang et al., 2011). The TS group also had increased activity in the sensorimotor cortex, putamen, pallidum and substantia nigra relative to controls; on the other hand, activity in the anterior cingulate and caudate was weaker in the TS group. Both the increase and decrease in activity in comparison the controls correlated with tic severity. The authors conclude that tics arise from a combination of "excessive activity in motor pathways and reduced activation in control portions of cortico-striato-thalamo-cortical circuits" (Wang et al., 2011). Researchers from the same group carried out an investigation using an eye blink inhibition task and found that the 51 TS participants showed more activations in the middle frontal gyrus and dorsal ACC (dACC), and deactivations in the superior frontal gyrus relative to 69 controls (Mazzone et al., 2010). Activations in the left middle frontal gyrus and left caudate correlated positively with tic severity. On the other hand, activation in the right inferolateral PFC, left IFG and right putamen correlated negatively with tic severity (Mazzone et al., 2010).

The left IFG was also implicated in a resting state fMRI study in which it had increased regional homogeneity during tic suppression in comparison to the free ticking state in 14 adults with TS (Ganos, Kahl, et al., 2014). In another resting state fMRI study, in which tic onset times were used as regressors to investigate brain activations prior to tic onset in 10 adults with TS, ACC and SMA were activated at 2 and 1 s prior to tic onset (Neuner et al., 2014). In an interesting fMRI study design, brain activations in 11 adult participants with TS who were asked to suppress ocular tics but not eye blinks, were compared to brain activations in 19 control participants who were asked to suppress eye blinks in a block design (van der Salm et al., 2018). Suppression of tics in the participants with TS was associated with increased activity in right anterior and dorsal anterior PFC, bilateral frontal eye fields, and right DLPFC. Suppression of eye blinks in the control participants was associated with activations in the bilateral premotor cortex, bilateral SMA, right IFG, right anterior PFC, insula bilaterally, left caudate and putamen. Compared to controls, participants with TS had higher activations in the right anterior PFC, right ACC, right DLPFC, left premotor cortex and frontal eye fields; in contrast, controls had higher activations in bilateral SMA, bilateral IFG and insula, and right putamen (van der Salm et al., 2018). These contrast analyses need to be interpreted bearing in mind that the two groups of participants performed different tasks.

In an interesting recent electroencephalography (EEG) study of 15 participants with TS (both children and adults) found an absence of event-related desynchronization prior to onset of tics (Morera Maiquez, Jackson, & Jackson, 2021). Prior to the onset of voluntary movement, the TS participants had event-related desynchronization in beta-band but not mu-band oscillation, in contrast to the 15 controls, who had event-related desynchronization in both beta-band and mu-band oscillations. The authors explain that beta-band desynchronization is involved in activation of movement-related neuronal populations, whereas mu-band oscillation are involved in inhibiting neuronal populations as part of movement selection. The implication is that tics originate from basal ganglia and paralimbic cortical areas, and their release is a result of a deficit of GABAergic cortical inhibition (Morera Maiquez et al., 2021). In another recent study using high-density EEG in 72 children with TS, tic suppression relative to resting state involved increased connectivity across widespread cortical areas, including multiple prefrontal and parietal areas (Morand-Beaulieu et al., 2021).

In a double-blind, randomized, sham-controlled, crossover real-time fMRI study, neurofeedback to control SMA blood-oxygen-level-dependent (BOLD) signal in 21 adolescents with TS resulted in significant tic reduction, although there was no effect on the ability to control the SMA (Sukhodolsky et al., 2020).

3.2 Cognitive control

In a task-based fMRI study, investigators evaluated the effects of age on brain activations during performance of the Stroop task in 66 participants with Tourette syndrome (children and adults) and 70 controls (Marsh, Zhu, Wang, Skudlarski, & Peterson, 2007). In contrast to healthy controls, TS participants did not have greater deactivations with age in the mesial PFC and ventral ACC, nor increased activations with age in the right inferolateral PFC. In addition, activations in the inferolateral PFC, mesial frontal gyrus, and DLPFC correlated inversely with task performance in the TS group, while activations in the right inferolateral PFC correlated positively with performance in the healthy controls; as well, activation in the ventromedial PFC (vmPFC) correlated positively with task performance in the TS group, and negatively in the control group. Finally, the magnitude of DLPFC activation correlated positively with tic symptom severity (Marsh et al., 2007). In an event-related fMRI study using a cognitive control task, 18 children with TS exhibited greater left DLPFC activations in comparison to age-matched 19 healthy controls (Baym, Corbett, Wright, & Bunge, 2008). Moreover, activations in premotor cortex, superior frontal gyrus and right IFG correlated with tic severity (Baym et al., 2008). In another cognitive control study, investigators used the Simon task to compare activations in 42 participants with TS and 37 controls in both children and adults (Raz et al., 2009). There was greater PFC activation in TS compared to healthy controls, and this was even more so in adults. In the TS group, activations in SMA and middle frontal gyrus, among others, correlated inversely with task performance. Activations in the left inferior prefrontal regions correlated with tic severity. Of interest, in the group of adults with TS, activations in the midfrontal gyrus following an incongruent stimulus were delayed relative to controls (Raz et al., 2009).

A study of cognitive control based on "task-maintenance" and "adaptive control" using a semantic task showed activity differences across multiple brain regions in 27 adolescents with TS and 27 controls (Church et al., 2009). In particular, activations associated with "adaptive control" were lower in a number of brain regions in the TS group, including in the middle frontal gyrus. Activations in several regions in the TS group, including a dorsal medial frontal region, appeared similar to adolescent participants in the control group. Activations associated with "task-maintenance" were higher in the TS compared to the control group in multiple regions, including the middle frontal gyrus, which had similar activation to child participants from a "developmental" dataset of typically-developing children and adults. According to the authors, these findings suggest "functional immaturity" of the control networks in TS (Church et al., 2009). In an investigation of neural responses using a reinforcement learning task, there was no difference in SMA activation between the 60 adults with TS and 50 healthy controls (Worbe et al., 2011). Of note, the TS participants with comorbid OCD and those that were receiving a dopamine antagonist medication had decreased activations in the vmPFC as well as poorer performance on the learning task (Worbe et al., 2011). Using near-infrared spectroscopy (NIRS), a study investigated prefrontal hemodynamic response during the Stroop color-word task and found that 10 children with TS had lower PFC activation in the DLPFC area compared to 10 age- and sex-matched control participants (Yamamuro et al., 2015).

A recent EEG study compared 60 children with a tic disorder to 35 controls during performance of a flanker task (Jurgiel et al., 2021). While performance was similar in both groups, there were neural differences during incongruent flanker trials. Specifically, there was attenuated activation of the ACC, and greater connectivity between the ACC and other front-parietal network hubs, namely midcingulate cortex and precuneus, in TS relative to control participants. These changes were associated with tic impairment, faster reaction time, and worse inhibitory performance, suggestive of adaptive changes resulting from chronic tic inhibition (Jurgiel et al., 2021).

3.3 Motor task switching

In an interesting multimodal imaging study, 13 child participants with TS had faster response times compared to 13 age- and gender-matched controls in a manual task-switching paradigm (Jackson, Parkinson, Jung, et al., 2011). This enhanced control over motor output in the participants with TS correlated inversely with tic severity. On diffusion-weighted imaging, there were differences between the TS ($n = 14$) and control ($n = 14$) group in the corpus callosum and forceps minor. White matter fractional anisotropy and mean diffusivity measures in these areas correlated with tic severity. In the control but not the TS group, corpus callosum fractional anisotropy was negatively correlated with task response times. In the TS but not the control group, forceps minor fractional anisotropy was negatively correlated with task response times. Finally, task-switching associated fMRI BOLD response in a PFC region of interest (ROI) immediately adjacent to the vicinity of the right forceps minor was increased in 10 TS participants relative to 15 controls; in the TS but not the control group, activation in this area was negatively correlated with task performance (Jackson, Parkinson, Jung, et al., 2011). In another fMRI study by the same group using the same task, 10 children with TS had less ROI activations compared to 10 controls in the SMA, ACC and inferolateral cortex (Jung et al., 2013). In this study, there was no significant difference in task performance between the TS and control groups. Of interest, activations in the SMA, ACC and inferior frontal cortex ROIs was positively correlated with performance in terms of response time (though this was not statistically significant); in the TS group only, activations in these ROIs were strongly correlated with the task-switching effect ("switch costs"), and these activations also correlated with tic severity (though this was not statistically significant) (Jung et al., 2013).

3.4 Inhibitory control

In a study using the stop-signal task, the group of 14 adults with TS showed greater activation compared to 15 controls in the dorsal premotor cortex during the "go" versus the "successful-stop" condition, a pattern that was opposite to that seen in the healthy controls (Ganos, Kühn, et al., 2014). On exploratory correlation analyses, activations in the SMA during the "successful-stop" versus "go" conditions correlated positively with motor tic frequency (Ganos, Kühn, et al., 2014). In another fMRI study using the stop-signal task, in adults with TS ($n = 18$), OCD ($n = 18$) and healthy controls ($n = 20$), there were no differences identified in PFC with three-group whole brain analysis (Fan et al., 2017). ROI analysis of activity during error processing showed higher activation in the TS relative to the other groups in the SMA (Fan et al., 2017). In a go/no-go fMRI study, 15 adults with TS had lower activations compared to 15 control participants during the "go" condition in the left dorsal premotor cortex and SMA

(Thomalla et al., 2014). In another, more recent go/no-go fMRI study, 23 adults with TS had increased activity relative to 21 control participants in the IFG, insula, and caudate (Rae et al., 2020). As well, there was a positive correlation between premonitory sensations and functional connectivity of the pre-SMA with the caudate, globus pallidus, and thalamus (Rae et al., 2020).

In an earlier fMRI go/no-go study, investigators used the task with and without an infusion of levodopa to study the role of dopamine in modulating task performance-related brain activity (Hershey et al., 2004); they found no differences between 8 adults with TS and 10 matched healthy control participants (Hershey et al., 2004). In a task-based fMRI study in 39 children with TS and 39 controls, Debes and colleagues found no significant differences in activation between the TS and control participant groups during performance of the Stroop, go/no-go and finger tapping tasks (Debes, Hansen, Skov, & Larsson, 2011).

In a recent fMRI study using the four-choice serial reaction time tasks, waiting motor impulsivity was increased in 36 unmedicated but not the 19 medicated adults with TS relative to 31 controls; this was correlated to tic severity, and also with higher connectivity between right OFC and caudate bilaterally, and right medial frontal gyrus and right IFG (Atkinson-Clement et al., 2021). In another recent fMRI study, this time using the stop signal task in children with TS, the 23 participants with comorbid ADHD, but not 28 without comorbid ADHD, had impaired response inhibition relative to the 52 controls (Openneer et al., 2021). Failed inhibition was associated with increased activity in IFG in participants with TS and comorbid ADHD relative to controls (Openneer et al., 2021).

3.5 Finger tapping

In finger-tapping study, 19 children with TS showed higher activations relative to 16 controls in the right medial frontal gyrus and left caudate but relatively lower activations in the right middle frontal gyrus and left precentral gyrus (SMA area) (Roessner et al., 2012). In a subsequent publication by the same group, 22 children with TS had less activations in left middle frontal gyrus and caudate relative to 22 controls during a finger tapping task with the dominant (right) hand, but greater activations in medial, superior frontal and precentral gyri, pre-SMA, DLPFC and ACC during finger tapping with the nondominant (left) hand (Roessner et al., 2013). In yet another finger-tapping investigation, 19 adult participants with TS showed increased left PFC activations relative to 18 controls during finger tapping with the nondominant (left) hand and during bimanual tapping (Werner et al., 2011). In addition, whereas control participants activated the ACC only during the bimanual tapping, the TS group recruited the area similarly in the dominant, nondominant, and bimanual finger tapping conditions (Werner et al., 2011). An investigation during a finger opposition motor task found the group of 24 adults with TS to have greater activations in superior and middle frontal gyri, and IFG relative to 24 controls during task execution, and in the superior and middle frontal gyri during imagination of the task (Zapparoli et al., 2016).

3.6 Prepulse inhibition

In an fMRI study of prepulse inhibition (PPI), 17 adult participants with TS had a reduced startle response relative to 16 healthy controls, and lower activations in multiple regions including the OFC, ventral lateral prefrontal cortex/anterior insula, and caudate (Zebardast et al., 2013). In the TS group, task-related activity in the caudate and in the OFC correlated positively with tic and obsessive-compulsive symptom severity scores, respectively (Zebardast et al., 2013). A combined electromyography/fMRI study also using PPI by Buse and colleagues found that the startle response was reduced in 22 adolescent participants with TS relative to 22 controls, as was activity in the middle frontal gyrus and caudate (Buse, Beste, Herrmann, & Roessner, 2016). In another paper by the same group, 17 adolescents with TS had more activation relative to 23 controls in the ACC in response to harmonic versus disharmonic musical chords, in contrast to healthy controls, who showed the opposite pattern (Buse, Beste, & Roessner, 2017). Further, activation during the harmonic chord sequence in the TS participants correlated with tic severity. In addition, the control participants had greater premotor cortex activation in the disharmonic versus the harmonic chords, whereas there was no such difference in the TS group (Buse et al., 2017).

3.7 Emotion and theory of mind

In an event-related fMRI study of neural responses to emotional faces, 19 adults with Tourette syndrome exhibited increased activation relative to 19 age- and sex-matched controls in the medial and dorsolateral aspects of the superior frontal gyrus, the IFG, and the anterior, middle and posterior cingulum, as well as the amygdala in response to any emotional facial expression (Neuner et al., 2010). An fMRI investigation compared 24 participants with TS and 24 age- and gender-matched healthy controls during performance of a theory of mind task (Eddy, Cavanna, Rickards, & Hansen, 2016). TS participants had lower activations compared to controls in the temporoparietal junction (TPJ), posterior cingulate cortex (PCC), and superior frontal gyrus (Eddy et al., 2016). In another investigation from the same study, this time using an emotion recognition task, the authors found that relative to the controls, the TS participants had increased activation in the TPJ, PCC and OFC (Eddy, Cavanna, & Hansen, 2017). In an interesting fMRI study, 21 adults with TS and 21 control participants were exposed to images of neutral and angry faces (Rae et al., 2018). TS participants had increased activation in IFG, and in the anterior insula, which showed task-related functional connectivity with middle frontal gyrus, ACC, pre-SMA, premotor cortex, primary motor cortex and putamen. Furthermore, insula connectivity with globus pallidus and thalamus correlated with tic severity, and connectivity with SMA correlated with premonitory sensations (Rae et al., 2018). In an fMRI study of eyeblink inhibition while viewing facial expressions in 40 adults with TS, tic severity was positively associated with activity in the superior frontal gyrus during blink inhibition while viewing angry facial expressions (Bhikram, Crawley, Arnold, Abi-Jaoude, & Sandor, 2021).

3.8 Behavioral therapy

An fMRI study compared eight adult patients with TS before and after Comprehensive Behavioral Intervention for Tics (CBIT), to eight control participants matched for age, gender, education and handedness before and after a 10 week waiting period, using the visuospatial priming task (Deckersbach et al., 2014). The authors used selective ROIs, though the rationale for their choices was not clear. There were no prime-specific condition differences identified. The investigators resorted to comparing a concatenation of all visuospatial priming task conditions (negative, positive, and neutral) contrasted to a fixation cross condition. However, given the disparate conditions of the task, it is not clear what the meaning is of such an analysis and related findings. Furthermore, such a contrast does not allow for the subtraction of nonspecific motor response effects. Bearing that in mind, the authors found an interaction between group and time, whereby there was a decrease in activation in the putamen from pre- to post-CBIT.

The investigators carried out multiple correlations—changes in and pre-treatment scores for several symptom rating scales with task-related activation changes from pre- to post-CBIT. They found a significant negative correlation between changes in tic severity score and changes in activation in the IFG (Deckersbach et al., 2014).

3.9 Deep brain stimulation

In a small study, five adult patients with severe medically refractory TS undergoing centromedian and parafascicularis complex (CMPf) thalamic deep brain stimulation (DBS) had intraoperative fMRI scans done (while under general anesthesia) with and without stimulation in a block paradigm (Jo et al., 2018). Simulation was associated with widespread BOLD signal changes, including decreased activity in the ACC, OFC, medial PFC, caudate, nucleus accumbens and amygdala, increased activity in primary motor and parietal regions, and both increased and decreased activity in the insula. BOLD signal changes in multiple areas correlated with reduction in tics, including ACC, OFC, insula, thalamus, amygdala and accumbens. The investigators also looked at BOLD signal changes and correlations based on tic type, i.e., vocal or motor. There were multiple such correlations across a wide range of brain regions. The authors argued that there were distinct BOLD signal changes and correlations for motor and vocal tics (Jo et al., 2018). Different results were obtained from a prior DBS study in which five adult patients with severe medically refractory TS underwent SPECT scanning at baseline and postoperatively (all under general anesthesia), as did six control participants (while awake) (Haense et al., 2016). The postoperative scans were carried out 3 months after either CMPf thalamus, GPi, or sham stimulation. At baseline, regional cerebral blood flow (rCBF) was decreased in the SMA and middle frontal cortex in the TS patients in comparison to controls. Relative to baseline, sham stimulation was followed by increase in SMA rCBF, and decrease in ACC, inferior, middle and superior frontal cortex, and striatum. GPi stimulation was associated with decrease in rCBF in inferior frontal cortex, striatum, and insula relative to baseline, and decrease in OFC, insula and striatum, and increase in superior and middle frontal cortex relative to sham stimulation. Thalamus CMPf stimulation was associated with rCBF decrease in inferior frontal cortex and increase in middle frontal cortex relative to baseline, and increase in SMA, middle and superior frontal cortex relative to sham stimulation. The authors suggest that the decreased rCBF in frontal regions in TS relative to controls could be reversed with DBS. It should be noted, however, that DBS in this study did not result in statistically significant improvement in tic symptoms relative to sham stimulation (Haense et al., 2016).

3.10 Cerebral blood flow and metabolism

Besides MRI, functional neuroimaging studies using other modalities also point to the PFC as being involved in TS. In an early F-18 fluorodeoxyglucose positron emission tomography (FDG-PET) study, 16 adult patients with TS had decreased metabolic activity relative to 16 age- and sex-matched controls in ventral prefrontal cortex, in particular OFC, and ventral striatum, and increased metabolism in the SMA and lateral premotor cortices (Braun et al., 1993); in addition, cerebral metabolic rate in the OFC was positively associated compulsive and impulsive behavioral symptoms in the TS patients (Braun et al., 1995). In a SPECT study, 38 children with TS had lower regional cerebral perfusion than 18 control participants in the DLPFC, OFC, cingulum and caudate (Diler, Reyhanli, Toros, Kibar, & Avci, 2002). In an FDG-PET study of glucose metabolism in 12 adults with TS and 12 age- (but not sex-) matched controls, principal components analyses across participant groups showed a pattern of resting metabolic activity that in the TS group was decreased in the OFC and striatum and increased in the premotor cortex relative to the controls (Pourfar et al., 2011). Additional analysis within the TS group showed that the six participants considered to have clinically significant OCD symptoms had a metabolic activity pattern that was relatively decreased in the ACC and DLPFC and increased in the primary motor cortex (Pourfar et al., 2011). In a $H_2{}^{15}O$-PET study in which participants were shown pictures with asymmetrically ordered objects, 14 patients with TS and symmetry behavior had larger increase in rCBF in the ACC relative to 10 controls, who had larger increase in dorsomedial PFC (de Vries et al., 2013). The participants with TS reported higher urge to correct the asymmetry relative to controls; the ratings correlated with OFC activity in the TS group, whereas they correlated negatively with DLPFC activation in the control group (de Vries et al., 2013).

3.11 Activation likelihood estimation

An ALE meta-analysis included 14 task-based neuroimaging studies that compared participants with TS to controls, entailing 25 experiments and 651 child and adult participants (Polyanska et al., 2017). Differences between TS and control participants were identified in lateral PFC, including IFG, middle, and superior frontal gyri, ACC, lateral premotor cortex and SMA. In addition, correlation with tic severity were analyzed across 7 studies comprising 378 participants and 8 experiments. Tic severity in participants with TS was found to correlate with activations in the SMA, lateral premotor cortex, and lateral PFC (middle frontal gyrus) (Polyanska et al., 2017).

3.12 Summary of imaging studies

Thus, cortical control regions have been frequently found to be implicated in TS (Ganos et al., 2013; Jackson et al., 2015; Stern et al., 2008; Worbe, Lehericy, et al., 2015). In particular, PFC regions are consistently involved in functional neuroimaging studies of tic suppression (Ganos, Kahl, et al., 2014; Peterson et al., 1998; van der Salm et al., 2018), tic generation (Bohlhalter et al., 2006; Hampson et al., 2009; Neuner et al., 2014; Wang et al., 2011), cognitive control (Baym et al., 2008; Church et al., 2009; Marsh et al., 2007; Raz et al., 2009; Yamamuro et al., 2015), and inhibitory control (Debes et al., 2011; Fan et al., 2017; Ganos, Kühn, et al., 2014; Hershey et al., 2004; Thomalla et al., 2014). As well, PFC regions have been consistently implicated in cerebral blood flow and metabolism studies in TS (Braun et al., 1995, 1993; de Vries et al., 2013; Diler et al., 2002; Pourfar et al., 2011). However, there are several inconsistent and even contradictory findings. In addition, multiple studies have not identified any differences between TS and control participants. A structural MRI VBM study did not find any differences between 38 boys with TS and 38 healthy boys matched for age and IQ (Roessner et al., 2009). A later study did not find any differences in gray or white matter between 24 children with TS and 18 controls using several methods including VBM, tract-based spatial-statistics analysis, fractional anisotropy, and apparent diffusion coefficient (Jeppesen et al., 2014). An earlier study using large deformation high dimensional brain mapping to study basal ganglia structure did not identify any differences between 15 adult TS and 15 control participants matched for age, gender, and handedness (Wang et al., 2007). In terms of fMRI studies, as described earlier, an investigation using the Stroop, go/no-go and finger tapping tasks in 39 children with TS and 39 controls did not identify brain differences between the groups (Debes et al., 2011). Similarly, an earlier fMRI study in which adult participants were administered levodopa during the go/no-go task did not show any differences between 8 participants with TS and 10 controls (Hershey et al., 2004). These divergent findings may be a result of a multitude of variables that differ across studies including sample size, demographic characteristics, tic characteristics including severity, illness duration, comorbid conditions, medication use, consideration of online tics, extent of control participant matching, study task, image acquisition and processing techniques, methods for data analysis, and scanner hardware factors.

Further complications come from considerations around interpretations of study results. In particular, it can be difficult to judge whether brain findings are causally related to clinical symptoms on the one hand, or if they are the result of compensatory or adaptive changes. Indeed, there are a number of studies in which participants with TS outperform healthy participants on measures of behavioral control (Brandt, Moczydlowski, et al., 2017; Brandt, Stock, Münchau, & Beste, 2017; Hovik, Plessen, Skogli, Andersen, & Øie, 2016; Jackson, Mueller, Hambleton, & Hollis, 2007; Jackson, Parkinson, Jung, et al., 2011; Jung, Jackson, Nam, Hollis, & Jackson, 2015; Mueller, Jackson, Dhalla, Datsopoulos, & Hollis, 2006; Tajik-Parvinchi & Sandor, 2011, 2012). For example, in a study described earlier, 13 children with TS had shorter reaction times than 13 gender- and age-matched controls during mixed trials of a manual task-switching paradigm (Jackson, Parkinson, Jung, et al., 2011). Subsequent neuroimaging showed widespread differences between the two groups in the white matter microstructure, in particular the forceps minor, and the TS group had higher task-related fMRI BOLD signal in a right PFC ROI adjacent to the forceps minor (Jackson, Parkinson, Jung, et al., 2011). The suggestion from these studies is that chronic efforts at tic symptom suppression may contribute to an enhanced capacity with certain control tasks.

Nevertheless, a meta-analysis that included 61 studies of neuropsychological tasks found that overall TS patients exhibited inhibitory deficits compared to control participants (Morand-Beaulieu et al., 2017). As well, meta-analyses of structural and task-based functional neuroimaging studies in TS implicate PFC control regions (Polyanska et al., 2017; Wan et al., 2021; Wen et al., 2021). Based on a body of literature including postmortem neuropathologic studies in patients with TS, as well as animal models of TS, it is believed that the basal ganglia, subcortical structures known for their role in habit formation and movement, are involved in the generation of tic symptoms (reviewed by Ganos and colleagues (Ganos et al., 2013)). Thus, the overall picture suggests that TS involves aberrant basal ganglia output escaping frontal cortical inhibition (Ganos et al., 2013; Jackson et al., 2015; Stern et al., 2008; Worbe, Lehericy, et al., 2015).

4. The PFC in TS

The DLPFC, IFG, ACC, and the SMA have been implicated in an ALE meta-analysis of task-based neuroimaging studies compared TS to control participants (Polyanska et al., 2017). Moreover, activations in the SMA and lateral PFC/middle frontal gyrus in TS participants were found to correlate with tic severity (Polyanska et al., 2017). The IFG has been involved in tic suppression, during which it had increased regional homogeneity compared to free ticking states (Ganos, Kahl, et al., 2014). An fMRI study of neural response to emotional faces showed the IFG to have increased activation in adult TS participants relative to controls (Neuner et al., 2010). In another such study in adults, TS participants had increased IFG and insula activation when viewing images of neutral and angry faces; moreover, insula connectivity with SMA correlated with premonitory sensations (Rae et al., 2018). In an fMRI study using a motor task-switching paradigm, IFG and SMA activation was decreased in children with TS compared to controls; in the TS group only, activations in these regions strongly correlated with task-switching costs (Jung et al., 2013). The SMA has been involved in fMRI studies of tic release (Bohlhalter et al., 2006; Hampson et al., 2009; Neuner et al., 2014). In a go/no-go fMRI study, SMA activation was lower in adult TS participants relative to controls (Thomalla et al., 2014). In an fMRI study using the stop-signal task, SMA activity during error processing was higher in adult TS participants in comparison to the OCD and control groups (Fan et al., 2017). In another stop-signal task fMRI study in adult participants, SMA activation during the "successful-stop" versus "go" conditions correlated positively with motor tic frequency (Ganos, Kühn, et al., 2014). A finger tapping fMRI study found the SMA to be decreased in children with TS compared to controls (Roessner et al., 2012); the children with TS also had decreased middle frontal gyrus activation when finger tapping with the dominant hand, whereas doing the task with the nondominant hand involved increased activation of the DLPFC relative to controls (Roessner et al., 2013). In a finger opposition fMRI study in adults, TS participants had increased middle frontal gyrus and IFG activation relative to controls (Zapparoli et al., 2016). In a study using PPI in adults, the TS group had decreased activation relative to controls in the ventral lateral PFC and OFC (Zebardast et al., 2013). In a PPI combined electromyography/fMRI study, middle frontal gyrus activity was decreased in adolescents with TS relative to control participants (Buse et al., 2016).

A number of functional neuroimaging studies using cognitive control paradigms show differences between TS and control participants in PFC control regions. In an fMRI study of the Stroop task in both children and adults, in contrast to controls, the TS group did not have greater deactivations with age in the vmPFC and ventral ACC, nor increased activations with age in the inferolateral PFC (Marsh et al., 2007). Further, activations in the inferolateral PFC and DLPFC correlated negatively with task performance in the TS group, while activations in the inferolateral PFC correlated positively with performance in the control group; moreover, activation in the vmPFC correlated positively with task performance in the TS group, and negatively in the control group. In addition, DLPFC activation correlated with tic symptom severity (Marsh et al., 2007). In another fMRI study using a cognitive control task, children with TS had increased DLPFC activity in comparison to controls; in addition, activation in the IFG correlated with tic severity (Baym et al., 2008). A study using the Simon task found that TS participants (children and adults) had higher PFC activation relative to controls; further, in the TS group, activity in the middle frontal gyrus and SMA correlated inversely with task performance (Raz et al., 2009). An fMRI study using a semantic task in adolescents showed that in comparison to controls, middle frontal gyrus activity in the TS group was lower during adaptive control, and higher during task maintenance (Church et al., 2009). In a NIRS study using the Stroop task, children with TS had lower PFC activity in the DLPFC area compared to the control group (Yamamuro et al., 2015).

Several fMRI studies have used an eyeblink suppression paradigm. An fMRI study using such a task in children and adults found activity in the middle frontal gyrus and dorsal ACC to be higher in the TS group relative to controls (Mazzone et al., 2010). Furthermore, activity in this area correlated positively with tic severity, whereas activity in the IFG correlated negatively with tic severity (Mazzone et al., 2010). An fMRI study compared suppression of ocular tics in adults with TS to suppression of eye blinks in healthy controls (van der Salm et al., 2018). The TS group had relatively increased activity in the DLPFC, but decreased activity in the IFG and SMA. It is noteworthy that in both of these studies (Mazzone et al., 2010; van der Salm et al., 2018), the TS participants had increased DLPFC activity relative to the controls. In fact, increased activity in PFC control areas in TS participants is a common finding across a number of the studies described here (Baym et al., 2008; Fan et al., 2017; Neuner et al., 2010; Rae et al., 2018; Raz et al., 2009; Roessner et al., 2013; Zapparoli et al., 2016). The general explanations for such findings is that these increases represent compensatory mechanisms in the TS group, particularly since in some cases they correlate negatively with task performance (Baym et al., 2008; Marsh et al., 2007), and positively with symptom severity (Ganos, Kühn, et al., 2014; Marsh et al., 2007; Mazzone et al., 2010; Rae et al., 2018; Raz et al., 2009).

The vmPFC and OFC also show variability across fMRI studies in TS, though perhaps less so than the other regions described above. In the PPI study described earlier, adults with TS had decreased activation in the OFC relative to controls (Zebardast et al., 2013); moreover, OFC activation correlated positively with obsessive-compulsive symptom severity (Zebardast et al., 2013). In the Stroop study, the TS group did not have greater deactivations with age in the vmPFC and ventral ACC, in contrast to controls (Marsh et al., 2007); further, activation in the vmPFC correlated positively with task performance in the TS group, and negatively in the control group (Marsh et al., 2007). In a study of reinforcement learning, adults with TS and comorbid OCD had decreased vmPFC and poorer task performance relative to controls (Worbe et al., 2011). In a study of emotion recognition, adults with TS had decreased OFC activation relative to controls (Eddy et al., 2017). In a DBS study, thalamic stimulation in adult patients with TS was associated with decreased activity in the medial PFC and OFC, with BOLD signal changes in the latter being associated with reduction in tics (Jo et al., 2018). Overall, the vmPFC and OFC do not feature as frequently in functional neuroimaging studies of TS relative to the SMA, IFG, and DLPFC (Polyanska et al., 2017), and may be more related to comorbid OCD (Worbe et al., 2011; Zebardast et al., 2013).

There is no clear explanation for the contrasting findings across the studies. There is much heterogeneity across studies, including around demographic and clinical characteristics, study design, and data analysis, among other factors. It should also be noted that there is wide variability in the control areas (DLPFC, IFG, SMA, others), aspects of the task (for example, the "go" condition, the "no-go" condition, or contrast between the two), and symptoms severity measures (motor, vocal, total, current, worst ever, other) used in the correlational analyses—this may have contributed to findings of uncertain reliability.

Interestingly, findings from nuclear imaging studies appear to be more consistent, with a tendency for decreased activity in TS in the PFC regions considered. Regional cerebral perfusion was found to be decreased in the DLPFC and OFC in a SPECT study of children with TS (Diler et al., 2002). In an FDG-PET study, adults with TS had resting metabolic activity patterns that was decreased in the OFC (Pourfar et al., 2011). In the same study, those with comorbid OCD had activity patterns that was decreased in the DLPFC (Pourfar et al., 2011). An earlier FDG-PET study in adults with TS found decreased metabolic activity the ventral PFC and OFC, but increased activity in the SMA (Braun et al., 1993). Furthermore, metabolic rate in the OFC correlated positively with compulsive and behavioral symptoms (Braun et al., 1995). In an $H_2^{15}O$-PET study, obsessive-compulsive symmetry symptom provocation correlated positively with OFC rCBF in adults with TS, whereas in the control group they correlated negatively with the DLPFC (de Vries et al., 2013). In a SPECT study of patients undergoing DBS, adults with TS had lower rCBF at baseline compared to controls in the SMA and middle frontal cortex (Haense et al., 2016). GPi stimulation was associated with decrease in rCBF in OFC and increase rCBF in middle frontal cortex relative to sham stimulation (Haense et al., 2016). Thalamic stimulation was associated with rCBF increase in SMA and middle and frontal cortex relative to sham stimulation (Haense et al., 2016).

The involvement of the PFC in TS, including in tic inhibition, should come as no surprise given the known role of these regions in various aspects of executive function, inhibition, and motor control (Grafton & Volz, 2019). The DLPFC is known to be involved in various executive functions, including working memory, cognitive flexibility, planning, and inhibition. It has been described as the highest cortical area in terms of motor planning, organization, and regulation. The dACC plays a role at the interface between sensory, motor and cognitive processing, and it interacts with the SMA in motor control (Asemi, Ramaseshan, Burgess, Diwadkar, & Bressler, 2015). Interesting, the dACC has been found to be active in an ALE of physiological urges (Jackson, Parkinson, Kim, Schüermann, & Eickhoff, 2011). The SMA is involved in internal cue-related selection of movement, response inhibition, and initiation and suppression of movement (Potgieser, de Jong, Wagemakers, Hoving, & Groen, 2014). As well, the pre-SMA has been found to be involved in exerting control over voluntary actions in situations of response conflict, and it is seen as part of a cognitive control network involving the IFG, striatum, and subthalamic nucleus (Nachev, Kennard, & Husain, 2008). The IFG is consistently involved in studies of impulse control, in particular, inhibition of prepotent response, likely via direct projections to the subthalamic nucleus in the hyperdirect pathway. The OFC has been found to be involved in sensory integration, decision-making, comparing expected and actual outcomes of an action, and adaptive learning.

Overall, it is clear that key control areas of the PFC are implicated in TS, though in task-based functional imaging studies, the variability in the findings may be related to multiple factors, including the possible role of compensatory mechanisms. Nevertheless, it is likely that the causes and neural mechanisms involved in TS are complex, varied, and may involve interactions among different systems. As such, the field would benefit from concerted efforts and collaborations to carry out multimodal studies with large samples and longitudinal design.

Conflict of Interest

The authors have no financial conflicts of interest to disclose.

References

Abi-Jaoude, E., Segura, B., Obeso, I., Cho, S. S., Houle, S., Lang, A. E., et al. (2015). Similar striatal D2/D3 dopamine receptor availability in adults with Tourette syndrome compared with healthy controls: A [(11) C]-(+)-PHNO and [(11) C]raclopride positron emission tomography imaging study. *Human Brain Mapping*, *36*(7), 2592–2601. https://doi.org/10.1002/hbm.22793.

Asemi, A., Ramaseshan, K., Burgess, A., Diwadkar, V. A., & Bressler, S. L. (2015). Dorsal anterior cingulate cortex modulates supplementary motor area in coordinated unimanual motor behavior. *Frontiers in Human Neuroscience*, *9*. https://doi.org/10.3389/fnhum.2015.00309.

Atkinson-Clement, C., Porte, C.-A., de Liege, A., Klein, Y., Delorme, C., Beranger, B., et al. (2021). Impulsive prepotent actions and tics in Tourette disorder underpinned by a common neural network. *Molecular Psychiatry*, *26*(7), 3548–3557. https://doi.org/10.1038/s41380-020-00890-5.

Atkinson-Clement, C., Porte, C.-A., de Liege, A., Wattiez, N., Klein, Y., Beranger, B., et al. (2020). Neural correlates and role of medication in reactive motor impulsivity in Tourette disorder. *Cortex; A Journal Devoted to the Study of the Nervous System and Behavior*, *125*, 60–72. https://doi.org/10.1016/j.cortex.2019.12.007.

Baym, C. L., Corbett, B. A., Wright, S. B., & Bunge, S. A. (2008). Neural correlates of tic severity and cognitive control in children with Tourette syndrome. *Brain: A Journal of Neurology*, *131*(Pt. 1), 165–179. https://doi.org/10.1093/brain/awm278.

Bhikram, T., Crawley, A., Arnold, P., Abi-Jaoude, E., & Sandor, P. (2021). Neuroimaging the emotional modulation of urge inhibition in Tourette syndrome. *Cortex; A Journal Devoted to the Study of the Nervous System and Behavior*, *135*, 341–351. https://doi.org/10.1016/j.cortex.2020.10.010.

Bohlhalter, S., Goldfine, A., Matteson, S., Garraux, G., Hanakawa, T., Kansaku, K., et al. (2006). Neural correlates of tic generation in Tourette syndrome: An event-related functional MRI study. *Brain: A Journal of Neurology*, *129*(Pt. 8), 2029–2037. https://doi.org/10.1093/brain/awl050.

Brandt, V. C., Moczydlowski, A., Jonas, M., Boelmans, K., Bäumer, T., Brass, M., et al. (2017). Imitation inhibition in children with Tourette syndrome. *Journal of Neuropsychology*. https://doi.org/10.1111/jnp.12132.

Brandt, V. C., Stock, A.-K., Münchau, A., & Beste, C. (2017). Evidence for enhanced multi-component behaviour in Tourette syndrome—An EEG study. *Scientific Reports*, 7(1), 7722. https://doi.org/10.1038/s41598-017-08158-9.

Braun, A. R., Randolph, C., Stoetter, B., Mohr, E., Cox, C., Vladar, K., et al. (1995). The functional neuroanatomy of Tourette's syndrome: An FDG-PET study. II: Relationships between regional cerebral metabolism and associated behavioral and cognitive features of the illness. *Neuropsychopharmacology: Official Publication of the American College of Neuropsychopharmacology*, *13*(2), 151–168. https://doi.org/10.1016/0893-133X(95)00052-F.

Braun, A. R., Stoetter, B., Randolph, C., Hsiao, J. K., Vladar, K., Gernert, J., et al. (1993). The functional neuroanatomy of Tourette's syndrome: An FDG-PET study. I. Regional changes in cerebral glucose metabolism differentiating patients and controls. *Neuropsychopharmacology: Official Publication of the American College of Neuropsychopharmacology*, *9*(4), 277–291. https://doi.org/10.1038/npp.1993.64.

Briguglio, M., Galentino, R., Michele, S. D., Dell'Osso, B., Fogassi, L., & Porta, M. (2021). The imitation game in children with Tourette syndrome: A lack of impulse control to mirror environmental stimuli. *Motor Control*, *1*(aop), 1–5. https://doi.org/10.1123/mc.2021-0064.

Budman, C. L., Bruun, R. D., Park, K. S., Lesser, M., & Olson, M. (2000). Explosive outbursts in children with Tourette's disorder. *Journal of the American Academy of Child and Adolescent Psychiatry*, *39*(10), 1270–1276. https://doi.org/10.1097/00004583-200010000-00014.

Budman, C. L., Rockmore, L., Stokes, J., & Sossin, M. (2003). Clinical phenomenology of episodic rage in children with Tourette syndrome. *Journal of Psychosomatic Research*, *55*(1), 59–65.

Buse, J., Beste, C., Herrmann, E., & Roessner, V. (2016). Neural correlates of altered sensorimotor gating in boys with Tourette syndrome: A combined EMG/fMRI study. *The World Journal of Biological Psychiatry: The Official Journal of the World Federation of Societies of Biological Psychiatry*, *17*(3), 187–197. https://doi.org/10.3109/15622975.2015.1112033.

Buse, J., Beste, C., & Roessner, V. (2017). Neural correlates of prediction violations in boys with Tourette syndrome: Evidence from harmonic expectancy. *The World Journal of Biological Psychiatry: The Official Journal of the World Federation of Societies of Biological Psychiatry*, 1–12. https://doi.org/10.1080/15622975.2016.1274052.

Buse, J., Schoenefeld, K., Münchau, A., & Roessner, V. (2013). Neuromodulation in Tourette syndrome: Dopamine and beyond. *Neuroscience and Biobehavioral Reviews*, *37*(6), 1069–1084. https://doi.org/10.1016/j.neubiorev.2012.10.004.

Channon, S., Crawford, S., Vakili, K., & Robertson, M. M. (2003). Real-life-type problem solving in Tourette syndrome. *Cognitive and Behavioral Neurology: Official Journal of the Society for Behavioral and Cognitive Neurology*, *16*(1), 3–15.

Channon, S., Pratt, P., & Robertson, M. M. (2003). Executive function, memory, and learning in Tourette's syndrome. *Neuropsychology*, *17*(2), 247–254.

Channon, S., Sinclair, E., Waller, D., Healey, L., & Robertson, M. M. (2004). Social cognition in Tourette's syndrome: Intact theory of mind and impaired inhibitory functioning. *Journal of Autism and Developmental Disorders*, *34*(6), 669–677.

Chen, K., Budman, C. L., Diego Herrera, L., Witkin, J. E., Weiss, N. T., Lowe, T. L., et al. (2013). Prevalence and clinical correlates of explosive outbursts in Tourette syndrome. *Psychiatry Research*, *205*(3), 269–275. https://doi.org/10.1016/j.psychres.2012.09.029.

Church, J. A., Wenger, K. K., Dosenbach, N. U. F., Miezin, F. M., Petersen, S. E., & Schlaggar, B. L. (2009). Task control signals in pediatric Tourette syndrome show evidence of immature and anomalous functional activity. *Frontiers in Human Neuroscience, 3*, 38. https://doi.org/10.3389/neuro.09.038.2009.

Conte, G., Valente, F., Fioriello, F., & Cardona, F. (2020). Rage attacks in Tourette syndrome and chronic tic disorder: A systematic review. *Neuroscience and Biobehavioral Reviews, 119*, 21–36. https://doi.org/10.1016/j.neubiorev.2020.09.019.

Debes, N. M. M. M., Hansen, A., Skov, L., & Larsson, H. (2011). A functional magnetic resonance imaging study of a large clinical cohort of children with Tourette syndrome. *Journal of Child Neurology, 26*(5), 560–569. https://doi.org/10.1177/0883073810387928.

Deckersbach, T., Chou, T., Britton, J. C., Carlson, L. E., Reese, H. E., Siev, J., et al. (2014). Neural correlates of behavior therapy for Tourette's disorder. *Psychiatry Research, 224*(3), 269–274. https://doi.org/10.1016/j.pscychresns.2014.09.003.

de Vries, F. E., van den Heuvel, O. A., Cath, D. C., Groenewegen, H. J., van Balkom, A. J. L. M., Boellaard, R., et al. (2013). Limbic and motor circuits involved in symmetry behavior in Tourette's syndrome. *CNS Spectrums, 18*(1), 34–42. https://doi.org/10.1017/S1092852912000703.

Diler, R. S., Reyhanli, M., Toros, F., Kibar, M., & Avci, A. (2002). Tc-99m-ECD SPECT brain imaging in children with Tourette's syndrome. *Yonsei Medical Journal, 43*(4), 403–410. https://doi.org/10.3349/ymj.2002.43.4.403.

Draganski, B., Martino, D., Cavanna, A. E., Hutton, C., Orth, M., Robertson, M. M., et al. (2010). Multispectral brain morphometry in Tourette syndrome persisting into adulthood. *Brain: A Journal of Neurology, 133*(Pt. 12), 3661–3675. https://doi.org/10.1093/brain/awq300.

Eddy, C. M., & Cavanna, A. E. (2017). Set-shifting deficits: A possible neurocognitive endophenotype for Tourette syndrome without ADHD. *Journal of Attention Disorders, 21*(10), 824–834. https://doi.org/10.1177/1087054714545536.

Eddy, C. M., Cavanna, A. E., & Hansen, P. C. (2017). Empathy and aversion: The neural signature of mentalizing in Tourette syndrome. *Psychological Medicine, 47*(3), 507–517. https://doi.org/10.1017/S0033291716002725.

Eddy, C. M., Cavanna, A. E., Rickards, H. E., & Hansen, P. C. (2016). Temporo-parietal dysfunction in Tourette syndrome: Insights from an fMRI study of Theory of Mind. *Journal of Psychiatric Research, 81*, 102–111. https://doi.org/10.1016/j.jpsychires.2016.07.002.

Eddy, C. M., Rickards, H. E., & Cavanna, A. E. (2012). Executive functions in uncomplicated Tourette syndrome. *Psychiatry Research, 200*(1), 46–48. https://doi.org/10.1016/j.psychres.2012.05.023.

Fahim, C., Yoon, U., Das, S., Lyttelton, O., Chen, J., Arnaoutelis, R., et al. (2010). Somatosensory-motor bodily representation cortical thinning in Tourette: Effects of tic severity, age and gender. *Cortex; A Journal Devoted to the Study of the Nervous System and Behavior, 46*(6), 750–760. https://doi.org/10.1016/j.cortex.2009.06.008.

Fan, S., Cath, D. C., van der Werf, Y. D., de Wit, S., Veltman, D. J., & van den Heuvel, O. A. (2017). Trans-diagnostic comparison of response inhibition in Tourette's disorder and obsessive-compulsive disorder. *The World Journal of Biological Psychiatry: The Official Journal of the World Federation of Societies of Biological Psychiatry*, 1–11. https://doi.org/10.1080/15622975.2017.1347711.

Ganos, C., Kahl, U., Brandt, V., Schunke, O., Bäumer, T., Thomalla, G., et al. (2014). The neural correlates of tic inhibition in Gilles de la Tourette syndrome. *Neuropsychologia, 65*, 297–301. https://doi.org/10.1016/j.neuropsychologia.2014.08.007.

Ganos, C., Kühn, S., Kahl, U., Schunke, O., Feldheim, J., Gerloff, C., et al. (2014). Action inhibition in Tourette syndrome. *Movement Disorders: Official Journal of the Movement Disorder Society, 29*(12), 1532–1538. https://doi.org/10.1002/mds.25944.

Ganos, C., Rocchi, L., Latorre, A., Hockey, L., Palmer, C., Joyce, E. M., et al. (2018). Motor cortical excitability during voluntary inhibition of involuntary tic movements. *Movement Disorders: Official Journal of the Movement Disorder Society*, *33*(11), 1804–1809. https://doi.org/10.1002/mds.27479.

Ganos, C., Roessner, V., & Münchau, A. (2013). The functional anatomy of Gilles de la Tourette syndrome. *Neuroscience and Biobehavioral Reviews*, *37*(6), 1050–1062. https://doi.org/10.1016/j.neubiorev.2012.11.004.

Grafton, S. T., & Volz, L. J. (2019). Chapter 13—From ideas to action: The prefrontal–premotor connections that shape motor behavior. In M. D'Esposito, & J. H. Grafman (Eds.), *Vol. 163. Handbook of clinical neurology* (pp. 237–255). Elsevier. https://doi.org/10.1016/B978-0-12-804281-6.00013-6.

Greene, D. J., Williams, A. C., III, Koller, J. M., Schlaggar, B. L., & Black, K. J. (2016). Brain structure in pediatric Tourette syndrome. *Molecular Psychiatry*. https://doi.org/10.1038/mp.2016.194.

Greimel, E., Wanderer, S., Rothenberger, A., Herpertz-Dahlmann, B., Konrad, K., & Roessner, V. (2011). Attentional performance in children and adolescents with tic disorder and co-occurring attention-deficit/hyperactivity disorder: New insights from a 2 × 2 factorial design study. *Journal of Abnormal Child Psychology*, *39*(6), 819–828. https://doi.org/10.1007/s10802-011-9493-7.

Haense, C., Müller-Vahl, K. R., Wilke, F., Schrader, C., Capelle, H. H., Geworski, L., et al. (2016). Effect of deep brain stimulation on regional cerebral blood flow in patients with medically refractory Tourette syndrome. *Frontiers in Psychiatry*, 7. https://doi.org/10.3389/fpsyt.2016.00118.

Hampson, M., Tokoglu, F., King, R. A., Constable, R. T., & Leckman, J. F. (2009). Brain areas coactivating with motor cortex during chronic motor tics and intentional movements. *Biological Psychiatry*, *65*(7), 594–599. https://doi.org/10.1016/j.biopsych.2008.11.012.

Hershey, T., Black, K. J., Hartlein, J., Braver, T. S., Barch, D. M., Carl, J. L., et al. (2004). Dopaminergic modulation of response inhibition: An fMRI study. *Brain Research. Cognitive Brain Research*, *20*(3), 438–448. https://doi.org/10.1016/j.cogbrainres.2004.03.018.

Hovik, K. T., Egeland, J., Isquith, P. K., Gioia, G., Skogli, E. W., Andersen, P. N., et al. (2017). Distinct patterns of everyday executive function problems distinguish children with Tourette syndrome from children with ADHD or autism spectrum disorders. *Journal of Attention Disorders*, *21*(10), 811–823. https://doi.org/10.1177/1087054714550336.

Hovik, K. T., Plessen, K. J., Skogli, E. W., Andersen, P. N., & Øie, M. (2016). Dissociable response inhibition in children with Tourette's syndrome compared with children with ADHD. *Journal of Attention Disorders*, *20*(10), 825–835. https://doi.org/10.1177/1087054713512371.

Jackson, G. M., Draper, A., Dyke, K., Pépés, S. E., & Jackson, S. R. (2015). Inhibition, disinhibition, and the control of action in Tourette syndrome. *Trends in Cognitive Sciences*, *19*(11), 655–665. https://doi.org/10.1016/j.tics.2015.08.006.

Jackson, G. M., Mueller, S. C., Hambleton, K., & Hollis, C. P. (2007). Enhanced cognitive control in Tourette syndrome during task uncertainty. *Experimental Brain Research*, *182*(3), 357–364. https://doi.org/10.1007/s00221-007-0999-8.

Jackson, S. R., Parkinson, A., Jung, J., Ryan, S. E., Morgan, P. S., Hollis, C., et al. (2011). Compensatory neural reorganization in Tourette syndrome. *Current Biology: CB*, *21*(7), 580–585. https://doi.org/10.1016/j.cub.2011.02.047.

Jackson, S. R., Parkinson, A., Kim, S. Y., Schüermann, M., & Eickhoff, S. B. (2011). On the functional anatomy of the urge-for-action. *Cognitive Neuroscience*, *2*(3–4), 227–243. https://doi.org/10.1080/17588928.2011.604717.

Jackson, S. R., Sigurdsson, H. P., Dyke, K., Condon, M., & Jackson, G. M. (2021). The role of the cingulate cortex in the generation of motor tics and the experience of the premonitory urge-to-tic in Tourette syndrome. *Journal of Neuropsychology, 15*(3), 340–362. https://doi.org/10.1111/jnp.12242.

Jeppesen, S. S., Debes, N. M., Simonsen, H. J., Rostrup, E., Larsson, H. B. W., & Skov, L. (2014). Study of medication-free children with Tourette syndrome do not show imaging abnormalities. *Movement Disorders: Official Journal of the Movement Disorder Society, 29*(9), 1212–1216. https://doi.org/10.1002/mds.25858.

Jeter, C. B., Patel, S. S., Morris, J. S., Chuang, A. Z., Butler, I. J., & Sereno, A. B. (2015). Oculomotor executive function abnormalities with increased tic severity in Tourette syndrome. *Journal of Child Psychology and Psychiatry, and Allied Disciplines, 56*(2), 193–202. https://doi.org/10.1111/jcpp.12298.

Jo, H. J., McCairn, K. W., Gibson, W. S., Testini, P., Zhao, C. Z., Gorny, K. R., et al. (2018). Global network modulation during thalamic stimulation for Tourette syndrome. *NeuroImage. Clinical, 18*, 502–509. https://doi.org/10.1016/j.nicl.2018.02.018.

Jung, J., Jackson, S. R., Nam, K., Hollis, C., & Jackson, G. M. (2015). Enhanced saccadic control in young people with Tourette syndrome despite slowed pro-saccades. *Journal of Neuropsychology, 9*(2), 172–183. https://doi.org/10.1111/jnp.12044.

Jung, J., Jackson, S. R., Parkinson, A., & Jackson, G. M. (2013). Cognitive control over motor output in Tourette syndrome. *Neuroscience and Biobehavioral Reviews, 37*(6), 1016–1025. https://doi.org/10.1016/j.neubiorev.2012.08.009.

Jurgiel, J., Miyakoshi, M., Dillon, A., Piacentini, J., Makeig, S., & Loo, S. K. (2021). Inhibitory control in children with tic disorder: Aberrant fronto-parietal network activity and connectivity. *Brain Communications, 3*(2), fcab067. https://doi.org/10.1093/braincomms/fcab067.

Kalanithi, P. S. A., Zheng, W., Kataoka, Y., DiFiglia, M., Grantz, H., Saper, C. B., et al. (2005). Altered parvalbumin-positive neuron distribution in basal ganglia of individuals with Tourette syndrome. *Proceedings of the National Academy of Sciences of the United States of America, 102*(37), 13307–13312. https://doi.org/10.1073/pnas.0502624102.

Kataoka, Y., Kalanithi, P. S. A., Grantz, H., Schwartz, M. L., Saper, C., Leckman, J. F., et al. (2010). Decreased number of parvalbumin and cholinergic interneurons in the striatum of individuals with Tourette syndrome. *The Journal of Comparative Neurology, 518*(3), 277–291. https://doi.org/10.1002/cne.22206.

Kong, L., Lv, B., Wu, T., Zhang, J., Fan, Y., Ouyang, M., et al. (2020). Altered structural cerebral cortex in children with Tourette syndrome. *European Journal of Radiology, 129*, 109119. https://doi.org/10.1016/j.ejrad.2020.109119.

Kurvits, L., Martino, D., & Ganos, C. (2020). Clinical features that evoke the concept of disinhibition in Tourette syndrome. *Frontiers in Psychiatry, 11*, 21. https://doi.org/10.3389/fpsyt.2020.00021.

Lin, Y.-J., Lai, M.-C., & Gau, S. S.-F. (2012). Youths with ADHD with and without tic disorders: Comorbid psychopathology, executive function and social adjustment. *Research in Developmental Disabilities, 33*(3), 951–963. https://doi.org/10.1016/j.ridd.2012.01.001.

Marsh, R., Zhu, H., Wang, Z., Skudlarski, P., & Peterson, B. S. (2007). A developmental fMRI study of self-regulatory control in Tourette's syndrome. *The American Journal of Psychiatry, 164*(6), 955–966. https://doi.org/10.1176/ajp.2007.164.6.955.

Martino, D., & Mink, J. W. (2013). Tic disorders. *Continuum (Minneapolis, Minn.), 19*(5 Movement Disorders), 1287–1311. https://doi.org/10.1212/01.CON.0000436157.31662.af.

Mazzone, L., Yu, S., Blair, C., Gunter, B. C., Wang, Z., Marsh, R., et al. (2010). An FMRI study of frontostriatal circuits during the inhibition of eye blinking in persons with Tourette syndrome. *The American Journal of Psychiatry, 167*(3), 341–349. https://doi.org/10.1176/appi.ajp.2009.08121831.

Mink, J. W. (2001). Neurobiology of basal ganglia circuits in Tourette syndrome: Faulty inhibition of unwanted motor patterns? *Advances in Neurology, 85*, 113–122.

Morand-Beaulieu, S., Grot, S., Lavoie, J., Leclerc, J. B., Luck, D., & Lavoie, M. E. (2017). The puzzling question of inhibitory control in Tourette syndrome: A meta-analysis. *Neuroscience and Biobehavioral Reviews, 80*, 240–262. https://doi.org/10.1016/j.neubiorev.2017.05.006.

Morand-Beaulieu, S., Wu, J., Mayes, L. C., Grantz, H., Leckman, J. F., Crowley, M. J., et al. (2021). Increased alpha-band connectivity during tic suppression in children with Tourette syndrome revealed by source electroencephalography analyses. *Biological Psychiatry. Cognitive Neuroscience and Neuroimaging*. S2451-9022(21)00140-3 https://doi.org/10.1016/j.bpsc.2021.05.001.

Morera Maiquez, B., Jackson, G. M., & Jackson, S. R. (2021). Examining the neural antecedents of tics in Tourette syndrome using electroencephalography. *Journal of Neuropsychology*. https://doi.org/10.1111/jnp.12245.

Mueller, S. C., Jackson, G. M., Dhalla, R., Datsopoulos, S., & Hollis, C. P. (2006). Enhanced cognitive control in young people with Tourette's syndrome. *Current Biology: CB, 16*(6), 570–573. https://doi.org/10.1016/j.cub.2006.01.064.

Müller-Vahl, K. R., Kaufmann, J., Grosskreutz, J., Dengler, R., Emrich, H. M., & Peschel, T. (2009). Prefrontal and anterior cingulate cortex abnormalities in Tourette syndrome: Evidence from voxel-based morphometry and magnetization transfer imaging. *BMC Neuroscience, 10*, 47. https://doi.org/10.1186/1471-2202-10-47.

Murphy, T., & Muter, V. (2012). Risk factors for comorbidity in ADHD and GTS: Looking beyond a single-deficit model. *Applied Neuropsychology. Child, 1*(2), 129–136. https://doi.org/10.1080/21622965.2012.703889.

Nachev, P., Kennard, C., & Husain, M. (2008). Functional role of the supplementary and pre-supplementary motor areas. *Nature Reviews. Neuroscience, 9*(11), 856–869. https://doi.org/10.1038/nrn2478.

Neuner, I., Kellermann, T., Stöcker, T., Kircher, T., Habel, U., Shah, J. N., et al. (2010). Amygdala hypersensitivity in response to emotional faces in Tourette's patients. *The World Journal of Biological Psychiatry: The Official Journal of the World Federation of Societies of Biological Psychiatry, 11*(7), 858–872. https://doi.org/10.3109/15622975.2010.480984.

Neuner, I., Werner, C. J., Arrubla, J., Stöcker, T., Ehlen, C., Wegener, H. P., et al. (2014). Imaging the where and when of tic generation and resting state networks in adult Tourette patients. *Frontiers in Human Neuroscience, 8*. https://doi.org/10.3389/fnhum.2014.00362.

Openneer, T. J. C., Forde, N. J., Akkermans, S. E. A., Naaijen, J., Buitelaar, J. K., Hoekstra, P. J., et al. (2020). Executive function in children with Tourette syndrome and attention-deficit/hyperactivity disorder: Cross-disorder or unique impairments? *Cortex, 124*, 176–187. https://doi.org/10.1016/j.cortex.2019.11.007.

Openneer, T. J. C., van der Meer, D., Marsman, J.-B. C., Forde, N. J., Akkermans, S. E. A., Naaijen, J., et al. (2021). Impaired response inhibition during a stop-signal task in children with Tourette syndrome is related to ADHD symptoms: A functional magnetic resonance imaging study. *The World Journal of Biological Psychiatry: The Official Journal of the World Federation of Societies of Biological Psychiatry, 22*(5), 350–361. https://doi.org/10.1080/15622975.2020.1813329.

Peterson, B. S., Skudlarski, P., Anderson, A. W., Zhang, H., Gatenby, J. C., Lacadie, C. M., et al. (1998). A functional magnetic resonance imaging study of tic suppression in Tourette syndrome. *Archives of General Psychiatry, 55*(4), 326–333.

Peterson, B. S., Staib, L., Scahill, L., Zhang, H., Anderson, C., Leckman, J. F., et al. (2001). Regional brain and ventricular volumes in Tourette syndrome. *Archives of General Psychiatry, 58*(5), 427–440.

Polyanska, L., Critchley, H. D., & Rae, C. L. (2017). Centrality of prefrontal and motor preparation cortices to Tourette syndrome revealed by meta-analysis of task-based neuroimaging studies. *NeuroImage. Clinical, 16*, 257–267. https://doi.org/10.1016/j.nicl.2017.08.004.

Potgieser, A. R. E., de Jong, B. M., Wagemakers, M., Hoving, E. W., & Groen, R. J. M. (2014). Insights from the supplementary motor area syndrome in balancing movement initiation and inhibition. *Frontiers in Human Neuroscience, 8*, 960. https://doi.org/10.3389/fnhum.2014.00960.

Pourfar, M., Feigin, A., Tang, C. C., Carbon-Correll, M., Bussa, M., Budman, C., et al. (2011). Abnormal metabolic brain networks in Tourette syndrome. *Neurology, 76*(11), 944–952. https://doi.org/10.1212/WNL.0b013e3182104106.

Rae, C. L., Critchley, H. D., & Seth, A. K. (2019). A Bayesian account of the sensory-motor interactions underlying symptoms of Tourette syndrome. *Frontiers in Psychiatry, 10*, 29. https://doi.org/10.3389/fpsyt.2019.00029.

Rae, C. L., Parkinson, J., Betka, S., Gouldvan Praag, C. D., Bouyagoub, S., Polyanska, L., et al. (2020). Amplified engagement of prefrontal cortex during control of voluntary action in Tourette syndrome. *Brain Communications, 2*(2), fcaa199. https://doi.org/10.1093/braincomms/fcaa199.

Rae, C. L., Polyanska, L., Gould van Praag, C. D., Parkinson, J., Bouyagoub, S., Nagai, Y., et al. (2018). Face perception enhances insula and motor network reactivity in Tourette syndrome. *Brain, 141*(11), 3249–3261. https://doi.org/10.1093/brain/awy254.

Rawji, V., Modi, S., Latorre, A., Rocchi, L., Hockey, L., Bhatia, K., et al. (2020). Impaired automatic but intact volitional inhibition in primary tic disorders. *Brain: A Journal of Neurology, 143*(3), 906–919. https://doi.org/10.1093/brain/awaa024.

Raz, A., Zhu, H., Yu, S., Bansal, R., Wang, Z., Alexander, G. M., et al. (2009). Neural substrates of self-regulatory control in children and adults with Tourette syndrome. *Canadian Journal of Psychiatry. Revue Canadienne de Psychiatrie, 54*(9), 579–588.

Roessner, V., Overlack, S., Baudewig, J., Dechent, P., Rothenberger, A., & Helms, G. (2009). No brain structure abnormalities in boys with Tourette's syndrome: A voxel-based morphometry study. *Movement Disorders: Official Journal of the Movement Disorder Society, 24*(16), 2398–2403. https://doi.org/10.1002/mds.22847.

Roessner, V., Wittfoth, M., August, J. M., Rothenberger, A., Baudewig, J., & Dechent, P. (2013). Finger tapping-related activation differences in treatment-naïve pediatric Tourette syndrome: A comparison of the preferred and nonpreferred hand. *Journal of Child Psychology and Psychiatry, and Allied Disciplines, 54*(3), 273–279. https://doi.org/10.1111/j.1469-7610.2012.02584.x.

Roessner, V., Wittfoth, M., Schmidt-Samoa, C., Rothenberger, A., Dechent, P., & Baudewig, J. (2012). Altered motor network recruitment during finger tapping in boys with Tourette syndrome. *Human Brain Mapping, 33*(3), 666–675. https://doi.org/10.1002/hbm.21240.

Sowell, E. R., Kan, E., Yoshii, J., Thompson, P. M., Bansal, R., Xu, D., et al. (2008). Thinning of sensorimotor cortices in children with Tourette syndrome. *Nature Neuroscience, 11*(6), 637–639. https://doi.org/10.1038/nn.2121.

Stern, E. R., Blair, C., & Peterson, B. S. (2008). Inhibitory deficits in Tourette's syndrome. *Developmental Psychobiology, 50*(1), 9–18. https://doi.org/10.1002/dev.20266.

Sukhodolsky, D. G., Landeros-Weisenberger, A., Scahill, L., Leckman, J. F., & Schultz, R. T. (2010). Neuropsychological functioning in children with Tourette syndrome with and without attention-deficit/hyperactivity disorder. *Journal of the American Academy of Child and Adolescent Psychiatry, 49*(11), 1155–1164. https://doi.org/10.1016/j.jaac.2010.08.008.

Sukhodolsky, D. G., Walsh, C., Koller, W. N., Eilbott, J., Rance, M., Fulbright, R. K., et al. (2020). Randomized, sham-controlled trial of real-time functional magnetic resonance imaging neurofeedback for tics in adolescents with Tourette syndrome. *Biological Psychiatry*, *87*(12), 1063–1070. https://doi.org/10.1016/j.biopsych.2019.07.035.

Tajik-Parvinchi, D. J., & Sandor, P. (2011). Smooth pursuit and fixation ability in children with Tourette syndrome. *Cognitive and Behavioral Neurology: Official Journal of the Society for Behavioral and Cognitive Neurology*, *24*(4), 174–186. https://doi.org/10.1097/WNN.0b013e31823f90eb.

Tajik-Parvinchi, D. J., & Sandor, P. (2012). Unique saccadic abilities associated with Tourette syndrome: Pure and comorbid groups a controlled study. *Journal of Obsessive-Compulsive and Related Disorders*, *1*(4), 283–293. https://doi.org/10.1016/j.jocrd.2012.07.004.

Thomalla, G., Jonas, M., Bäumer, T., Siebner, H. R., Biermann-Ruben, K., Ganos, C., et al. (2014). Costs of control: Decreased motor cortex engagement during a Go/NoGo task in Tourette's syndrome. *Brain*, *137*(1), 122–136. https://doi.org/10.1093/brain/awt288.

van der Salm, S. M. A., van der Meer, J. N., Cath, D. C., Groot, P. F. C., van der Werf, Y. D., Brouwers, E., et al. (2018). Distinctive tics suppression network in Gilles de la Tourette syndrome distinguished from suppression of natural urges using multimodal imaging. *NeuroImage. Clinical*, *20*, 783–792. https://doi.org/10.1016/j.nicl.2018.09.014.

Wan, X., Zhang, S., Wang, W., Su, X., Li, J., Yang, X., et al. (2021). Gray matter abnormalities in Tourette syndrome: A meta-analysis of voxel-based morphometry studies. *Translational Psychiatry*, *11*(1), 287. https://doi.org/10.1038/s41398-021-01394-8.

Wang, L., Lee, D. Y., Bailey, E., Hartlein, J. M., Gado, M. H., Miller, M. I., et al. (2007). Validity of large-deformation high dimensional brain mapping of the basal ganglia in adults with Tourette syndrome. *Psychiatry Research*, *154*(2), 181–190. https://doi.org/10.1016/j.pscychresns.2006.08.006.

Wang, Z., Maia, T. V., Marsh, R., Colibazzi, T., Gerber, A., & Peterson, B. S. (2011). The neural circuits that generate tics in Tourette's syndrome. *The American Journal of Psychiatry*, *168*(12), 1326–1337. https://doi.org/10.1176/appi.ajp.2011.09111692.

Wen, F., Yan, J., Yu, L., Wang, F., Liu, J., Li, Y., et al. (2021). Grey matter abnormalities in Tourette syndrome: An activation likelihood estimation meta-analysis. *BMC Psychiatry*, *21*(1), 184. https://doi.org/10.1186/s12888-021-03187-1.

Werner, C. J., Stöcker, T., Kellermann, T., Bath, J., Beldoch, M., Schneider, F., et al. (2011). Altered motor network activation and functional connectivity in adult Tourette's syndrome. *Human Brain Mapping*, *32*(11), 2014–2026. https://doi.org/10.1002/hbm.21175.

Wittfoth, M., Bornmann, S., Peschel, T., Grosskreutz, J., Glahn, A., Buddensiek, N., et al. (2012). Lateral frontal cortex volume reduction in Tourette syndrome revealed by VBM. *BMC Neuroscience*, *13*, 17. https://doi.org/10.1186/1471-2202-13-17.

Worbe, Y., Gerardin, E., Hartmann, A., Valabrégue, R., Chupin, M., Tremblay, L., et al. (2010). Distinct structural changes underpin clinical phenotypes in patients with Gilles de la Tourette syndrome. *Brain: A Journal of Neurology*, *133*(Pt. 12), 3649–3660. https://doi.org/10.1093/brain/awq293.

Worbe, Y., Lehericy, S., & Hartmann, A. (2015). Neuroimaging of tic genesis: Present status and future perspectives. *Movement Disorders: Official Journal of the Movement Disorder Society*, *30*(9), 1179–1183. https://doi.org/10.1002/mds.26333.

Worbe, Y., Marrakchi-Kacem, L., Lecomte, S., Valabregue, R., Poupon, F., Guevara, P., et al. (2015). Altered structural connectivity of cortico-striato-pallido-thalamic networks in Gilles de la Tourette syndrome. *Brain: A Journal of Neurology*, *138*(Pt. 2), 472–482. https://doi.org/10.1093/brain/awu311.

Worbe, Y., Palminteri, S., Hartmann, A., Vidailhet, M., Lehéricy, S., & Pessiglione, M. (2011). Reinforcement learning and Gilles de la Tourette syndrome: Dissociation of clinical phenotypes and pharmacological treatments. *Archives of General Psychiatry*, *68*(12), 1257–1266. https://doi.org/10.1001/archgenpsychiatry.2011.137.
Wright, A., Rickards, H., & Cavanna, A. E. (2012). Impulse-control disorders in Gilles de la Tourette syndrome. *The Journal of Neuropsychiatry and Clinical Neurosciences*, *24*(1), 16–27. https://doi.org/10.1176/appi.neuropsych.10010013.
Wylie, S. A., Claassen, D. O., Kanoff, K. E., van Wouwe, N. C., & van den Wildenberg, W. P. M. (2016). Stopping manual and vocal actions in Tourette's syndrome. *The Journal of Neuropsychiatry and Clinical Neurosciences*, appineuropsych15110387. https://doi.org/10.1176/appi.neuropsych.15110387.
Yamamuro, K., Ota, T., Iida, J., Nakanishi, Y., Uratani, M., Matsuura, H., et al. (2015). Prefrontal dysfunction in pediatric Tourette's disorder as measured by near-infrared spectroscopy. *BMC Psychiatry*, *15*, 102. https://doi.org/10.1186/s12888-015-0472-3.
Yaniv, A., Benaroya-Milshtein, N., Steinberg, T., Ruhrrman, D., Apter, A., & Lavidor, M. (2017). Specific executive control impairments in Tourette syndrome: The role of response inhibition. *Research in Developmental Disabilities*, *61*, 1–10. https://doi.org/10.1016/j.ridd.2016.12.007.
Zapparoli, L., Porta, M., Gandola, M., Invernizzi, P., Colajanni, V., Servello, D., et al. (2016). A functional magnetic resonance imaging investigation of motor control in Gilles de la Tourette syndrome during imagined and executed movements. *The European Journal of Neuroscience*, *43*(4), 494–508. https://doi.org/10.1111/ejn.13130.
Zebardast, N., Crowley, M. J., Bloch, M. H., Mayes, L. C., Wyk, B. V., Leckman, J. F., et al. (2013). Brain mechanisms for prepulse inhibition in adults with Tourette syndrome: Initial findings. *Psychiatry Research*, *214*(1), 33–41. https://doi.org/10.1016/j.pscychresns.2013.05.009.

SECTION III

The structure and function of neural circuits: Data from neuroimaging and electrophysiology

CHAPTER TEN

Somatomotor cortical mapping in Tourette syndrome using neuro-navigated transcranial magnetic stimulation

Caitlin M. Smith[a,b], Hilmar P. Sigurdsson[c], Katherine Dyke[a], Rosa Sanchez Panchuelo[d], Susan T. Francis[b,d], Georgina M. Jackson[e], and Stephen R. Jackson[a,b,e,*]

[a]School of Psychology, University of Nottingham, University Park, Nottingham, United Kingdom
[b]Beacon of Excellence in Precision Imaging, University of Nottingham, United Kingdom
[c]Brain and Movement Research Group, Clinical Ageing Research Unit, Translational and Clinical Research Institute, Newcastle University, Newcastle upon Tyne, United Kingdom
[d]Sir Peter Mansfield Imaging Centre, School of Physics and Astronomy, University of Nottingham, University Park, Nottingham, United Kingdom
[e]Institute of Mental Health, School of Medicine, University of Nottingham, University Park, Nottingham, United Kingdom
[*]Corresponding author: e-mail address: stephen.jackson@nottingham.ac.uk

Contents

International Review of Movement Disorders, Volume 3
ISSN 2666-7878
https://doi.org/10.1016/bs.irmvd.2021.11.008

1. Introduction

Tourette syndrome (TS) is a neurological disorder of childhood onset characterized by the occurrence of vocal and motor tics (Leckman, 2002). Tics are repetitive, stereotyped behaviors that occur with a limited duration (Cohen, Leckman, & Bloch, 2013). Motor tics can be simple or complex in appearance, ranging from repetitive movements to coordinated action sequences. Verbal tics can consist of repetitive sounds, words or utterances, the production of inappropriate or obscene utterances, or the repetition of another's words. Tics occur in bouts, typically many times in a single day, and are the most common form of movement disorder in children. Note, tics can be temporarily suppressed, but individuals often find this very uncomfortable and vary in their ability to achieve this.

TS has been associated with: dysfunction within cortical-striatal-thalamic-cortical (CSTC) brain circuits that are implicated in the selection of movements and in habit learning (Albin & Mink, 2006); impaired operation of gamma-aminobutyric acid (GABA) inhibitory signaling within both striatum and cortical motor areas (Albin & Mink, 2006; Gilbert et al., 2004; Kalanithi et al., 2005; Lerner et al., 2012; Orth, Münchau, & Rothwell, 2008; Orth & Rothwell, 2009); and hyper-excitability of limbic and sensorimotor regions that may contribute to the occurrence of tics (Heise et al., 2010; Vaccarino, Kataoka-Sasaki, & Lennington, 2013). The role of GABA signaling in the maintenance of cortical somatomotor representations will be discussed in greater detail below.

The volitional nature of tics has been a topic of debate (Cavanna, Black, Hallett, & Voon, 2017). While many refer to tics as involuntary movements (e.g., Ganos et al., 2015), others have argued that tics are instead voluntary, and occur in response to so-called premonitory urges (Leckman, Bloch, Scahill, & King, 2006). Thus, the majority of individuals with TS report that their tics are often preceded by premonitory sensory/urge phenomena (PU) that are described as uncomfortable cognitive or bodily sensations that occur prior to the execution of a tic and are experienced as a strong urge for motor discharge (Cohen et al., 2013). Individuals who experience PU often report that: these experiences are more bothersome than their tics; that expressing their tics give them relief from, and temporarily abolishes, their PU; and that they would not exhibit tics if they did not experience PU. For this reason, it has been proposed that PU should be considered as the driving force behind the occurrence of tics, and that tics are a learnt response to the experience of PU (Cavanna et al., 2017; Leckman et al., 2006).

2. Alterations in somatosensory function in Tourette syndrome

There is mounting evidence that TS may be associated with the altered processing of somatosensory information, including altered interoception (Tinaz, Malone, Hallett, & Horovitz, 2015).

2.1 Premonitory urge phenomena in Tourette syndrome

As noted above, the majority (~90%) of individuals with TS report that their tics are often *preceded* by PU, which are of particular clinical importance because they form the core component of behavioral therapies that are currently used in the treatment of tic disorders (Cohen et al., 2013). Importantly, it has been suggested that PU arise in part from increased sensitivity to somatic stimulation (Belluscio, Jin, Watters, Lee, & Hallett, 2011; Jackson, Parkinson, Kim, Schüermann, & Eickhoff, 2011). Consistent with this proposal, individuals with TS often report heightened sensitivity to external sensory stimuli, such as irritation due to clothing fabrics, that causes distress (Belluscio et al., 2011; Cohen & Leckman, 1992). Surprisingly, given the theoretical and clinical importance of somatosensation to the pathophysiology of TS; our understanding of somatosensory abnormalities in TS and their relationship to tics, and tic suppression is currently limited.

Both cross-sectional and longitudinal data have demonstrated that reports of urge experiences increase with age and are experienced by over 80–90% of those with TS in adolescence/adulthood (Banaschewski, Woerner, & Rothenberger, 2003; Cohen & Leckman, 1992; Draper, Jackson, Morgan, & Jackson, 2016; Kwak, Dat Vuong, & Jankovic, 2003; Leckman, Walker, & Cohen, 1993; Sutherland Owens, Miguel, & Swerdlow, 2011). However, it is thought unlikely that PU are due to TS symptom progression, and may instead reflect language, cognitive and introspective development. For instance, when evaluating the global PU measure, namely, Premonitory Urge for Tic Scale (PUTS), it was evident that PUTS demonstrated lower temporal stability and internal consistency in children under 10 years old, whereas PUTS in older individuals showed stable and consistent properties (Steinberg et al., 2010; Woods, Piacentini, Himle, & Chang, 2005). Therefore, it is suggested that the reported increase in PU from the age of 10 is likely to reflect cognitive development and the enhanced awareness of urges (Banaschewski et al., 2003; Li et al., 2019). For instance, PU are associated with higher IQ scores and therefore may reflect a greater verbal and expressive ability (Reese et al., 2014; Sambrani, Jakubovski, & Müller-Vahl, 2016). As a result, the increased reporting of urges may be due to a developmental shift in introspective, cognitive and linguistic development, which are necessary to recognize and accurately describe emotional states and behaviors (Lane & Schwartz, 1987; Sutherland Owens et al., 2011). This development may allow children to recognize their urges and as a result, can allow for accurate differentiation between types of behaviors. For example, tic behaviors may become more recognized as compulsive behaviors as a result of urge awareness, rather than as involuntary behaviors (Kane, 1994). This increase of urges across ages is therefore likely to reflect cognitive development for the awareness and expression of urges and their characteristics.

An alternative perspective is, however, that rather than view tics as being *caused* by the presence of PU, we might take an opposing view that PU are in fact experienced when movements such as tics are effectively *withheld, suppressed* or *postponed.* Anecdotal evidence for this proposal comes from the very many reports from individuals with TS who state that the urge to tic builds up and becomes increasingly uncomfortable when they are successful in suppressing or withholding their tics, and empirical evidence consistent with this view comes from demonstrations that suppressing relatively automatic movements such as blinking (Berman, Horovitz, Morel, & Hallett, 2012; Botteron et al., 2019) or yawning (Brown et al., 2017) leads to a clear increase in self-reported urge experiences. Viewed from this perspective, the developmental time course for reporting PU may not reflect a developmental shift in introspective, cognitive and linguistic development, so much as a developmental increase in the ability to voluntarily control our actions, including the increased ability to suppress, withhold or delay action.

As noted above, the precise nature and causes of PU remain largely unknown and subject to debate; however, there is evidence to indicate that PU overlap with other forms of urge phenomena. For instance, the insula and cingulate cortex, which have been heavily associated with TS and PU experiences, align with the brain areas involved in the experience of other forms of urge such as the urge to swallow, the urge to yawn, or the urge to micturate, in typically developing individuals (Jackson et al., 2011; O'Neill, Piacentini, & Peterson, 2019). Furthermore, functional imaging studies have demonstrated that the insular and cingulate cortices exhibit significant increases in activity directly prior to the execution of tics, and may therefore reflect the occurrence of PU, as different brain areas (e.g., sensorimotor cortex) increase their activity at tic onset; suggesting separate mechanisms behind PU and tic generation (Bohlhalter et al., 2006; Neuner et al., 2014). Separation in function between PU and tic generation is also seen *within* the insula and cingulate cortices, with anatomical data demonstrating that motor tic severity and PU scores are associated with different anatomical regions within both of these structures (Jackson et al., 2020; Jackson, Sigurdsson, Dyke, Condon, & Jackson, 2021).

2.2 Alterations interoceptive awareness in Tourette syndrome

A further example of altered somatosensory processing experienced by individuals with TS is that of altered interoceptive awareness or interoception. Interoception reflects an individual's awareness of their internal bodily processes, and is reportedly increased in TS, as evidenced by self-reported measures (Eddy, Rickards, & Cavanna, 2014; Ganos et al., 2015; Rae, Larsson, Garfinkel, & Critchley, 2019). This may indicate those with TS are more aware of their internal bodily sensations and thus have a greater capacity to experience PU as a consequence. Consistent with this proposal, interoceptive awareness scores have been demonstrated to predict PU severity (Ganos et al., 2015; Rae et al., 2019—although see Eddy et al., 2014; Pile, Lau, Topor, Hedderly, & Robinson, 2018 for alternative findings). However, although self-reported measures suggest heightened interoceptive awareness in TS, objective interoceptive detection tasks, such as heartbeat tracking, have demonstrated, paradoxically, lower detection accuracy to interoceptive signals compared to controls (Ganos et al., 2015; Rae et al., 2019). This discrepancy, or interoceptive prediction error, might suggest that there is an increased higher-order sensitivity to bodily signals, but noisier detection of peripheral afferent signals. In other words, those with TS may be both oversensitive and yet less precise to internal bodily signals (Rae et al., 2019). Alternatively, reduced interoceptive detection ability may reflect a compensatory interoception downregulation in order to suppress PU, and may be an alternative explanation for why interoceptive awareness is related to PU (Ganos et al., 2015). Further investigation into the role of interoceptive prediction error in TS and its relationship with PU is essential. However, due to the mismatch between self-reported and objective measures of interoception, it is most likely that this discrepancy arises from abnormal central processing mechanisms.

2.3 Alterations exteroceptive awareness in Tourette syndrome

It has been proposed that individuals with TS may also be hypersensitive to external somatosensory stimuli, i.e., exteroception (e.g., Belluscio et al., 2011). For instance, hypersensitivity to sensory stimuli, such as clothing materials, lights, and sounds, has been (self) reported in ~70% of individuals with TS and can cause considerable discomfort (Belluscio et al., 2011; Cohen et al., 2013; Cohen & Leckman, 1992). Moreover, individuals with TS are more likely to report over-inclusion/hyper-awareness of, and feeling "flooded" and easily distracted by, external stimuli, all of which are enhanced under stress and anxiety (Sutherland Owens et al., 2011). This is especially the case with faint, nonsalient and repetitive stimuli which are perceived as more disruptive and bothersome than intense stimuli in individuals with TS compared to typically developing individuals (Belluscio et al., 2011). However, once again, objective quantitative somatosensory testing has demonstrated that afferent detection thresholds to thermal, tactile and pain stimuli were not significantly different between individuals with TS and matched controls (Schunke et al., 2016). Thus, any hypersensitivity to external somatosensory stimuli is unlikely to reflect an underlying deficit in basic somatosensory processing mechanisms. Instead, somatosensory abnormalities in TS, whether interoception or exteroception, may reflect abnormal higher-order sensorimotor processing mechanisms. Nonetheless, studies involving quantitative testing of sensory function in TS are very few in number and considerable further investigation is required.

3. GABA signaling may determine the neurophysiological response to somatic stimulation

3.1 Physiological mechanisms implicated in cortical mapping

The mechanisms that give rise to the stability and dynamic properties of somatomotor representations are currently unclear. However, one key mechanism implicated in the mediation of sensorimotor representations is the inhibitory neurotransmitter γ-aminobutyric acid (GABA) and its role in implementing surround inhibition (Beck & Hallett, 2011; Jacobs & Donoghue, 1991; Jones, 1993; Kolasinski et al., 2017; Sigurdsson, Jackson, Kim, Dyke, & Jackson, 2020; Sigurdsson, Molloy, & Jackson, 2020).

GABA is the most abundant inhibitory neurotransmitter in the cortex, acting through GABAA and GABAB receptors. It is synthesized from glutamine via glutamate and exists as a neurotransmitter and in metabolic pools within cell bodies (Rae, 2014; Rae, Nasrallah, Griffin, & Balcar, 2009). Surround inhibition mediated through GABA signaling, specifically GABAA, is a physiological mechanism that is thought to focus neuronal activity (Beck & Hallett, 2011; Sillito, 1975; Sohn & Hallett, 2004), improve the discrimination of somatosensory events, and aid the efficient selection of movements (Beck & Hallett, 2011). Consistent with this view, it has been shown that injections of GABAA-antagonist, bicuculline, within the motor cortex results in the fusion of motor representations between adjacent muscles and weakens sensory map plasticity and receptive field tuning (Foeller, Celikel, & Feldman, 2005; Jacobs & Donoghue, 1991; Schneider, Devanne, Lavoie, & Capaday, 2002). Furthermore, increased concentration of GABA, measured within the somatomotor cortex using magnetic resonance spectroscopy (MRS), is associated with more selective tuning functions within sensorimotor cortex (SMC) and enhanced somatosensory perception (Kolasinski et al., 2017). However, it is important to note that while human studies illustrate an association between GABA levels measured using MRS (MRS-GABA) and neural representations of cortical activity (e.g., neural tuning function), it should be recognized that MRS-GABA is highly unlikely to reflect phasic GABA neurotransmitter signaling, and in particular the synaptic inhibition that has been linked to surround inhibition. Instead, it is likely that MRS-GABA may reflect tonic levels of extra-synaptic GABA that contribute to the longer-term inhibitory tone of a brain region (Dyke et al., 2017; Stagg et al., 2011; Tremblay et al., 2013).

3.2 Altered somatomotor maps in focal hand dystonia

Focal hand dystonia (FHD) is a neurological condition characterized by the inappropriate co-contraction of muscles resulting in uncoordinated movements and abnormal hand postures (Meunier, Lehéricy, Garnero, & Vidailhet, 2003; Schabrun, Stinear, Byblow, & Ridding, 2009). FHD is associated with altered cortical somatomotor representations, as determined by motor mapping studies conducted using transcranial magnetic stimulation (TMS) techniques to measure the size and extent of individual cortical muscle representations. These studies have demonstrated that the TMS motor (muscle) maps obtained from FHD patients differ from those of healthy controls. Specifically, the maps of FHD patients exhibit larger areas and reduced inter-map distances between hand muscle representations (Byrnes et al., 1998; Schabrun et al., 2009). Other evidence also indicates altered somatosensory and motor representations in individuals with FHD (Bara-Jimenez, Catalan, Hallett, & Gerloff, 1998; Butterworth et al., 2003; Byl, McKenzie, & Nagarajan, 2000; Elbert et al., 1998; Elbert, Pantev, Wienbruch, Rockstroh, & Taub, 1995; Meunier et al., 2003; Sohn & Hallett, 2004). Specifically, neuroimaging studies of individuals with FHD have also consistently demonstrated abnormalities in the cortical sensory system in FHD (e.g., Butterworth et al., 2003; Elbert et al., 1995). One recent advance has been to use ultra-high field (7 Tesla) functional MRI techniques to quantify the somatomotor cortical maps in FHD and to associate alterations in cortical somatomotor representation with behavioral measures of somatosensation including: temporal discrimination; amplitude thresholding; and spatial acuity assessment (Packenham et al., 2019).

Packenham et al. (2019) developed a traveling-wave approach to map the brain activations associated with the movements of individual digits (i.e., digits D1-D5) of each hand, or somatosensory stimulation of individual digits, in both neurological healthy individuals and patients with FHD. Individuals executed movements of digits 1–5 in sequence, or were stimulated on digits 1–5 in sequence, and the activations associated with each digit were modeled. This approach allows cortical somatosensory and/or somatomotor maps to be precisely quantified in individual patients using noninvasive brain imaging and associated with behavioral alterations in motor or somatosensory function.

A second recent advance has been the use of positron emission tomography (PET) imaging to investigate alterations in GABA in patients with FHD (Gallea et al., 2018). Gallea and colleagues measured the density of GABA-A receptors in a PET imaging study using [11C] Flumazenil—a benzodiazepine antagonist that binds to GABA-A receptors—in a group of individuals with FHD affecting their right hand and a group of neurologically healthy controls. They demonstrated that GABA binding, as indexed by the binding potential of Flumazenil, was significantly reduced in the right cerebellum and the left sensorimotor cortex in individuals with FHD. Additional evidence, has also demonstrated reduced GABAA-mediated short interval intracortical inhibition (SICI), which has been proposed to contribute to surround inhibition (Beck & Hallett, 2011), in individuals with FHD (Beck et al., 2008; Stinear & Byblow, 2004).

Taken together the above evidence suggests that cortical somatomotor maps are significantly altered in individuals with FHD, that these alterations are associated with behavioral impairments in hand sensorimotor function, and that alterations in both sensorimotor maps and behavioral measures of hand function, are likely linked to dysfunction in GABA signaling. Furthermore, we suggest that many of the abnormalities in GABA signaling observed in FHD are also reported for individuals with TS as outlined below.

4. Evidence for alterations in GABA signaling in Tourette syndrome

There are multiple lines of evidence to indicate that TS is associated with altered GABAergic function (Clarke, Lee, & Eapen, 2012; Ramamoorthi & Lin, 2011). First, postmortem investigations of the brains of a small number of individuals with TS has demonstrated a substantial decrease (~50%) in the number of GABAergic interneurons found within the striatum (Kalanithi et al., 2005), and a later study, using positron emission tomography (PET) reported converging evidence for this finding, by demonstrating that there are widespread alterations in GABAA receptor binding in the brains of individuals with TS, including decreases in GABAA receptor binding within the ventral striatum, globus pallidus and the thalamus (Lerner et al., 2012). By contrast, studies using MRS to measure GABA concentration differences in individuals with TS compared to typically developing controls have yielded mixed results. Some have reported that MRS-GABA measured within the sensorimotor cortex (SMC) of young children with TS is reduced compared to typically developing controls and also correlated with motor tic severity (Puts et al., 2015). However, a recent study by the same research group, using a larger sample, failed to replicate these findings and concluded that MRS-GABA concentrations in the primary sensorimotor cortex, SMA, and insula were in fact comparable between individuals with TS and controls (He et al., 2021). Nonetheless, they did report that reduced MRS-GABA levels within the SMA were associated with more severe and more frequent PU (He et al., 2021). Furthermore, an ultra-high-field (7 Tesla) MRS study had previously reported that MRS-GABA levels within the SMA region were increased in individuals with TS relative to matched controls, and that MRS-GABA levels within the SMA in TS were associated with alterations in both motor gain function (i.e., the increase in motor excitability ahead of a voluntary movement) and motor tic severity (Draper et al., 2014).

Second, studies that have investigated the effects of striatal disinhibition in rodents and nonhuman primates, have consistently demonstrated that the localized administration of a microinjection of GABAA-antagonist can produce tic-like movements (Bronfeld, Israelashvili, & Bar-Gad, 2013; Bronfeld, Yael, Belelovsky, & Bar-Gad, 2013; Matsumura, Sawaguchi, Oishi, Ueki, & Kubota, 1991; McCairn, Iriki, & Isoda, 2013; McCairn & Isoda, 2013). Moreover, the site of these microinjections determines the nature of behavior/tic exhibited. For instance, injection of a GABAA antagonist into focal somatotopic areas of the striatum is demonstrated to produce tic-like movements in specific body parts within minutes, lasting for up to 2h (Bronfeld, Yael, et al., 2013). Importantly, investigations into striatal disinhibition using GABAA antagonists have demonstrated the involvement of other neural regions for the generation of tic-like behaviors. Specifically, after microinjections of striatal GABAA antagonist were administered to nonhuman primates, electrophysiological recordings—local field potential (LFP) spikes—were recorded at multiple sites including the striatum, pallidum, cerebellum, and primary motor cortex (McCairn et al., 2013). LFP spikes within the striatum were observed prior to tic-like behaviors but also in the intervals between tics (these it was suggested these might be linked to the experience of PU, McCairn & Isoda, 2013). Conversely, LFP spikes in the cerebellum and motor cortex were always and only present immediately prior to tic execution, suggesting these sites may gate the occurrence of tic execution (McCairn et al., 2013).

Finally, TMS studies of physiological inhibition within the primary motor cortex have demonstrated reduced $GABA_A$ receptor dependent physiological inhibition in TS, for both short-interval intra-hemispheric inhibition (SICI) (Orth, 2009) and interhemispheric inhibition (IHI) protocols (Bäumer et al., 2010). While SICI may not necessarily be reduced in children with TS (Moll et al., 1999), a reduction in SICI has been consistently demonstrated in adults with TS (Heise et al., 2010; Orth, 2009; Orth et al., 2008; Orth, Amann, Robertson, & Rothwell, 2005; Ziemann, Paulus, & Rothenberg, 1997). Moreover, SICI measurements have been shown to be correlated with both motor tics and ADHD symptoms, with lower levels of SICI associated with greater tic severity and more severe ADHD symptoms (Gilbert et al., 2004; Gilbert, Sallee, Zhang, Lipps, & Wassermann, 2005).

4.1 Measuring cortical somatomotor representations using neuro-navigated TMS

We recently commenced investigating how cortical somatomotor representations might be altered in TS using a combination of ultra-high-field fMRI neuroimaging techniques; physiological TMS; and neuro-navigated TMS mapping of cortical motor representations (Sigurdsson, Jackson, et al., 2020). For the TMS mapping study of cortical motor representations (as illustrated in Fig. 1), data were collected from 16 individuals with TS (5 females, aged 20.4 ± 7.1 years, range: 13.7–35.1 years) and 14 typically developing controls (4 females, aged 17.4 ± 4.0, years range: 12.2–24.8). Single pulses of super-threshold TMS were delivered at multiple random locations within a grid centered over the left sensorimotor cortex of each individual, and care was taken to sample from within the entire grid (Fig. 1). During stimulation of each site, electromyographic (EMG) recordings were collected simultaneously from five different targeted muscles contralateral to the stimulation: the FDI (hand); deltoid (shoulder); orbicularis oris (lips); orbicularis oculi (eye); and masseter (jaw) and the amplitude of the TMS-induced motor evoked electrical potential (MEP) was calculated to produce a spatial map of TMS-induced muscle activity for each muscle (Figs. 1 and 2A). Note, the orbicularis oris, orbicularis oculi, and masseter muscles were selected for measurement, as these facial muscles are likely to be involved during tic production in many individuals with TS and are also often associated with the experience of PU (Ganos et al., 2012; Kwak et al., 2003; Schunke et al., 2016). By contrast, the deltoid and FDI muscles are less likely to be associated with tics and PU. The FDI muscle was also selected as it is the muscle most often used to measure SICI. These maps were then co-registered with an anatomical MRI for each individual and the maps were translated from scalp coordinated into brain coordinates (see Sigurdsson, Jackson, et al., 2020 for details).

Several parameters were calculated to describe each map. First, the *center-of-gravity (COG)* for each muscle map was calculated, where the COG is the mean position in a map, based on weighted mean intensity of each position. Second, a Euclidian distance (ED) measure was calculated in 2D space to describe the distance between the COGs each pair of muscle representations. Third, we calculated an estimate of the variability of the area of the map for each targeted muscle representation. The variability is derived from the eigenvalues of the covariance matrix and generated by fitting a 95% confidence interval ellipse to the peak activity. This

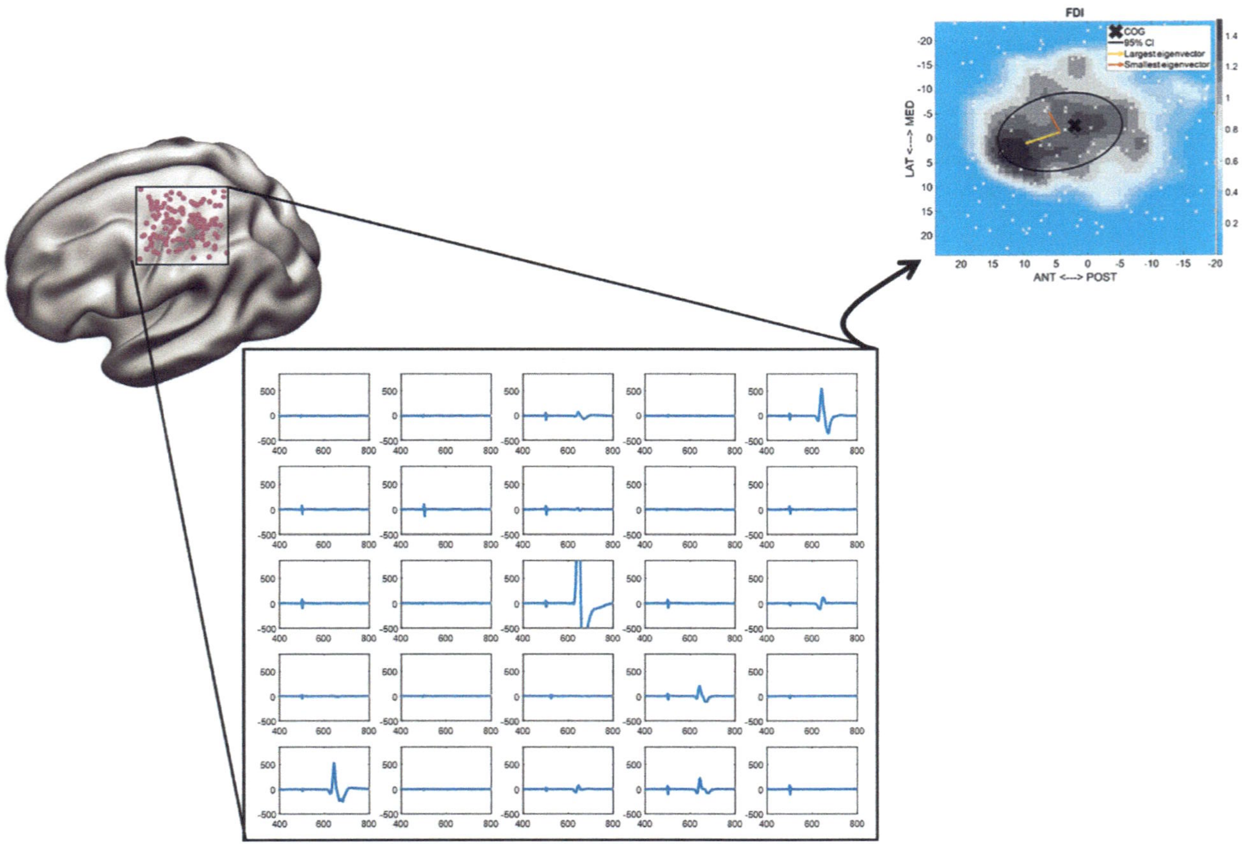

Fig. 1 Illustrates the neuro-navigated TMS approach to computing individual muscle maps. Super-threshold TMS pulses are delivered at random locations within a grid located over the sensorimotor cortex and TMS-induced electrical potentials (MEPs) are recorded simultaneously from multiple targeted muscles. From these recordings, a map of muscle response to TMS is constructed and then translated from scalp coordinates into brain coordinates through co-registration with that individual's anatomical MRI scan.

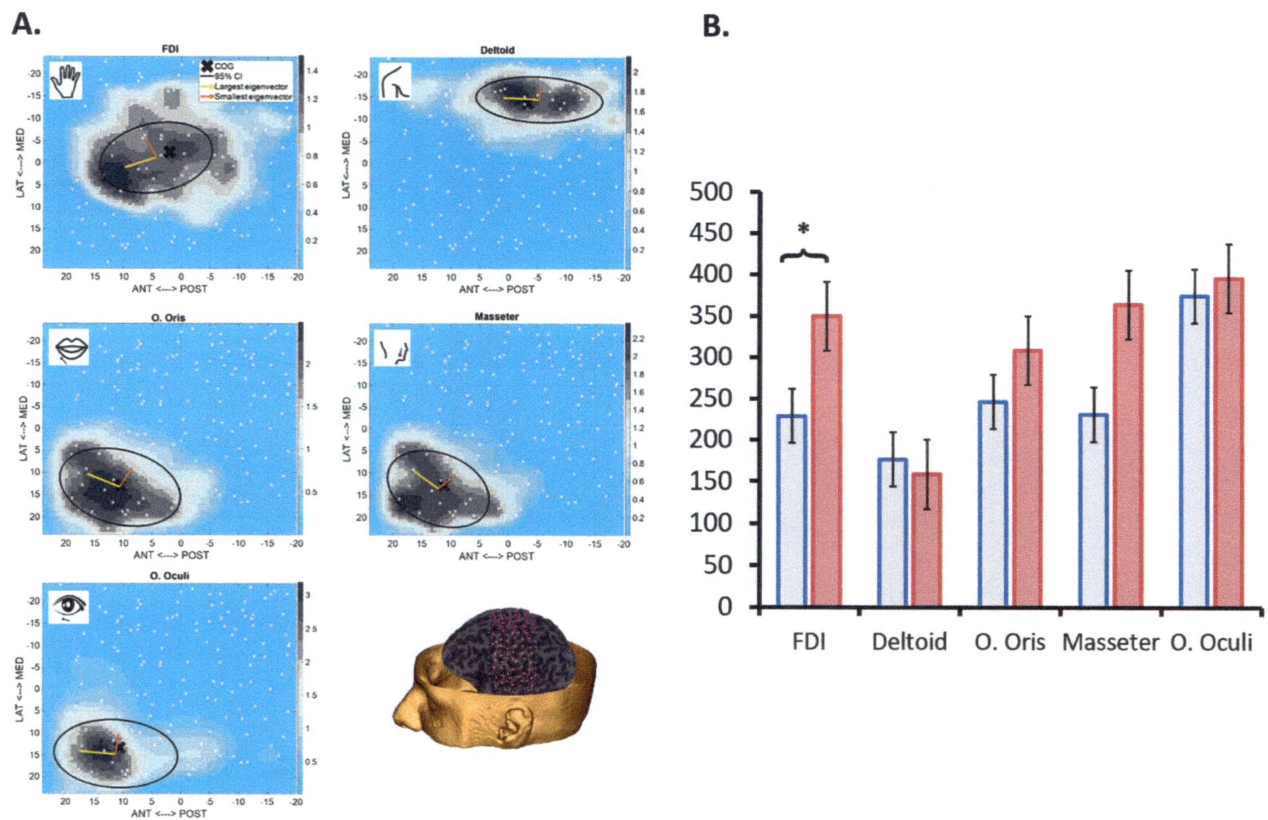

Fig. 2 (A) Illustrates representative muscle maps for each targeted muscle and the location of the 95% CI ellipse in each case. (B) Shows the results of a between-group comparison of the mean area of the 95% CI ellipse for each muscle map. The difference in area for the FDI muscle was statistically significant ($P < 0.05$, FDR-corrected).

results in an approximation of the *area* covered by the representation in a map (see Fig. 2A). Between-group statistical comparison of these map parameters were conducted to examine group differences and correlation analyses were conducted to investigate associations between physiological measures (e.g., SICI) or clinical measures of tic severity and PU, and map parameters.

4.1.1 Between group differences in the area of muscle maps

Fig. 2B shows the average area of each muscle map for the TS group and controls. Inspection of this figure indicates that the area of the muscle representation was reduced in the TS group for several muscles (e.g., FDI [hand], O. Oris [lip], and Masseter [jaw]) but not others (e.g., deltoid [shoulder] and O. oculi [eye]). In the case of the FDI muscle, this difference was statistically significant.

4.1.2 Relationship between muscle area and GABA-mediated inhibition

In addition to using neuro-navigated TMS to generate muscle maps for each individual, we used physiological TMS methods to measure the efficacy of $GABA_A$-mediated physiological inhibition (SICI) for the FDI muscle for all participants. The relationship between individual's FDI SICI measurement and the area of each individual's FDI map is illustrated in Fig. 3B. Inspection of this figure shows that magnitude of SICI inhibition (indicated by the reduction in MEP amplitude for SICI [i.e., paired-pulse trials] compared to single-pulse unconditioned trials) was significantly associated with the area of area of the FDI muscle. Specifically, increased $GABA_A$-mediated inhibition was associated with an increase in the area of the FDI muscle map. Given that the area of the FDI muscle map was significantly reduced in the TS group and the extent of the map area is positively associated with increased SICI, this finding appears consistent with previous reports of decreased $GABA_A$-mediated physiological inhibition in TS [as indexed by SICI] (Orth, 2009). Our current research is investigating whether map parameters are associated with MRS-GABA levels within cortical sensorimotor areas and to what extent any differences in map parameters observed in individuals with TS can be explained by alterations in MRS-GABA levels.

4.1.3 Association between map parameters and premonitory urges

In addition to conducting between-group statistical comparison of muscle map parameters, we also conducted correlational analyses between map

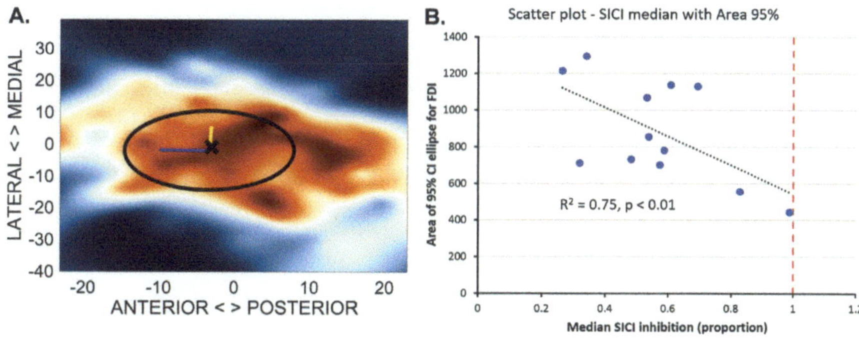

Fig. 3 (A) Representative muscle map for the FDI muscle illustrating the 95% CI ellipse (black ellipse) and the large (blue) and small (yellow) Eigenvectors from a group of 12 healthy young adults. (B) Scatter plot illustrating the association between median SICI inhibition value measured from the FDI muscle and the area of the 95% CI ellipse for the FDI muscle. The red dotted line illustrates the point in which there is no difference in MEP amplitude between conditioned (i.e., SICI) and unconditioned (i.e., baseline) trials. Thus, decreasing values that are less than 1 represent increasing levels of inhibition. *Data from Sigurdsson, H. P., Molloy, K., & Jackson, S. R. (2020). Somatomotor cortical representations are predicted by levels of short-interval intracortical inhibition. BioRxiv. https://doi.org/10.1101/2020.04.24.057331.*

parameters for the TS individuals and their clinical measures. Fig. 4 shows spider plots that illustrate the relationship between different map parameters and PU scores, as measured using PUTS scale. Inspection of this figure shows that while the 2D location of the center-of-gravity (COG) was not associated with PU severity, both the area and the distance between individual maps were significantly associated with PU severity. Specifically, the *area* of the muscle map was significantly *negatively* correlated with PU severity for the Orbicularis Oculi (eye) and Masseter (jaw) muscles, *positively* correlated with PU severity for the FDI (hand) muscle, and uncorrelated with PU severity for the Deltoid (shoulder) and Orbicularis Oris (lip) muscles. Similarly, the Euclidian distance between muscle maps was significantly associated with PU severity for the FDI–O. Oris, FDI–Masseter, and FDI–O. Oculi muscle pairs. Importantly, reduced distance between muscle pairs was associated with increased PU score. Again, no relationship was seen for the Deltoid muscle.

Taken together, these data indicate that the quantitative mapping of individual somatomotor representations is possible using both neuroimaging (fMRI) and neuro-navigated TMS methods and that this approach can be used to investigate the pathophysiology of TS, including the role played by GABA-ergic signaling in somatomotor impairment in TS.

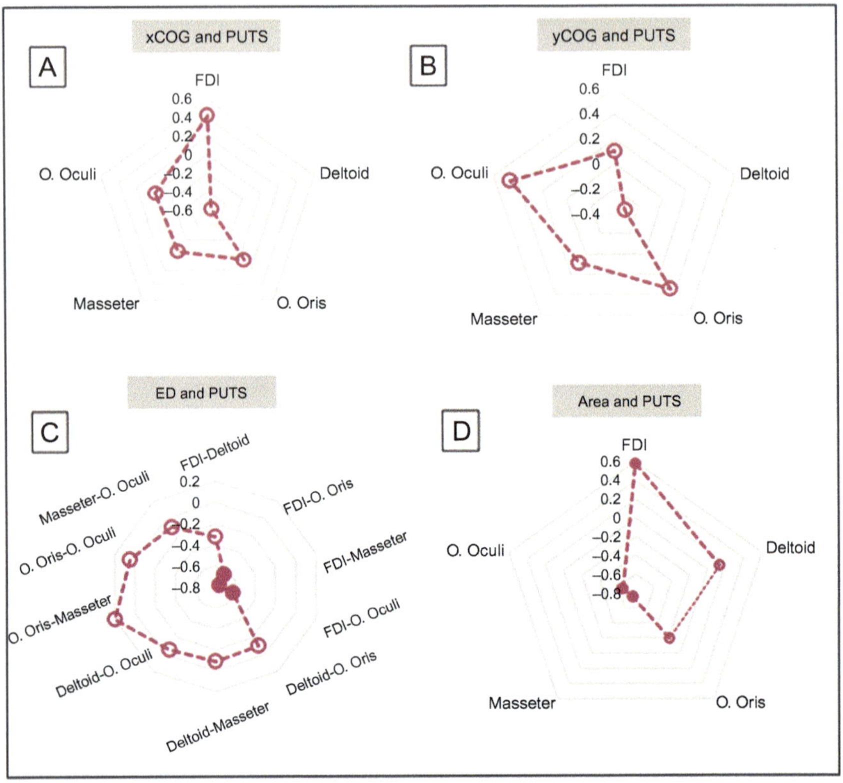

Fig. 4 Spider plots showing the statistical association between muscle map parameters such as center-of-gravity (COG), Euclidian distance (ED), and area, and PU severity as measured by the PUTS scale. (A) Association between xCOG (anterior-posterior) and PU score for each muscle. (B) Association between yCOG (lateral-medial) and PU score for each muscle. (C) Association between ED and PU score for each muscle. (D) Association between area (95% CI ellipse) and PU score for each muscle. Note that only filled circles represent statistically significant correlations.

References

Albin, R. L., & Mink, J. W. (2006). Recent advances in Tourette syndrome research. *Trends in Neurosciences*, *29*(3), 175–182. https://doi.org/10.1016/j.tins.2006.01.001.

Banaschewski, T., Woerner, W., & Rothenberger, A. (2003). Premonitory sensory phenomena and suppressibility of tics in Tourette syndrome: Developmental aspects in children and adolescents. *Developmental Medicine and Child Neurology*, *45*(10), 700–703. https://doi.org/10.1017/S0012162203001294.

Bara-Jimenez, W., Catalan, M. J., Hallett, M., & Gerloff, C. (1998). Abnormal somatosensory homunculus in dystonia of the hand. *Annals of Neurology*, *44*(5), 828–831. https://doi.org/10.1002/ana.410440520.

Bäumer, T., Thomalla, G., Kroeger, J., Jonas, M., Gerloff, C., Hummel, F. C., et al. (2010). Interhemispheric motor networks are abnormal in patients with Gilles de la Tourette syndrome. *Movement Disorders*, *25*(16), 2828–2837. https://doi.org/10.1002/mds.23418.

Beck, S., & Hallett, M. (2011). Surround inhibition in the motor system. *Experimental Brain Research, 210*(2), 165–172. https://doi.org/10.1007/s00221-011-2610-6.
Beck, S., Richardson, S. P., Shamim, E. A., Dang, N., Schubert, M., & Hallett, M. (2008). Short intracortical and surround inhibition are selectively reduced during movement initiation in focal hand dystonia. *Journal of Neuroscience, 28*(41), 10363–10369. https://doi.org/10.1523/JNEUROSCI.3564-08.2008.
Belluscio, B. A., Jin, L., Watters, V., Lee, T. H., & Hallett, M. (2011). Sensory sensitivity to external stimuli in Tourette syndrome patients. *Movement Disorders, 26*(14), 2538–2543. https://doi.org/10.1002/mds.23977.
Berman, B. D., Horovitz, S. G., Morel, B., & Hallett, M. (2012). Neural correlates of blink suppression and the buildup of a natural bodily urge. *NeuroImage, 59*(2), 1441–1450. https://doi.org/10.1016/j.neuroimage.2011.08.050.
Bohlhalter, S., Goldfine, A., Matteson, S., Garraux, G., Hanakawa, T., Kansaku, K., et al. (2006). Neural correlates of tic generation in Tourette syndrome: An event-related functional MRI study. *Brain, 129*(8), 2029–2037. https://doi.org/10.1093/brain/awl050.
Botteron, H. E., Richards, C. A., Nishino, T., Ueda, K., Acevedo, H. K., Koller, J. M., et al. (2019). The urge to blink in Tourette syndrome. *Cortex, 120*, 556–566. https://doi.org/10.1016/j.cortex.2019.07.010.
Bronfeld, M., Israelashvili, M., & Bar-Gad, I. (2013). Pharmacological animal models of Tourette syndrome. *Neuroscience & Biobehavioral Reviews, 37*(6), 1101–1119. https://doi.org/10.1016/j.neubiorev.2012.09.010.
Bronfeld, M., Yael, D., Belelovsky, K., & Bar-Gad, I. (2013). Motor tics evoked by striatal disinhibition in the rat. *Frontiers in Systems Neuroscience*, 7, 1–10. https://doi.org/10.3389/fnsys.2013.00050.
Brown, B. J., Kim, S., Saunders, H., Bachmann, C., Thompson, J., Ropar, D., et al. (2017). A neural basis for contagious yawning. *Current Biology, 27*(17), 2713–2717.e2. https://doi.org/10.1016/j.cub.2017.07.062.
Butterworth, S., Francis, S., Kelly, E., McGlone, F., Bowtell, R., & Sawle, G. V. (2003). Abnormal cortical sensory activation in dystonia: An fMRI study. *Movement Disorders, 18*(6), 673–682. https://doi.org/10.1002/mds.10416.
Byl, N. N., McKenzie, A., & Nagarajan, S. S. (2000). Differences in somatosensory hand organization in a healthy flutist and a flutist with focal hand dystonia: A case report. *Journal of Hand Therapy, 13*(4), 302–309. https://doi.org/10.1016/S0894-1130(00)80022-8.
Byrnes, M. L., Thickbroom, G. W., Wilson, S. A., Sacco, P., Shipman, J. M., Stell, R., et al. (1998). The corticomotor representation of upper limb muscles in writer's cramp and changes following botulinum toxin injection. *Brain, 121*(5), 977–988. https://doi.org/10.1093/brain/121.5.977.
Cavanna, A. E., Black, K. J., Hallett, M., & Voon, V. (2017). Neurobiology of the premonitory urge in Tourette's syndrome: Pathophysiology and treatment implications. *The Journal of Neuropsychiatry and Clinical Neurosciences, 29*(2), 95–104. https://doi.org/10.1176/appi.neuropsych.16070141.
Clarke, R. A., Lee, S., & Eapen, V. (2012). Pathogenetic model for Tourette syndrome delineates overlap with related neurodevelopmental disorders including autism. *Translational Psychiatry, 2*(9), 1–13. https://doi.org/10.1038/tp.2012.75.
Cohen, A. J., & Leckman, J. F. (1992). Sensory phenomena associated with Gilles de la Tourette's syndrome. *Journal of Clinical Psychiatry, 53*(9), 319–323.
Cohen, S. C., Leckman, J. F., & Bloch, M. H. (2013). Clinical assessment of Tourette syndrome and tic disorders. *Neuroscience & Biobehavioral Reviews, 37*(6), 997–1007. https://doi.org/10.1016/j.neubiorev.2012.11.013.
Draper, A., Jackson, G. M., Morgan, P. S., & Jackson, S. R. (2016). Premonitory urges are associated with decreased grey matter thickness within the insula and sensorimotor

cortex in young people with Tourette syndrome. *Journal of Neuropsychology, 10*(1), 143–153. https://doi.org/10.1111/jnp.12089.

Draper, A., Stephenson, M. C., Jackson, G. M., Pépés, S., Morgan, P. S., Morris, P. G., et al. (2014). Increased GABA contributes to enhanced control over motor excitability in Tourette syndrome. *Current Biology, 24*(19), 2343–2347. https://doi.org/10.1016/j.cub.2014.08.038.

Dyke, K., Pépés, S. E., Chen, C., Kim, S., Sigurdsson, H. P., Draper, A., et al. (2017). Comparing GABA-dependent physiological measures of inhibition with proton magnetic resonance spectroscopy measurement of GABA using ultra-high-field MRI. *NeuroImage, 152*, 360–370. https://doi.org/10.1016/j.neuroimage.2017.03.011.

Eddy, C. M., Rickards, H. E., & Cavanna, A. E. (2014). Physiological awareness is negatively related to inhibitory functioning in Tourette syndrome. *Behavior Modification, 38*(2), 319–335. https://doi.org/10.1177/0145445513504431.

Elbert, T., Candia, V., Altenmüller, E., Rau, H., Sterr, A., Rockstroh, B., et al. (1998). Alteration of digital representations in somatosensory cortex in focal hand dystonia. *Neuroreport, 9*(16), 3571–3575. https://doi.org/10.1097/00001756-199811160-00006.

Elbert, T., Pantev, C., Wienbruch, C., Rockstroh, B., & Taub, E. (1995). Increased cortical representation of the fingers of the left hand in string players. *Science, 270*(5234), 305–307. https://doi.org/10.1126/science.270.5234.305.

Foeller, E., Celikel, T., & Feldman, D. E. (2005). Inhibitory sharpening of receptive fields contributes to whisker map plasticity in rat somatosensory cortex. *Journal of Neurophysiology, 94*(6), 4387–4400. https://doi.org/10.1152/jn.00553.2005.

Gallea, C., Herath, P., Voon, V., Lerner, A., Ostuni, J., Saad, Z., et al. (2018). Loss of inhibition in sensorimotor networks in focal hand dystonia. *NeuroImage: Clinical, 17*, 90–97. https://doi.org/10.1016/j.nicl.2017.10.011.

Ganos, C., Garrido, A., Navalpotro-Gómez, I., Ricciardi, L., Martino, D., Edwards, M. J., et al. (2015). Premonitory urge to tic in tourette's is associated with interoceptive awareness. *Movement Disorders, 30*(9), 1198–1202. https://doi.org/10.1002/mds.26228.

Ganos, C., Kahl, U., Schunke, O., Kühn, S., Haggard, P., Gerloff, C., et al. (2012). Are premonitory urges a prerequisite of tic inhibition in Gilles de la Tourette syndrome? *Journal of Neurology, Neurosurgery & Psychiatry, 83*(10), 975–978. https://doi.org/10.1136/jnnp-2012-303033.

Gilbert, D. L., Bansal, A. S., Sethuraman, G., Sallee, F. R., Zhang, J., Lipps, T., et al. (2004). Association of cortical disinhibition with tic, ADHD, and OCD severity in Tourette syndrome. *Movement Disorders, 19*(4), 416–425. https://doi.org/10.1002/mds.20044.

Gilbert, D. L., Sallee, F. R., Zhang, J., Lipps, T. D., & Wassermann, E. M. (2005). Transcranial magnetic stimulation-evoked cortical inhibition: A consistent marker of attention-deficit/hyperactivity disorder scores in Tourette syndrome. *Biological Psychiatry, 57*(12), 1597–1600. https://doi.org/10.1016/j.biopsych.2005.02.022.

He, J. L., Mikkelsen, M., Huddleston, D. A., Crocetti, D., Cecil, K. M., Singer, H. S., et al. (2021). Frequency and intensity of premonitory urges-to-tic in Tourette syndrome is associated with SMA GABA+ levels. *PsyArXiv*. https://doi.org/10.31234/osf.io/j8uwc.

Heise, K. F., Steven, B., Liuzzi, G., Thomalla, G., Jonas, M., Muller-Vahl, K., et al. (2010). Altered modulation of intracortical excitability during movement preparation in Gilles de la Tourette syndrome. *Brain, 133*, 580–590. https://doi.org/10.1093/brain/awp299.

Jackson, S. R., Loayza, J., Crighton, M., Sigurdsson, H. P., Dyke, K., & Jackson, G. M. (2020). The role of the insula in the generation of motor tics and the experience of the premonitory urge-to-tic in Tourette syndrome. *Cortex, 126*, 119–133. https://doi.org/10.1016/j.cortex.2019.12.021.

Jackson, S. R., Parkinson, A., Kim, S. Y., Schüermann, M., & Eickhoff, S. B. (2011). On the functional anatomy of the urge-for-action. *Cognitive Neuroscience, 2*(3–4), 227–243. https://doi.org/10.1080/17588928.2011.604717.

Jackson, S. R., Sigurdsson, H. P., Dyke, K., Condon, M., & Jackson, G. M. (2021). The role of the cingulate cortex in the generation of motor tics and the experience of the premonitory urge-to-tic in Tourette syndrome. *Journal of Neuropsychology*, *15*(3), 340–362. https://doi.org/10.1111/jnp.12242.

Jacobs, K., & Donoghue, J. (1991). Reshaping the cortical motor map by unmasking latent intracortical connections. *Science*, *251*(4996), 944–947. https://doi.org/10.1126/science.2000496.

Jones, E. G. (1993). Gabaergic neurons and their role in cortical plasticity in primates. *Cerebral Cortex*, *3*(5), 361–372. https://doi.org/10.1093/cercor/3.5.361-a.

Kalanithi, P. S. A., Zheng, W., Kataoka, Y., DiFiglia, M., Grantz, H., Saper, C. B., et al. (2005). Altered parvabumin-positive neuron distribution in basal ganglia of individuals with Tourette syndrome. *PNAS*, *102*(37), 13307–13312.

Kane, M. J. (1994). Premonitory urges as "attentional tics" in Tourette's syndrome. *Journal of the American Academy of Child & Adolescent Psychiatry*, *33*(6), 805–808.

Kolasinski, J., Logan, J. P., Hinson, E. L., Manners, D., Divanbeighi Zand, A. P., Makin, T. R., et al. (2017). A mechanistic link from GABA to cortical architecture and perception. *Current Biology*, *27*(11), 1685–1691. https://doi.org/10.1016/j.cub.2017.04.055.

Kwak, C., Dat Vuong, K., & Jankovic, J. (2003). Premonitory sensory phenomenon in Tourette's syndrome. *Movement Disorders*, *18*(12), 1530–1533. https://doi.org/10.1002/mds.10618.

Lane, R. D., & Schwartz, G. E. (1987). Levels of emotional awareness: A cognitive-developmental theory and its application to psychopathology. *American Journal of Psychiatry*, *144*(2), 133–143. https://doi.org/10.1176/ajp.144.2.133.

Leckman, J. F. (2002). Tourette's syndrome. *The Lancet*, *360*(9345), 1577–1586. https://doi.org/10.1016/S0140-6736(02)11526-1.

Leckman, J. F., Bloch, M. H., Scahill, L., & King, R. A. (2006). Tourette Syndrome: The self under siege. *Journal of Child Neurology*, *21*(8), 642–649. https://doi.org/10.1177/08830738060210081001.

Leckman, J. F., Walker, D. E., & Cohen, D. J. (1993). Premonitory urges in Tourette's syndrome. *American Journal of Psychiatry*, *150*(1), 98–102. https://doi.org/10.1176/ajp.150.1.98.

Lerner, A., Bagic, A., Simmons, J. M., Mari, Z., Bonne, O., Xu, B., et al. (2012). Widespread abnormality of the gamma-aminobutyric acid-ergic system in Tourette syndrome. *Brain*, *135*(Pt. 6), 1926–1936. https://doi.org/10.1093/brain/aws104.

Li, Y., Wang, F., Liu, J., Wen, F., Yan, C., Zhang, J., et al. (2019). The correlation between the severity of premonitory urges and tic symptoms: A meta-analysis. *Journal of Child and Adolescent Psychopharmacology*, *29*(9), 652–658. https://doi.org/10.1089/cap.2019.0048.

Matsumura, M., Sawaguchi, T., Oishi, T., Ueki, K., & Kubota, K. (1991). Behavioral deficits induced by local injection of bicuculline and muscimol into the primate motor and premotor cortex. *Journal of Neurophysiology*, *65*(6), 1542–1553. https://doi.org/10.1152/jn.1991.65.6.1542.

McCairn, K. W., Iriki, A., & Isoda, M. (2013). Global dysrhythmia of cerebro-basal ganglia-cerebellar networks underlies motor tics following striatal disinhibition. *Journal of Neuroscience*, *33*(2), 697–708. https://doi.org/10.1523/JNEUROSCI.4018-12.2013.

McCairn, K. W., & Isoda, M. (2013). Pharmacological animal models of tic disorders. *International Review of Neurobiology*, *112*, 179–209. https://doi.org/10.1016/B978-0-12-411546-0.00007-X.

Meunier, S., Lehéricy, S., Garnero, L., & Vidailhet, M. (2003). Dystonia: Lessons from brain mapping. *The Neuroscientist*, *9*(1), 76–81. https://doi.org/10.1177/1073858402239593.

Moll, G. H., Wischer, S., Heinrich, H., Tergau, F., Paulus, W., & Rothenberger, A. (1999). Deficient motor control in children with tic disorder: Evidence from transcranial magnetic stimulation. *Neuroscience Letters*, *272*(1), 37–40. https://doi.org/10.1016/S0304-3940(99)00575-3.

Neuner, I., Werner, C. J., Arrubla, J., Stöcker, T., Ehlen, C., Wegener, H. P., et al. (2014). Imaging the where and when of tic generation and resting state networks in adult Tourette patients. *Frontiers in Human Neuroscience*, *8*, 1–16. https://doi.org/10.3389/fnhum.2014.00362.

O'Neill, J., Piacentini, J. C., & Peterson, B. S. (2019). Cingulate role in Tourette syndrome. In B. A. Vogt (Ed.), *Handbook of clinical neurology* (pp. 165–221). Elsevier. https://doi.org/10.1016/B978-0-444-64196-0.00011-X.

Orth, M. (2009). Transcranial magnetic stimulation in Gilles de la Tourette syndrome. *Journal of Psychosomatic Research*, *67*(6), 591–598. https://doi.org/10.1016/j.jpsychores.2009.07.014.

Orth, M., Amann, B., Robertson, M. M., & Rothwell, J. C. (2005). Excitability of motor cortex inhibitory circuits in Tourette syndrome before and after single dose nicotine. *Brain*, *128*(Pt. 6), 1292–1300. https://doi.org/10.1093/brain/awh473.

Orth, M., Münchau, A., & Rothwell, J. C. (2008). Corticospinal system excitability at rest is associated with tic severity in Tourette syndrome. *Biological Psychiatry*, *64*(3), 248–251. https://doi.org/10.1016/j.biopsych.2007.12.009.

Orth, M., & Rothwell, J. C. (2009). Motor cortex excitability and comorbidity in Gilles de la Tourette syndrome. *Journal of Neurology, Neurosurgery and Psychiatry*, *80*(1), 29–34. https://doi.org/10.1136/jnnp.2008.149484.

Packenham, D., Asghar, M., Glover, P., O'Neill, G. C., Sengupta, A., Schluppeck, D., et al. (2019). Assessing somatotopic and mototopic organisation in focal hand dystonia using high-resolution 7T fMRI [abstract]. *Proceedings of the International Society for Magnetic Resonance in Medicine*, *27*, 0361 (https://index.mirasmart.com/ISMRM2019/PDFfiles/0361.html).

Pile, V., Lau, J. Y. F., Topor, M., Hedderly, T., & Robinson, S. (2018). Interoceptive accuracy in youth with tic disorders: Exploring links with premonitory urge, anxiety and quality of life. *Journal of Autism and Developmental Disorders*, *48*(10), 3474–3482. https://doi.org/10.1007/s10803-018-3608-8.

Puts, N. A. J., Harris, A. D., Crocetti, D., Nettles, C., Singer, H. S., Tommerdahl, M., et al. (2015). Reduced GABAergic inhibition and abnormal sensory symptoms in children with Tourette syndrome. *Journal of Neurophysiology*, *114*(2), 808–817. https://doi.org/10.1152/jn.00060.2015.

Rae, C. D. (2014). A guide to the metabolic pathways and function of metabolites observed in human brain 1H magnetic resonance spectra. *Neurochemical Research*, *39*(1), 1–36. https://doi.org/10.1007/s11064-013-1199-5.

Rae, C. L., Larsson, D. E. O., Garfinkel, S. N., & Critchley, H. D. (2019). Dimensions of interoception predict premonitory urges and tic severity in Tourette syndrome. *Psychiatry Research*, *271*, 469–475. https://doi.org/10.1016/j.psychres.2018.12.036.

Rae, C. D., Nasrallah, F. A., Griffin, J. L., & Balcar, V. J. (2009). Now I know my ABC. A systems neurochemistry and functional metabolomic approach to understanding the GABAergic system. *Journal of Neurochemistry*, *109*(Suppl. 1), 109–116. https://doi.org/10.1111/j.1471-4159.2009.05803.x.

Ramamoorthi, K., & Lin, Y. (2011). The contribution of GABAergic dysfunction to neurodevelopmental disorders. *Trends in Molecular Medicine*, *17*(8), 452–462. https://doi.org/10.1016/j.molmed.2011.03.003.

Reese, H. E., Scahill, L., Peterson, A. L., Crowe, K., Woods, D. W., Piacentini, J., et al. (2014). The premonitory urge to tic: Measurement, characteristics, and correlates in older adolescents and adults. *Behavior Therapy*, *45*(2), 177–186. https://doi.org/10.1016/j.beth.2013.09.002.

Sambrani, T., Jakubovski, E., & Müller-Vahl, K. R. (2016). New insights into clinical characteristics of Gilles de la Tourette syndrome: Findings in 1032 patients from a single German center. *Frontiers in Neuroscience*, *10*, 1–13. https://doi.org/10.3389/fnins.2016.00415.

Schabrun, S. M., Stinear, C. M., Byblow, W. D., & Ridding, M. C. (2009). Normalizing motor cortex representations in focal hand dystonia. *Cerebral Cortex, 19*(9), 1968–1977. https://doi.org/10.1093/cercor/bhn224.

Schneider, C., Devanne, H., Lavoie, B. A., & Capaday, C. (2002). Neural mechanisms involved in the functional linking of motor cortical points. *Experimental Brain Research, 146*(1), 86–94. https://doi.org/10.1007/s00221-002-1137-2.

Schunke, O., Grashorn, W., Kahl, U., Schöttle, D., Haggard, P., Münchau, A., et al. (2016). Quantitative sensory testing in adults with Tourette syndrome. *Parkinsonism & Related Disorders, 24,* 132–136. https://doi.org/10.1016/j.parkreldis.2016.01.006.

Sigurdsson, H. P., Jackson, S. R., Kim, S., Dyke, K., & Jackson, G. M. (2020). A feasibility study for somatomotor cortical mapping in Tourette syndrome using neuronavigated transcranial magnetic stimulation. *Cortex, 129,* 175–187. https://doi.org/10.1016/j.cortex.2020.04.014.

Sigurdsson, H. P., Molloy, K., & Jackson, S. R. (2020). Somatomotor cortical representations are predicted by levels of short-interval intracortical inhibition. *BioRxiv*. https://doi.org/10.1101/2020.04.24.057331.

Sillito, A. M. (1975). The contribution of inhibitory mechanisms to the receptive field properties of neurones in the striate cortex of the cat. *Journal of Physiology, 250,* 305–329.

Sohn, Y., & Hallett, M. (2004). Surround inhibition in human motor system. *Experimental Brain Research, 158*(4), 397–404. https://doi.org/10.1007/s00221-004-1909-y.

Stagg, C. J., Bestmann, S., Constantinescu, A. O., Moreno Moreno, L., Allman, C., Mekle, R., et al. (2011). Relationship between physiological measures of excitability and levels of glutamate and GABA in the human motor cortex. *The Journal of Physiology, 589*(23), 5845–5855. https://doi.org/10.1113/jphysiol.2011.216978.

Steinberg, T., Shmuel Baruch, S., Harush, A., Dar, R., Woods, D., Piacentini, J., et al. (2010). Tic disorders and the premonitory urge. *Journal of Neural Transmission, 117*(2), 277–284. https://doi.org/10.1007/s00702-009-0353-3.

Stinear, C. M., & Byblow, W. D. (2004). Elevated threshold for intracortical inhibition in focal hand dystonia. *Movement Disorders, 19*(11), 1312–1317. https://doi.org/10.1002/mds.20160.

Sutherland Owens, A. N., Miguel, E. C., & Swerdlow, N. R. (2011). Sensory gating scales and premonitory urges in Tourette syndrome. *The Scientific World Journal, 11,* 736–741. https://doi.org/10.1100/tsw.2011.57.

Tinaz, S., Malone, P., Hallett, M., & Horovitz, S. G. (2015). Role of the right dorsal anterior insula in the urge to tic in Tourette syndrome. *Movement Disorders, 30*(9), 1190–1197. https://doi.org/10.1002/mds.26230.

Tremblay, S., Beaulé, V., Proulx, S., de Beaumont, L., Marjańska, M., Doyon, J., et al. (2013). Relationship between transcranial magnetic stimulation measures of intracortical inhibition and spectroscopy measures of GABA and glutamate + glutamine. *Journal of Neurophysiology, 109*(5), 1343–1349. https://doi.org/10.1152/jn.00704.2012.

Vaccarino, F. M., Kataoka-Sasaki, Y., & Lennington, J. B. (2013). Cellular and molecular pathology in Tourette syndrome. In D. Martino, & J. F. Leckman (Eds.), *Tourette Syndrome* (pp. 205–220). Oxford University Press. https://doi.org/10.1093/med/9780199796267.003.0010.

Woods, D. W., Piacentini, J. C., Himle, M. B., & Chang, S. (2005). Premonitory urge for tics scale (PUTS): Initial psychometric results and examination of the premonitory urge phenomenon in youths with tic disorders. *Journal of Developmental & Behavioral Pediatrics, 26*(6), 397–403. https://doi.org/10.1097/00004703-200512000-00001.

Ziemann, U., Paulus, W., & Rothenberg, A. (1997). Decreased motor inhibition in Tourette's disorder: Evidence from transcranial magnetic stimulation. *American Journal of Psychiatry, 154,* 1277–1284.

CHAPTER ELEVEN

Control of actions and tics: A psychophysiological perspective

Simon Morand-Beaulieu*, Rebecca P. Jordan, Elinor Baldwin, Julia Zhong, and Denis G. Sukhodolsky*
Child Study Center, Yale University School of Medicine, New Haven, CT, United States
*Corresponding authors: e-mail address: simon.morand-beaulieu@mcgill.ca; denis.sukhodolsky@yale.edu

Contents

1. Introduction

A distinctive feature of Tourette syndrome (TS), compared to other movement disorders, is that in most individuals, tics can be voluntarily suppressed for short periods of time. The fact that one can gain control over their tics is the foundation of an important area of research in TS. Many researchers in the field of TS have pursued efforts to better understand this mechanism, with the ultimate hope that it could lead to improved treatments. Many treatment options for TS, such as exposure and response prevention (Verdellen, Hoogduin, & Keijsers, 2007), as well as Comprehensive Behavioral Intervention for Tics (CBIT; Piacentini et al., 2010), aim to enhance this ability by teaching specific skills. However, it is also possible

International Review of Movement Disorders, Volume 3
ISSN 2666-7878
https://doi.org/10.1016/bs.irmvd.2022.01.003

that mechanisms involved in improved tic control through behavior therapy differ from those of self-initiated tic suppression. While the brain regions and circuits involved in the generation of tics should be similar across individuals, there might be more than a single way to act on these regions and circuits in order to decrease tic frequency or severity.

In the TS literature, the role played by cognitive control in tic suppression has been the subject of many studies. A recent meta-analysis of 61 cognitive control studies suggested slightly reduced performance in neurocognitive tasks of inhibitory control in individuals with TS (Morand-Beaulieu et al., 2017). As noted in a commentary by Jackson and Jackson (2020), it has been assumed for a long time that the inability to suppress tics stemmed from impaired volitional control or response inhibition mechanisms. However, this remains speculative, and it is not clear that reduced performance during a cognitive control task is associated with the mechanisms involved in tic suppression, or even more so with the mechanisms that give rise to tics in the first place. It is also important to note that ADHD, which co-occurs frequently with TS, contributes significantly to decreased performance in neurocognitive tasks of inhibitory control (Morand-Beaulieu et al., 2017; Sukhodolsky, Landeros-Weisenberger, Scahill, Leckman, & Schultz, 2010).

In parallel, and despite a general tendency for studies to show reduced behavioral performance during inhibitory control tasks, some studies have found enhanced response inhibition capacities among individuals with TS (Jung, Jackson, Parkinson, & Jackson, 2013; Mueller, Jackson, Dhalla, Datsopoulos, & Hollis, 2006). It has been suggested that this could represent an adaptive mechanism that emerges in individuals from years of constantly trying to suppress tics. In the same vein, cognitive control has been suggested as one of the putative mechanisms through which behavior therapy may reduce tic severity (Essoe, Ramsey, Singer, Grados, & McGuire, 2021). However, it is still unclear whether cognitive control actually plays a role in CBIT, given that baseline measures of cognitive control failed to predict treatment outcome in large randomized controlled trials of CBIT in children (Chang et al., 2018) and adults (Abramovitch et al., 2017) with TS. Nevertheless, a recent study found a link between a measure of inhibition/switching during the Stroop task and CBIT outcome (McGuire et al., 2021). More work is still required to better understand how cognitive control is related to meaningful tic reduction following behavioral therapy.

While response inhibition/cognitive control tasks offer an interesting insight into control abilities in TS, it is still unclear how they relate to tic suppression mechanisms. In this chapter, we will review the evidence from psychophysiological studies of response inhibition and tic suppression in order to better understand the similarities and differences in the mechanisms involved in both processes.

2. Control of actions: Inhibitory control in Tourette syndrome

2.1 Psychophysiological correlates of inhibitory control in Tourette syndrome

It is well known that response inhibition involves the prefrontal cortex. Extensive work has shown that the right inferior frontal gyrus is heavily involved in stopping behavior, and is part of a response inhibition network that also includes the pre-supplementary motor area (pre-SMA) and the subthalamic nucleus (Aron, Behrens, Smith, Frank, & Poldrack, 2007; Aron, Durston, et al., 2007; Aron, Robbins, & Poldrack, 2014). In adults with TS, greater involvement of the prefrontal cortex during voluntary inhibition in cognitive control tasks has been reported in many EEG and fMRI studies (Johannes et al., 2001; Morand-Beaulieu, O'Connor, Blanchet, & Lavoie, 2018; Morand-Beaulieu, O'Connor, Sauvé, Blanchet, & Lavoie, 2015; Rae et al., 2020; Serrien, Orth, Evans, Lees, & Brown, 2005; Thibault, O'Connor, Stip, & Lavoie, 2009).

In an early EEG study of response inhibition in TS, Johannes et al. (2001) assessed the ERP correlates of inhibitory control in 10 adults with TS. They assessed an effect known as the "NoGo Anteriorization" (NGA), which reflects the fact that the P3b amplitude has a more frontal distribution during NoGo or Stop trials than during Go trials. They found that the P3b to Stop trials was even more frontally distributed in adults with TS than in controls. A similar effect was observed in three later studies, which all used the same stimulus-response compatibility task that also included a NoGo condition (Morand-Beaulieu et al., 2018, 2015; Thibault et al., 2009). In the first of these three studies, Thibault et al. (2009) observed that the P3b to NoGo stimuli was more frontally shifted in 15 adults with TS compared with 20 controls. However, at this point there were no statistics to support this observation. A few years later, the frontal shift of the NoGo P3b was confirmed and replicated in 20 adults with TS compared with 20 controls (Morand-Beaulieu et al., 2015). A further study added 6 new participants to that TS group and compared it with a different group of controls and was also able to replicate this effect (Morand-Beaulieu et al., 2018). In the same vein, an early EEG study of TS assessed coherence during a Go/NoGo task in 9 adults with TS and found increased coherence between electrodes spanning the sensorimotor areas and the frontal cortex during the NoGo condition, relative to the Go condition (Serrien et al., 2005). However, this study had a small sample size and assessed sensor-level functional connectivity, increasing the risk of spurious connectivity due to volume conduction. A recent study aimed to circumvent these issues by assessing functional connectivity during a cognitive control task in a large sample of 60 children with TS and by using cortical-source functional connectivity algorithms (Jurgiel et al., 2021). Overall, children with TS exhibited lower spectral power in the ACC, but higher causal information flow between the ACC and other midline central and posterior regions, despite similar levels of task accuracy. Whole brain connectivity analyses highlighted the midcingulate and precuneus as fronto-parietal network hubs whose connections with several motor and sensory areas were atypical in TS. Attenuated ACC activation and greater information flow between and among nodes within the fronto-parietal attention network were significantly associated with tic impairment, faster reaction times, and worse inhibitory performance, suggesting it may be a putative neural adaption among individuals with TS.

Studies using fMRI to assess the brain correlates of cognitive control have suggested similar findings. For instance, Marsh, Zhu, Wang, Skudlarski, and Peterson (2007) examined neural activation in regulatory regions during the Stroop test in children and adults with and without TS and looked at effects of age and Stroop test performance on brain activation. Stroop test performance was associated with different areas of the brain in TS and controls. In TS, poorer performance was associated with greater activation of frontostriatal regions, whereas in controls, more activation of the right inferolateral prefrontal cortex accompanied better performance. The two groups also differed in their correlation of age with magnitude of brain activation during the task, specifically in regions associated with the default mode network: TS showed decreased activation in the ventral prefrontal and posterior cingulate cortices with increased age. These results suggest that individuals with TS show greater difficulty in actively suppressing neural activity/regulating during the more active task. Additionally, they struggle to engage mental processes that become active when freed from the demands of more challenging cognitive tasks. Furthermore, since Stroop test performance was unimpacted in TS, there may be a compensatory process occurring where greater activation of the frontostriatal systems helps individuals with TS maintain task performance.

Two further studies showed increased activation of the right prefrontal cortex in children (Jackson et al., 2011) and adults (Rae et al., 2020) with TS during tasks involving cognitive control. Jackson et al. (2011) assessed the functional and structural correlates of cognitive control during a manual task-switching paradigm in children with TS and typically developing controls. Enhanced cognitive control was found in children with TS during the task. The authors then used diffusion-weighted imaging (DWI) to see whether this enhanced cognitive control was accompanied by white matter changes in TS. Compared with typically developing controls, children with TS had reduced fractional anisotropy and increased mean diffusivity in the corpus callosum and forceps minor, which connects the lateral and medial areas of the prefrontal cortex. Further, this white matter microstructure was associated with tic severity. The authors believe changes in cortical white matter most likely reflect a functional adaptation to TS rather than a core symptom of the disorder. But it is possible that white matter differences could alternatively reflect individual variation in connectivity that might confer on some individuals with TS an increased capacity to suppress tics. Performance on the behavioral task (reaction time) was strongly associated with the white matter of corpus collosum in typically developing controls but not in children with TS. In the TS group, reaction times were associated with the white matter microstructure of the forceps minor instead. This suggests that for the TS group, the normal role performed by the corpus callosum in resolving the selection of motor outputs may be reduced in favor of an increased role for the prefrontal cortex. The experimenters followed up DWI analyses with fMRI to see if there were significant differences in brain activation during the manual task-switching paradigm. Compared with typically developing controls, children with TS had greater brain activation in the right prefrontal cortex during high conflict incongruent-switch trials relative to incongruent-repeat trials. In the TS group only, the fMRI BOLD response was very strongly and positively linearly associated with RT. Authors concluded that for the TS group only, enhanced cognitive control of motor output is accompanied by adaptive functional and structural changes in the prefrontal cortex. Rae et al. (2020) used a modified Go/NoGo task to uncover neural mechanisms by which adults with TS control actions. The researchers reported no group differences in task accuracy or reaction times between adults with TS and age-matched controls. They also reported that the neural processes by which adults with TS choose to act upon or withhold movements was similar to that of controls, as activity in the cardinal prefrontal and motor regions (including preSMA area) was anatomically similar in both groups. However, adults with TS demonstrated greater activity in the right inferior frontal gyrus across all task conditions.

Another inhibitory control study also highlighted the role played by the right prefrontal cortex in TS. It was found that in medicated adults with TS, gray matter signal in the right inferior frontal gyrus was correlated with stop-signal reaction times in an emotional stop-signal task (Atkinson-Clement et al., 2020). Stop signal reaction times were also associated with higher connectivity between the inferior frontal gyrus and the superior temporal gyrus, and with lower connectivity between the inferior frontal gyrus and the cerebellum. While these results were restricted to those on medication, it is possible that these individuals had increased tic severity (before medication), such that they would represent more severe cases of TS than unmedicated individuals.

Whereas this pattern of frontal overactivation has been reported across many studies using a wide range of experimental paradigms, it is important to note that some studies did not find such frontal overactivation associated with inhibitory control in TS. Notably, Debes, Hansen, Skov, and Larsson (2011) found no difference in brain activation between children with TS and typically developing controls during a Stroop or a Go/NoGo task. Also, Thomalla et al. (2014) only reported decreased brain activation in the left primary motor cortex and secondary motor areas during movement execution (Go trials) but not during response inhibition (NoGo trials) in adults with TS. Another study found increased right inferior parietal cortex activation in adults with TS relative to controls, but no difference over prefrontal regions (Fan et al., 2018). While many EEG studies reported frontal overactivation in the P3b time window for NoGo stimuli (Johannes et al., 2001; Morand-Beaulieu et al., 2018, 2015; Thibault et al., 2009), two recent studies using the stop-signal (Schüller et al., 2018) or the stop-change task (Brandt, Stock, Munchau, & Beste, 2017) reported decreased frontal P3b in adults with TS. Yet, in the study by Schüller et al. (2018), a decreased frontal P3b was observed across condition and was not specific to the stop condition.

In TS studies that reported a frontal overactivation associated with inhibitory control, it has often been interpreted as the source of inhibitory control of motor outputs, suggesting that it may also play a role in the control of tics. However, it is not clear whether this brain pattern is indeed adaptive.

For instance, it was found that the NoGo P300 over frontal electrodes was negatively correlated with reductions in tic severity after cognitive-psychophysiological (CoPs) therapy, and that it was also positively correlated with endpoint tic severity (Morand-Beaulieu, 2019, 2018). This suggests that adults with TS presenting with increased frontal activation during response inhibition respond less to this type of treatment and continue to show elevated tic severity after therapy. A study from Jung et al. (2013) argues that the role played by the prefrontal cortex in TS is not one of inhibitory control of motor outputs, but instead, that it is the source of facilitatory signals that may operate to bias response competition downstream within brain areas more directly linked to motor planning or motor execution. In support of this theory, authors showed that for children with TS, increased task difficulty was associated with increased activation of the inferior frontal cortex and the supplementary motor area. Tic severity was also positively associated with brain activation in these two regions, as well as with activation in the primary motor cortex and the anterior cingulate cortex. Thus, results run counter to the general idea that increased prefrontal cortex activation leads to increased inhibition of motor responses but is more consistent with recent reports that regions such as supplementary motor area may bias suppression mechanisms elsewhere instead of being the direct source of suppression.

In addition to the study of Jung et al. (2013), other studies have highlighted the important role of the supplementary motor area in inhibitory processes in TS. Ganos et al. (2014) conducted an fMRI study of inhibitory control in adults with TS using the stop-signal task. They found that tic frequency was positively correlated with activations in supplementary motor area during the Stop vs Go contrasts. In agreement with Jung et al. (2013), they concluded that the supplementary motor area seems to act as a global inhibition mechanism in TS and might be used to simultaneously stop tics and voluntary actions. Another study that also used a stop-signal task similarly reported worse cognitive control performance in medicated adults with TS to be associated with increased gray matter signal in the supplementary motor area (Atkinson-Clement et al., 2020). An EEG study also showed that lateralized readiness potentials (LRP), which are partly generated by the supplementary motor area, had an increased amplitude in adults with TS relative to controls during a stimulus-response compatibility task (Morand-Beaulieu et al., 2018). This increased LRP amplitude was also correlated with the severity of phonic tics. Finally, it was found that error-related brain activity in the supplementary motor area was increased during a cognitive control task in adults with TS (Fan et al., 2018).

2.2 Inhibitory control in TS: The impact of co-occurring conditions

Co-occurring conditions are common in TS. It is said that up to 90% of individuals with TS also present with a co-occurring condition, the most frequent being ADHD and OCD (Bitsko et al., 2014; Freeman et al., 2000; Hirschtritt et al., 2015). Co-occurring ADHD is known to have an impact on inhibitory control in TS (Morand-Beaulieu et al., 2017) and could, thus, explain some inconsistencies between studies. The best way to disentangle the effects of ADHD (or any other co-occurring condition) is by using a 2 × 2 factorial design, where we have a group with TS-only, a group with ADHD-only, a group with both conditions (TS + ADHD) and a group with neither (controls) (Rothenberger & Heinrich, 2021). This design was used in an ERP study that assessed how TS and ADHD may act separately or additively on electrophysiological markers of inhibitory controls (Shephard, Jackson, & Groom, 2016). In this study, ERPs were recorded while children with TS-only, ADHD-only, TS + ADHD, and typically developing controls performed a Go/NoGo task. Analyses revealed that reduced P3b amplitude was associated with the ADHD factor, rather than the TS factor, suggesting that children with ADHD-only and TS + ADHD had a smaller P3b amplitude than those with TS-only and typically developing controls. Recently, our research team aimed to expand this previous work by including a larger sample of children with TS and/or ADHD, and by adding time-frequency analyses on top of ERPs (Morand-Beaulieu et al., 2022). Similar to the findings of Shephard et al. (2016), we found that children with ADHD-only had reduced task performance and frontal midline theta, as well as delayed latency of N2 and P3 ERPs, while those with TS-only did not differ from typically developing controls on any measures. However, we found unexpected results regarding children with TS + ADHD. Contrary to our expectations and to the findings of Shephard et al. (2016), children in this group showed similar EEG markers of inhibitory control to that of typically developing controls and did not significantly differ from them in terms of task performance either. It is possible that this difference could be explained by the severity of tic symptoms across studies. In our study, the group of children with TS + ADHD did not differ from the group of children with TS-only in terms of tic severity as assessed with the YGTSS. However, in the study of Shephard et al. (2016), children with TS + ADHD not only had more ADHD symptoms than children with TS-only, but they also had increased TS symptoms. This difference in symptoms between studies may in part explain the observed difference in EEG markers between them. It is also possible that some ADHD symptoms in children with TS could be related to TS-associated factors, such as reduced attention due to attempts at suppressing tics or by being distracted by one's own tics (Erenberg, 2005).

Eichele et al. (2016, 2017) also conducted an ERP study comparing children with TS, ADHD, and typically developing controls. In this study, they assessed the development of cognitive control using a modified Eriksen flanker task at two assessments separated by 5 years. At the first assessment, TS showed higher amplitudes of an early P3 component across experimental conditions and in separate trial outcomes. However, the early P3 showed a significant increase for all groups, suggesting that it develops earlier in children with TS compared with controls, but later converges over development. While the TS group was small ($n = 17$), the authors still performed subgroup analyses to compare children with TS-only ($n = 7$) to those with TS + ADHD ($n = 10$). Similar to our findings, the authors did not find major differences between these subgroups on behavioral or electrophysiological measures of cognitive control.

However, an fMRI study has reported differences between children with TS without ADHD (TS-ADHD) and those with TS + ADHD during a stop-signal task (Openneer et al., 2021). This study reported increased stop-signal reaction times during the task, as well as atypical neural patterns during failed inhibition in the TS + ADHD group but not in the TS-ADHD group, indicating that co-occurring ADHD was the driving force behind these differences. For all children, neural activation was found in areas associated with response inhibition, including the inferior and prefrontal gyrus, insula, and temporal gyrus. However, only in children with TS + ADHD was increased activation found in the right inferior frontal gyrus and the left insula during failed inhibition. This overactivation could represent either the effects of ADHD symptoms in TS during difficult conditions or the combined effects of controlling tics and associated premonitory urges. Increased activity of the middle & superior temporal gyrus was also found for children with TS + ADHD during failed inhibition, which could underscore the possibility of an immature inhibitory network in children with TS +ADHD, leading to cognitive difficulties.

Many factors could explain these discrepancies between studies, such as the level of cognitive demand implied by different tasks, the age of participants, or the severity of TS and/or ADHD symptoms. In this vein, the study of Rae et al. (2020) revealed that greater ADHD severity was associated with greater connectivity between the inferior frontal gyrus & premotor cortex during Choose-NoGo trials. This suggests that adults with TS with worse ADHD symptoms require stronger leverage of prefrontal resources to motor cortices in order to withhold actions.

3. Control of tics: Brain correlates of tic suppression

3.1 Tic suppression vs resting-state contrasts

Most individuals with TS are able to suppress their tics for short periods of time. However, to the best of our knowledge, there have only been eight studies to date that tried to understand the physiological processes involved in tic suppression. The first such study was conducted by Peterson, Skudlarski, Anderson, et al. (1998), who assessed the brain correlates of tic suppression in 22 adults with TS using fMRI. In contrast with periods where they were allowed to tic freely, adults with TS showed increased and decreased activations in various areas of the brain during tic suppression. In most subcortical areas (bilateral putamen, thalamus, and globus pallidus), there were reduced activations during tic suppression. The right caudate nucleus was the only subcortical structure showing increased activation during tic suppression. At the cortical level, there were increased brain activations in the right frontal cortex, bilateral temporal gyri, right anterior cingulate cortex, and bilateral occipital cortices. There were also decreased activations in the right posterior cingulate cortex, the bilateral hippocampus and cuneus, as well as the left sensorimotor and inferior parietal cortices. Correlational analyses also showed that baseline tic severity (outside of the scanner), was negatively associated with increased/decreased activations of subcortical structures (as well as sensorimotor and inferior parietal cortices), such that participants showing the most important changes in activation during the tic suppression task were those with the least severe tics. While this study showed that many brain structures were activated/deactivated during tic suppression, the increased activation of the right frontal cortex has been used for a long time to support the relationship between tic suppression and inhibitory control mechanisms. Frontal cortex involvement in tic suppression is also supported by findings of increased regional homogeneity in the left inferior frontal gyrus during tic suppression (Ganos et al., 2014). While the right inferior frontal gyrus is known to play a major role in inhibitory control, as one of the key nodes of the stop network (Aron et al., 2014), the role played by the left inferior frontal gyrus in inhibition is less clear.

Another study, which was conducted by Serrien et al. (2005), is also often referenced to support the parallel between inhibitory control and tic suppression. In this study, coherence analyses were conducted to assess the functional connectivity between pairs of EEG electrodes while 9 adults suppressed their tics. The authors found that EEG alpha coherence between all electrode pairs was increased during tic suppression. Notably, there were four connections (F3-C3, FCz-C3, FCz-C4, F4-C4) where coherence was more significantly increased during tic suppression relative to rest. According to the authors, these connections represented increased functional connectivity between the sensorimotor areas and the prefrontal and mesial frontal cortex. As mentioned earlier in this chapter, the authors also assessed the same connections during a Go/NoGo task and found that coherence was increased during the NoGo in adults with TS. This led to the conclusion that response inhibition and tic suppression may occur through the same fronto-mesial network. Another study also performed similar analyses, but in 9 children with Tourette syndrome (Hong et al., 2013). Here, the authors used partial directed coherence to analyze effective connectivity between 18 scalp electrodes. Authors compared tic suppression and rest for each possible connection (306 connections) and five frequency bands (delta, theta, alpha, beta, gamma) using paired-sample *t*-tests and found increased coherence in the theta band from electrodes F3 and C3 to T4. In the beta band, they reported increased coherence from F3 and F8 to T3. However, these results did not survive correction for multiple comparisons and thus their meaningfulness may be limited. Additionally, they found increased delta power at F7 and increased theta power at Fp1 during tic suppression, but these results did not survive correction for multiple comparisons either. While these two latter studies reported increased connectivity during tic suppression compared to rest, they both share methodological shortcomings including small samples and limited spatial resolution. The latter issue is mostly attributable to the small number of electrodes used in these studies and to problems associated with volume conduction when assessing connectivity at the sensor level. It is, therefore, complicated to interpret the brain structures involved in the connection highlighted in both studies.

We tackled these issues in one of our recent studies (Morand-Beaulieu et al., 2021), where we aimed to replicate and extend the earlier findings of Serrien et al. (2005) of increased alpha-band connectivity during tic suppression. While their study was conducted in 9 adults with TS, we recruited a large sample of 72 children with TS. Additionally, we assessed functional connectivity during tic suppression on source-reconstructed EEG, a technique that overcomes some limitations of sensor-level functional connectivity, such as the risk of spurious connectivity due to volume conduction and the limited capacity for consideration of the brain regions involved. Our study revealed increased alpha-band functional connectivity during tic suppression compared with a resting-state when tics were not suppressed. Our finding of increased connectivity in the alpha band is consistent to the seminal findings of Serrien et al. (2005). However, our approach allowed for a better refinement of the brain regions involved in these connections and consequently suggested potentially different conclusions. Our analyses revealed a subnetwork of 28 nodes that were connected through 29 edges. Within that subnetwork, there were three hubs, i.e., brain regions that were connected to many others in the same subnetwork. These hubs were the right superior frontal gyrus, the left precuneus, and the left banks of the superior temporal sulcus (Fig. 1).

In the Desikan-Killiany atlas that we used for brain parcellations, the anatomically-defined superior frontal gyrus encompasses the functionally-defined dorsomedial prefrontal cortex (Desikan et al., 2006). Both the SFG and the precuneus have been identified by previous fMRI investigations as part of the default mode network (Buckner, Andrews-Hanna, & Schacter, 2008; Utevsky, Smith, & Huettel, 2014), suggesting that this large-scale brain network may be involved in voluntary tic suppression. It may also suggest that brain mechanisms involved in tic suppression differ from those of response inhibition. This hypothesis is supported by a recent fMRI study of tic suppression, which contrasted tic suppression in adults with TS to eyeblink inhibition in controls (van der Salm et al., 2018). In controls, eyeblink inhibition mainly activated the sensorimotor circuit, including notably the right ventrolateral prefrontal cortex, the premotor cortex, and the supplementary motor area. This circuit is typically involved in response inhibition. In adults with TS, tic suppression mainly activated the superior prefrontal gyrus, the anterior and dorsolateral prefrontal cortex, and the dorsal anterior cingulate cortex. Based on these findings, the authors' interpretation was that in controls, the "classical stop network" is recruited to inhibit eyeblinks, while individuals with TS recruits the "limbic network" to suppress tics. This latter study beautifully demonstrated that voluntary tic suppression could involve a mechanism that is distinct from that of response inhibition.

Despite different approach and interpretations, these results may be consistent with those of our study of source-EEG connectivity during tic

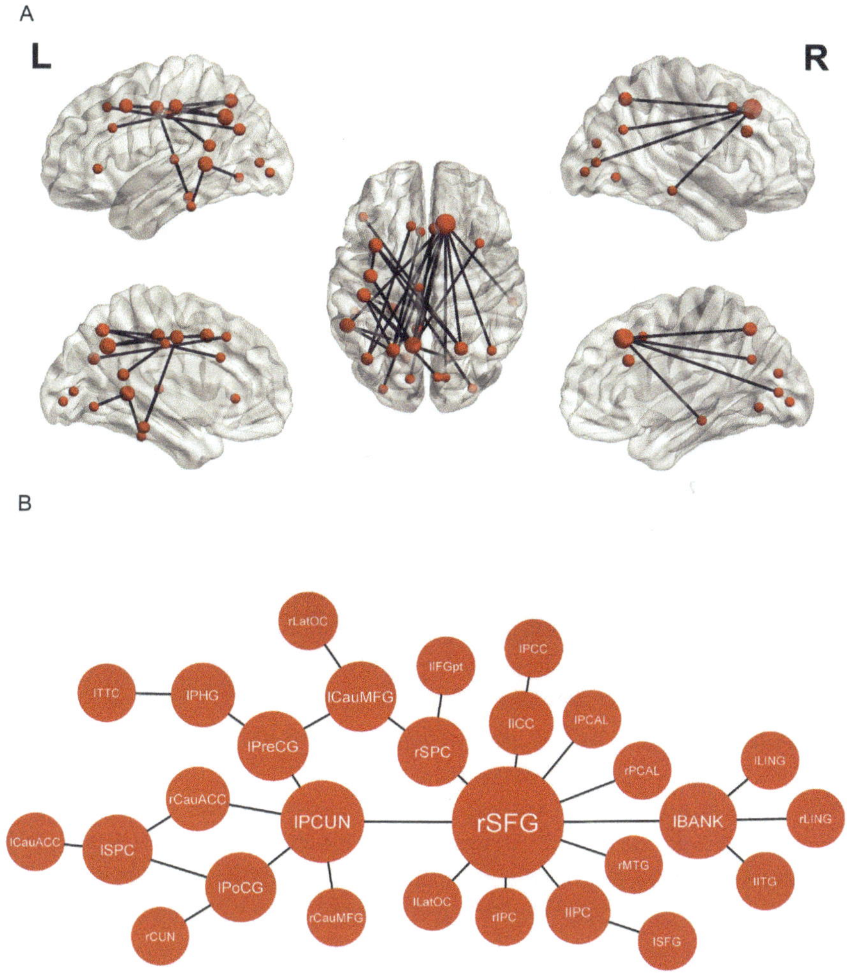

Fig. 1 Subnetwork of functional connectivity in the alpha band during tic suppression. Nodes depicted in this figure showed increased connectivity during tic suppression in comparison to rest. Node size corresponds to connectivity degree (number of connections per node). (A) Hubs in this network were the right superior frontal gyrus (10 connections, degree z-score: 4.2), the left precuneus (5 connections, degree z-score: 1.5), and the left banks of superior temporal sulcus (4 connections, degree z-score: 1.01). (B) Graphical representation of the alpha-band subnetwork. Note: "l" & "r" denotes the lateralization of the nodes; CauACC: caudal anterior cingulate cortex, CauMFG: caudal middle frontal gyrus, CUN: cuneus, ICC: isthmus of the cingulate cortex, IFGpt: pars triangularis of the inferior frontal gyrus, IPC: inferior parietal cortex, ITG: inferior temporal gyrus, LatOC: lateral occipital cortex, LING: lingual gyrus, MTG: middle temporal gyrus, PCAL: pericalcarine cortex, PCC: posterior cingulate cortex, PCUN: precuneus, PHG: parahippocampal gyrus, PoCG: postcentral gyrus, PreCG: precentral gyrus, SFG: superior frontal gyrus, SPC: superior parietal cortex, TTC: transverse temporal cortex. *Reproduced from Morand-Beaulieu, S., Wu, J., Mayes, L.C., Grantz, H., Leckman, J.F., Crowley, M.J., Sukhodolsky, D.G., 2021. Increased alpha-band connectivity during tic suppression in children with Tourette syndrome revealed by source EEG analyses. Biological Psychiatry: Cognitive Neuroscience and Neuroimaging.*

suppression in children. van der Salm et al. (2018) argued that the pattern of activations during tic suppression suggested that it relied on the limbic network. However, the existence of the limbic network has been a subject of debate in the neuroscience literature in the recent years (LeDoux, 2012; Pessoa & Hof, 2015). In a recent review paper, Uddin, Yeo, and Spreng (2019) proposed a taxonomy of large-scale brain networks that is based mostly on neuroanatomical rather than functional labels. For instance, networks known as the DMN or the fronto-parietal control network in the cognitive domain are, respectively, named the medial and lateral frontoparietal networks in this taxonomy. Furthermore, in the Uddin classification, this limbic network is subsumed by the medial frontoparietal network (or DMN). Therefore, this classification appears to be the key to connect our results with those of van der Salm et al. (2018). In their study, the brain regions activated during tic suppression were mostly midline frontal areas. These regions also somewhat overlapped with one of the core hubs of increased connectivity during tic suppression in our study in children with TS. Additionally, in the study from van der Salm et al. (2018), brain regions activated during eyeblink inhibition in healthy controls are consistent with those that are included in the lateral frontoparietal network. Indeed, the inferior frontal gyrus/ventrolateral prefrontal cortex, which was one of the largest clusters activated during eyeblink inhibition, is a core region of the lateral frontoparietal network.

Why increased DMN connectivity would occur during tic suppression remains to be elucidated. It is possible that individuals with TS engage in self-referential thinking while they are instructed to suppress their tics. Self-referential thinking has been shown to involve the DMN (Cavanna & Trimble, 2006; Gusnard, Akbudak, Shulman, & Raichle, 2001; Sheline et al., 2009). In this vein, a recent study reported increased connectivity between the DMN and basal ganglia and hypothesized that it could represent a brain correlate of self-referential thinking related to tics (Ramkiran, Heidemeyer, Gaebler, Shah, & Neuner, 2019). Thus, children in our study might have been more focused on premonitory urges and incoming tics, relying on self-referential processing to suppress tics, which would result in increased DMN connectivity.

Additionally, the sensorimotor network (SMN), which stands in a reciprocal relationship with the DMN (Martino et al., 2016; Northoff, Hirjak, Wolf, Magioncalda, & Martino, 2021) is known to be involved in tic generation (Bohlhalter et al., 2006; Neuner et al., 2014). This reciprocal relationship means that when activity in the DMN goes up, activity in the SMN goes down, and vice versa. Thus, increased DMN connectivity during tic suppression could balance and reduced hyperactive sensorimotor network activity, resulting in reduced tic frequency.

3.2 Task-based studies of tic suppression

While all of these previous studies contrasted two "resting state" conditions—one where tics were suppressed and one where participants were allowed to tic freely—a recent EEG study assessed tic suppression during a motor task (Zapparoli, Macerollo, Joyce, Martino, & Kilner, 2019). This study showed that while adults with TS were ticcing freely, they did not show the same modulation in EEG beta power that was observed in controls during finger movement (decrease in beta power). However, when they were asked to suppress their tics, their decrease in beta power during finger movement was larger than during the tic freely condition. In other words, the modulation of beta oscillations during finger movement during tic suppression was similar to that of controls. According to the authors, this suggests that tic suppression operates by normalizing abnormal beta oscillations over the sensorimotor cortex, at least during the context of a finger movement task.

However, it is possible that individuals with TS use different strategies to suppress tics in different contexts, and thus different mechanisms may be involved in tic suppression depending on context. In studies where individuals with TS are asked to suppress tics during a resting-state session, there are no other instructions than to suppress tics. Therefore, participants may either be intensively focused on their tics and premonitory urges or may let their mind wander. These experimentations may reflect some real-life situations such as sitting at home or in a waiting room, where no other actions are expected or required. Conversely, the experimental manipulation in the study of Zapparoli et al. (2019) may reflect what happens in a classroom or at work, when tics may be suppressed while other tasks are performed. Therefore, future studies aiming to understand the neural correlates of tics suppression should assess these in various contexts and tasks that are close proxies to the everyday life of individuals with TS.

3.3 Neurodevelopmental trajectory of tic suppression

Previous studies have shown that tic suppression follows a developmental trajectory. In a pooled analysis of 9 behavioral studies of tic suppression, Conelea et al. (2018) found that tic suppressibility was positively correlated with age. Only two studies have assessed the brain correlates of tic suppression in children (Hong et al., 2013; Morand-Beaulieu et al., 2021), thus making it hard to understand how brain development may impact tic suppression. In our recent study, we found that connectivity within a specific subnetwork, in which the right superior frontal gyrus was the main hub, was positively correlated with age during tic suppression but not during rest (Morand-Beaulieu et al., 2021). Therefore, in that subnetwork, connectivity did not differ between rest and tic suppression in younger children, but older children and adolescents showed increased connectivity relative to rest. However, it remains unclear whether this increased connectivity in older children and adolescents plays a role in tic suppression capacities.

3.4 Impact of co-occurring conditions on brain correlates of tic suppression

Behavioral studies of tic suppression suggest that co-occurring conditions do not impact tic suppression Despite two small studies suggesting a possible negative impact of attention problems on tic suppression (Himle & Woods, 2005; Müller-Vahl, Riemann, & Bokemeyer, 2014), a recent large study reported no impact of ADHD on tic suppressibility (Conelea et al., 2018). Additionally, the latter study reported no impact of OCD or anxiety disorders as well. Among the few psychophysiological studies that have assessed the impact of co-occurring conditions on brain correlates of tic suppression, none have reported a significant effect of either ADHD or OCD (Morand-Beaulieu et al., 2021; Peterson et al., 1998).

4. Conclusions and future directions

In this chapter, we reviewed the psychophysiological literature on response inhibition and tic suppression in children and adults with TS. While seminal psychophysiological studies of tic suppression suggested similar mechanisms for response inhibition and tic suppression, more recent studies revealed that these processes are probably more different than previously hypothesized. However, tic suppression remains understudied and poorly understood. It is something that most individuals with TS do on an almost-daily basis but how it occurs remains somewhat unclear. It is an interesting and intriguing feature of TS and something that distinguishes it from most other movement disorders. A better understanding of this phenomenon may potentially allow us to enhance tic suppression capacity, such as through behavioral treatments.

One way to further our understanding of tic suppression would be to develop new paradigms to study it. Current paradigms that have been used in many psychophysiological studies, such as a contrast between tic suppression and a "resting-state," may lack ecological validity. For instance, it would be interesting to contrast brain activity in settings where tics are more likely to be suppressed, such as the classroom or the workplace, and settings where active tic suppression is less likely, such as being home alone.

Furthermore, future psychophysiological studies of tic suppression should recruit larger and more diverse samples. Current evidence does not show a gender effect (Conelea et al., 2018), but the number of girls and women recruited in tic suppression studies is likely too small to identify any potential gender and/or sex effects. Understanding how neurodevelopment affects tic suppression is also very important and future studies should additionally aim to assess tic suppression at different developmental stages.

References

Abramovitch, A., Hallion, L. S., Reese, H. E., Woods, D. W., Peterson, A., Walkup, J. T., et al. (2017). Neurocognitive predictors of treatment response to randomized treatment in adults with tic disorders. *Progress in Neuro-Psychopharmacology & Biological Psychiatry*, *74*, 9–14.

Aron, A. R., Behrens, T. E., Smith, S., Frank, M. J., & Poldrack, R. A. (2007). Triangulating a cognitive control network using diffusion-weighted magnetic resonance imaging (MRI) and functional MRI. *The Journal of neuroscience: The official journal of the Society for Neuroscience*, *27*(14), 3743–3752.

Aron, A. R., Durston, S., Eagle, D. M., Logan, G. D., Stinear, C. M., & Stuphorn, V. (2007). Converging evidence for a Fronto-basal-ganglia network for inhibitory control of action and cognition. *The Journal of Neuroscience*, *27*(44), 11860–11864.

Aron, A. R., Robbins, T. W., & Poldrack, R. A. (2014). Inhibition and the right inferior frontal cortex: One decade on. *Trends in Cognitive Sciences*, *18*(4), 177–185.

Atkinson-Clement, C., Porte, C.-A., de Liege, A., Wattiez, N., Klein, Y., Beranger, B., et al. (2020). Neural correlates and role of medication in reactive motor impulsivity in Tourette disorder. *Cortex*, *125*, 60–72.

Bitsko, R. H., Holbrook, J. R., Visser, S. N., Mink, J. W., Zinner, S. H., Ghandour, R. M., et al. (2014). A national profile of Tourette syndrome, 2011–2012. *Journal of Developmental & Behavioral Pediatrics*, *35*(5).

Bohlhalter, S., Goldfine, A., Matteson, S., Garraux, G., Hanakawa, T., Kansaku, K., et al. (2006). Neural correlates of tic generation in Tourette syndrome: An event-related functional MRI study. *Brain*, *129*(Pt 8), 2029–2037.

Brandt, V. C., Stock, A. K., Munchau, A., & Beste, C. (2017). Evidence for enhanced multi-component behaviour in Tourette syndrome - an EEG study. *Scientific Reports*, 7(1), 7722.

Buckner, R. L., Andrews-Hanna, J. R., & Schacter, D. L. (2008). The Brain's default network. *Annals of the New York Academy of Sciences*, *1124*(1), 1–38.

Cavanna, A. E., & Trimble, M. R. (2006). The precuneus: A review of its functional anatomy and behavioural correlates. *Brain*, *129*(3), 564–583.

Chang, S. W., McGuire, J. F., Walkup, J. T., Woods, D. W., Scahill, L., Wilhelm, S., et al. (2018). Neurocognitive correlates of treatment response in children with Tourette's disorder. *Psychiatry Research, 261*, 464–472.

Conelea, C. A., Wellen, B., Woods, D. W., Greene, D. J., Black, K. J., Specht, M., et al. (2018). Patterns and predictors of tic suppressibility in youth with tic disorders. *Frontiers in Psychiatry, 9*, 188.

Debes, N. M., Hansen, A., Skov, L., & Larsson, H. (2011). A functional magnetic resonance imaging study of a large clinical cohort of children with Tourette syndrome. *Journal of Child Neurology, 26*(5), 560–569.

Desikan, R. S., Ségonne, F., Fischl, B., Quinn, B. T., Dickerson, B. C., Blacker, D., et al. (2006). An automated labeling system for subdividing the human cerebral cortex on MRI scans into gyral based regions of interest. *NeuroImage, 31*(3), 968–980.

Eichele, H., Eichele, T., Bjelland, I., Hovik, M. F., Sorensen, L., van Wageningen, H., et al. (2016). Performance monitoring in medication-naive children with Tourette syndrome. *Frontiers in Neuroscience, 10*, 50.

Eichele, H., Eichele, T., Marquardt, L., Adolfsdottir, S., Hugdahl, K., Sorensen, L., et al. (2017). Development of performance and ERPs in a flanker task in children and adolescents with Tourette dyndrome-a follow-up study. *Frontiers in Neuroscience, 11*, 305.

Erenberg, G. (2005). The relationship between tourette syndrome, attention deficit hyperactivity disorder, and stimulant medication: A critical review. *Seminars in Pediatric Neurology, 12*(4), 217–221.

Essoe, J. K. Y., Ramsey, K. A., Singer, H. S., Grados, M., & McGuire, J. F. (2021). Mechanisms underlying behavior therapy for Tourette's disorder. *Current Developmental Disorders Reports, 8*(3), 161–174.

Fan, S., Cath, D. C., van der Werf, Y. D., de Wit, S., Veltman, D. J., & van den Heuvel, O. A. (2018). Trans-diagnostic comparison of response inhibition in Tourette's disorder and obsessive-compulsive disorder. *The World Journal of Biological Psychiatry, 19*(7), 527–537.

Freeman, R. D., Fast, D. K., Burd, L., Kerbeshian, J., Robertson, M. M., & Sandor, P. (2000). An international perspective on Tourette syndrome: Selected findings from 3,500 individuals in 22 countries. *Developmental Medicine and Child Neurology, 42*(7), 436–447.

Ganos, C., Kahl, U., Brandt, V., Schunke, O., Baumer, T., Thomalla, G., et al. (2014). The neural correlates of tic inhibition in Gilles de la Tourette syndrome. *Neuropsychologia, 65*, 297–301.

Ganos, C., Kuhn, S., Kahl, U., Schunke, O., Feldheim, J., Gerloff, C., et al. (2014). Action inhibition in Tourette syndrome. *Movement Disorders, 29*(12), 1532–1538.

Gusnard, D. A., Akbudak, E., Shulman, G. L., & Raichle, M. E. (2001). Medial prefrontal cortex and self-referential mental activity: Relation to a default mode of brain function. *Proceedings of the National Academy of Sciences, 98*(7), 4259.

Himle, M. B., & Woods, D. W. (2005). An experimental evaluation of tic suppression and the tic rebound effect. *Behaviour Research and Therapy, 43*(11), 1443–1451.

Hirschtritt, M. E., Lee, P. C., Pauls, D. L., Dion, Y., Grados, M. A., Illmann, C., et al. (2015). Lifetime prevalence, age of risk, and genetic relationships of comorbid psychiatric disorders in Tourette syndrome. *JAMA Psychiatry, 72*(4), 325–333.

Hong, H. J., Sohn, H., Cha, M., Kim, S., Oh, J., Chu, M. K., et al. (2013). Increased Frontomotor oscillations during tic suppression in children with Tourette syndrome. *Journal of Child Neurology, 28*(5), 615–624.

Jackson, G. M., & Jackson, S. R. (2020). Rethinking the nature of inhibitory control deficits in Tourette syndrome. *Brain, 143*(3), 721–722.

Jackson, S. R., Parkinson, A., Jung, J., Ryan, S. E., Morgan, P. S., Hollis, C., et al. (2011). Compensatory neural reorganization in Tourette syndrome. *Current Biology, 21*(7), 580–585.

Johannes, S., Wieringa, B. M., Mantey, M., Nager, W., Rada, D., Muller-Vahl, K. R., et al. (2001). Altered inhibition of motor responses in Tourette syndrome and obsessive-compulsive disorder. *Acta Neurologica Scandinavica, 104*(1), 36–43.

Jung, J., Jackson, S. R., Parkinson, A., & Jackson, G. M. (2013). Cognitive control over motor output in Tourette syndrome. *Neuroscience and Biobehavioral Reviews, 37*(6), 1016–1025.

Jurgiel, J., Miyakoshi, M., Dillon, A., Piacentini, J., Makeig, S., & Loo, S. K. (2021). Inhibitory control in children with tic disorder: Aberrant fronto-parietal network activity and connectivity. *Brain Communications, 3*(2).

LeDoux, J. E. (2012). Evolution of human emotion: A view through fear. *Progress in Brain Research, 195*, 431–442.

Marsh, R., Zhu, H., Wang, Z., Skudlarski, P., & Peterson, B. S. (2007). A developmental fMRI study of self-regulatory control in Tourette's syndrome. *The American Journal of Psychiatry, 164*(6), 955–966.

Martino, M., Magioncalda, P., Huang, Z., Conio, B., Piaggio, N., Duncan, N. W., et al. (2016). Contrasting variability patterns in the default mode and sensorimotor networks balance in bipolar depression and mania. *Proceedings of the National Academy of Sciences of the United States of America, 113*(17), 4824–4829.

McGuire, J. F., Sturm, A., Ricketts, E. J., Montalbano, G. E., Chang, S., Loo, S. K., et al. (2021). Cognitive control processes in behavior therapy for youth with Tourette's disorder. *Journal of Child Psychology and Psychiatry*.

Morand-Beaulieu, S. (2019). *Étude des capacités d'inhibition, des processus moteurs, et de l'impact de la thérapie cognitive-comportementale sur le fonctionnement cérébral des patients atteints du syndrome de Gilles de la Tourette*. Université de Montréal.

Morand-Beaulieu, S., Grot, S., Lavoie, J., Leclerc, J. B., Luck, D., & Lavoie, M. E. (2017). The puzzling question of inhibitory control in Tourette syndrome: A meta-analysis. *Neuroscience and Biobehavioral Reviews, 80*, 240–262.

Morand-Beaulieu, S., O'Connor, K. P., Blanchet, P. J., & Lavoie, M. E. (2018). Electrophysiological predictors of cognitive-behavioral therapy outcome in tic disorders. *Journal of Psychiatric Research, 105*, 113–122.

Morand-Beaulieu, S., O'Connor, K. P., Sauvé, G., Blanchet, P. J., & Lavoie, M. E. (2015). Cognitive-behavioral therapy induces sensorimotor and specific electrocortical changes in chronic tic and Tourette's disorder. *Neuropsychologia, 79*(Part B), 310–321.

Morand-Beaulieu, S., Smith, S. D., Ibrahim, K., Wu, J., Leckman, J. F., Crowley, M. J., et al. (2022). Electrophysiological signatures of inhibitory control in children with Tourette syndrome and attention-deficit/hyperactivity disorder. *Cortex, 147*, 157–168.

Morand-Beaulieu, S., Wu, J., Mayes, L. C., Grantz, H., Leckman, J. F., Crowley, M. J., et al. (2021). Increased alpha-band connectivity during tic suppression in children with Tourette syndrome revealed by source EEG analyses. *Biological Psychiatry: Cognitive Neuroscience and Neuroimaging*..

Mueller, S. C., Jackson, G. M., Dhalla, R., Datsopoulos, S., & Hollis, C. P. (2006). Enhanced cognitive control in young people with Tourette's syndrome. *Current Biology, 16*(6), 570–573.

Müller-Vahl, K. R., Riemann, L., & Bokemeyer, S. (2014). Tourette patients' misbelief of a tic rebound is due to overall difficulties in reliable tic rating. *Journal of Psychosomatic Research, 76*(6), 472–476.

Neuner, I., Werner, C. J., Arrubla, J., Stocker, T., Ehlen, C., Wegener, H. P., et al. (2014). Imaging the where and when of tic generation and resting state networks in adult Tourette patients. *Frontiers in Human Neuroscience, 8*, 362.

Northoff, G., Hirjak, D., Wolf, R. C., Magioncalda, P., & Martino, M. (2021). All roads lead to the motor cortex: Psychomotor mechanisms and their biochemical modulation in psychiatric disorders. *Molecular Psychiatry, 26*(1), 92–102.

Openneer, T. J. C., van der Meer, D., Marsman, J.-B. C., Forde, N. J., Akkermans, S. E. A., Naaijen, J., et al. (2021). Impaired response inhibition during a stop-signal task in children with Tourette syndrome is related to ADHD symptoms: A functional magnetic resonance imaging study. *The World Journal of Biological Psychiatry, 22*(5), 350–361.

Pessoa, L., & Hof, P. R. (2015). From Paul Broca's great limbic lobe to the limbic system. *The Journal of Comparative Neurology, 523*(17), 2495–2500.

Peterson, B. S., Skudlarski, P., Anderson, A. W., et al. (1998). A functional magnetic resonance imaging study of tic suppression in tourette syndrome. *Archives of General Psychiatry, 55*(4), 326–333.

Piacentini, J., Woods, D. W., Scahill, L., Wilhelm, S., Peterson, A. L., Chang, S., et al. (2010). Behavior therapy for children with Tourette disorder: A randomized controlled trial. *JAMA: The Journal of the American Medical Association, 303*(19), 1929–1937.

Rae, C. L., Parkinson, J., Betka, S., Gouldvan Praag, C. D., Bouyagoub, S., Polyanska, L., et al. (2020). Amplified engagement of prefrontal cortex during control of voluntary action in Tourette syndrome. *Brain Communications, 2*(2), fcaa199.

Ramkiran, S., Heidemeyer, L., Gaebler, A., Shah, N. J., & Neuner, I. (2019). Alterations in basal ganglia-cerebello-thalamo-cortical connectivity and whole brain functional network topology in Tourette's syndrome. *NeuroImage: Clinical, 24*, 101998.

Rothenberger, A., & Heinrich, H. (2021). Electrophysiology echoes brain dynamics in children and adolescents with Tourette syndrome-a developmental perspective. *Frontiers in Neurology, 12*, 587097.

Schüller, T., Gruendler, T. O. J., Huster, R., Baldermann, J. C., Huys, D., Ullsperger, M., et al. (2018). Altered electrophysiological correlates of motor inhibition and performance monitoring in Tourette's syndrome. *Clinical Neurophysiology, 129*(9), 1866–1872.

Serrien, D. J., Orth, M., Evans, A. H., Lees, A. J., & Brown, P. (2005). Motor inhibition in patients with Gilles de la Tourette syndrome: Functional activation patterns as revealed by EEG coherence. *Brain, 128*(Pt 1), 116–125.

Sheline, Y. I., Barch, D. M., Price, J. L., Rundle, M. M., Vaishnavi, S. N., Snyder, A. Z., et al. (2009). The default mode network and self-referential processes in depression. *Proceedings of the National Academy of Sciences, 106*(6), 1942.

Shephard, E., Jackson, G. M., & Groom, M. J. (2016). The effects of co-occurring ADHD symptoms on electrophysiological correlates of cognitive control in young people with Tourette syndrome. *Journal of Neuropsychology, 10*(2), 223–238.

Sukhodolsky, D. G., Landeros-Weisenberger, A., Scahill, L., Leckman, J. F., & Schultz, R. T. (2010). Neuropsychological functioning in children with Tourette syndrome with and without attention-deficit/hyperactivity disorder. *Journal of the American Academy of Child and Adolescent Psychiatry, 49*(11), 1155–1164.

Thibault, G., O'Connor, K. P., Stip, E., & Lavoie, M. E. (2009). Electrophysiological manifestations of stimulus evaluation, response inhibition and motor processing in Tourette syndrome patients. *Psychiatry Research, 167*(3), 202–220.

Thomalla, G., Jonas, M., Baumer, T., Siebner, H. R., Biermann-Ruben, K., Ganos, C., et al. (2014). Costs of control: Decreased motor cortex engagement during a go/NoGo task in Tourette's syndrome. *Brain, 137*(Pt 1), 122–136.

Uddin, L. Q., Yeo, B. T. T., & Spreng, R. N. (2019). Towards a universal taxonomy of macro-scale functional human brain networks. *Brain Topography, 32*(6), 926–942.

Utevsky, A. V., Smith, D. V., & Huettel, S. A. (2014). Precuneus is a functional core of the default-mode network. *The Journal of neuroscience: The official journal of the Society for Neuroscience, 34*(3), 932–940.

van der Salm, S. M. A., van der Meer, J. N., Cath, D. C., Groot, P. F. C., van der Werf, Y. D., Brouwers, E., et al. (2018). Distinctive tics suppression network in Gilles de la Tourette syndrome distinguished from suppression of natural urges using multimodal imaging. *NeuroImage: Clinical, 20*, 783–792.

Verdellen, C. W., Hoogduin, C. A., & Keijsers, G. P. (2007). Tic suppression in the treatment of Tourette's syndrome with exposure therapy: The rebound phenomenon reconsidered. *Movement Disorders*, *22*(11), 1601–1606.

Zapparoli, L., Macerollo, A., Joyce, E. M., Martino, D., & Kilner, J. M. (2019). Voluntary tic suppression and the normalization of motor cortical beta power in Gilles de la Tourette syndrome: An EEG study. *European Journal of Neuroscience*, *50*(12), 3944–3957.

CHAPTER TWELVE

Integrating psychophysiology into cognitive-behavioral treatment for the Gilles de la Tourette syndrome

Marc E. Lavoie[a,*], Marie-Ange Perrault[b], and Julie Leclerc[c]

[a]Département de Psychiatrie et Addictologie, Laboratoire de Psychophysiologie Cognitive et Sociale, Groupe d'étude sur les Troubles tic, d'Accumulation Compulsive et Obsessionnel-Compulsif (GE-tic-tac-toc), Centre de recherche de l'Institut Universitaire en Santé Mentale de Montréal, Université de Montréal, Montréal, Québec, Canada

[b]Département de Psychologie, Laboratoire de Psychophysiologie Cognitive et Sociale, Groupe d'étude sur les Troubles tic, d'Accumulation Compulsive et Obsessionnel-Compulsif (GE-tic-tac-toc), Centre de recherche de l'Institut Universitaire en Santé Mentale de Montréal, Université de Montréal, Montréal, Québec, Canada

[c]Département de Psychologie, Groupe d'étude sur les Troubles tic, d'Accumulation Compulsive et Obsessionnel-Compulsif (GE-tic-tac-toc), Centre de recherche de l'Institut Universitaire en Santé Mentale de Montréal, Université du Québec à Montréal, Montréal, Québec, Canada

*Corresponding author: e-mail address: marc.lavoie@umontreal.ca

Contents

International Review of Movement Disorders, Volume 3
ISSN 2666-7878
https://doi.org/10.1016/bs.irmvd.2021.11.004

Abbreviations

CBIT Comprehensive Behavioral Intervention for Tics
ERP exposure and response prevention
TS Tourette syndrome

1. Brief historical context

The first chapter of this book gives a substantive and thorough portrait of TS's historical context, but it is essential to highlight the roots of the contemporary neurobiological models underlying this syndrome. Originally known references to tics probably go back to medieval times. In the fifteenth century, two Dominican monks reported the case of a priest who could not help but grimace and emit vocalizations whenever he was praying (Kramer & Sprenger, 1948). Later in 1825, Jean-Marc Gaspard Itard described tics in a systematic way (Itard, 1825). The latter reports the case of a 26-year-old French noblewoman, the Countess de Dampierre, who presented involuntary convulsive spasms and contortions at the shoulders, neck, and face level. Shortly afterward, he also reported the presence of "spasms affecting the organs of voice and speech." He also notes the presence of strange screams and senseless words in the absence of a circumscribed mental disorder. This initial description paved the way for the next generation of clinician interested in that syndrome. The Gilles de la Tourette syndrome is named after the French neurologist Georges Gilles de la Tourette, who, in 1885, described again in the case of the Countess de Dampierre, now aged 86 years old, who continued to make abrupt movements and sounds also known as tics. The same year, Tourette described eight other patients with motor and vocal tics (Gilles de la Tourette, 1885), consistent with similar observations from American clinicians done 1 year later (Dana & Wilkin, 1886). In a doctoral dissertation published under the supervision of Jean-Martin Charcot, Jacques Catrou documented 26 other cases (Catrou, 1890). Subsequently, few systematic investigations, clinical observations, or etiological developments were undertaken during the first half of the 20th century. Instead, a psychoanalytic explanation prevailed in the etiology and the treatment during this period, with little or no notable empirical support (Ascher, 1948; Ferenczi, 1921; Mahler, 1944, 1949; Mahler, Luke, & Daltroff, 1945).

However, in the 1960s, an experimental drug treatment (i.e., haloperidol) surfaced for tics (Seignot, 1961).[a] These results encouraged clinical trials in the United States, which further supported the beneficial effects of neuroleptics (Corbett, Mathews, Connell, & Shapiro, 1969; Shapiro, 1970; Shapiro & Shapiro, 1968). These seminal investigations instigated the race to find an effective pharmacological treatment and, therefore, the search for a neurobiological etiology, relegating to the background, the psychotherapeutic or any cognitive and behavioral approach as well (Shapiro, 1970; Shapiro et al., 1976; Shapiro & Shapiro, 1971). At that time, it was claimed that clinicians prematurely attribute etiology of tics to psychological factors and thus tend to overprescribe psychological treatments, which in their opinion' "*illustrates some of the dangers of excessive, overelaborate, and premature psychological theorizing in place of careful clinical observation*," (Shapiro & Shapiro, 1971; page 159–160). If we replace that view in context, the prevailing psychoanalytic approach and the behavioral interpretation were often considered embedded within the same family of conceptual approaches.

Nonetheless, at about the same time, Azrin and Nunn (1973) were among the first to suggest that tics could be prevented with a behavioral approach entitled habit reversal. Their book "Habit Reversal in a day," printed in 1977, suggested that tics could be prevented by activating a response antagonistic to the tic and tensing the opposite muscles with a counter-conditioning process. In practice, the tic could be controlled until it is countermanded, and the person practiced this antagonist response when they detected tic onset. This therapy allows connecting the physiological response build-up often observed before tic onset and the awareness to muscle tension. Fortunately, the integration of the psychological and pharmacological approaches to reduce tics and other behaviors are now standard, and psychological as well as neurobiological approaches are considered complementary. Therapeutic interventions must target tic symptoms and coping strategies that can modify the unique impact that TS symptoms may have on an individual's well-being (Peterson & Cohen, 1998).

[a] However, it is interesting to note that Rickards, Hartley, and Robertson (1997) reported that the single case involved in the Seignot's paper had been subject to a frontal lobectomy prior to the treatment with haloperidol.

1.1 Recent advances and evolution in psychotherapy for the treatment of Tics

The habit reversal model considers tics to be a normal reaction, but it represents an accentuated response in terms of frequency and is maintained by a conditioning process. This treatment includes training in awareness, progressive muscle relaxation, and the development of an incompatible response to tic (Azrin & Nunn, 1973; Frank & Cavanna, 2013). Contingency management and the generalization of achievements are also part of the process (Azrin & Peterson, 1988). Although effective (Deckersbach, Rauch, Buhlmann, & Wilhelm, 2006; Wilhelm et al., 2003), these results were collected during experiments with relatively small numbers of participants from various populations affected with chronic tics, TS, or habit disorders. As it evolved, habit reversal is no longer widely used alone, as it has been incorporated into a new treatment model, the so-called Comprehensive Behavioral Intervention for Tics or CBIT that is now more widely studied (Woods et al., 2008). These therapies take different forms depending on the conceptualization of tics. CBIT includes the basic principles of habit reversal, to which psychological education and functional intervention steps have been added. Thus, this model considers the historical and tic-like elements that influence the variability of their expression (Capriotti et al., 2015; Himle et al., 2014). This type of behavioral therapy was evaluated in a multi-site randomized controlled trial, which followed 126 children between 9 and 17 years old diagnosed with TS or chronic tics (Piacentini et al., 2010). In this study, all children were randomly assigned to eight behavioral therapy sessions for 10 weeks or to educational support therapy sessions. The behavioral therapy sessions significantly decreased tic symptoms compared to the support therapy (53% vs 19% respectively) with, in addition, lasting effects for 6 months in 87% of cases. Its effectiveness has also been demonstrated in adults (Wilhelm et al., 2012), where the treatment sessions significantly decreased the total tic score compared to the support therapy (26% vs 11% respectively). This treatment has a lasting effect for 6 months in 80% of cases compared with 25% improvement in the support and education sessions. Currently, the American Academy of Neurology Guidelines recommend the CBIT as a first-line treatment, either alone or in conjunction with medication if necessary (Pringsheim, Holler-Managan, et al., 2019; Pringsheim, Okun, et al., 2019). These guidelines recommend that clinicians prescribe CBIT delivered over a teleconference or secure voice over internet protocol delivery systems if face-to-face options are unavailable in a patient care center. If CBIT is unavailable, other behavioral interventions for tics may be acceptable, such as exposure and response prevention (ERP), where the client tolerates the urge to tic while resisting the action of ticking (Verdellen, van de Griendt, Hartmann, Murphy, & ESSTS Guidelines Group, 2011). This intervention targets the physiological dynamics that underlie tics. It is based on the principle that tics are triggered and maintained by a negative reinforcement process where the tic allows the neutralization of an unpleasant sensation and the reduction of muscle tension (Evers & van de Wetering, 1994; Hoogduin, Verdellen, & Cath, 1997; Verdellen, Keijsers, Cath, & Hoogduin, 2004). Exposure and response prevention, which shows interesting effects, aims to identify premonitory sensations and then tolerate them while resisting the tic manifestation (Verdellen et al., 2011). These behavioral approaches addressed the tic when it has occurred or when it is about to occur. Although this treatment seems effective, especially with severe tics not amenable to other approaches, not all people with tic disorder experience premonitory urges (Woods & Himle, 2004). There is less data on exposure and response prevention than CBIT in the scientific literature, but a controlled trial seems to suggest a similar average decrease in tics between the two interventions (Verdellen et al., 2004). However, the constraint of ERP is that it is often emotionally challenging for the child to resist the tics because the therapeutic process does not operate much on the thoughts triggering tics.

1.2 The step toward the integration of behavioral, cognitive and physiological components for a more comprehensive treatment

A more recent and integrative intervention model targets both the cognitive, behavioral, and psychophysiological factors inherent in triggering and maintaining tics (O'Connor, 2002; O'Connor, Lavoie, & Schoendorff, 2017). This type of intervention aims at uniting several approaches within a psychophysiological model. Psychophysiology is a domain of neuroscience that seeks to comprehend how a person's mental state (psychology) and biological (physiology) responses interact to affect one another (Lang, McTeague, & Bradley, 2014). Because tics fluctuate in frequency, intensity, and location, the purpose of identifying the cognitive processes and the physiological state is of crucial clinical interest. Our findings with tic disorder patients confirmed the identification of consistent patterns in motor and cognitive processing. These were reflected through behavioral (O'Connor, Brisebois, Brault, Robillard, & Loiselle, 2003; O'Connor & Gareau, 1994; O'Connor, Lavoie, Robert, Stip, & Borgeat, 2005), neuropsychological

(Lavoie, Thibault, Stip, & O'Connor, 2007; Morand-Beaulieu et al., 2017; O'Connor, Lavoie, Stip, Borgeat, & Laverdure, 2008), and electrophysiological measures (Lavoie, Imbriglio, Stip, & O'Connor, 2011; Morand-Beaulieu & Lavoie, 2019). For instance, earlier studies showed that adults with tic disorders exhibit a mode of action planning that increases sensory-motor activation and muscle tension that can be addressed directly in a specific treatment (O'Connor, Lavoie, Blanchet, & St-Pierre-Delorme, 2016; O'Connor et al., 2008). These discoveries have led to the elaboration of a Cognitive-behavioral Psychophysiological (CoPs) model for the treatment of adults (O'Connor et al., 2017) and children (Leclerc, Forget, & O'Connor, 2008; Leclerc, O'Connor, Nolin, Valois, & Lavoie, 2016) with habit and tic disorders.

The CoPs approach is multimodal and targets underlying sensorimotor processes rather than tics per se. Within that framework, the tics temporarily release muscle tension through a sensory-motor self-regulation process, but this tends to strengthen the tics cycle in the medium term. Hence, tics represent an endpoint where, proximally, a high muscle contraction level triggers the tic, but distal causes are traced back to how the person planned telic action and managed the interplay of inhibition/facilitation in planning action. This therapy aims at a global cognitive and behavioral restructuration of anticipation and appraisals in high-risk tic situations, leading to reeducation related to movement planning and, consequently, a reduction in tics. Studies show good results in reducing tics in both adults (O'Connor et al., 2001, 2009, 2016) and children (Leclerc, O'Connor, et al., 2016; Leclerc, Valois, et al., 2016). In addition, the main outcome measures on Tourette global scale and measures of mood revealed that CoPs treatment for tic disorders is effective either in combination with medication or alone (O'Connor et al., 2009).

Further results from our team showed a clinically significant reduction of tic frequency following CoPs (Cohen's d - 1.43–2.34), maintained at the 6-month follow-up (O'Connor et al., 2016). That therapy was later adapted for children and adolescents and demonstrated similar effectiveness in decreasing tic severity (Leclerc, O'Connor, et al., 2016; Leclerc, Valois, et al., 2016) and improved quality of life (Nolin & Leclerc, 2021). Our open-label trial with adults confirmed a clinically significant reduction in tic symptoms with a large effect size (Cohen's d - 2.36–2.81) following CoPs treatment in chronic tic and TS groups patients (O'Connor et al., 2016). Despite these promising numbers, 35% of individuals with chronic tic and 25% of the group with TS responded sub-optimally with less than 35% improvement to CoPs treatment, even with medication for some of them. These results are comparable to those obtained from the CBIT obtained in children (Piacentini et al., 2010) and adults (Wilhelm et al., 2012) that led to significant symptoms decrease in both populations. In earlier studies, a low rate of positive response in adults compared with children was typically seen. To explain the discrepancies between children's and adults' results, it is proposed that emerging adults with enduring tics may represent a more pervasive form of the disorder (Wilhelm et al., 2012; Woods et al., 2011). However, these adults also develop symptoms through a period where the cerebral regions implied in tic generations are still in development, particularly between 15 and 25 years of age (Langen, Durston, Kas, van Engeland, & Staal, 2011; Lavoie, Leclerc, & O'Connor, 2013). Thus, it is reasonable to hypothesize that the insertion of treatment at that period should have a larger impact in younger patients with more durable long-term effects. Therefore, it is essential to improve our psychological treatments to be more beneficial in those emerging adult patients who respond less well to therapy.

Over the years, studies with adults that have focused on CoPs treatment have shown motor improvements associated with better coordination and control during movement execution (O'Connor et al., 2005). Other related studies revealed an induced change in brain activity following CoPs treatment (Morand-Beaulieu, O'Connor, Blanchet, & Lavoie, 2018; Morand-Beaulieu et al., 2016; Morand-Beaulieu, O'Connor, Sauvé, Blanchet, & Lavoie, 2015). For the first time, specific modifications on the activation of the premotor cortex and SMA (Lavoie et al., 2011; Morand-Beaulieu et al., 2018, 2015) showed that this treatment induces a change in sensorimotor activation patterns that would tend to normalize following therapy and allow, among other things, to develop better motor control.

2. Broad etiological concept and recent data on the neuropsychology and physiology of motor performance in Tic disorders

TS has been initially considered to be of "nervous origin" even in the absence of an explanatory physiological model of underlying symptoms at that time (Gilles de la Tourette, 1885). More recent models brought that if the mechanisms responsible for this syndrome could be mainly ascribed to dopaminergic dysfunctions, provoking a higher level of cerebral activation than normal, it is thus plausible to postulate that this type of imbalance is responsible for provoking these motor inhibition symptoms (Hallett, 2015; van Wouwe et al., 2017). Dopamine is a neurotransmitter primarily associated with motor control as well as with rewards and pleasure. Although anti-dopaminergic drugs are widely used in treating TS, there has been an increasing interest in adrenergic alpha agonists, histaminergic agents, cannabidiol, and neurosurgical methods comprising deep brain stimulation (Gunduz & Okun, 2016) that will be covered in other chapters of this book. So, it is expected that complex interactions between different mechanisms are responsible for generating the observed characteristic symptoms. It is possible that these complex networks, associated with motor skills, are particularly involved in TS. Consistent with that hypothesis, past findings revealed that motor performance differed significantly on the Purdue Pegboard Tests in tic disorders patients (O'Connor et al., 2008; Sukhodolsky, Landeros-Weisenberger, Scahill, Leckman, & Schultz, 2010), which represent a test of fine motor dexterity. Interestingly, other studies found a correlation between the Purdue Pegboard Test score and dopamine D1 binding potential (Wang et al., 1998) and better response to dopaminergic medication (Smits et al., 2017). Furthermore, reduced performances with the dominant hand on the Purdue Pegboard test predicted worse adulthood tic severity and correlated with tic severity at the time of childhood assessment (Bloch, Sukhodolsky, Leckman, & Schultz, 2006). It is also noteworthy that cognitive-behavioral treatment improved motor performances at the Purdue pegboard in tic disorders compared to a waitlist control, and this improvement was positively correlated to clinical outcome measures (O'Connor et al., 2008).

This clinical evolution raises the hypothesis of a possible alteration in motor dexterity and motor inhibition processing. Therefore, if an inhibitory control is associated with TS, the next step would be to investigate the functional relationship between motor inhibition skills and the cerebral structures involved. This parallel between inhibitory control and cerebral processes is important to investigate further the impact of brain development in the evolving expression of symptoms. At the neuroanatomical level, neuroimaging studies have pinpointed the contribution of the frontal motor cortex and deeper subcortical regions such as the basal ganglia, which have been structurally affected among children and adults afflicted with TS (Peterson et al., 2003). These deep nuclei project themselves via the thalamus and then to different cortex regions that seem to be affected in TS, such as the prefrontal and somatosensory cortex (Thomalla et al., 2009). More recent papers addressing brain connectivity showed widespread structural white matter connectivity deficits analyzed by probabilistic fiber tractography in TS patients. Lower connectivity values were found specifically in tracts connecting the SMA with basal ganglia (pre-SMA–putamen, SMA–putamen) and frontal cortico-cortical circuits (Cheng et al., 2014). In addition, structural connectivity of frontal brain networks involved in executive functions such as planning and controlling motor actions was reduced and associated with tic severity. These authors hypothesized that TS could be considered a neurodevelopmental disorder of brain immaturity.

These breakthroughs provide an excellent topography of the potential malfunctioning structures in TS. However, it does not give us clues whether these affected structures can operate normally or not under certain demands. Despite these alterations of motor cortex excitability at rest demonstrated in the TS population, there has been little research into changes in cortical motor activity during voluntary movements and tic generations. To address that question, Neuner et al. (2014) revealed the chronometry of brain activity before and during tics with certain spatial and temporal precision. For instance, two seconds before a tic, the supplementary motor area, the ventral primary motor cortex, the primary sensorimotor cortex, and the parietal operculum exhibited activation. Then, at one second before the tic generation, the anterior cingulate, putamen, insula, amygdala, cerebellum, and the extrastriatal-visual cortex showed activation. Finally, the thalamus, central operculum, primary motor, and somatosensory cortices exhibited activation with tic onset. Analysis of resting-state data resulted in network strength in SMA that was also active before tic occurrence, correlated significantly with tic severity.

2.1 A cerebral motor over-activation hypothesis

Hence, data obtained from neuropsychology, electrophysiology, and brain imaging showed the SMA seems to be a structure consistently affected structurally and functionally in TS patients. The SMA forms a cluster within the dorsomedial frontal cortex with the supplementary eye field and the pre-supplementary motor area. According to Nachev, Kennard, and Husain (2008), the SMA exhibits sensitivity to tasks with action contingencies. This process illustrated by differences in SMA activity between' self-initiated and "externally triggered" actions, by movement sequences, by well-learned and poorly learned actions, and by switching between motor action possibilities. Damage to this structure disrupts behavior in complex ways, affecting not just the commission but also the omission of actions, and the effects are broadly reflective of the neurophysiological properties of the region. Consistent with these findings, there is often a problem in planning and executing motor action in patients with tic disorders. One of the first fMRI studies investigating motor functions in TS patients showed heightened activation in the premotor cortex and SMA during a finger-tapping task (Biswal et al., 1998). However, this pattern of activation could depend on a nonselective overactivity of the motor system or on a problem in modulating effort. To address that question, Serrien et al. (2002) showed that the SMA of the TS patients has small or greatly reduced activation when executing a manipulative task compared with a baseline condition. Cortical areas involved in movement preparation were continuously activated. This pattern suggested that the constant activation of SMA might explain the involuntary urges to move, preventing accurate planning of voluntary behavior.

Regarding goal-directed behavior during a motor countermanding task, O'Connor et al. (2005) examined electrophysiological indices of preparation (readiness potential) and execution (movement-associated potential) during automated and controlled reaction time in adults with a chronic tic disorder, habit disorder, and neurotypical control. The patients with chronic tics showed reduced readiness potential and no correlation between reaction times and cerebral motor preparation responses. This could further confirm that adults with chronic tics may not modulate cortical motor activation optimally in planning and executing both automated and controlled responses. Furthermore, this suggests that chronic tic disorders involve heightened activation, which could impede initiation and the control of complex motor actions. Brain imaging findings also seem to point towards greater activation of bilateral frontostriatal regions in TS, which accompanied poorer performance on the Stroop, a well-known task of cognitive interference inhibition (Marsh, Zhu, Wang, Skudlarski, & Peterson, 2007). This finding implied that greater activation of the frontostriatal system helps maintain task performance in individuals with TS. This result was partly confirmed by combining neuropsychological evaluation and electrophysiology (Thibault, O'Connor, Stip, & Lavoie, 2009), which revealed cerebral motor response onset faster in response to incompatible stimuli and faster interference processing with the Stroop test in the TS compared with paired adult controls. There was also a larger frontal activation during inhibition of motor responses in the TS group. This is consistent with increased activation in the direct pathway through the basal ganglia and increased activation in the prefrontal cortex and the subthalamic nucleus observed during a similar inhibition control task in the TS population (Baym, Corbett, Wright, & Bunge, 2008). These data are convergent with a probable cortical motor over-activation hypothesis of TS involving faster motor program selection in processing conflicting stimulus-response configuration, which could involve suboptimal functioning in some basal ganglia structures and other related structures. But we can also introduce the issue of whether these results could be contrasted across cortical activation and inhibition. A study using magnetoencephalography event-related desynchronization (ERDm) and event-related synchronization (ERSm) during self-paced finger movement tasks was proposed with TS patients (Franzkowiak et al., 2010). ERDm represents cortical motor activation, whereas ERS most likely indicates its inhibition. In TS patients, ERDm was significantly increased in the contralateral hemisphere before and during movements, whereas ERSm following movement termination was increased in the ipsilateral motor cortex (M1), which was inversely correlated with tic severity. These results support the hypothesis that during voluntary movement, motor cortical reactivity is altered in TS patients. The pattern of increased activation ERD before and during movement execution followed by increased inhibition ERS after movement termination could suggest abnormally increased motor cortical activation, possibly driving stronger inhibition (Franzkowiak et al., 2010). The same research group pushed further and performed a brain connectivity analysis to identify the circuit responsible for this activation and found greater connectivity between the primary motor cortex and the SMA in patients with TS (Franzkowiak et al., 2012). These results are coherent with a functional neuroimaging study that also identified such connectivity between these two regions during the occurrence of tics (Hampson, Tokoglu, King, Constable, & Leckman, 2009).

In the light of these findings, it appears that the stronger this inhibition is, the better symptoms appear to be controlled, and again, this could have crucial importance for cognitive-behavioral treatment. There could be a compensatory reorganization at the level of frontoparietal circuits, which would result from the overactivity of the CSTC network. This reorganization would allow patients to have better control over their tics but cause some delay during an experimental task (Thomalla et al., 2014). Consistent with that hypothesis, another study also discovered a particular pattern of brain activity in patients with TS during a stop-signal task, where activation of the dorsal premotor cortex was more prominent during the presentation of Go stimuli than during the presentation of Stop stimuli. Interestingly, a significant positive correlation was present between motor tic frequency and SMA activations during the Stop than the Go condition in patients (Ganos et al., 2014). Thus, it seems that in patients with tics, the SMA acts as the overall mechanism of inhibition. Again, the SMA is probably a key relay of inhibitory processes mediating both suppressions of tics and inhibition of voluntary action. Thus, the more a person tends to have tics, the more resources are needed to properly inhibit them. It is thus conceivable that common mechanisms could be involved in both the inhibition of tics and voluntary movements. This could also suggest a modulation deficit not only in motor response inhibition but also in cognitive control.

To sum up, data obtained from multimodal measures showed that the SMA is a crucial structure affected both structurally and functionally in TS patients. This could explain in part a problem in planning and execution of motor action. The presence of constant activation of SMA might explain the involuntary urges to move, preventing accurate planning of voluntary behavior. Moreover, this suggests the involvement of a state of heightened activation, which could impede initiation and the control of complex motor actions. In addition, brain-imaging findings implied that greater activation of the frontal and striatal system might help to maintain task performance in individuals with TS. Increased activation in the direct pathway through the basal ganglia and increased activation in the prefrontal cortex and the subthalamic nucleus during motor inhibition tasks represent an important network underlying tics. There is also a pattern of abnormally increased motor cortical activation, possibly driving stronger inhibition, consistent with greater connectivity between the primary motor cortex and the SMA. It appears that the stronger this inhibition is, the better symptoms appear to be controlled, and we purport that this represents a key ingredient to therapeutic interventions. With that proviso, the next phase is to propose an integrative model with a component of motor training that allows the improvement of overactive and inhibitory control. These initial baseline factors may relate to high sensorimotor activation, elevated muscle contraction and so complement the application of behavioral principles.

2.2 Psychotherapy and developmental cerebral plasticity: Toward a perfect window of opportunity to intervene

From a clinical standpoint, tic disorders are characterized by their inherent fluctuating nature over time, and their developmental trajectory needs to be considered. Through longitudinal studies, certain hypotheses have underlined cerebral anomalies associated with symptoms persistency in adulthood. Originally, Peterson (2001) proposed that because it is present in every age group, the hypometabolism of the caudate nucleus could constitute a core feature of TS. Moreover, the volume decrease of the putamen, the internal globus pallidus and prefrontal areas, as well as the increase in the volume of premotor areas, are uniquely present among adults, suggesting an association with evolving mechanisms contributing to the maintenance or inhibition of symptoms at least among adult subgroups with significant symptoms persisting during adulthood. Among these individuals, there seems to be a failure of cerebral plasticity mechanisms that allows compensating the presence of tics by overactivation of a motor inhibition process. Unlike adults, TS children have a larger orbitofrontal volume (Peterson et al., 2001; Spessot & Peterson, 2006; Spessot, Plessen, & Peterson, 2004), constituting adaptive plasticity in response to the tic expression, which, in turn, would help to inhibit them more easily. A study using transcranial magnetic stimulation to induce plasticity confirms that aberrant synaptic plasticity in TS is related to reducing long-term consolidation of motor skills, which is correlated to symptom severity (Brandt et al., 2014). With the maturation of the prefrontal cortex during adolescence, this mechanism could gain strength and explain the symptom decrease during adolescence and early adulthood. Among adults with persisting symptoms, this prefrontal compensation could not occur or at least in the least efficient way. The decrease in the putamen and the globus pallidus volume, and thus the increase in the volume of the premotor frontal area, would only be secondary to this compensation. These neurodevelopmental observations are compatible with the CoPs model previously described (O'Connor, 2002, 2005; O'Connor et al., 2009, 2017). If the evolution and fluctuation of symptoms are related to a type of cerebral plasticity, the adolescence or young adulthood could represent a time window of opportunity to intervene. Thus, we propose that psychotherapy could improve symptoms, as well as favor cognitive and neurophysiological changes, with better outcome in that relatively early developmental period. The corresponding post-therapy normalization of cerebral and motor function has been observed previously (Lavoie et al., 2011; Morand-Beaulieu et al., 2015, 2016, 2018) and allows to postulate that motor and physiological processes are related to psychotherapeutic outcome.

3. Integrating the knowledge of neurobiology and physiology for cognitive-behavioral treatments and comprehensive biofeedback

If we purport that psychotherapy improves symptoms and that clinical improvement is related to psychophysiological changes during a given developmental stage, we suggest that the experimental discoveries could be regrouped into four main hypotheses. These findings propose that in tic disorders, there is, first, an *overactive style of planning* preventing optimal preparation for action. Secondly, there is a *dysregulation in arousal/inhibition* processes, particularly under circumstances where regulation is open-looped, controlled, and has unpredictable parameters. Third, there are *high levels of motor activation* that create tensions and frustrations and are likely to evoke tics. Finally, there is the possibility to address the cognitive *psychophysiological sources of motor activation* to reduce background tension and prevent tic onset.

Whereas traditional habit reversal targets the tic-implicated muscle solely in a competing response, an important additional component in our program is the modification of excessive overall motor activation by targeting cognitive and behavioral/physiological sources creating tensions. We were able to demonstrate that the temporal pattern of tic generation follows the cortico-striatal-thalamocortical circuit and that cortical structures precede subcortical activation. The analysis of spontaneous tic fluctuations highlights the role of cortical premotor structures. Several findings corroborate the notion of tic disorders as a syndrome in which abnormal resting-state network activity might contribute to the generation of tics. These circuit loops are involved in multiple functions such as emotions, motivation, personality, disinhibition, executive functions, motor planning, and the control of different muscles as well. They probably represent one of the best candidate networks underlying the expression of tics and constitute an interesting focus for treatment, especially biofeedback. The technique of biofeedback typically allows a patient to gain voluntary control over physiological responses by making them explicit through real-time visual or auditory feedback. Patients are typically able to learn how to adjust these physiological processes consciously. That technique incorporates an array of cognitive, behavioral, or psychophysiological therapies that have been gaining popularity in the management of neurologic and psychiatric conditions over the last decades. Such biofeedback is achieved through a combination of variations in attentional focus (cognition) and adjustments in muscle tension or heart rates and depth of breathing (physiology), for instance.

The initial attempts at biofeedback for TS used electroencephalography, targeting the enhancement of the 12–14 Hz rhythm that emerges over the sensorimotor area (Tansey, 1986). That technique was reported as successful in two patients with TS who showed a dramatic reduction in tic frequency. The biofeedback sessions were administered for 30 min once a week. However, the overall length of the treatment seemed individually tailored as one of the patients benefited from biofeedback after 5 weeks, while the other patient began to feel the benefit after 18 weeks. Another investigation using a muscle biofeedback protocol using electromyograms was recorded at rest, during a baseline period, and at post biofeedback training in nine subjects suffering from chronic tic disorder. In that study, patients received equivalent 2–4 biofeedback-training sessions aimed at enhancing their ability to control levels of muscle contraction in both affected and non-affected sites. Sixty percent of patients reported a clinically significant decrease in tic frequency, and the reflex-like quality of tic muscles was normalized by biofeedback training (O'Connor, Gareau, & Borgeat, 1995). More recently, a single case study with sensorimotor theta EEG neurofeedback training was delivered to a 17-year-old male with TS. The patient showed a significant tic reduction after twenty-two sessions suggesting that sensorimotor decrease might account for some clinical effects (Messerotti Benvenuti, Buodo, Leone, & Palomba, 2011). Theta is an EEG oscillation that underlies various aspects of cognition and behavior, including learning, memory and emotional regulation (Ertl, Hildebrandt, Ourina, Leicht, & Mulert, 2013). Yet, in these studies of biofeedback, given the lack of outcome measures and follow-up with these patients, nonspecific effects of treatment cannot be excluded as explanatory factors in their improvement.

The element of outcome measures is particularly relevant because a relationship is frequently described between tic frequency and states of bodily arousal and focused attention. One of the first proof of concept, in applying biofeedback systematically to TS, significantly showed tic frequency positively correlating with sympathetic arousal (Nagai, Cavanna, & Critchley, 2009). More precisely, they observed significantly lower tics during relaxation biofeedback compared to arousal biofeedback. These findings suggested that the conjunction of focused attention to task and reduced peripheral sympathetic tone inhibit tic expression. More recently, the same team conducted a randomized controlled trial of electrodermal biofeedback training in 21 patients with TS (Nagai et al., 2014). After training patients for three sessions a week over 4 weeks, they observed a significant reduction in tic frequency and improved indices of subjective well-being in both the active-biofeedback and sham-feedback groups. Nevertheless, the active-treatment group did not noticeably learn to reduce their sympathetic electrodermal tone using biofeedback. According to the authors, the main confounding factor appeared to be the 30 min duration of the training sessions, which made it hard for patients to sustain a reduction in sympathetic tone when their tics themselves were generating competing phasic electrodermal arousal responses.

As described here, only a few studies systematically applied biofeedback to the treatment of chronic tics, and only one was realized within a randomized controlled protocol. Challenge remains high, whereas some biofeedback experiments obtained encouraging results, while others showed more equivocal findings. This can be ascribed at least in part to the limited number of sessions in earlier studies, thus restraining the potential advantage of that technique. A lack of follow-up and outcome measures during and after biofeedback is also an important point to consider assessing the durable effects of the active treatment. But then, a more crucial point lies in the fact that the most promising results seem to point toward more directed biofeedback, addressing motor aspects related to activation of SMA and somatosensory frontal regions (Hampson et al., 2011). These considerations appear to be fundamental for improving specialized interventions for tic disorders. Despite evidence of effectiveness, some people with tics are treatment-resistant to psychotherapy. However, current gold-standard psychotherapy does not directly address the learning of self-controlled psychophysiological strategies. To overcome this limitation, we propose a solution by integrating psychophysiology in the psychologist's office, which provides an additional means to support current effective psychotherapy.

With that in mind, it is possible to integrate biofeedback measures within the CoPs therapy and focus on the biofeedback of motor control to improve clinical outcomes. Our previous results from Perrault et al. (Perrault, 2019; Perrault, Morand-Beaulieu, O'Connor, Blanchet, & Lavoie, 2018) revealed that the effective regulation of muscle tension before therapy was a sensitive predictor to treatment outcomes. To integrate these results, we conducted a pilot RCT focusing on biofeedback and motor regulation with nine mildly severe TS adult patients, and our results showed promising pre-post improvement in symptoms following biofeedback + CoPs (47%) compared with CoPs alone (33%). In addition, our pilot results showed a reduction of EMG activation across sessions and an accrual of time on biofeedback across sessions accompanied by a gradual reduction of the observed tic duration across sessions.

These considerations appear to be fundamental for the improvement of specialized interventions for TS in the near future. Integrating the psychophysiological model in the psychologist's office offers possibilities to support cognitive and behavioral treatments through learning self-controlled strategies. An important element not present in the earlier type of biofeedback is the limited account of aspects associated with active motor processing. The first attempts at motor biofeedback for TS used electroencephalography targeting enhancement of the motor rhythm that emerges over the sensorimotor area (Tansey, 1986). However, more recent studies have successfully shown that motor biofeedback enables them to gain control of activity in specific parts of their brain and alter functional connectivity between cerebral areas. Hampson et al. (2011) used biofeedback from real-time functional magnetic resonance imaging data to train healthy subjects to control activity in their SMA. Although a significant increase over the SMA during biofeedback was not found, participants were able to exert significant control over that region in later biofeedback sessions. When looking at functional connectivity to the SMA, the strength of functional connectivity with subcortical regions was reduced after biofeedback. These authors suggested that motor biofeedback has great potential to allow patients to develop greater conscious control over SMA activity by reducing the influence of cortical-striatal-thalamocortical loops. This implies that we need an active motor task that could activate these structures, to regain control and diminish tics. For research purposes, this fMRI biofeedback protocol is interesting to document the neurobiological as well as the functional levels. However, to push further that question, it is important to transfer this knowledge to the clinician offices. To do so, we are now proposing to step further by integrating biofeedback measures within the therapy and focusing on motor control to improve clinical outcomes.

3.1 Best practices for a biofeedback protocol embedded within psychotherapy sessions

In our literature review, we proposed a hypothesis of a cortical motor over-activation involving faster motor program selection and suboptimal functioning of the supplementary motor area. The question is: Can we intervene to impact these structures and improve motor functioning? The subsequent stage is to determine whether a treatment program combining psychotherapy and biofeedback focusing on motor regulation could be superior in efficacy to psychotherapy alone. Since tic disorders are considered neurodevelopmental, the other crucial question is to investigate an emerging adult population through a period where the cerebral regions implied in tic generations are still in development and probably more receptive to treatment. Notably, the investigation of the emerging adult populations was often neglected. Over 326 studies were detected using the keywords "Tourette" and "clinical trials" in PubMed. Among them, only 10% were executed in young adults of the 19–24 years age range. Thus, for future research, the importance to investigate that age range lie in the fact that it is clearly understudied and potentially corresponds to a window of opportunity where the motor cortex is still malleable to therapeutic change.

To evaluate whether the addition of biofeedback to psychotherapy is useful, we propose the use of a superiority trial. A superiority trial is designed to show that a new treatment is better than active control. In this type of psychotherapeutic trial, trained psychologists should deliver treatment sessions on a weekly one-to-one basis for a period of 10 weeks, plus 4-week home practice sessions should administer all therapies concurrently. Typically, the program should be progressive through seven major steps, including awareness training, constructing a high and low-risk tic profile, muscle discrimination, modifying background sensorimotor activation through modification of habit style of planning action, cognitive, meta-cognitive, and behavioral restructuration of anticipation and appraisals in risk tic situations, and preventing relapse (see O'Connor, Lavoie and Schoendorff (2017) for a manualized intervention). For the biofeedback sessions, the designated therapist for each participant should integrate the biofeedback training at the beginning of the standard therapy session. The biofeedback normally takes the form of computer-generated graphics visually presented on a computer monitor. Each biofeedback session should also begin with a 5 min resting baseline followed by an active biofeedback session for 15 min, followed by 5 min of resting baseline (total of 25 min + 5 min installation). Our pilot data showed good tolerance to that biofeedback duration before each psychotherapeutic sessions.

In an active biofeedback session, changes in the participant's muscular activity in the intended direction result in movement through an animated balance with a silver ball. Animated graphical feedback of this sort is particularly effective in capturing the attention and interactive engagement of participants (Fig. 1). The main goal will be to contract the arm muscle to keep the ball in the middle of the balance for 5 min for each arm and then 5 min with both arms. In this kind of protocol, another group should be randomly assigned to a sham biofeedback condition. A sham condition, also called placebo therapy, constitutes an inactive treatment or procedure that is intended to mimic as closely as a possible therapy in a clinical trial. In that condition, the patients view the same computer-generated graphics like those used in the active-biofeedback condition and try to modulate their EMG activity level using the evolving sequence of animations. In contrast to the active biofeedback group, the animation should change direction unrelated to the muscle contraction of the patient. Thus, the sham-control patients could not learn direct biofeedback control. To avoid frustrating the patient with a

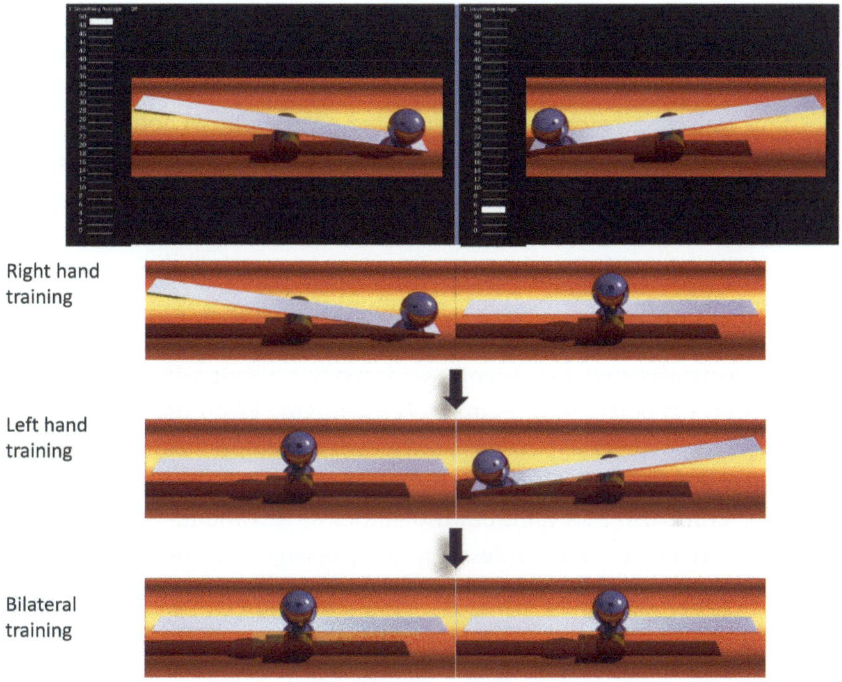

Fig. 1 Illustration of a biofeedback image to contract the arm muscle to keep the ball in the middle of the balance during 5 min for the left arm, 5 min for the right arm and then 5 min with both arms. In this depiction, the total EMG scale values are between 0 and 50 μV, and when the ball remain on the center of the balance, this represent a value of 25 μV with a response variance of ±15 μV at the first session corresponding to an EMG threshold criterion between 10 and 40 μV. The criterion could be adjusted gradually at each session incrementally when biofeedback is reached 80% of the time during the total session.

lack of apparent feedback responses, we propose to change the threshold criterion to obtain correct feedback by chance (less than 50% of the trial with a reduced response variance tolerance). The biofeedback criterion and variables are very important because these criteria decide the magnitude of muscle tension to attain correct biofeedback. In a typical scenario, the total EMG scale values are between 0 and 50 μV, and the ball at the center of balance represents a value of 25 μV with a response variance tolerance of ±15 μV at the first session corresponding to an EMG threshold criterion between 10 and 40 μV. The criterion could be adjusted gradually at each session incrementally when biofeedback is reached 80% of the time during the 15 min session. The duration in seconds, in which the patient holds the ball in the middle of the balance (EMG between 15 and 40 μV), could be entered as a variable for each hand and then with both hands. The differential tensions in the involved muscles during biofeedback (left arm, right arm, both arms) across sessions could also represent variables of interest. We suggest recording muscle tension of each extensor Carpi Ulnaris of the forearm and the upper trapezius separately with dry nickel-plated electrodes. The recording of the upper trapezius activity is not being related to the biofeedback signal and only represents a control to measure other muscles' involvement, non-related to the task. In addition to these measures, it is also important to report on-task tic monitoring. During both biofeedback sessions, participants should be monitored using a digital camera to record tics that could be later identified and quantified, for instance, with the Rush Video-Based Tic Rating Scale (Goetz, Pappert, Louis, Raman, & Leurgans, 1999) filled post-hoc by two independent raters to assess the symptom evolution during treatment.

Another interesting component not often assessed in the field of biofeedback is the comparison of different modalities of treatment. When comparing different treatment modalities, it is also important to control for sources of bias to ensure the independence of clinical evaluators and verify treatment integrity, therapeutic alliance, and treatment acceptability. To protect against sources of bias, it is important to control that none of the researchers will be directly involved in supervision or therapy. Ideally, they can share responsibility for the training and clinical supervision with an expert co-supervisor. Nevertheless, to ensure the credibility of modalities (e.g., biofeedback plus therapy vs. therapy alone), each treatment should be presented by the appropriate therapist with an allegiance to the approach as an empirically validated intervention with an introductory package outlining the approach and illustrating its validity. The *independence of evaluators* is also important, and they should ideally conduct clinical evaluations on four measurement times (pre-treatment, post-treatment, 3-month follow-up, and at a 6-month follow-up) to ensure treatment robustness. It is important to mention that evaluators must be blind to the participant treatment modality and should not receive information on participant history and psychopathology. In addition, the evaluator should be separated from the therapists by geographical location if possible. Another important part is to assess *treatment integrity* with a valid and reliable scale. For instance, the Implementation of the Treatment Integrity Procedures Scale could be proposed to evaluate treatment integrity, including establishing, assessing, and reporting integrity, as well as therapist treatment adherence and therapist proficiency (Perepletchikova, 2011). A series of measures should also assess motivation and the therapy process (credibility, expectancy, and perception of therapeutic relation) following recommendations.

The *therapeutic alliance* should also be assessed by the working alliance inventory, a 12-item scale comprising subscales of bonding, tasking, and goals (Mellado et al., 2017). Objective measures of *treatment tolerability* are counts of treatment refusers and dropouts across treatments. A refuser is defined as a participant who was eligible for the study but declined to participate after randomization or before the 1st session. A dropout is defined as a participant who withdraws participation after the beginning of the first treatment session. The reason for refusal or dropout must also be documented to verify a possible bias of selection.

Another critical research question to take into account is how other factors such as biopsychosocial variables mediate the treatment outcome. It is generally recognized that primary and secondary outcomes have the power to measure the impact of treatment, while process variables must identify other mechanisms taking place in psychotherapy that could lead to these outcomes. For that purpose, the addition of process variables such as the psychosocial (comorbidity, self-esteem, and quality of life), neurocognitive (motor function and response inhibition), and physiological (stress) variables is crucial to see how these mechanisms mediate clinical changes.

4. Conclusion

Tourette Syndrome has been studied intensively and comprehensively for the last 60 years, but it is still considered a complex and widely unknown neuropsychiatric disorder. We can undoubtedly assume that past investigations have made progress in the treatment of this syndrome, but many questions remain. For instance, why are symptoms fluctuating? Why have many patients failed to respond to current treatment? Do symptoms simply disappear or become successfully inhibited in adults? These questions can only be tackled with a multimodal approach combining neurology, psychology, and neuroscience. A unidimensional approach precludes integrating the various levels of cerebral functioning. Structured clinical interviews are useful to follow up on clinical states and traits, but they can produce superficial or indirect evidence on the syndrome. An archetypal approach from a single neurobiological model is unlikely since this pathology is too intricate. A cognitive-behavioral model has the potential to relate impairment with the clinical expression of the illness that will affect therapeutic approaches. However, it delivers little or indirect facts about the cerebral origins of the syndrome. Neuropsychology offers the potential for valid interpretations about discrete anomalies, but these interpretations are essentially based on our knowledge of focal lesions, not on functional disorders. In complement, novel brain imaging techniques are appropriate for identifying localized metabolic abnormalities. Despite this, it is restrained by its low temporal resolution that often fails to explain fast stream cognitive processes. Moreover, electrophysiological techniques (EEG and event-related potentials) unlocked the door to track in real-time the cerebral activation underlying rapid cognitive processes. Then again, the activity recorded over the scalp might also reflect more profound subcortical activity, which can be only extrapolated or modeled through multiple generating sources at best. Another crucial challenge is to connect multiple sources of information to behavioral and neurocognitive expressions of chronic tic symptoms. The benefit of a multifaceted approach will be to extract a complete profile allowing prediction of symptom development and treatment success prediction. From a clinical standpoint, individualized therapeutic action should include the improvement of motor symptomatology and include cognitive strategies to deal with tics.

In conclusion, two issues seem fundamental for the development of psychological treatment for TS by integrating psychophysiological technology as an instrument of treatment. These new possibilities can support cognitive and behavioral restructuring through learning self-controlled strategies. Second, the dissemination of study results on alternate interventions or other front lines must be done. Finally, treatments for TS symptoms, empirically acknowledged to be effective, should be known by the public and be more accessible.

References

Ascher, E. (1948). Psychodynamic considerations in Gilles de la Tourette's disease, maladie des tics, with a report of 5 cases and discussion of the literature. *American Journal of Psychiatry*, *105*(4), 267–276.

Azrin, N. H., & Nunn, R. G. (1973). Habit-reversal: A method of eliminating nervous habits and tics. *Behaviour Research and Therapy*, *11*(4), 619–628. https://doi.org/10.1016/0005-7967(73)90119-8.

Azrin, N. H., & Peterson, A. L. (1988). Habit reversal for the treatment of Tourette syndrome. *Behavior Research and Therapy*, *26*(4), 347–351.

Baym, C. L., Corbett, B. A., Wright, S. B., & Bunge, S. A. (2008). Neural correlates of tic severity and cognitive control in children with Tourette syndrome. *Brain: A Journal of Neurology*, *131*(1), 165–179. https://doi.org/10.1093/brain/awm278.

Biswal, B., Ulmer, J. L., Krippendorf, R. L., Harsch, H. H., Daniels, D. L., Hyde, J. S., et al. (1998). Abnormal cerebral activation associated with a motor task in Tourette syndrome. *American Journal of Neuroradiology*, *19*(8), 1509–1512.

Bloch, M. H., Sukhodolsky, D. G., Leckman, J. F., & Schultz, R. T. (2006). Fine-motor skill deficits in childhood predict adulthood tic severity and global psychosocial functioning in Tourette's syndrome. *Journal of Child Psychology and Psychiatry*, *47*(6), 551–559. https://doi.org/10.1111/j.1469-7610.2005.01561.x.

Brandt, V. C., Niessen, E., Ganos, C., Kahl, U., Baumer, T., & Munchau, A. (2014). Altered synaptic plasticity in Tourette's syndrome and its relationship to motor skill learning. *PLoS One [Electronic Resource]*, *9*(5), e98417. https://doi.org/10.1371/journal.pone.0098417.

Capriotti, M. R., Piacentini, J. C., Himle, M. B., Ricketts, E. J., Espil, F. M., Lee, H. J., et al. (2015). Assessing environmental consequences of ticcing in youth with chronic tic disorders: The tic accommodation and reactions scale. *Child Health Care*, *44*(3), 205–220. https://doi.org/10.1080/02739615.2014.948164.

Catrou, J. (1890). Étude sur la maladie des tics convulsifs. (Thèse pour le doctorat en médecine), Hôpital Pitié-Salpetrière, Paris. Retrieved from www.bium.univ-paris5.fr/histmed/medica/cote?TPAR1890x129.

Cheng, B., Braass, H., Ganos, C., Treszl, A., Biermann-Ruben, K., Hummel, F. C., et al. (2014). Altered intrahemispheric structural connectivity in Gilles de la Tourette syndrome. *NeuroImage Clinical*, *4*, 174–181. https://doi.org/10.1016/j.nicl.2013.11.011.

Corbett, J. A., Mathews, A. M., Connell, P. H., & Shapiro, D. A. (1969). Tics and Gilles de la Tourette's syndrome: A follow-up study and critical review. *British Journal of Psychiatry*, *115*(528), 1229–1241.

Dana, C. L., & Wilkin, W. P. (1886). On convulsive tic with explosive disturbances of speech (so-called Gilles de la Tourette's Disease). *Journal of Nervous & Mental Disease*, *13*(7), 407–412.

Deckersbach, T., Rauch, S., Buhlmann, U., & Wilhelm, S. (2006). Habit reversal versus supportive psychotherapy in Tourette's disorder: A randomized controlled trial and predictors of treatment response. *Behaviour Research and Therapy*, *44*(8), 1079–1090. https://doi.org/10.1016/j.brat.2005.08.007.

Ertl, M., Hildebrandt, M., Ourina, K., Leicht, G., & Mulert, C. (2013). Emotion regulation by cognitive reappraisal—The role of frontal theta oscillations. *Neuroimage*, *81*, 412–421. https://doi.org/10.1016/j.neuroimage.2013.05.044.

Evers, R. A., & van de Wetering, B. J. (1994). A treatment model for motor tics based on a specific tension-reduction technique. *Journal of Behavior Therapy and Experimental Psychiatry*, *25*(3), 255–260.

Ferenczi, S. (1921). Psycho-analytical observation on tic'. *International Journal of Psycho-Analysis*, 1–30.

Frank, M., & Cavanna, A. E. (2013). Behavioural treatments for Tourette syndrome: An evidence-based review. *Behavioural Neurology*, *27*(1), 105–117. https://doi.org/10.1155/2013/134863.

Franzkowiak, S., Pollok, B., Biermann-Ruben, K., Sudmeyer, M., Paszek, J., Jonas, M., et al. (2010). Altered pattern of motor cortical activation-inhibition during voluntary movements in Tourette syndrome. *Movement Disorders*, *25*(12), 1960–1966. https://doi.org/10.1002/mds.23186.

Franzkowiak, S., Pollok, B., Biermann-Ruben, K., Sudmeyer, M., Paszek, J., Thomalla, G., et al. (2012). Motor-cortical interaction in Gilles de la Tourette syndrome. *PLoS One*, 7(1), e27850. https://doi.org/10.1371/journal.pone.0027850.

Ganos, C., Kuhn, S., Kahl, U., Schunke, O., Feldheim, J., Gerloff, C., et al. (2014). Action inhibition in Tourette syndrome. *Movement Disorders*, *29*(12), 1532–1538. https://doi.org/10.1002/mds.25944.

Gilles de la Tourette, G. (1885). Étude sur une affection nerveuse caractérisée par de l'incoordination motrice accompagnée d'écholalie et de coprolalie. *Archives de Neurologie*, *9*(19-42), 158–200.

Goetz, C. G., Pappert, E. J., Louis, E. D., Raman, R., & Leurgans, S. (1999). Advantages of a modified scoring method for the rush video-based tic rating scale. *Movement Disorders*, *14*(3), 502–506.

Gunduz, A., & Okun, M. S. (2016). A review and update on Tourette syndrome: Where is the field headed? *Current Neurology & Neuroscience Reports*, *16*(4), 37. https://doi.org/10.1007/s11910-016-0633-x.

Hallett, M. (2015). Tourette syndrome: Update. *Brain & Development*, *37*(7), 651–655. https://doi.org/10.1016/j.braindev.2014.11.005.

Hampson, M., Scheinost, D., Qiu, M., Bhawnani, J., Lacadie, C. M., Leckman, J. F., et al. (2011). Biofeedback of real-time functional magnetic resonance imaging data from the supplementary motor area reduces functional connectivity to subcortical regions. *Brain Connect*, *1*(1), 91–98. https://doi.org/10.1089/brain.2011.0002.

Hampson, M., Scheinost, D., Qiu, M., Bhawnani, J., Lacadie, C. M., Leckman, J. F., et al. (2011). Biofeedback of real-time functional magnetic resonance imaging data from the supplementary motor area reduces functional connectivity to subcortical regions. *Brain Connect*, *1*(1), 91–98. https://doi.org/10.1089/brain.2011.0002.

Hampson, M., Tokoglu, F., King, R. A., Constable, R. T., & Leckman, J. F. (2009). Brain areas coactivating with motor cortex during chronic motor tics and intentional movements. *Biological Psychiatry*, *65*(7), 594–599.

Himle, M. B., Capriotti, M. R., Hayes, L. P., Ramanujam, K., Scahill, L., Sukhodolsky, D. G., et al. (2014). Variables associated with Tic exacerbation in children with chronic tic disorders. *Behavior Modification*, *38*(2), 163–183. https://doi.org/10.1177/0145445514531016.

Hoogduin, K., Verdellen, C., & Cath, D. C. (1997). Exposure and response prevention in the treatment of Gilles de la Tourette's syndrome: Four case studies. *Clinical Psychology and Psychotherapy*, *4*(2), 125–135.

Itard, J. (1825). Mémoire sur quelques fonctions involontaires des appareils de la locomotion, de la préhension et de la voix. *Archives Générales de Médecine*, *8*, 385–407.

Kramer, H., & Sprenger, J. (1948). *Malleus maleficarum*. London: Pushkin Press.

Lang, P. J., McTeague, L. M., & Bradley, M. M. (2014). Pathological anxiety and function/dysfunction in the brain's fear/defense circuitry. *Restorative Neurology & Neuroscience*, *32*(1), 63–77. https://doi.org/10.3233/RNN-139012.

Langen, M., Durston, S., Kas, M. J., van Engeland, H., & Staal, W. G. (2011). The neurobiology of repetitive behavior: ...and men. *Neuroscience and Biobehavioral Reviews*, *35*(3), 356–365. S0149-7634(10)00029-1 [pii]10.1016/j.neubiorev.2010.02.005.

Lavoie, M. E., Imbriglio, T. V., Stip, E., & O'Connor, K. P. (2011). Neurocognitive changes following cognitive-behavioral treatment in Tourette Syndrome and chronic tic disorder. *International Journal of Cognitive Therapy*, *4*(1), 34–50. https://doi.org/10.1521/ijct.2011.4.1.34.

Lavoie, M. E., Leclerc, J., & O'Connor, K. P. (2013). Bridging neuroscience and clinical psychology: Cognitive behavioral and psychophysiological models in the evaluation and treatment of Gilles de la Tourette syndrome. *Neuropsychiatry (London)*, *3*(1), 75–87. https://doi.org/10.2217/npy.12.70.

Lavoie, M. E., Thibault, G., Stip, E., & O'Connor, K. P. (2007). Memory and executive functions in adults with Gilles de la Tourette syndrome and chronic tic disorder. *Cognitive Neuropsychiatry*, *12*(2), 165–181. https://doi.org/10.1080/13546800600826371.

Leclerc, J., Forget, J., & O'Connor, K. P. (2008). *Quand le corps fait à sa tête - Le syndrome de Gilles de la Tourette*. Montréal: Multimondes.

Leclerc, J. B., O'Connor, K. P., Nolin, G. J., Valois, P., & Lavoie, M. E. (2016). The effect of a new therapy for children with tics targeting underlying cognitive, behavioral, and physiological processes. *Frontiers in Psychiatry*, 7, 135. https://doi.org/10.3389/fpsyt.2016.00135.

Leclerc, J. B., Valois, P., Gabrielle, J., Bombardier, M., Ouellette, S., & O'Connor, K. P. (2016). A therapy for tics in children managing underlying processes: A pilot study. *Journal of Developmental and Physical Disabilities*, *28*, 581–593. https://doi.org/10.1007/s10882-016-9496-y.

Mahler, M. S. (1944). Tics and impulsions in children: A study of motility. *The Psychoanalytic Quarterly*, *xiii*, 430–444.

Mahler, M. S. (1949). A psychoanalytic evaluation of tic in psychopathology of children; symptomatic and tic syndrome. In A. Freud, H. Hartmann, & E. Kris (Eds.), *Vol. 3/4. The psychoanalytic study of the child* (pp. 279–310). Oxford, England: International Universities Press.

Mahler, M. S., Luke, J. A., & Daltroff, W. (1945). Clinical and follow-up study of the tic syndrome in children. *American Journal of Orthopsychiatry*, *xv*, 631–647.

Marsh, R., Zhu, H., Wang, Z., Skudlarski, P., & Peterson, B. S. (2007). A developmental fMRI study of self-regulatory control in Tourette's syndrome. *The American Journal of Psychiatry*, *164*(6), 955–966. https://doi.org/10.1176/appi.ajp.164.6.955.

Mellado, A., Suarez, N., Altimir, C., Martinez, C., Perez, J., Krause, M., et al. (2017). Disentangling the change-alliance relationship: Observational assessment of the therapeutic alliance during change and stuck episodes. *Psychotherapy Research*, *27*(5), 595–607. https://doi.org/10.1080/10503307.2016.1147657.

Messerotti Benvenuti, S., Buodo, G., Leone, V., & Palomba, D. (2011). Neurofeedback training for tourette syndrome: An uncontrolled single case study. *Applied Psychophysiology and Biofeedback*, *36*(4), 281–288. https://doi.org/10.1007/s10484-011-9169-7.

Morand-Beaulieu, S., & Lavoie, M. E. (2019). Cognitive and motor event-related potentials in Tourette syndrome and tic disorders: A systematic review. *Clinical Neurophysiology*, *130*(6), 1041–1057. https://doi.org/10.1016/j.clinph.2018.10.022.

Morand-Beaulieu, S., Leclerc, J. B., Valois, P., Lavoie, M. E., O'Connor, K. P., & Gauthier, B. (2017). A review of the neuropsychological dimensions of Tourette syndrome. *Brain Science*, 7(8). https://doi.org/10.3390/brainsci7080106.

Morand-Beaulieu, S., O'Connor, K. P., Blanchet, P. J., & Lavoie, M. E. (2018). Electrophysiological predictors of cognitive-behavioral therapy outcome in tic disorders. *Journal of Psychiatric Research*, *105*, 113–122. https://doi.org/10.1016/j.jpsychires.2018.08.020.

Morand-Beaulieu, S., O'Connor, K. P., Richard, M., Sauve, G., Leclerc, J. B., Blanchet, P. J., et al. (2016). The impact of a cognitive-behavioral therapy on event-related potentials in patients with Tic disorders or body-focused repetitive behaviors. *Frontiers in Psychiatry*, 7, 81. https://doi.org/10.3389/fpsyt.2016.00081.

Morand-Beaulieu, S., O'Connor, K. P., Sauvé, G., Blanchet, P. J., & Lavoie, M. E. (2015). Cognitive-behavioral therapy induces sensorimotor and specific electrocortical changes in chronic tic and Tourette's disorder. *Neuropsychologia*, *79*(Pt B), 310–321. https://doi.org/10.1016/j.neuropsychologia.2015.05.024.

Nachev, P., Kennard, C., & Husain, M. (2008). Functional role of the supplementary and pre-supplementary motor areas. *Nature Reviews. Neuroscience*, *9*(11), 856–869. https://doi.org/10.1038/nrn2478.

Nagai, Y., Cavanna, A., & Critchley, H. D. (2009). Influence of sympathetic autonomic arousal on tics: Implications for a therapeutic behavioral intervention for Tourette syndrome. *Journal of Psychosomatic Research*, *67*(6), 599–605. https://doi.org/10.1016/j.jpsychores.2009.06.004.

Nagai, Y., Cavanna, A. E., Critchley, H. D., Stern, J. J., Robertson, M. M., & Joyce, E. M. (2014). Biofeedback treatment for Tourette syndrome: A preliminary randomized controlled trial. *Cognitive and Behavioral Neurology*, *27*(1), 17–24. https://doi.org/10.1097/WNN.0000000000000019.

Neuner, I., Werner, C. J., Arrubla, J., Stocker, T., Ehlen, C., Wegener, H. P., et al. (2014). Imaging the where and when of tic generation and resting state networks in adult Tourette patients. *Frontiers in Human Neuroscience*, *8*, 362. https://doi.org/10.3389/fnhum.2014.00362.

Nolin, G., & Leclerc, J. (2021). La qualité de vie d'enfants avec le syndrome de Gilles de la Tourette (The quality of life of children with Tourette disorder). *Revue québécoise de psychologie*, *42*(1), 171–194.

O'Connor, K. P. (2002). A cognitive-behavioral/psychophysiological model of tic disorders. *Behaviour Research and Therapy, 40*, 1113–1142.
O'Connor, K. P. (2005). Testing the cognitive-psychophysiological model: validation of a style of planning action (STOP) as a discriminator between tic disorder, obsessive-compulsive disorder and generalized anxiety. *Cognitive-Behavioral Management of Tic Disorders*, 65–73. Chichester: John-Wiley and Sons.
O'Connor, K. P., Brault, M., Robillard, S., Loiselle, J., Borgeat, F., & Stip, E. (2001). Evaluation of a cognitive-behavioural program for the management of chronic tic and habit disorders. *Behavioral Research Therapy, 39*(6), 667–681. https://doi.org/10.1016/s0005-7967(00)00048-6.
O'Connor, K., Gareau, D., & Borgeat, F. (1995). Muscle control in chronic tic disorders. *Biofeedback and Self-Regulation, 20*(2), 111–122. https://doi.org/10.1007/BF01720968.
O'Connor, K., Brisebois, H., Brault, M., Robillard, S., & Loiselle, J. (2003). Behavioral activity associated with onset in chronic tic and habit disorder. *Behavior Research and Therapy, 41*(2), 241–249.
O'Connor, K. P., & Gareau, D. (1994). *Tics et problèmes de tension musculaire.* Montréal: Les Éditions de l'Homme.
O'Connor, Laverdure, A., Taillon, A., Stip, E., Borgeat, F., & Lavoie, M. (2009). Cognitive behavioral management of Tourette's syndrome and chronic tic disorder in medicated and unmedicated samples. *Behaviour Research and Therapy, 47*(12), 1090–1095.
O'Connor, K., Lavoie, M., Blanchet, P., & St-Pierre-Delorme, M. E. (2016). Evaluation of a cognitive psychophysiological model for management of tic disorders: An open trial. *The British Journal of Psychiatry, 209*(1), 76–83. https://doi.org/10.1192/bjp.bp.114.154518.
O'Connor, Lavoie, M. E., Robert, M., Stip, E., & Borgeat, F. (2005). Brain-behavior relations during motor processing in chronic tic and habit disorder. *Cognitive and Behavioral Neurology, 18*(2), 79–88. https://doi.org/10.1097/01.wnq.0000151131.06699.af.
O'Connor, K. P., Lavoie, M. E., & Schoendorff, B. (2017). *Managing Tic and Habit Disorders: A Cognitive Psychophysiological Treatment Approach with Acceptance Strategies.* Wiley-Blackwell.
O'Connor, K. P., Lavoie, M. E., Stip, E., Borgeat, F., & Laverdure, A. (2008). Cognitive-behaviour therapy and skilled motor performance in adults with chronic tic disorder. *Neuropsychological Rehabilitation, 18*(1), 45–64. https://doi.org/10.1080/09602010701390835.
Perepletchikova, F. (2011). On the topic of treatment integrity. *Clinical Psychology (New York), 18*(2), 148–153. https://doi.org/10.1111/j.1468-2850.2011.01246.x.
Perrault, M.-A. (2019). *Modulation de l'activité électromyographique de surface à la suite d'un traitement cognitif-comportemental et psychophysiologique chez des patients atteints du syndrome de Gilles de la Tourette ou de tics chroniques persistants (Master thesis Thèses et mémoires électroniques de l'Université de Montréal [17622]).* Montréal: Université de Montréal. Retrieved from http://hdl.handle.net/1866/22737 Available from Faculté des arts et des sciences – Département de psychologie - Thèses et mémoires [835].
Perrault, M.-A., Morand-Beaulieu, S., O'Connor, K. P., Blanchet, P. J., & Lavoie, M. E. (2018). In *Normalization of the muscle tension induced by cognitive behavioral therapy amongst subjects living with Gilles de la Tourette syndrome Paper presented at the 29th International Congress of Applied Psychology (ICAP). (Montréal, Canada).*
Peterson, B. S. (2001). Neuroimaging studies of Tourette syndrome: A decade of progress. *Advances in Neurology, 85*, 179–196.
Peterson, B. S., & Cohen, D. J. (1998). The treatment of Tourette's syndrome: Multimodal, developmental intervention. *Journal of Clinical Psychiatry, 59*(Suppl. 1), 62–72. discussion 73–64.
Peterson, B. S., Staib, L., Scahill, L., Zhang, H., Anderson, C., Leckman, J. F., et al. (2001). Regional brain and ventricular volumes in Tourette syndrome. *Archives of General Psychiatry, 58*(5), 427–440.

Peterson, B. S., Thomas, P., Kane, M. J., Scahill, L., Zhang, H., Bronen, R., et al. (2003). Basal Ganglia volumes in patients with Gilles de la Tourette syndrome. *Archives of General Psychiatry*, *60*(4), 415–424.

Piacentini, J., Woods, D. W., Scahill, L., Wilhelm, S., Peterson, A. L., Chang, S., et al. (2010). Behavior therapy for children with Tourette disorder: A randomized controlled trial. *Journal of the American Medical Association*, *303*(19), 1929–1937.

Pringsheim, T., Holler-Managan, Y., Okun, M. S., Jankovic, J., Piacentini, J., Cavanna, A. E., et al. (2019). Comprehensive systematic review summary: Treatment of tics in people with Tourette syndrome and chronic tic disorders. *Neurology*, *92*(19), 907–915. https://doi.org/10.1212/WNL.0000000000007467.

Pringsheim, T., Okun, M. S., Muller-Vahl, K., Martino, D., Jankovic, J., Cavanna, A. E., et al. (2019). Practice guideline recommendations summary: Treatment of tics in people with Tourette syndrome and chronic tic disorders. *Neurology*, *92*(19), 896–906. https://doi.org/10.1212/WNL.0000000000007466.

Rickards, H., Hartley, N., & Robertson, M. M. (1997). Seignot's paper on the treatment of Tourette's syndrome with haloperidol. *History of Psychiatry*, *8*(31, Pt. 3), 433–436. https://doi.org/10.1177/0957154X9700803109.

Seignot, J. N. (1961). A case of tic of Gilles de la Tourette cured by R 1625. *Annales de Medecine Psychologique (Paris)*, *119*(1), 578–579.

Serrien, D. J., Nirkko, A. C., Loher, T. J., Lovblad, K. O., Burgunder, J. M., & Wiesendanger, M. (2002). Movement control of manipulative tasks in patients with Gilles de la Tourette syndrome. *Brain*, *125*(Pt 2), 290–300.

Shapiro, A. K. (1970). Gilles de la Tourette's syndrome. Dangers of premature psychologic diagnosis. *New York State Journal of Medicine*, *70*(17), 2210–2214.

Shapiro, A. K., & Shapiro, E. (1968). Treatment of Gilles de la Tourette's syndrome with haloperidol. *British Journal of Psychiatry*, *114*(508), 345–350.

Shapiro, A. K., & Shapiro, E. (1971). Clinical dangers of psychological theorizing. The Gilles de la Tourette syndrome. *The Psychiatric Quarterly*, *45*(2), 159–171.

Shapiro, A. K., Shapiro, E. S., Bruun, R. D., Sweet, R., Wayne, H., & Solomon, G. (1976). Gilles de la Tourette's syndrome: Summary of clinical experience with 250 patients and suggested nomenclature for tic syndromes. *Advances in Neurology*, *14*, 277–283.

Smits, E. J., Tolonen, A. J., Cluitmans, L., van Gils, M., Zietsma, R. C., Borgemeester, R. W., et al. (2017). Graphical tasks to measure upper limb function in patients with Parkinson's disease: Validity and response to dopaminergic medication. *IEEE Journal of Biomedical and Health Informatics*, *21*(1), 283–289. https://doi.org/10.1109/JBHI.2015.2503802.

Spessot, A. L., & Peterson, B. S. (2006). Tourette's syndrome: A multifactorial, developmental psychopathology. *Developmental Psychopathology*, *3*. Risk, disorder, and adaptation (2nd ed.).

Spessot, A. L., Plessen, K. J., & Peterson, B. S. (2004). Neuroimaging of developmental psychopathologies: The importance of self-regulatory and neuroplastic processes in adolescence. *Annals of the New York Academy of Sciences*, *1021*, 86–104.

Sukhodolsky, D. G., Landeros-Weisenberger, A., Scahill, L., Leckman, J. F., & Schultz, R. T. (2010). Neuropsychological functioning in children with Tourette syndrome with and without attention-deficit/hyperactivity disorder. *Journal of the American Academy of Child & Adolescent Psychiatry*, *49*(11), 1155–1164.

Tansey, M. A. (1986). A simple and a complex tic (Gilles de la Tourette's syndrome): Their response to EEG sensorimotor rhythm biofeedback training. *International Journal of Psychophysiology*, *4*(2), 91–97.

Thibault, G., O'Connor, K. P., Stip, E., & Lavoie, M. E. (2009). Electrophysiological manifestations of stimulus evaluation, response inhibition and motor processing in Tourette syndrome patients. *Psychiatry Research*, *167*(3), 202–220. https://doi.org/10.1016/j.psychres.2008.03.021.

Thomalla, G., Jonas, M., Bäumer, T., Siebner, H. R., Biermann-Ruben, K., Ganos, C., et al. (2014). Costs of control: Decreased motor cortex engagement during a Go/NoGo task in Tourette's syndrome. *Brain: A Journal of Neurology*, *137*(1), 122–136. https://doi.org/10.1093/brain/awt288.

Thomalla, G., Siebner, H. R., Jonas, M., Baumer, T., Biermann-Ruben, K., Hummel, F., et al. (2009). Structural changes in the somatosensory system correlate with tic severity in Gilles de la Tourette syndrome. *Brain*, *132*(Pt 3), 765–777.

van Wouwe, N. C., Pallavaram, S., Phibbs, F. T., Martinez-Ramirez, D., Neimat, J. S., Dawant, B. M., et al. (2017). Focused stimulation of dorsal subthalamic nucleus improves reactive inhibitory control of action impulses. *Neuropsychologia*, *99*, 37–47. https://doi.org/10.1016/j.neuropsychologia.2017.02.016.

Verdellen, C. W., Keijsers, G. P., Cath, D. C., & Hoogduin, C. A. (2004). Exposure with response prevention versus habit reversal in Tourettes's syndrome: A controlled study. *Behavior and Research Therapy*, *42*(5), 501–511.

Verdellen, C., van de Griendt, J., Hartmann, A., Murphy, T., & ESSTS Guidelines Group. (2011). European clinical guidelines for Tourette syndrome and other tic disorders. Part III: Behavioural and psychosocial interventions. *European Child & Adolescent Psychiatry*, *20*(4), 197–207. https://doi.org/10.1007/s00787-011-0167-3.

Wang, Y., Chan, G. L., Holden, J. E., Dobko, T., Mak, E., Schulzer, M., et al. (1998). Age-dependent decline of dopamine D1 receptors in human brain: A PET study. *Synapse*, *30*(1), 56–61. https://doi.org/10.1002/(SICI)1098-2396(199809)30:1<56::AID-SYN7>3.0.CO;2-J.

Wilhelm, S., Deckersbach, T., Coffey, B. J., Bohne, A., Peterson, A. L., & Baer, L. (2003). Habit reversal versus supportive psychotherapy for Tourette's disorder: A randomized controlled trial. *American Journal of Psychiatry*, *160*(6), 1175–1177. https://doi.org/10.1176/appi.ajp.160.6.1175.

Wilhelm, S., Peterson, A. L., Piacentini, J., Woods, D. W., Deckersbach, T., Sukhodolsky, D. G., et al. (2012). Randomized trial of behavior therapy for adults with Tourette syndrome. *Archives of General Psychiatry*, *69*(8), 795–803. https://doi.org/10.1001/archgenpsychiatry.2011.1528.

Woods, D. W., & Himle, M. B. (2004). Creating tic suppression: comparing the effects of verbal instruction to differential reinforcement. *Journal of Applied Behavioral Analysis*, *37*(3), 417–420.

Woods, D. W., Himle, M. B., Miltenberger, R. G., Carr, J. E., Osmon, D. C., Karsten, A. M., et al. (2008). Durability, negative impact, and neuropsychological predictors of tic suppression in children with chronic tic disorder. *Journal of Abnormal Child Psychology*, *36*(2), 237–245. https://doi.org/10.1007/s10802-007-9173-9.

Woods, D. W., Piacentini, J. C., Scahill, L., Peterson, A. L., Wilhelm, S., Chang, S., et al. (2011). Behavior therapy for tics in children: Acute and long-term effects on psychiatric and psychosocial functioning. *Journal of Child Neurology*, *26*(7), 858–865. doi:0883073810397046 [pii]. https://doi.org/10.1177/0883073810397046.

CHAPTER THIRTEEN

The functional neuroimaging of Tourette syndrome and obsessive-compulsive disorder

Tracy Bhikram*
Department of Psychiatry, University Health Network, University of Toronto, Toronto, ON, Canada
*Corresponding author: e-mail address: tracy.bhikram@mail.utoronto.ca

Contents

Tourette syndrome (TS) is a chronic neurodevelopmental disorder that traverses the disciplines of psychiatry and neurology. The disorder is characterized by the presence of tics, which are sudden, repetitive, stereotyped movements or vocalizations that can be temporarily, but not indefinitely suppressed. The majority of patients with TS describe uncomfortable

International Review of Movement Disorders, Volume 3
ISSN 2666-7878
https://doi.org/10.1016/bs.irmvd.2021.12.005

sensations that precede their tics, often referred to as premonitory urges or sensory phenomena (SP); they are usually described as arising internally and building until the tic is released (Cox, Seri, & Cavanna, 2018). Once the tic is expressed, the unpleasant characteristics of these sensations are temporarily resolved, leading to a sense of relief or gratification, which in turn is thought to perpetuate ticcing behavior. Thus, a negative reinforcement cycle develops, supporting the hypothesis that tic expression is a voluntary response to an involuntary SP (Bullen & Hemsley, 1983; Crossley & Cavanna, 2013).

Studies have shown that 86–90% of patients with TS have comorbid psychopathologies, which often can have the greatest impact on quality of life, therefore adding to the clinical burden of TS and affecting its overall prognosis (Rizzo, Gulisano, Pellico, Cali, & Curatolo, 2014). Obsessive-compulsive disorder (OCD) is one of the most common comorbidities in TS, with 27–50% of TS patients having an OCD diagnosis and obsessive-compulsive symptoms (OCS) affecting 50–80% of patients with TS (Freeman et al., 2000; Grados & Mathews, 2008; Hirschtritt et al., 2015; Lebowitz et al., 2012). In contrast, while 20–40% of patients with OCD report having a history of tics, only 16–25% of patients with OCD, mostly children and adolescents, have a TS diagnosis (Goodman, Storch, Geffken, & Murphy, 2006). While OCD is characterized by the presence of either obsessions, compulsions, or most commonly, both, it shares many phenomenological similarities with TS as they both involve involuntary or intrusive repetitive behaviors that are commonly reinforced by immediate distress reduction, and symptoms can be exacerbated by stress and heightened emotions (Eddy & Cavanna, 2014; Ferrão, Miguel, & Stein, 2009).

Meta-analyses have grouped OCS into different subtypes according to their thematic content, including those consisting of contamination/cleaning symptoms, forbidden thoughts and checking compulsions, and symmetry/ordering symptoms (Bloch, Landeros-Weisenberger, Rosario, Pittenger, & Leckman, 2008). There are differences in the OCS profile of patients with TS and those with OCD only; TS + OCD patients are less likely to have contamination and cleaning symptoms, and more likely to exhibit symptoms relating to the need for symmetry, violent/sexual images and somatic obsessions (Diniz et al., 2006; Holzer et al., 1994; Petter, Richter, & Sandor, 1998), though such evidence has since been deemed controversial (Kloft, Steinel, & Kathmann, 2018). The symmetry/ordering and forbidden thoughts/aggression dimensions in particular have been shown to be associated with global tic severity in children and across all ages, respectively (Kano et al., 2015). In addition to also being more common in TS + OCD than in "classical" OCD, ordering, counting and tic-like compulsions are less frequently preceded by obsessions and are usually not performed to relieve anxiety, further differentiating them from other OCS (de Vries et al., 2016). Such differences have caused some investigators to argue that tic-related OCD should be a distinct entity from non-tic related OCD, especially since it has an earlier age of onset, a male preponderance, and is usually associated with a family history of either OCD or tics (de Vries et al., 2016; Kloft et al., 2018).

An additional similarity in the symptomatology of TS and OCD is the presence of SP, which has also been found to be common among OCD patients (Ferrão et al., 2012; Rosario et al., 2009). These SP are perceived by patients as being uncomfortable and/or aversive and include the same types of SP reported in TS (e.g., feelings of incompleteness, bodily sensations, urges and just right perceptions) (Miguel et al., 2000). Though SP are phenomenologically different from cognitive obsessions, they serve similar functions by prompting the performance of compulsions until the obsessions/SP are relieved, which in turn, negatively reinforces the compulsions (Belloch et al., 2016).

Many of the symptoms of TS can be conceptualized as a failure of neural mechanisms to suppress unwanted and abnormal behaviors, such as tics, OCS, hyperactivity, rage and other emotional problems. This has prompted numerous investigations into the neurobiological correlates of TS, though there can be some uncertainty whether reported findings relate to the core symptoms of TS (i.e., tics and SP) or associated comorbidities. Teasing apart such distinctions is important since determining the neurobiological profile of the different symptom domains helps to answer key etiological questions about their associated disorders, which in turn, is crucial for the discovery and development of neurobiological markers and new approaches to treatment. The similarities and differences between TS and OCD in particular, and between the different symptoms that characterize them, raise questions as to their underlying neurobiology and whether they share neurobiological pathways that give rise to similar phenotypes. Below, functional neuroimaging findings relating to the symptom domains of TS and OCD will be reviewed with a special emphasis on comorbid TS + OCS, though there is a paucity of such research.

1. Tics

1.1 Tourette syndrome

The neural correlates of tic genesis were investigated in an early event-related positron emission tomography (PET) study that utilized time-synchronized audio and video taping of TS patients while they allowed their tics to occur spontaneously. The authors found that tic occurrence was highly correlated with increased activity in a set of neocortical (supplementary motor area [SMA], dorsolateral prefrontal cortex, premotor cortex, primary motor cortex, somatosensory cortex), paralimbic (insula, anterior cingulate cortex [ACC]) and subcortical regions (putamen and caudate) (Stern et al., 2000). These findings were largely mirrored in a subsequent PET study in which Lerner et al., compared an active ticcing paradigm with the control condition of stage 2 sleep to minimize tics and SP. They reported activation of multiple areas including the insula, ACC, SMA and the cerebellum, which the authors hypothesized to be the regions responsible for tic release and anticipation (Lerner et al., 2009).

Another method that has been utilized to study the neural mechanisms responsible for tics is by comparing spontaneous tics in TS patients to intentional tic-like movements in healthy controls. Hampton et al., used such a paradigm and compared the cross correlations of brain activity in the primary motor cortex with activity in the SMA, somatosensory cortex and subcortical nuclei between patients and controls using functional magnetic resonance imaging (fMRI). Similar patterns of neural activity were detected for both tics in TS patients and voluntary tic-like movements in controls across the entire motor system, except patients displayed stronger correlations of activity between the SMA and the primary motor cortex, highlighting the central role of the SMA for tic generation (Hampson, Tokoglu, King, Constable, & Leckman, 2009). Wang et al. also compared the activity of neuronal networks during spontaneous (in TS patients) and simulated tics (in healthy controls and TS patients). Patients displayed stronger activity in the somatosensory and premotor cortices, putamen, amygdala and hippocampus which correlated positively with tic severity. The authors suggested that the findings indicate that tics engage the same circuits that support the expression of normal voluntary movement, but does so to a greater extent (Wang et al., 2011). Patients also displayed reduced activity in the ACC, caudate and parietal operculum, which negatively correlated with tic severity, causing the authors to conclude that tics are caused by the combined effects of excessive activity in motor pathways and reduced activity of control regions. Additionally, within the patient group, greater activity in the putamen, parietal operculum, somatosensory cortex and amygdala was observed during spontaneous tics than during the voluntary imitations of tics. The authors contend that these regions represent the occurrence of SP, with greater activity of the amygdala representing either the emotional discomfort associated with premonitory urges or the relief experienced after tic expression (Wang et al., 2011).

1.2 OCD and TS + OCD

There have not yet been studies that have investigated the functional neurobiology of tic occurrence in patients with comorbid TS + OCD or in OCD patients who exhibit tic-like compulsions. Such studies would be informative as it would help to determine if the neurobiology of tics can be differentiated between TS, OCD and TS + OCD.

2. Sensory phenomena and urge suppression

The different types of SP experienced in OCD are similar to those experienced in TS (Miguel et al., 2000), however important differences have been found between them. Brandt and colleagues reported that the time-course of SP—including the build-up, relief after compulsion/tic performance, and subsequent return to baseline—is different in OCD than in TS (Brandt et al., 2016, 2018). This raises questions as to whether SP in TS + OCD varies from that in TS and OCD, and more specifically, if they involve different neurobiological substrates. However, there is a lack of neuroimaging research on the role of comorbid OCD in TS, especially with regards to SP; those that have been published are summarized below.

2.1 Tourette syndrome

To better delineate the neural systems related to SP and tic activity, studies have separately investigated neuronal activity both preceding and during tic activity. Using event-related fMRI, Bohlhalter specifically addressed this distinction by analyzing activation patterns at tic onset separately from the patterns occurring 2 seconds (s) prior to its onset, a time period relevant to the SP that precedes tics. Concomitant with the subjective experience of premonitory urges, the ACC, insula, SMA and parietal operculum were activated 2 s prior to tic onset. In contrast, at tic onset, the cerebellum, superior parietal lobule and sensorimotor areas were activated (Bohlhalter, 2006). In a more recent study, Neuner and colleagues attempted to refine these results by using time bins of 2 s prior to tic onset, 1 s prior, and at tic onset. At 2 s prior to tic onset, activations were detected in the SMA, sensorimotor cortex, and parietal operculum, whereas activation of a network consisting of the ACC, putamen, insula, amygdala and cerebellum was detected 1 s prior to tic onset. Activations of the thalamus, primary motor cortex and somatosensory cortex were observed at tic onset (Neuner et al., 2014). These findings support the results from many of the previously discussed studies, specifically the involvement of the sensory, motor and limbic cortices, and that urges and tics involve distinct neural systems. However, perhaps more importantly, these findings suggest that cortical activity both precedes and follows subcortical activity during tic generation.

Functional imaging studies using tic suppression versus free tic expression paradigms, in general, point to abnormal activity both in the basal ganglia and the frontal cortex. In an early fMRI study, Peterson and colleagues studied subjects with TS under conditions of spontaneous tic expression in contrast to periods of voluntary tic suppression. They found that tic suppression was accompanied by increased activity in the right frontal cortex and caudate, while decreased activity was detected in the globus pallidus, putamen and thalamus (Peterson et al., 1998). The severity of motor tics was correlated with activity in the basal ganglia and thalamus, suggesting that these components might be more closely related to the generation of tics. Ganos et al. procured different results when comparing free ticcing to tic inhibition states using regional homogeneity analyses. During tic inhibition, regional homogeneity increased in the inferior frontal gyrus (IFG) and was strongly correlated with the behavioral capacity to inhibit tics, but there was no relationship with measures of premonitory urge severity (Ganos et al., 2014). In fact, Ganos has suggested that urge and tic generation might have a distinct neurophysiological system from that of tic control and suppressibility. Ganos' findings were supported by Tinaz et al., who reported that functional connectivity between the insula and SMA correlated with urge severity, whereas activity of the IFG was associated with the ability to successfully inhibit tics both inside and outside the scanner (Tinaz et al., 2014). The regions found to be significantly activated in the above studies may in fact be loci involved in tic and urge generation, or alternatively, they may represent the regions involved in the attentional and internal monitoring required for tic suppression (Frey & Albin, 2006).

Blink suppression has also been used as a model for studying tic suppression because it requires participants to suppress a semi-involuntary behavior that can be assessed both in patients and healthy controls. Mazzone et al., contrasted brain activity during blocks of blink inhibition and free blinking between 66 healthy controls and 51 patients with TS. They found increased activation in the middle frontal gyrus, the ACC and in the middle and superior temporal gyri in the patient group (Mazzone et al., 2010). The authors reasoned that the hyperactivity observed in these regions may represent a disturbance in the ability to suppress somatic sensations associated with urges in TS. In addition, the magnitude of activation in the inferolateral prefrontal cortex, superior temporal gyrus, IFG, and putamen correlated inversely with tic severity; whereas activity in the caudate, middle frontal and temporal gyri positively correlated with tic severity (Mazzone et al., 2010).

In an interesting study, van der Salm and colleagues compared blink inhibition in controls to ocular tic suppression (e.g., eye rolling, squinting, twitching, etc.) in TS patients (van der Salm et al., 2018). Specifically, while patients completed alternating fMRI blocks of tic suppression and release, healthy controls completed alternating blocks of blink inhibition and rest. They found that during suppression, patients exhibited increased activity in the ACC, premotor cortex, superior frontal gyrus and orbitofrontal cortex (OFC) when compared to controls. Contrastingly, controls exhibited greater activity than patients in the SMA, insula, putamen, IFG, cerebellum, and inferior parietal cortex, suggesting that different suppression strategies are utilized in patients and controls (van der Salm et al., 2018). However, limitations of this study included the fact that patients and controls inhibited different behaviors, which may in part explain some of the observed group differences, and that the relatively short length of the suppression blocks (25 s) may not have fully allowed for the sufficient build-up of urges.

2.2 OCD

Even though SP are commonly experienced by OCD patients, the relationship between neural activity and SP in OCD has rarely been investigated. In a study by Brown and colleagues, 19 patients with OCD watched videos depicting body movements/sensations (e.g., a paintbrush stroking the palm of a hand) or control videos (e.g., the back of a paint brush sliding across a table) (Brown et al., 2018). Investigators found that SP severity was positively associated with activity in the insula, somatosensory cortex, posterior temporal cortex, OFC and dorsal prefrontal cortex (Brown et al., 2018). These associations remained significant even after controlling for other symptoms, comorbidities and medication usage, suggesting that SP are behaviorally and neurally dissociable from other clinical characteristics of OCD. However, this study was limited as the authors did not assess whether the task actually provoked SP in the patients. This is especially pertinent since watching videos of body movements/sensations may not be a good representation of the many types of SP in OCD, such as urges and feelings of incompleteness. Nevertheless, the reported results are interesting because for the most part, these regions are distinct from those found to be associated with SP in TS, which suggests that while SP in OCD and TS may be experientially similar, they may not be neurobiologically similar. Alternatively, such differences may instead be due to the use of different experimental paradigms.

Such a limitation was rectified by Stern and colleagues who used a similar blink inhibition paradigm as the previously outlined studies in TS; 46 OCD patients and 23 healthy control participants completed alternating blocks of blink inhibition and free blinking, though for the analyses, the blink inhibition blocks were divided so that the first half of the blocks were analyzed separately from the second half (Stern et al., 2020). The authors found that when compared to controls, OCD patients experienced more "escape blinks" (i.e., blinks occurring during the inhibition blocks) during both the first and second halves of the inhibition blocks. During the first half of the inhibition blocks, OCD patients displayed greater activity than controls in three large clusters comprising of the (1) cingulate, somatosensory cortex, precuneus, inferior parietal cortex, parahippocampal gyrus, (2) insula, IFG, caudate, putamen, amygdala, and (3) the cerebellum. During the second half of the inhibition blocks, patients displayed greater activity in the posterior cingulate, precuneus, inferior parietal cortex, and parahippocampal gyrus when compared to controls. However, none of the selected ROIs were found to correlate with either OCS or SP severity (Stern et al., 2020). Notably, many of these regions are different from those found to be significant in the blink and ocular tic suppression studies conducted in TS.

2.3 TS + OCD

Only one study has explicitly investigated how the neural profile of urge suppression in TS is affected by the severity of tics and OCS, as well as emotion, since heightened emotional states can exacerbate both tic and OCS severity. Forty patients with TS and 20 healthy controls completed alternating blocks of blink inhibition and free blinking while viewing pictures of angry and neutral facial expressions. While there were no significant differences between the patient group and the controls in the amount of blinks that occurred during the blocks, patients exhibited greater activity in the superior temporal gyrus and midcingulate while inhibiting blinks (Bhikram, Crawley, Arnold, Abi-Jaoude, & Sandor, 2021). These results are similar to those reported by Mazzone and colleagues who found increased activity in the superior temporal gyrus and cingulate of TS patients when compared to controls during blink inhibition (Mazzone et al., 2010). More interestingly, within the patient group, differences between neural associations with tic and OCS severity were highlighted depending on whether or not the emotional content of the blocks were taken into consideration (Bhikram et al., 2021). Specifically, OCS severity was inversely associated with activity in the OFC and insula, regardless of the emotional condition of the blocks, whereas there were no associations with tic severity scores. These significant findings with OCS severity involve the same regions found to be associated with SP severity in a previous study in which OCD patients viewed body-focused videos meant to mimic SP (Brown et al., 2018).

However, when the emotional context of the blocks were considered, tic severity was positively associated with activity in the superior frontal gyrus while there were no significant associations with OCS severity (Bhikram et al., 2021). Patients with worse tics may have required greater superior frontal gyrus recruitment to monitor urge inhibition and task performance when faced with emotional stimuli since this region has been shown to be involved in processing emotional stimuli (Neuner et al., 2010) and in self-awareness during sensorimotor processing (Goldberg, Harel, & Malach, 2006). These results further dissociate tics from OCS in TS even though they are phenomenologically similar. Furthermore, these results highlight the important role emotions have on the symptomatology of TS, with tics frequently becoming worse during times of anger and excitement, and episodic behavioral outbursts and anger control problems affecting 25–70% of TS patients (Budman, Rockmore, Stokes, & Sossin, 2003; Kadesjo & Gillberg, 2000). The regions reported by this study as being significantly associated with the different symptom domains of TS may represent compensatory cortical activity needed to keep tics and OCS under control, or alternatively, could represent networks that are responsible for the manifestation of urges to perform tics and compulsions (Bhikram et al., 2021).

3. Obsessive-compulsive symptoms

3.1 Tourette syndrome

Symptom provocation tasks or paradigms (SPPs) are used to develop and test theories about the neural correlates of OCS by exposing individuals to stimuli that are directly related to concerns typically reported in OCD, with the goal of inducing such symptoms during neuroimaging. While these studies are commonly conducted in OCD patient samples, as reviewed below, only one such neuroimaging study has been conducted in a sample of patients with primarily "pure" TS even though OCS are extremely common in TS populations.

In a PET study, de Vries and colleagues had 14 TS patients and 10 controls view pictures of symmetrically (baseline condition) and asymmetrically (provocation condition) ordered objects. During the provocation condition, patients reported similar levels of distress as controls, but experienced significantly greater urges to rearrange the objects (de Vries et al., 2013). Within the patient group, increased cerebral blood flow in the ACC, IFG and SMA was observed during the provocation condition as compared to baseline. However, when compared to controls, TS patients showed significantly larger increases of cerebral blood flow in the ACC and smaller increases in the occipital cortex, motor cortex and dorsal-medial prefrontal cortex (de Vries et al., 2013). Interestingly, these regions are very similar to those reported in tic generation and tic suppression studies conducted in TS populations, and are for the most part different from the results obtained during traditional symptom provocation in OCD. The authors speculated that increased cerebral blood flow in the ACC, IFG and SMA represents the attention capture of a salient stimulus, the urge to move, and an inhibitory response to prevent the movement (de Vries et al., 2013). This is an important study as it investigated the common occurrence of OCS in TS, but it was not without limitations. Firstly, only two of the patients had comorbid OCD, therefore the results are not representative of TS + OCD. Secondly, only symmetry symptoms were investigated; while symmetry-related OCS are common in TS, they are not the only type of OCS in TS.

3.2 OCD

Numerous SPP studies conducted with "classical OCD" samples have grouped together patients regardless of their OCS dimensions and a variety of techniques have been used to provoke symptoms, including having participants view individually tailored photographs or hold sealed tubes containing stimuli. Many of these studies have produced largely similar results, that is, increased activity in the basal ganglia, thalamus, cingulate and OFC (Adler et al., 2000; McGuire et al., 1994; Rauch et al., 1994; Simon, Adler, Kaufmann, & Kathmann, 2014; Simon, Kaufmann, Musch, Kischkel, & Kathmann, 2010), with one study also reporting increased skin conductance in OCD patients in response to OCD triggers, demonstrating activation of the autonomic nervous system (Simon, Kaufmann, Kniesche, Kischkel, & Kathmann, 2013).

Other investigators have opted to examine the neural correlates of the different types of OCS separately or dimensionally, with most reporting somewhat different neural profiles for each of the investigated symptom dimensions. Researchers using PET have reported that checking and symmetry symptoms correlated with increased and decreased cerebral blood flow in the striatum, respectively; whereas washing symptoms correlated with blood flow in the OFC and ACC (Rauch et al., 1998). Different results were found in an fMRI study; Phillips reported that when viewing contamination/washing related pictures, "washers" demonstrated activity in the insula and occipital lobe, whereas "checkers" demonstrated activation in frontostriatal regions and the thalamus (Phillips et al., 2000). To avoid dividing patients into mutually exclusive subgroups, Mataix-Cols et al., used a dimensional approach to investigate the neural correlates of washing and checking symptoms within the patient group. The authors found that washing symptoms were correlated with greater activity in the insula, ventrolateral prefrontal cortex and temporal pole, whereas checking severity was associated with greater activity in the putamen, pallidum and thalamus. The authors concluded that each symptom dimension might reflect the dysregulation of highly conserved and partially overlapping neural systems that serve to detect, appraise, and respond to potential threats (Mataix-Cols et al., 2004).

Among other regions, these studies further implicate the OFC, ACC, thalamus and striatum in OCD, findings which have been supported by the results of multiple meta-analyses. The first reported that symptom provocation was associated with activity in numerous cortical and subcortical regions, including the OFC, ACC, dorsolateral prefrontal cortex, precuneus, pallidum, thalamus, hippocampus and superior temporal gyrus (Rotge et al., 2008). Another more recent meta-analysis included SPP studies as well as other more traditional emotional processing investigations and produced mostly similar results; compared with healthy participants, patients showed increased activation in the amygdala, OFC, ACC putamen, ventromedial prefrontal cortex, middle temporal gyrus and occipital regions during emotional processing (Thorsen et al., 2018). They also reported that OCD severity was associated with increased activity in the ACC. The above implicated regions are thought to be involved in emotion, visual attention and cognitive control functions. Specifically, it has been hypothesized that the OFC and ACC plays a major role in the cognitive processes underlying the emergence of OCS, while frontal and parietal regions are involved in patients' attempts to turn their attention away from their obsessive thoughts (Rotge et al., 2008).

3.3 TS+OCD

A single study has investigated the neurobiological profile of OCS provocation in patients with comorbid TS and OCD. TS patients with moderate/severe OCS ($n=19$), TS patients with absent/mild OCS ($n=21$) and healthy controls ($n=20$) completed fMRI scans while viewing blocks of OCS-provoking pictures relating to washing, checking and symmetry symptoms, as well as neutral scenes. When comparing the groups, severe OCS patients exhibited less activity than controls in the insula for both the checking and washing contrasts. TS patients with severe OCS also displayed less activity than TS patients with mild OCS in the SMA, somatosensory cortex and supramarginal gyri for the washing contrast. There were, however, no significant differences between any of the groups for the symmetry contrasts (Bhikram, Crawley, Arnold, Abi-Jaoude, & Sandor, 2020). Interestingly, the above significant regions do not include those typically found to be hyperactive in OCD patients during SPP tasks, including the OFC, ACC, thalamus and basal ganglia, which further suggests that OCS in TS may by characterized by mostly distinct brain regions.

When examining the entire patient group for associations with OCS severity, inverse associations were found with activity in the primary motor cortex, superior temporal gyrus, supramarginal gyrus, IFG and in visual processing areas during the checking blocks (Bhikram, Arnold, Crawley, Abi-Jaoude, & Sandor, 2020; Bhikram, Crawley, et al., 2020). For the symmetry blocks, OCS severity was negatively associated with activity in the primary motor cortex, SMA, insula, midcingulate, IFG, and the supramarginal gyrus. Finally, negative associations were found between OCS severity and activity in the precuneus and supramarginal gyrus for the washing condition. Generally, activity in the supramarginal gyrus, the precuneus,

SMA and IFG were common in multiple provocation conditions, and along with the regions found to be significant when comparing the different groups, paint a picture as to the neural correlates of OCS in TS (Bhikram, Arnold, et al., 2020; Bhikram, Crawley, et al., 2020). Aberrant activity in the supramarginal gyrus/precuneus, SMA and IFG may interfere with the ability of patients to shift their attention away from distressing stimuli, may contribute to the SP that elicits compulsion performance, and may impair top-down inhibitory cortical control mechanisms, respectively.

In contrast to the associations with OCS, there was only one significant association with tic severity in this study; higher severity scores were associated with greater activity in the ACC for the symmetry contrasts (Bhikram, Arnold, et al., 2020; Bhikram, Crawley, et al., 2020), an interesting finding especially since symmetry-related OCS are one of the most common OCS dimensions in TS (Kloft et al., 2018). An overactive ACC may result in excessive error signals that are interpreted as SP, resulting in the elicitation of tics and symmetry-related compulsions due to the region's role in producing sensory symptoms and in error monitoring (Allman, Hakeem, Erwin, Nimchinsky, & Hof, 2001; Bush, Luu, & Posner, 2000).

4. Resting state functional connectivity studies

4.1 TS

Another factor that remains unknown in TS is how functional connectivity of resting state networks influence the generation of tics. Examining the temporal correlations between the activities of remote brain regions while participants do not engage in any particular task continues to be a growing area of interest in TS. An early PET study reported a broad range of functional connectivity differences in patients with TS when compared to controls. Specifically, Jeffries et al., found increased striatal activity correlated with activity in the insula, OFC, cerebellum, and motor cortices, which fits with the idea that TS is a disorder that crosses over the boundary between motor and emotional functions (Jeffries et al., 2002; Rickards, 2009). Tinaz and colleagues reported increased functional connectivity in patients with TS between the sensorimotor cortex and the anterior insula, regions previously found to be associated with sensations and urges. These results suggest increased awareness of bodily sensations in TS, possibly leading to SP (Tinaz et al., 2014). More recently, the same authors observed decreased connectivity in the dorsomedial frontal nodes and increased connectivity in the striatothalamic and anterior insular nodes in a TS group compared to controls. Moreover, severity of premonitory urges was associated with greater connectivity between the insula and the SMA, regions found to be active in urge provocation paradigms (Tinaz, Malone, Hallett, & Horovitz, 2015). A meta-analysis of three resting state studies reported that the brain regions that consistently showed abnormal functional connectivity in TS were the SMA, lingual gyrus and the thalamus (Zapparoli, Porta, & Paulesu, 2015).

Neuner and peers investigated resting state networks in TS patients and found over 20 neuroanatomically meaningful maps of functional connectivity, including a preserved default mode network (DMN) (Neuner et al., 2014). Other studies, however, reported increased connectivity of the DMN which negatively correlated with tic severity (Fan et al., 2018). In healthy controls, the DMN is more intensely active under resting conditions and deactivated during task participation. The DMN consists of the PCC, medial prefrontal cortex and the inferior parietal cortex (bilaterally), and has been found to be altered in other movement disorders (Delaveau et al., 2010). Anomalies in other resting state networks including the frontoparietal network (FPN), sensorimotor and salience networks have also been reported in TS patients (Fan et al., 2018; Wen et al., 2018). Specifically, Fan et al., reported that when compared to healthy controls, TS patients exhibited weaker connectivity of the FPN, but stronger coupling between the DMN and left FPN, which the authors hypothesized serves to compensate for altered connectivity within those networks (Fan et al., 2018).

4.2 OCD

Within the last decade, studies examining resting state functional connectivity and neural networks in OCD have become increasingly prevalent. Li and colleagues measured the functional connectivity between the anterior prefrontal cortex and the rest of the cortex. They reported that OCD patients had stronger connectivity between the prefrontal cortex and the insula and cingulate cortex, which the authors hypothesized represents abnormalities in cognitive control. The authors also reported lower connectivity between the prefrontal cortex and the pre- and postcentral gyri, which was interpreted as reflecting lower motor inhibition capacity (Li et al., 2012). Studies using the striatum as a seed region of interest have reported very similar results; OCD patients exhibit stronger functional connectivity between the striatum and prefrontal regions, including the OFC, but with conflicting evidence as to whether the results correlate with measures of OCS severity (Fitzgerald et al., 2011; Harrison et al., 2009; Sakai et al., 2011).

A recent study conducted a meta-analysis of resting state studies in OCD to determine functional connectivity patterns within and between common neural networks. They found that relative to healthy controls, subjects with OCD exhibited hypoconnectivity within the FPN, DMN and the salience network, which is thought to have an important role in detecting salient stimuli but also in modulating attentional and cognitive resources for such stimuli (Gursel, Avram, Sorg, Brandl, & Koch, 2018; Seeley et al., 2007). The authors hypothesized that dysfunctions in these three core networks may explain difficulties in switching attention between thoughts and goal directed behaviors, and should thus be regarded as a central mechanism underlying OCD behavior (Gursel et al., 2018). Interestingly, the authors did not find any significant connectivity differences between the groups within the limbic network, considering regions in the limbic network are commonly reported as being abnormally active during functional task-based studies in OCD patients (Menzies et al., 2008; Rotge et al., 2008).

Researchers have also investigated if there are relationships between functional connectivity and the different OCS dimensions. Jang et al., described how connectivity between the PCC and medial frontal cortex, regions in the DMN, correlated positively and negatively with washing and checking symptoms, respectively (Jang et al., 2010). In a subsequent, larger study, investigators reported that patients with greater aggression-related symptoms showed increased connectivity between the striatum and ventromedial prefrontal cortex, and lower connectivity between the striatum and amygdala, possibly reflecting heightened fear and threat estimation processes that occur in the harm/aggression OCS subtype (Harrison et al., 2013). On the other hand, patients with more sexual/religious symptoms displayed greater connectivity between the ventral striatum and the insula, which may reflect self-directed disgust that is often experienced by individuals with this symptom subtype. There were no significant functional connectivity associations with either the contamination/cleaning or symmetry/ordering subtypes (Harrison et al., 2013).

Overall, the main findings suggest that OCD patients exhibit increased connectivity between the striatum and regions in the frontal cortex, which has also been reported in TS populations (Jeffries et al., 2002). However, important differences between the two clinical populations exist. OCD patients have hypoconnectivity of the DMN, whereas a preserved or hyperconnected DMN is observed in TS. As well, TS patients exhibit greater connectivity between the striatum and the sensorimotor cortex, which for the most part, is not commonly reported in OCD patients

(Neuner et al., 2014; Tinaz et al., 2014). Considering the differences in functional connectivity that have been reported in TS and OCD subjects at rest, questions arise as to what resting state functional connectivity looks like in TS + OCD populations; several studies that have pursued such research avenues are summarized below.

4.3 TS + OCD

In a resting state fMRI study, Worbe demonstrated that a group of 55 adult TS patients with comorbid OCD, in contrast to 27 matched controls, had immature connectivity patterns and increased functional integration characterized by shorter path lengths between areas, an increased number and strength of functional connections between regions, and fewer "hub" regions (Worbe et al., 2012). "Hubs" are regions rich in connections thought to facilitate the fast and efficient transfer of information over long distances within the brain (Bullmore & Sporns, 2009). Furthermore, connectivity changes in limbic and associative networks were correlated with OCS, but not with overall tic severity (Worbe et al., 2012). In a different study of children with TS and comorbid OCD and/or attention deficit/hyperactivity disorder (ADHD), the investigators found underconnectivity of a fronto-parietal functional system involved in adaptive control and in a cingulo-opercular task-maintenance system. Both of these networks resembled the maps of younger children, suggesting deviant brain maturation (Church et al., 2009).

Poufar et al., identified metabolic patterns that differentiated TS patients with prominent OCS from those without, as well as patterns that differentiated TS patients from controls. They found that compared to controls, the TS related metabolic pattern was characterized by increased premotor and cerebellar metabolism that covaried with decreased striatal and orbitofrontal metabolism (Pourfar et al., 2011). Perhaps more interestingly, within the patient group, the OCD related metabolic pattern was characterized by metabolic reductions in the ACC and dlPFC which covaried with increases in activity of the primary motor cortex and the precuneus, indicating that OCD in TS is associated with a different metabolic pattern than general TS (Pourfar et al., 2011). However, important limitations of this study included very small sample sizes in the TS subgroups, and that tic severity was not comparable between the TS patients with and without severe OCS, which may have ultimately influenced the results.

A subsequent study addressed some of these limitations and examined resting state functional connectivity in 40 TS patients and 20 healthy controls to determine how key regions are associated with both the tic and OCS phenotype in TS. Specifically, seeds were placed in the SMA and putamen—regions hypothesized to be involved in the etiology of TS; the OFC and caudate—regions believed to be involved in OCD pathophysiology; and the insula, which has been shown to be related to both disorders. When comparing the patient group to the controls, patients exhibited stronger connectivity between the insula and inferior and middle temporal gyri, between the OFC and the ACC, and between the putamen and the superior and middle temporal gyri, as well as weaker connectivity between the caudate and the cerebellum and calcarine sulcus (Bhikram, Arnold, et al., 2020). When parsing out the symptom domains of TS + OCD, tic severity was positively associated with connectivity between the caudate and the superior occipital gyrus, as well as between the putamen and a cluster spanning the somatosensory cortex and primary motor cortex, key regions of the sensorimotor circuit. Dysfunctions in this circuit are at the core of neurobiological models hypothesized to be responsible for TS, as greater connectivity of the sensorimotor cortex and putamen may impair control mechanisms leading to more severe tics and associated SP (Mercadante, Rosario-Campos, Quarantini, & Sato, 2004). This increased connectivity could also reflect instances in which sensory information is coupled to motor responses, leading to aberrant behavior, such as tics and SP. OCS severity, on the other hand, was negatively associated with connectivity between the caudate and the precuneus, potentially reflecting the inability of patients with greater OCS to distract themselves from obsessive and distressing thoughts while trying to let their minds wander during the resting state task (Bhikram, Arnold, et al., 2020; Bhikram, Crawley, et al., 2020), since the precuneus is thought to play a role in self-generated thoughts and the reallocation of attentional resources across sensory modalities (Cavanna & Trimble, 2006). OCS severity was also negatively associated with connectivity between the SMA and thalamus; the thalamus is major site for the SMA's projections and both are frequently implicated in the occurrence of tics (Alexander & Crutcher, 1990). Such findings could potentially help to explain the phenomenological similarities between tics and OCS, and supports the theory that OCD in TS + OCD may be somewhat neurobiologically distinct from pure OCD.

Such conclusions were also supported by another recent study which investigated functional connectivity in a variety of networks by comparing pediatric patients with TS, OCD, and TS + OCD between each other, and with healthy controls. Tikoo and colleagues found that when compared to healthy controls, the pure TS and TS + OCD groups displayed connectivity patterns involving almost identical regions, which included increased connectivity between regions in the basal ganglia network (BGN), cerebellar network (CBN), DMN and sensorimotor network, as well as reduced connectivity in the FPN and the salience network (Tikoo et al., 2020). Contrastingly, pure OCD patients displayed increased connectivity in all of the above six networks when compared to controls, with slightly different regions involved than found in the TS subgroups. However, no differences in the orbitofrontal network was detected in any of the patient groups. Interestingly, when the patient groups were compared to each other, OCD patients displayed increased connectivity in the FPN and CBN group when compared separately to both the TS and TS + OCD groups. The striking similarities between the TS and TS + OCD groups provide further evidence that TS + OCS may be neurobiological more similar to TS than to OCD, a hypothesis that has been suggested by other researchers (Coffey et al., 1998). This was further evidenced by the correlations with clinical severity measures. Both the pure TS and TS + OCD groups displayed significant positive associations with tic severity scores and connectivity in the basal ganglia and DMN; tics were also negatively associated with connectivity in the CBN and FPN in the TS groups and TS + OCD groups, respectively. Strikingly, there were no significant associations with OCS scores in the TS + OCD group even though such associations were found in the pure OCD group, where connectivity measures in the CBN and FPN were positively associated with OCS (Tikoo et al., 2020).

In a subsequent paper, the same authors then conducted functional connectivity analyses using the dentate gyrus (DG) of the cerebellum as the seed of interest and the same pediatric diagnostic groupings. When compared to controls, each of the TS, TS + OCD and OCD groups displayed decreased connectivity between the DG and the somatosensory cortex, primary motor cortex, inferior temporal gyrus and thalamus (Tikoo et al., 2021). However, there were also differences between the groups when compared to controls; both the TS and the TS + OCD groups displayed decreased connectivity with the prefrontal cortex when compared to controls, whereas OCD patients displayed increased connectivity with the prefrontal cortex and the OFC. When the patient groups were directly compared to one another, both the TS and the TS + OCD groups displayed higher DG connectivity with the primary motor cortex and somatosensory cortex when compared to the OCD group; TS + OCD patients also displayed increased and decreased connectivity with the prefrontal cortex and the primary motor cortex, respectively, when compared to the pure TS group (Tikoo et al., 2021). More differences were found between the patient groups when assessing for correlations with clinical measures. Specifically, OCS scores were positively associated with DG connectivity between the prefrontal cortex and OFC in the OCD group, whereas for the TS group, tic severity was negatively and positively correlated with DG connectivity between the PFC and left lobule VI of the cerebellum. Moreover, the complexities associated with comorbid TS + OCD were further highlighted by the results; even though TS + OCD patients displayed several regions of aberrant connectivity with the DG and had comparable tic and OCS severity scores as the "pure" TS and OCD groups respectively, correlations between tic or OCS severity and DG connectivity could not be gleamed in the TS + OCD subgroup (Tikoo et al., 2021).

5. Neurobiological models

The etiology of both TS and OCD are hypothesized to involve dysfunctions in the frontostriatal circuits that connect the basal ganglia to the frontal cortex, referred to as cortico-striato-thalamo-cortical (CSTC) circuits. Each circuit is composed of loops which serves to exist in equilibrium; an excitatory direct pathway and an inhibitory indirect pathway (Saxena & Rauch, 2000). Imbalances between the pathways are thought to be at the etiological core of both TS and OCD, either due to excessive activity in the direct pathway or deficits in the indirect pathway, ultimately resulting in the hyperactivity of cortical regions (Saxena & Rauch, 2000).

5.1 Tourette syndrome

In general, the regions that are found to be consistently implicated in studies of TS include the SMA, ACC, insula, as well as numerous frontal and motor areas. The functions of these regions further implicate their potential roles in TS. The ACC and insula have been found to be involved in various different natural urges in healthy controls, and stimulation of the SMA results in unusual sensations and in the anticipation of movement (Fried et al., 1991; Jackson, Parkinson, Kim, Schuermann, & Eickhoff, 2011); as such, both regions may contribute to the occurrence of SP. As well, stimulation of the putamen or medial frontal lobes can produce motor and vocal outputs that appear similar to tics (Leckman & Riddle, 2000; Rothenberger & Roessner, 2013), whereas abnormal activity in regions such as the IFG may represent deficits in cortical control areas responsible for monitoring and regulating neural activity and behavior associated with other regions.

Regarding the hypothesized dysfunctional CSTC circuits, TS and its accompanying tics are thought to be a result of abnormalities in the sensorimotor circuit, with the main cortical projections emanating from the SMA, but also involving projections from the motor and somatosensory cortices, to the putamen and the dorsolateral portion of the caudate (Mercadante et al., 2004). The occurrence of tics is thought to be due to altered inhibitory activity of GABAergic projections from the striatum to the globus pallidus (Mercadante et al., 2004). Indeed, studies in TS have reported 30–60% reductions in GABAergic interneurons, reflecting inhibitory loss and possible deficits in the inhibition of unwanted motor patterns. Such inhibitory deficits could result in the increased excitability of the motor pathways converging on the SMA, primary motor and somatosensory cortices, regions that are thought to drive the generation of tics.

Specifically, the SMA, which is believed to be involved in movement initiation, becomes overstimulated by the excess of unfiltered sensory inputs coming from the thalamus, resulting in SP that precede and accompany tics. Greater activity of this circuit could in turn require greater activation of other CSTC circuits to suppress and modulate the overactivity of the motor pathway (Mazzone et al., 2010). After the SMA and motor cortex processes and translates preparatory motor signals into SP and tics, they would likely relay signals to regions responsible for top-down cortical control and action monitoring, such as the prefrontal cortex, to strategically compensate for unwanted tics and SP. Moreover, more generalized CSTC dysfunction may account for alterations across neuroanatomical systems, predisposing patients with TS to the heterogeneity of symptoms and comorbidities commonly observed in TS (Ganos, Roessner, & Munchau, 2013).

5.2 OCD

Most of the reviewed functional neuroimaging studies conducted in OCD patients generally converge to show hyperactivity within the OFC, caudate, ACC and the thalamus. These regions display elevated metabolism and activity when compared to controls, and during the provocation of symptoms. Importantly, there are consistent foci of functional abnormalities in other areas as well, including the prefrontal and parietal cortices, indicating that more distributed large-scale brain systems may be involved in OCD (Menzies et al., 2008). The OFC is thought to play a role in ascribing and monitoring changes in reward value, including in the awareness of expected rewards and the probability such rewards will occur. It also plays a role in reversal learning, especially in detecting changes in the motivational value of a stimulus and in changing behaviors when it is adaptive to do so (Menzies et al., 2008). The ACC, on the other hand, is involved in monitoring conflicts and detecting errors, as well as in the regulation of affective responses (Allman et al., 2001; Bush et al., 2000).

In OCD, the involvement of the orbitofrontal loop of the CSTC circuit is thought to be implicated, involving projections from the OFC to the head of the caudate and ventral striatum, then to the mediodorsal thalamus and finally returning to the OFC (Burguiere, Monteiro, Mallet, Feng, & Graybiel, 2015). Researchers have modified the model of this circuit to also include inputs from other limbic areas, such as the ACC, that are extensively connected to the OFC and are thought to be important in the regulation of affective states and emotional perception (Menzies et al., 2008). Each of these regions work in close cooperation but also have their own roles to play, the OFC as a monitor of appropriate behavior, and the ACC as an action monitor and regulator (Nakao, Okada, & Kanba, 2014). According to this model, the primary dysfunction is within the caudate, leading to inefficient gating at the level of the thalamus, thereby resulting in hyperactivity of the OFC and ACC that in turn causes obsessions. Dysfunctional processing in the ACC has been hypothesized to result in deficits in error monitoring systems, generating excessive error signals that are manifested as obsessions and SP, which then translates into corrective actions such as tics and compulsions (Schwartz, 1998). Hyperactivity of the OFC can result in rigid and inflexible behaviors, thus playing a role in the cycle that reinforces obsessions and compulsions, and their accompanying anxiety. Compulsions are performed to recruit the inefficient striatum to finally achieve thalamic gating (Del Casale et al., 2011).

5.3 TS + OCD

Early characterization of the CSTC circuits emphasized their parallel and segregated nature, however, it is now believed that the loops are much more integrated within the striatum and thalamus, and are therefore not fully segregated (Milad & Rauch, 2012). Instead, it has been suggested that the loops are better described as spirals, with information passing down from one loop to the next so that cortical outflow can instead affect wider parts of the striatum than previously thought (Haber & Knutson, 2010). Haber has suggested that it is in the basal ganglia that information can move across functional circuits. Specifically, reciprocal connections linking regions associated with similar functions and non-reciprocal connections that link regions that are associated with different cortico-basal ganglia connections, send feedback and feedforward signals allowing the transfer of information between circuits (Haber & Knutson, 2010). Importantly, this may help to explain comorbid symptomatology in TS, such as TS + OCD. Heimer expanded on this, he postulated that a general flow of information passes from ventromedial to dorsolateral parts of the striatum, consistent with the idea of information flow from motivational to motor parts of the CSTC circuitry (Heimer, Van Hoesen, Trimble, & Zahm, 2008). TS + OCD may then be explained by overlapping circuits emanating from impairments at related anatomical sites, with the spectrum of symptoms explained by topographical dysfunctions specifically in the striatum (Levy, 2011).

Indeed, the neuroimaging literature investigating tics and SP in samples of patients with TS + OCD for the most part support the findings of those conducted in traditional samples of TS patients. However, for the studies investigating OCS in TS + OCD, results both support and diverge from the regions consistently found to be associated with "classical OCD." For example, aberrant activity of the insula, ACC, parietal regions and the IFG were reported in the studies of TS + OCD, and are also commonly reported in OCD investigations with more traditional patients; thus, these regions appear to be common neural substrates of both comorbid OCD in TS and traditional OCD.

More interestingly, regions not commonly reported to be involved in OCD were sometimes found to be significant in TS + OCD patients, including the sensorimotor cortex. This could potentially explain the differences in the symptomatology between the OCS in TS and in "pure" OCD, as the former are generally more sensory-based whereas traditional OCS are commonly more cognitive-based (Diniz et al., 2006; Rosario et al., 2009). Additionally, the OFC was not consistently found to be associated with OCS in TS + OCD even though it is one of the most implicated regions in studies of traditional OCD patients (Menzies et al., 2008; Whiteside, Port, & Abramowitz, 2004), such that dysfunctions in the orbitofrontal loop of the CSTC circuit are at the core of neurobiological models of OCD. Differences between the importance of the OFC in OCD and TS + OCD may in part be explained by the functions of this region, as it is important in affective processing and thus may play a role in the anxiety that accompanies obsessions (Wood & Ahmari, 2015). The symptomatology of

OCS in TS, however, is not always characterized by anxiety and obsessions, as compulsions are instead often preceded by SP, and therefore, the OFC may not be as heavily involved in TS + OCD. These differences provide evidence of the neural distinction between pure OCD and OCD in TS, supporting the theory that TS + OCD does not always involve typical OCD regions.

When scrutinizing the regions found to be associated with tics and OCS in TS + OCD, the symptom domains may best be described as having partially overlapping, but distinct neural substrates. The similarities in the regions associated with tics, OCS and SP, such as the cingulate cortex and IFG, provide possible explanations as to why OCS are so common in TS, and at times, so similar. Perhaps faulty activity of such regions produces both tics and OCS, or manifests as SP that precedes them, with the full symptom profile of each disorder ultimately dependent on the other regions showing uniquely aberrant activity. More research into the neural correlates of tics and OCS in TS is greatly needed to further understand how they contribute to the neurobiological profile of TS + OCD.

References

Adler, C. M., McDonough-Ryan, P., Sax, K. W., Holland, S. K., Arndt, S., & Strakowski, S. M. (2000). fMRI of neuronal activation with symptom provocation in unmedicated patients with obsessive compulsive disorder. *Journal of Psychiatric Research, 34*(4–5), 317–324.

Alexander, G. E., & Crutcher, M. D. (1990). Functional architecture of basal ganglia circuits: Neural substrates of parallel processing. *Trends in Neurosciences, 13*(7), 266–271. https://doi.org/10.1016/0166-2236(90)90107-l.

Allman, J. M., Hakeem, A., Erwin, J. M., Nimchinsky, E., & Hof, P. (2001). The anterior cingulate cortex. The evolution of an interface between emotion and cognition. *Annals of the New York Academy of Sciences, 935*, 107–117.

Belloch, A., Fornacs, G., Carrasco, A., López-Solá, C., Alonso, P., & Menchan, J. M. (2016). Incompleteness and not just right experiences in the explanation of obsessive-compulsive disorder. *Psychiatry Research, 236*, 1–8. https://doi.org/10.1016/j.psychres.2016.01.012.

Bhikram, T., Arnold, P., Crawley, A., Abi-Jaoude, E., & Sandor, P. (2020). The functional connectivity profile of tics and obsessive-compulsive symptoms in Tourette syndrome. *Journal of Psychiatric Research, 123*, 128–135. https://doi.org/10.1016/j.jpsychires.2020.01.019.

Bhikram, T., Crawley, A., Arnold, P., Abi-Jaoude, E., & Sandor, P. (2020). Examining the functional activity of different obsessive-compulsive symptom dimensions in Tourette syndrome. *NeuroImage. Clinical, 26*, 102198. https://doi.org/10.1016/j.nicl.2020.102198.

Bhikram, T., Crawley, A., Arnold, P., Abi-Jaoude, E., & Sandor, P. (2021). Neuroimaging the emotional modulation of urge inhibition in Tourette syndrome. *Cortex; A Journal Devoted to the Study of the Nervous System and Behavior, 135*, 341–351. https://doi.org/10.1016/j.cortex.2020.10.010.

Bloch, M. H., Landeros-Weisenberger, A., Rosario, M. C., Pittenger, C., & Leckman, J. F. (2008). Meta-analysis of the symptom structure of obsessive-compulsive disorder. *American Journal of Psychiatry, 165*(12), 1532–1542. https://doi.org/10.1176/appi.ajp.2008.08020320.

Bohlhalter, S. (2006). Neural correlates of tic generation in Tourette syndrome: An event-related functional MRI study. *Brain, 129*(8), 2029–2037.
Brandt, V. C., Beck, C., Sajin, V., Baaske, M. K., Baumer, T., Beste, C., et al. (2016). Temporal relationship between premonitory urges and tics in Gilles de la Tourette syndrome. *Cortex; A Journal Devoted to the Study of the Nervous System and Behavior, 77*, 24–37. https://doi.org/10.1016/j.cortex.2016.01.008.
Brandt, V. C., Hermanns, J., Beck, C., Baumer, T., Zurowski, B., & Munchau, A. (2018). The temporal relationship between premonitory urges and covert compulsions in patients with obsessive-compulsive disorder. *Psychiatry Research, 262*, 6–12. https://doi.org/10.1016/j.psychres.2018.01.041.
Brown, C., Shahab, R., Collins, K., Fleysher, L., Goodman, W. K., Burdick, K. E., et al. (2018). Functional neural mechanisms of sensory phenomena in obsessive-compulsive disorder. *Journal of Psychiatric Research, 109*, 68–75. https://doi.org/10.1016/j.jpsychires.2018.11.018.
Budman, C. L., Rockmore, L., Stokes, J., & Sossin, M. (2003). Clinical phenomenology of episodic rage in children with Tourette syndrome. *Journal of Psychosomatic Research, 55*(1), 59–65. https://doi.org/10.1016/s0022-3999(02)00584-6.
Bullen, J. G., & Hemsley, D. R. (1983). Sensory experience as a trigger in Gilles de la Tourette's syndrome. *Journal of behavior therapy and experimental psychiatry*. https://doi.org/10.1016/0005-7916(83)90048-4.
Bullmore, E., & Sporns, O. (2009). Complex brain networks: Graph theoretical analysis of structural and functional systems. *Nature Reviews. Neuroscience, 10*(3), 186–198. https://doi.org/10.1038/nrn2575.
Burguiere, E., Monteiro, P., Mallet, L., Feng, G., & Graybiel, A. M. (2015). Striatal circuits, habits, and implications for obsessive-compulsive disorder. *Current Opinion in Neurobiology, 30*, 59–65. https://doi.org/10.1016/j.conb.2014.08.008.
Bush, G., Luu, P., & Posner, M. I. (2000). Cognitive and emotional influences in anterior cingulate cortex. *Trends in Cognitive Sciences, 4*(6), 215–222. https://doi.org/10.1016/s1364-6613(00)01483-2.
Cavanna, A. E., & Trimble, M. R. (2006). The precuneus: A review of its functional anatomy and behavioural correlates. *Brain, 129*(Pt. 3), 564–583. https://doi.org/10.1093/brain/awl004.
Church, J. A., Fair, D. A., Dosenbach, N. U., Cohen, A. L., Miezin, F. M., Petersen, S. E., et al. (2009). Control networks in paediatric Tourette syndrome show immature and anomalous patterns of functional connectivity. *Brain, 132*(Pt. 1), 225–238. https://doi.org/10.1093/brain/awn223.
Coffey, B. J., Fau, M. E., Biederman, J., O'Sullivan, R., Savage, C. R., Phillips, K., et al. (1998). Tourette's disorder with and without obsessive-compulsive disorder in adults: Are they different? *Journal of Nervous and Mental Disease, 186*, 201–206.
Cox, J. H., Seri, S., & Cavanna, A. E. (2018). Sensory aspects of Tourette syndrome. *Neuroscience & Biobehavioral Reviews, 88*, 170–176. https://doi.org/10.1016/j.neubiorev.2018.03.016.
Crossley, E., & Cavanna, A. E. (2013). Sensory phenomena: Clinical correlates and impact on quality of life in adult patients with Tourette syndrome. *Psychiatry Research, 209*(3), 705–710. https://doi.org/10.1016/j.psychres.2013.04.019.
Delaveau, P., Salgado-Pineda, P., Fossati, P., Witjas, T., Azulay, J. P., & Blin, O. (2010). Dopaminergic modulation of the default mode network in Parkinson's disease. *European Neuropsychopharmacology: The Journal of the European College of Neuropsychopharmacology, 20*(11), 784–792. https://doi.org/10.1016/j.euroneuro.2010.07.001.
Del Casale, A., Kotzalidis, G. D., Rapinesi, C., Serata, D., Ambrosi, E., Simonetti, A., et al. (2011). Functional neuroimaging in obsessive-compulsive disorder. *Neuropsychobiology, 64*(2), 61–85.

de Vries, F. E., Cath, D. C., Hoogendoorn, A. W., van Oppen, P., Glas, G., Veltman, D. J., et al. (2016). Tic-related versus tic-free obsessive-compulsive disorder: Clinical picture and 2-year natural course. *The Journal of Clinical Psychiatry*, 77(10), e1240–e1247. https://doi.org/10.4088/JCP.14m09736.

de Vries, F. E., van den Heuvel, O. A., Cath, D. C., Groenewegen, H. J., van Balkom, A. J., Boellaard, R., et al. (2013). Limbic and motor circuits involved in symmetry behavior in Tourette's syndrome. *CNS Spectrums*, *18*(1), 34–42. https://doi.org/10.1017/S1092852912000703.

Diniz, J. B., Rosario-Campos, M. C., Hounie, A. G., Curi, M., Shavitt, R. G., Lopes, A. C., et al. (2006). Chronic tics and Tourette syndrome in patients with obsessive-compulsive disorder. *Journal of Psychiatric Research*, *40*(6), 487–493. https://doi.org/10.1016/j.jpsychires.2005.09.002.

Eddy, C. M., & Cavanna, A. E. (2014). Tourette syndrome and obsessive compulsive disorder: Compulsivity along the continuum. *Journal of Obsessive-Compulsive and Related Disorders*, *3*(4), 363–371. https://doi.org/10.1016/j.jocrd.2014.04.003.

Fan, S., van den Heuvel, O. A., Cath, D. C., de Wit, S. J., Vriend, C., Veltman, D. J., et al. (2018). Altered functional connectivity in resting state networks in Tourette's disorder. *Frontiers in Human Neuroscience*, *12*, 363. https://doi.org/10.3389/fnhum.2018.00363.

Ferrão, Y. A., Miguel, E., & Stein, D. J. (2009). Tourette's syndrome, trichotillomania, and obsessive-compulsive disorder: How closely are they related? *Psychiatry Research*, *170*(1), 32–42.

Ferrão, Y. A., Shavitt, R. G., Prado, H., Fontenelle, L. F., Malavazzi, D. M., de Mathis, M. A., et al. (2012). Sensory phenomena associated with repetitive behaviors in obsessive-compulsive disorder: An exploratory study of 1001 patients. *Psychiatry Research*, *197*(3), 253–258.

Fitzgerald, K. D., Welsh, R. C., Stern, E. R., Angstadt, M., Hanna, G. L., Abelson, J. L., et al. (2011). Developmental alterations of frontal-striatal-thalamic connectivity in obsessive-compulsive disorder. *Journal of the American Academy of Child and Adolescent Psychiatry*, *50*(9). https://doi.org/10.1016/j.jaac.2011.06.011. 938–948.e3.

Freeman, R. D., Fast, D. K., Burd, L., Kerbeshian, J., Robertson, M. M., & Sandor, P. (2000). An international perspective on Tourette syndrome: Selected findings from 3500 individuals in 22 countries. *Developmental Medicine & Child Neurology*, *42*(7), 436–447. https://doi.org/10.1017/S0012162200000839.

Frey, K. A., & Albin, R. L. (2006). Neuroimaging of Tourette syndrome. *Journal of Child Neurology*, *21*(8), 672–677.

Fried, I., Katz, A., McCarthy, G., Sass, K. J., Williamson, P., Spencer, S. S., et al. (1991). Functional organization of human supplementary motor cortex studied by electrical stimulation. *The Journal of Neuroscience : The Official Journal of the Society for Neuroscience*, *11*(11), 3656–3666.

Ganos, C., Kahl, U., Brandt, V., Schunke, O., Baumer, T., Thomalla, G., et al. (2014). The neural correlates of tic inhibition in Gilles de la Tourette syndrome. *Neuropsychologia*, *65*, 297–301. https://doi.org/10.1016/j.neuropsychologia.2014.08.007.

Ganos, C., Roessner, V., & Munchau, A. (2013). The functional anatomy of Gilles de la Tourette syndrome. *Neuroscience and Biobehavioral Reviews*, *37*(6), 1050–1062. https://doi.org/10.1016/j.neubiorev.2012.11.004.

Goldberg, I. I., Harel, M., & Malach, R. (2006). When the brain loses its self: Prefrontal inactivation during sensorimotor processing. *Neuron*, *50*(2), 329–339. https://doi.org/10.1016/j.neuron.2006.03.015.

Goodman, W. K., Storch, E. A., Geffken, G. R., & Murphy, T. K. (2006). Obsessive-compulsive disorder in Tourette syndrome. *Journal of Child Neurology*, *21*(8), 704–714. https://doi.org/10.1177/08830738060210081201.

Grados, M. A., & Mathews, C. A. (2008). Latent class analysis of Gilles de la Tourette syndrome using comorbidities: Clinical and genetic implications. *Biological Psychiatry*, *64*(3), 219–225. https://doi.org/10.1016/j.biopsych.2008.01.019.

Gursel, D. A., Avram, M., Sorg, C., Brandl, F., & Koch, K. (2018). Frontoparietal areas link impairments of large-scale intrinsic brain networks with aberrant fronto-striatal interactions in OCD: A meta-analysis of resting-state functional connectivity. *Neuroscience and Biobehavioral Reviews*, *87*, 151–160. https://doi.org/10.1016/j.neubiorev.2018.01.016.

Haber, S. N., & Knutson, B. (2010). The reward circuit: Linking primate anatomy and human imaging. *Neuropsychopharmacology: Official Publication of the American College of Neuropsychopharmacology*, *35*(1), 4–26. https://doi.org/10.1038/npp.2009.129.

Hampson, M., Tokoglu, F., King, R. A., Constable, R. T., & Leckman, J. F. (2009). Brain areas coactivating with motor cortex during chronic motor tics and intentional movements. *Biological Psychiatry*, *65*(7), 594–599. https://doi.org/10.1016/j.biopsych.2008.11.012.

Harrison, B. J., Pujol, J., Cardoner, N., Deus, J., Alonso, P., López-Solà, M., et al. (2013). Brain corticostriatal systems and the major clinical symptom dimensions of obsessive-compulsive disorder. *Biological Psychiatry*, *73*(4), 321–328.

Harrison, B. J., Soriano-Mas, C., Pujol, J., Ortiz, H., Lopez-Sola, M., Hernandez-Ribas, R., et al. (2009). Altered corticostriatal functional connectivity in obsessive-compulsive disorder. *Archives of General Psychiatry*, *66*(11), 1189–1200. https://doi.org/10.1001/archgenpsychiatry.2009.152.

Heimer, L., Van Hoesen, G. W., Trimble, M., & Zahm, D. S. (2008). The anatomy of the basal forebrain. In L. Heimer, G. W. Van Hoesen, M. Trimble, & D. S. Zahm (Eds.), *Vol. 3. Anatomy of neuropsychiatry* (pp. 27–67). San Diego: Academic Press. https://doi.org/10.1016/B978-012374239-1.50006-3. Retrieved from http://www.sciencedirect.com/science/article/pii/B9780123742391500063.

Hirschtritt, M. E., Lee, P. C., Pauls, D. L., Dion, Y., Grados, M. A., Illmann, C., et al. (2015). Lifetime prevalence, age of risk, and genetic relationships of comorbid psychiatric disorders in Tourette syndrome. *JAMA Psychiatry*, *72*(4), 325.

Holzer, J. C., Goodman, W. K., McDougle, C. J., Baer, L., Boyarsky, B. K., Leckman, J. F., et al. (1994). Obsessive-compulsive disorder with and without a chronic tic disorder. A comparison of symptoms in 70 patients. *British Journal of Psychiatry*, *164*(4), 469–473.

Jackson, S. R., Parkinson, A., Kim, S. Y., Schuermann, M., & Eickhoff, S. B. (2011). On the functional anatomy of the urge-for-action. *Cognitive Neuroscience*, *2*(3–4), 227–243. https://doi.org/10.1080/17588928.2011.604717.

Jang, J. H., Kim, J., Jung, W. H., Choi, J., Jung, M. H., Lee, J., et al. (2010). Functional connectivity in fronto-subcortical circuitry during the resting state in obsessive-compulsive disorder. *Neuroscience Letters*, *474*(3), 158–162. https://doi.org/10.1016/j.neulet.2010.03.031.

Jeffries, K. J., Schooler, C., Schoenbach, C., Herscovitch, P., Chase, T. N., & Braun, A. R. (2002). The functional neuroanatomy of Tourette's syndrome: An FDG PET study III: Functional coupling of regional cerebral metabolic rates. *Neuropsychopharmacology : Official Publication of the American College of Neuropsychopharmacology*, *27*(1), 92–104. https://doi.org/10.1016/S0893-133X(01)00428-6.

Kadesjo, B., & Gillberg, C. (2000). Tourette's disorder: Epidemiology and comorbidity in primary school children. *Journal of the American Academy of Child and Adolescent Psychiatry*, *39*(5), 548–555. https://doi.org/10.1097/00004583-200005000-00007.

Kano, Y., Kono, T., Matsuda, N., Nonaka, M., Kuwabara, H., Shimada, T., et al. (2015). The impact of tics, obsessive-compulsive symptoms, and impulsivity on global functioning in Tourette syndrome. *Psychiatry Research*, *226*(1), 156–161. https://doi.org/10.1016/j.psychres.2014.12.041.

Kloft, L., Steinel, T., & Kathmann, N. (2018). Systematic review of co-occurring OCD and TD: Evidence for a tic-related OCD subtype? *Neuroscience and Biobehavioral Reviews*, *95*, 280–314. https://doi.org/10.1016/j.neubiorev.2018.09.021.

Lebowitz, E. R., Motlagh, M. G., Katsovich, L., King, R. A., Lombroso, P. J., Grantz, H., et al. (2012). Tourette syndrome in youth with and without obsessive compulsive disorder and attention deficit hyperactivity disorder. *European Child & Adolescent Psychiatry, 21*(8), 451–457. https://doi.org/10.1007/s00787-012-0278-5.

Leckman, J. F., & Riddle, M. A. (2000). Tourette's syndrome: When habit-forming systems form habits of their own? *Neuron, 28*(2), 349–354. https://doi.org/10.1016/s0896-6273(00)00114-8.

Lerner, A., Bagic, A., Hanakawa, T., Boudreau, E. A., Pagan, F., Mari, Z., et al. (2009). Involvement of insula and cingulate cortices in control and suppression of natural urges. *Cerebral Cortex (New York, N.Y.: 1991), 19*(1), 218–223.

Levy, F. (2011). Cortico-striatal-thalamic-cortical (CSTC) circuits, Tourette's disorder, and OCD. *The ADHD Report, 19*(2). https://doi.org/10.1521/adhd.2011.19.2.8. 8–9, 15.

Li, P., Li, S., Dong, Z., Luo, J., Han, H., Xiong, H., et al. (2012). Altered resting state functional connectivity patterns of the anterior prefrontal cortex in obsessive-compulsive disorder. *Neuroreport, 23*(11), 681–686. https://doi.org/10.1097/WNR.0b013e328355a5fe.

Mataix-Cols, D., Wooderson, S., Lawrence, N., Brammer, M. J., Speckens, A., & Phillips, M. L. (2004). Distinct neural correlates of washing, checking, and hoarding symptom dimensions in obsessive-compulsive disorder. *Archives of General Psychiatry, 61*(6), 564–576.

Mazzone, L., Yu, S., Blair, C., Gunter, B. C., Wang, Z., Marsh, R., et al. (2010). An FMRI study of frontostriatal circuits during the inhibition of eye blinking in persons with Tourette syndrome. *American Journal of Psychiatry, 167*(3), 341–349.

McGuire, P. K., Bench, C. J., Frith, C. D., Marks, I. M., Frackowiak, R. S., & Dolan, R. J. (1994). Functional anatomy of obsessive-compulsive phenomena. *British Journal of Psychiatry, 164*(4), 459–468. https://doi.org/10.1192/bjp.164.4.459.

Menzies, L., Chamberlain, S. R., Laird, A. R., Thelen, S. M., Sahakian, B. J., & Bullmore, E. T. (2008). Integrating evidence from neuroimaging and neuropsychological studies of obsessive-compulsive disorder: The orbitofronto-striatal model revisited. *Neuroscience and Biobehavioral Reviews, 32*(3), 525–549. https://doi.org/10.1016/j.neubiorev.2007.09.005.

Mercadante, M. T., Rosario-Campos, M. C., Quarantini, L. C., & Sato, F. P. (2004). The neurobiological bases of obsessive-compulsive disorder and Tourette syndrome. [As bases neurobiologicas do transtorno obsessivo-compulsivo e da sindrome de Tourette]. *Jornal de Pediatria, 80*(2 Suppl), S35–S44.

Miguel, E. C., do Rosário-Campos, M. C., Prado, H. S., do Valle, R., Rauch, S. L., Coffey, B. J., et al. (2000). Sensory phenomena in obsessive-compulsive disorder and Tourette's disorder. *The Journal of Clinical Psychiatry, 61*(2), 150–156. quiz 157.

Milad, M. R., & Rauch, S. L. (2012). Obsessive-compulsive disorder: Beyond segregated cortico-striatal pathways. *Trends in Cognitive Sciences, 16*(1), 43–51. https://doi.org/10.1016/j.tics.2011.11.003.

Nakao, T., Okada, K., & Kanba, S. (2014). Neurobiological model of obsessive-compulsive disorder: Evidence from recent neuropsychological and neuroimaging findings. *Psychiatry and Clinical Neurosciences, 68*(8), 587–605. https://doi.org/10.1111/pcn.12195.

Neuner, I., Kellermann, T., Stocker, T., Kircher, T., Habel, U., Shah, J. N., et al. (2010). Amygdala hypersensitivity in response to emotional faces in Tourette's patients. *The World Journal of Biological Psychiatry: The Official Journal of the World Federation of Societies of Biological Psychiatry, 11*(7), 858–872. https://doi.org/10.3109/15622975.2010.480984.

Neuner, I., Werner, C. J., Arrubla, J., Stocker, T., Ehlen, C., Wegener, H. P., et al. (2014). Imaging the where and when of tic generation and resting state networks in adult Tourette patients. *Frontiers in Human Neuroscience, 8*, 362. https://doi.org/10.3389/fnhum.2014.00362.

Peterson, B. S., Skudlarski, P., Anderson, A. W., Zhang, H., Gatenby, J. C., Lacadie, C. M., et al. (1998). A functional magnetic resonance imaging study of tic suppression in Tourette syndrome. *Archives of General Psychiatry*, *55*(4), 326–333.

Petter, T., Richter, M. A., & Sandor, P. (1998). Clinical features distinguishing patients with Tourette's syndrome and obsessive-compulsive disorder from patients with obsessive-compulsive disorder without tics. *The Journal of Clinical Psychiatry*, *59*(9), 456–459.

Phillips, M. L., Marks, I. M., Senior, C., Lythgoe, D., O'Dwyer, A. M., Meehan, O., et al. (2000). A differential neural response in obsessive-compulsive disorder patients with washing compared with checking symptoms to disgust. *Psychological Medicine*, *30*(5), 1037–1050.

Pourfar, M., Feigin, A., Tang, C. C., Carbon-Correll, M., Bussa, M., Budman, C., et al. (2011). Abnormal metabolic brain networks in Tourette syndrome. *Neurology*, *76*(11), 944–952.

Rauch, S. L., Dougherty, D. D., Shin, L. M., Alpert, N. M., Manzo, P., Leahy, L., et al. (1998). Neural correlates of factor-analyzed OCD symptom dimensions: A PET study. *CNS Spectrums*, *3*(7), 37–43. https://doi.org/10.1017/S1092852900006167.

Rauch, S. L., Jenike, M. A., Alpert, N. M., Baer, L., Breiter, H. C., Savage, C. R., et al. (1994). Regional cerebral blood flow measured during symptom provocation in obsessive-compulsive disorder using oxygen 15-labeled carbon dioxide and positron emission tomography. *Archives of General Psychiatry*, *51*(1), 62–70.

Rickards, H. (2009). Functional neuroimaging in Tourette syndrome. *Journal of Psychosomatic Research*, *67*(6), 575–584.

Rizzo, R., Gulisano, M., Pellico, A., Cali, P. V., & Curatolo, P. (2014). Tourette syndrome and comorbid conditions: A spectrum of different severities and complexities. *Journal of Child Neurology*, *29*(10), 1383–1389. https://doi.org/10.1177/0883073814534317.

Rosario, M. C., Prado, H. S., Borcato, S., Diniz, J. B., Shavitt, R. G., Hounie, A. G., et al. (2009). Validation of the University of São Paulo Sensory Phenomena Scale: Initial psychometric properties. *CNS Spectrums*, *14*(6), 315–323.

Rotge, J., Guehl, D., Dilharreguy, B., Cuny, E., Tignol, J., Bioulac, B., et al. (2008). Provocation of obsessive-compulsive symptoms: A quantitative voxel-based meta-analysis of functional neuroimaging studies. *Journal of Psychiatry & Neuroscience: JPN*, *33*(5), 405–412.

Rothenberger, A., & Roessner, V. (2013). Functional neuroimaging investigations of motor networks in Tourette syndrome. *Behavioural Neurology*, *27*(1), 47–55. https://doi.org/10.3233/BEN-120281.

Sakai, Y., Narumoto, J., Nishida, S., Nakamae, T., Yamada, K., Nishimura, T., et al. (2011). Corticostriatal functional connectivity in non-medicated patients with obsessive-compulsive disorder. *European Psychiatry: The Journal of the Association of European Psychiatrists*, *26*(7), 463–469. https://doi.org/10.1016/j.eurpsy.2010.09.005.

Saxena, S., & Rauch, S. L. (2000). Functional neuroimaging and the neuroanatomy of obsessive-compulsive disorder. *The Psychiatric Clinics of North America*, *23*(3), 563–586.

Schwartz, J. M. (1998). Neuroanatomical aspects of cognitive-behavioural therapy response in obsessive-compulsive disorder. An evolving perspective on brain and behaviour. *British Journal of Psychiatry. Supplement*, (35), 38–44.

Seeley, W. W., Menon, V., Schatzberg, A. F., Keller, J., Glover, G. H., Kenna, H., et al. (2007). Dissociable intrinsic connectivity networks for salience processing and executive control. *The Journal of Neuroscience: The Official Journal of the Society for Neuroscience*, *27*(9), 2349–2356. https://doi.org/10.1523/JNEUROSCI.5587-06.2007.

Simon, D., Adler, N., Kaufmann, C., & Kathmann, N. (2014). Amygdala hyperactivation during symptom provocation in obsessive-compulsive disorder and its modulation by distraction. *NeuroImage. Clinical*, *4*, 549–557. https://doi.org/10.1016/j.nicl.2014.03.011.

Simon, D., Kaufmann, C., Kniesche, R., Kischkel, E., & Kathmann, N. (2013). Autonomic responses and neural-cardiac coupling during individually tailored symptom provocation in obsessive-compulsive disorder. *Journal of Anxiety Disorders*, *27*(7), 635–644. https://doi.org/10.1016/j.janxdis.2013.08.005.

Simon, D., Kaufmann, C., Musch, K., Kischkel, E., & Kathmann, N. (2010). Fronto-striato-limbic hyperactivation in obsessive-compulsive disorder during individually tailored symptom provocation. *Psychophysiology*, *47*(4), 728–738. https://doi.org/10.1111/j.1469-8986.2010.00980.x.

Stern, E. R., Brown, C., Ludlow, M., Shahab, R., Collins, K., Lieval, A., et al. (2020). The buildup of an urge in obsessive-compulsive disorder: Behavioral and neuroimaging correlates. *Human Brain Mapping*, *41*(6), 1611–1625. https://doi.org/10.1002/hbm.24898.

Stern, E., Silbersweig, D. A., Chee, K. Y., Holmes, A., Robertson, M. M., Trimble, M., et al. (2000). A functional neuroanatomy of tics in Tourette syndrome. *Archives of General Psychiatry*, *57*(8), 741–748. https://doi.org/10.1001/archpsyc.57.8.741.

Thorsen, A. L., Hagland, P., Radua, J., Mataix-Cols, D., Kvale, G., Hansen, B., et al. (2018). Emotional processing in obsessive-compulsive disorder: A systematic review and meta-analysis of 25 functional neuroimaging studies. *Biological Psychiatry. Cognitive Neuroscience and Neuroimaging*, *3*(6), 563–571. https://doi.org/10.1016/j.bpsc.2018.01.009.

Tikoo, S., Cardona, F., Tommasin, S., Gianni, C., Conte, G., Upadhyay, N., et al. (2020). Resting-state functional connectivity in drug-naive pediatric patients with Tourette syndrome and obsessive-compulsive disorder. *Journal of Psychiatric Research*, *129*, 129–140. https://doi.org/10.1016/j.jpsychires.2020.06.021.

Tikoo, S., Suppa, A., Tommasin, S., Gianni, C., Conte, G., Mirabella, G., et al. (2021). The cerebellum in drug-naive children with Tourette syndrome and obsessive-compulsive disorder. *Cerebellum (London, England)*. https://doi.org/10.1007/s12311-021-01327-7.

Tinaz, S., Belluscio, B. A., Malone, P., van der Veen, J. W., Hallett, M., & Horovitz, S. G. (2014). Role of the sensorimotor cortex in Tourette syndrome using multimodal imaging. *Human Brain Mapping*, *35*(12), 5834–5846. https://doi.org/10.1002/hbm.22588.

Tinaz, S., Malone, P., Hallett, M., & Horovitz, S. G. (2015). Role of the right dorsal anterior insula in the urge to tic in Tourette syndrome. *Movement Disorders: Official Journal of the Movement Disorder Society*, *30*(9), 1190–1197. https://doi.org/10.1002/mds.26230.

van der Salm, S. M. A., van der Meer, J. N., Cath, D. C., Groot, P. F. C., van der Werf, Y. D., Brouwers, E., et al. (2018). Distinctive tics suppression network in Gilles de la Tourette syndrome distinguished from suppression of natural urges using multimodal imaging. *NeuroImage. Clinical*, *20*, 783–792. https://doi.org/10.1016/j.nicl.2018.09.014.

Wang, Z., Maia, T., Marsh, R., Colibazzi, T., Gerber, A., & Peterson, B. (2011). The neural circuits that generate tics in Tourette syndrome. *American Journal of Psychiatry*, *168*(12), 1326–1337.

Wen, H., Liu, Y., Rekik, I., Wang, S., Chen, Z., Zhang, J., et al. (2018). Combining disrupted and discriminative topological properties of functional connectivity networks as neuroimaging biomarkers for accurate diagnosis of early Tourette syndrome children. *Molecular Neurobiology*, *55*(4), 3251–3269. https://doi.org/10.1007/s12035-017-0519-1.

Whiteside, S. P., Port, J. D., & Abramowitz, J. S. (2004). A meta-analysis of functional neuroimaging in obsessive-compulsive disorder. *Psychiatry Research*, *132*(1), 69–79. https://doi.org/10.1016/j.pscychresns.2004.07.001.

Wood, J., & Ahmari, S. E. (2015). A framework for understanding the emerging role of corticolimbic-ventral striatal networks in OCD-associated repetitive behaviors. *Frontiers in Systems Neuroscience*, *9*, 171. https://doi.org/10.3389/fnsys.2015.00171.

Worbe, Y., Malherbe, C., Hartmann, A., Pelegrini-Issac, M., Messe, A., Vidailhet, M., et al. (2012). Functional immaturity of cortico-basal ganglia networks in Gilles de la Tourette syndrome. *Brain*, *135*(Pt. 6), 1937–1946. https://doi.org/10.1093/brain/aws056.

Zapparoli, L., Porta, M., & Paulesu, E. (2015). The anarchic brain in action: The contribution of task-based fMRI studies to the understanding of Gilles de la Tourette syndrome. *Current Opinion in Neurology*, *28*(6), 604–611. https://doi.org/10.1097/WCO.0000000000000261.

CHAPTER FOURTEEN

Co-occurring ADHD symptoms and electrophysiological correlates of cognitive control in Tourette syndrome

Elizabeth Shephard[a,b,*], Madeleine J. Groom[c], and Georgina M. Jackson[c]
[a]Institute of Psychiatry, Psychology & Neuroscience (IoPPN), King's College London, London, United Kingdom
[b]Department of Psychiatry, Faculdade de Medicina FMUSP, University of Sao Paulo, Sao Paulo, Brazil
[c]Academic Unit of Mental Health & Clinical Neurosciences, School of Medicine, Institute of Mental Health, University of Nottingham, Nottingham, United Kingdom
[*]Corresponding author: e-mail address: elizabeth.1.shephard@kcl.ac.uk

Contents

1. Introduction

A large proportion of individuals with Tourette syndrome (TS) have co-occurring diagnoses or symptoms of other neurodevelopmental or psychiatric conditions. For instance, a large multi-site study reported that 85.7% of 1,374 children and adults with TS living in the United Kingdom, North America and the Netherlands met diagnostic criteria for at least

International Review of Movement Disorders, Volume 3
ISSN 2666-7878
https://doi.org/10.1016/bs.irmvd.2021.11.002

one other psychiatric condition; 57.7% met diagnostic criteria for two or more co-occurring conditions (Hirschtritt et al., 2015). The most common co-occurring condition in TS was attention-deficit/hyperactivity disorder (ADHD), with 54.3% of the participants meeting diagnostic criteria for this condition. The same pattern of findings has been reported in many other studies of TS (Cravedi et al., 2017; Sambrani, Jakubovski, & Müller-Vahl, 2016).

Co-occurring ADHD is associated with a range of additional or more severe impairments in TS. For example, individuals with TS and co-occurring ADHD have been reported to experience lower quality of life and greater functional difficulties in school, home and social settings than individuals with TS without ADHD (Eddy et al., 2012; Lebowitz et al., 2012; Ricketts et al., 2021). Functional impairments have also been shown to correlate with co-occurring ADHD symptoms over and above tic severity (e.g., Gorman et al., 2010). In addition, co-occurring ADHD is associated with poorer response to behavioral tic therapies in children and adults with TS (McGuire et al., 2014). Finally, co-occurring ADHD is associated with impairments in neurocognitive processes in TS, including visuomotor function (Sukhodolsky, Landeros-Weisenberger, Scahill, Leckman, & Schultz, 2010), emotion processing (Drury et al., 2012) and reinforcement learning (Shephard, Jackson, & Groom, 2016a).

Arguably one of the most important neurocognitive functions in TS is cognitive control. As defined in previous chapters, cognitive control is an umbrella term for a set of self-regulatory processes that monitor our ongoing behavior and implement modifications to ensure it remains consistent with our goals in the context of changing internal and external demands (Botvinick, Braver, Barch, Carter, & Cohen, 2001; Houghton & Tipper, 1996). Specific cognitive control processes include inhibition (the ability to withhold inappropriate motor behaviors and suppress interfering information), attention (the ability to focus on specific aspects of the environment in a selective and sustained manner), performance monitoring (the ability to monitor our own behavior), flexibility (the ability to flexibly adapt our behavior in response to internal and external stimuli) and working memory (the ability to hold and manipulate information in mind) (Botvinick et al., 2001; Houghton & Tipper, 1996). These self-regulatory processes are particularly important in TS because they enable the control of tics during temporary tic suppression and during behavioral therapy for tics (McGuire et al., 2021; Woods et al., 2008). It has also been proposed that increasing cognitive control over tics during adolescence contributes to the typical remitting trajectory of tics in late adolescence and adulthood (Jackson, Draper, Dyke, Pépés, & Jackson, 2015). Recent longitudinal work investigating developmental changes in cognitive control and tics supports this proposal (Yaniv et al., 2018).

Given the central role of cognitive control in tic control in TS, it is crucial to understand how co-occurring conditions affect these self-regulatory processes. Outside of the context of TS, ADHD is strongly associated with impairments in cognitive control (discussed in Section 2 below). It is therefore likely that co-occurring ADHD may introduce cognitive control difficulties in TS, which might in turn affect tic control, response to behavioral tic therapies and tic remission. In this chapter, we discuss the effects of co-occurring ADHD on cognitive control in individuals with TS. We focus specifically on electrophysiological correlates of cognitive control, i.e. indices of electrophysiological neural activity associated with different cognitive control processes measured with electroencephalography (EEG) (described in chapter "Control of actions and tics: An electrophysiological perspective" by Simon Morand-Beaulieu and Denis Sukhodolsky). Electrophysiological correlates are more sensitive to alterations than are behavioral measures and more directly reflect the neurocognitive processes involved in cognitive control. Electrophysiology is also particularly appropriate for investigating fast and adaptive neurocognitive functions such as cognitive control and tic control because, unlike other techniques of measuring neural activity such as functional magnetic resonance imaging (fMRI), it measures neural activity in the millisecond time-range and can therefore capture moment-to-moment changes in cognitive control processes.

2. Electrophysiological correlates of cognitive control in ADHD

ADHD is characterized by developmentally inappropriate and impairing symptoms of inattention, hyperactivity and impulsivity (American Psychiatric Association, 2013). These symptoms are strongly associated with impairments in cognitive control processes, with meta-analyses of behavioral measures reporting the largest effect sizes for impaired inhibition and working memory (Shephard et al., 2021; Willcutt, Doyle, Nigg, Faraone, & Pennington, 2005). EEG studies have consistently reported reduced amplitudes of electrophysiological correlates of cognitive control in ADHD, indicative of weaker or less efficient engagement of neural regions involved in cognitive control.

For example, during experimental tasks such as the Go/Nogo that require the withholding of a prepotent motor response, children and adults with ADHD show reduced amplitude of the P3 event-related potential component compared to participants without ADHD (see Kaiser et al., 2020 for a recent meta-analysis). The P3 is a positive deflection that occurs around 300-600 ms in the averaged electrophysiological activity in response to an event of interest (event-related potential, ERP) (Falkenstein, Hoormann, & Hohnsbein, 1999). In task contexts such as the Go/Nogo, the P3 is larger in amplitude at fronto-central scalp regions on trials that require withholding of the motor response (Nogo trials) compared to trials that require execution of the motor response (Go trials) (Falkenstein et al., 1999; Groom & Cragg, 2015). The enhancement of P3 amplitude on Nogo trials is believed to reflect the engagement of frontal cortical regions involved in inhibition (Falkenstein et al., 1999; Groom & Cragg, 2015). Thus, reduced P3 amplitude in ADHD during these tasks is interpreted as reflecting reduced or inefficient engagement of neural regions subserving inhibition (Kaiser et al., 2020).

Another set of electrophysiological correlates that are frequently reported to be reduced in amplitude in ADHD are the N2 ERP component, evoked oscillatory theta power and the error-related negativity (ERN) ERP component, all of which are believed to index processes of conflict and error monitoring (Cavanagh & Frank, 2014). The N2 is a negative deflection at ~200–400 ms in the ERP waveform that increases in amplitude over fronto-central scalp electrodes when task conditions elicit conflict (Donkers & Van Boxtel, 2004; Groom & Cragg, 2015). For example, in the Flanker task a central stimulus points in one direction and flanking stimuli point in the opposite direction; the participant must overcome the conflicting directional information introduced by the flankers and respond to the direction of the central stimulus. In the same time-range as the N2, evoked (i.e. stimulus-locked) theta oscillations increase in amplitude over fronto-central scalp during task conditions that elicit conflict; changes in the phase and/or amplitude of such theta activity are thought to contribute to the N2 component (Cavanagh & Frank, 2014). The ERN is a negative deflection at ~100 ms following an erroneous response and shows a similar fronto-central topography as the N2 (Gehring, Goss, Coles, Meyer, & Donchin, 1993). The ERN is also thought to be driven at least partly by changes in phase/amplitude of evoked theta activity (Cavanagh & Frank, 2014). Evidence indicates that the N2, evoked fronto-central theta and the ERN are generated by the anterior cingulate cortex (ACC) and reflect performance monitoring functions of this region, including error and conflict monitoring (Botvinick et al., 2001; Cavanagh & Frank, 2014). Reductions in amplitude of these components have been reported in children and adults with ADHD, indicative of impaired neural mechanisms of performance monitoring (Bellato et al., 2021; Groom et al., 2008; Groom et al., 2010).

Several other electrophysiological correlates of cognitive control have been implicated in ADHD. These include the error positivity (Pe), a positive deflection around 300–500 ms at centro-parietal electrodes that follows the ERN to erroneous responses and likely reflects evaluative processes (Falkenstein, Hoormann, Christ, & Hohnsbein, 2000). The Pe has consistently been reported as reduced in amplitude in children and adults with ADHD (Bellato et al., 2021; Kaiser et al., 2020). Reduced amplitude of the centro-parietal P3 in response to visual or auditory cue and target stimuli during cognitive control tasks has also been reported in ADHD (Tye et al., 2014) and likely reflects difficulties with attention, working memory updating and preparatory processes (Donchin & Coles, 1988; Kok, 2001).

Taken together, the findings reviewed above suggest that co-occurring ADHD symptoms are likely to negatively affect a range of cognitive control processes in TS. It is also important to highlight that cognitive control impairments in ADHD onset early in life. For instance, poorer inhibitory control, attention, flexibility and working memory are found as early as age 2–3 years in children who, later in childhood, meet diagnostic criteria for ADHD or who have elevated ADHD symptoms (Shephard et al., 2021). Thus, in addition to the broad range of cognitive control impairments associated with ADHD that may affect children with TS, it is also the case that these impairments are likely to be well-established before the onset of tics in mid-childhood.

3. Co-occurring ADHD symptoms and electrophysiological correlates of cognitive control in TS

3.1 Comparisons of cognitive control between individuals with TS, TS and co-occurring ADHD, ADHD and controls

Despite the importance of cognitive control in TS, the frequency of co-occurring ADHD symptoms and the evidence from the ADHD literature that these symptoms are associated with widespread difficulties in cognitive control, few studies have examined effects of co-occurring ADHD on cognitive control in TS. Even fewer studies have used methodological designs that are capable of attributing cognitive control difficulties in TS specifically to co-occurring ADHD. The strongest methodological design in this regard involves comparing participants with TS to those with TS and co-occurring ADHD (TS + ADHD), ADHD without TS and control participants without TS or ADHD. This four-group design enables the identification of cognitive control difficulties that are related to ADHD in both participants with TS + ADHD and ADHD, and cognitive control alterations that are related to TS with and without ADHD. In designs that compare participants with TS + ADHD to participants with TS only or to participants with ADHD only, it is difficult to attribute cognitive control alterations to ADHD specifically in the TS + ADHD group.

To our knowledge, only two published studies have investigated effects of co-occurring ADHD on electrophysiological correlates of cognitive control in TS using a four-group design. Yordanova, Heinrich, Kolev, and Rothenberger (2006) investigated evoked theta power to attended and non-attended auditory target stimuli during an auditory attention task in children with TS, TS + ADHD, ADHD and controls. Children with TS + ADHD showed significantly increased evoked theta power for attended targets at fronto-central electrodes in the first 200 ms following the auditory stimulus compared to all other groups. In addition, children with ADHD and TS + ADHD showed significantly increased fronto-central evoked theta power in a later time-window (200–450 ms post-stimulus) for non-attended targets compared to children with TS and controls. This later theta power was correlated with spontaneous theta power recorded during a resting-state condition, which was also significantly elevated in the TS + ADHD and ADHD groups. The authors concluded that because this late theta component was maximal for task-irrelevant non-attended stimuli rather than task-relevant attended stimuli and because of the similarity in the pattern of activity to spontaneous theta activity, the increased late evoked theta reflected a return to the baseline of increased (noisy) spontaneous theta activity after the identification of the non-attended target as a task-irrelevant stimulus. This conclusion is consistent with other work in ADHD reporting atypically increased spontaneous theta activity during rest (Tye, Rijsdijk, & McLoughlin, 2014) and increased neural noise in the theta frequency during cognitive control (Pertermann, Bluschke, Roessner, & Beste, 2019).

In contrast, since the increased early theta component in TS + ADHD was specific to task-relevant attended target stimuli, Yordanova et al. (2006) concluded that this increase reflected inappropriately enhanced engagement of cognitive control processes to achieve adequate task performance. Of note, there were no group differences in behavioral performance measures. The authors further suggested that this increased theta specific to the TS + ADHD group indicated the presence of unique cognitive control impairments in this group compared to the ADHD and TS groups. However, it could also be argued that the pattern of findings across both early and late theta responses simply reflected the impairing effects of ADHD on cognitive control in TS and that children with TS + ADHD were somewhat able to engage in compensatory adjustments to cope with those impairments (indicated by the increased early theta component). This proposal is in line with previous work showing that children with TS without co-occurring ADHD show compensatory enhancements of cognitive control processes, which likely result from the constant need to control their tics (Jackson et al., 2011; Jackson, Mueller, Hambleton, & Hollis, 2007; Mueller, Jackson, Dhalla, Datsopoulos, & Hollis, 2006).

The other EEG study that used a four-group design (Shephard, Jackson, & Groom, 2016b) compared electrophysiological correlates of cognitive control measured during a Go/Nogo task between children and adolescents with TS, TS + ADHD, ADHD and controls. This study reported significantly reduced amplitude of the P3 during Nogo trials and significantly reduced amplitudes of the ERN and Pe components on error trials in participants with ADHD (ADHD and TS + ADHD) compared to participants without ADHD (TS and controls). In contrast, there were no alterations in ERP markers of cognitive control associated with TS. Consistent with the ERP findings, participants with ADHD and TS + ADHD showed significantly poorer inhibitory performance and less evidence of adjusting performance following errors than participants with TS and controls. Together, these findings provide straightforward support for the proposal that co-occurring ADHD symptoms result in ADHD-related impairments in key cognitive control processes (inhibition and performance monitoring) in TS. It is important to note, however, that the sample sizes for each of the participant groups in this study were modest (<20 per group), which was also the case in the Yordanova et al. (2006) study (<15 per group). Replications of these findings in larger samples are therefore needed.

Still, behavioral studies using the four-group design have supported the electrophysiological findings from Shephard et al. (2016b) and Yordanova et al. (2006). For instance, Greimel et al. (2011) examined behavioral measures of sustained and divided attention, flexibility and inhibition in children and adolescents with TS, TS + ADHD, ADHD and controls. Similar to

Shephard et al.'s (2016b) findings and the late evoked theta findings of Yordanova et al. (2006), Greimel et al. (2011) found significantly poorer performance across all behavioral measures of cognitive control in participants with ADHD (ADHD and TS + ADHD groups) compared to participants without ADHD (TS and controls). Interestingly, Greimel et al. also found better flexibility in participants with TS (TS, TS + ADHD) compared to those without TS (ADHD, controls), which is consistent with previous findings of enhanced cognitive control in TS without co-occurring ADHD (Jackson et al., 2007; Jackson et al., 2011; Mueller et al., 2006).

In one of the largest samples reported to date, Sukhodolsky et al. (2010) compared children and adolescents with TS ($n = 56$), TS + ADHD ($n = 45$), ADHD ($n = 64$) and controls ($n = 71$) on behavioral measures of sustained attention and inhibition. The participants with ADHD without TS had significantly poorer performance on inhibition and attention measures than controls, while participants with TS + ADHD showed no impairments in inhibition but had equally poor sustained attention performance as the ADHD group. Participants with TS without ADHD did not differ from controls. Sukhodolsky et al. (2010) concluded that co-occurring ADHD symptoms may be responsible for cognitive control difficulties in TS. These authors also suggested that the intact performance in the TS group may have reflected compensatory enhancements in cognitive control associated with tic control. Whether the less severe impairments in cognitive control found in the TS + ADHD group compared to the ADHD group also reflected tic control-related enhancements is unclear.

In summary, the findings from electrophysiological and behavioral studies using a four-group design have reported clear indications that co-occurring ADHD symptoms introduce impairments in cognitive control in TS. However, due to the sparsity of the studies and the range of different cognitive control processes and behavioral/electrophysiological correlates examined across the different studies, it is difficult to draw conclusions about which specific aspects of cognitive control are most affected by co-occurring ADHD in TS. Further work aiming to replicate the existing findings in larger samples will be crucial to better understand the nature of cognitive control impairments related to ADHD symptoms in TS.

3.2 Effects of co-occurring ADHD symptoms on altered electrophysiological correlates of cognitive control in TS

A handful of other studies (Eichele et al., 2016; Lange, Seer, Müller-Vahl, & Kopp, 2017; Loo et al., 2019) have taken a different approach to investigating effects of co-occurring ADHD on cognitive control in TS. Specifically, these studies first compared electrophysiological correlates of cognitive control between participants with TS and control participants to identify alterations in these neurocognitive processes related to TS. Subsequently, these studies investigated whether co-occurring diagnoses or symptoms of ADHD in the TS group moderated the TS-related alterations in cognitive control. This methodological approach is therefore different to that of the four-group design used in the studies described above in that the full range of potential negative effects of ADHD on cognitive control in TS is not examined. Rather, the focus is on understanding how ADHD affects electrophysiological correlates of cognitive control processes that are specifically altered in TS.

In the earliest of these studies, Eichele et al. (2016) investigated differences in several ERP components including the P3, ERN and Pe measured during a Flanker task between a group of children with TS (with and without co-occurring ADHD), a group of children with ADHD (without TS) and a group of control children. All ERP components were measured at central electrode sites rather than fronto-central or centro-parietal regions. The authors found that P3 amplitudes were larger in the TS group than in the ADHD and control groups, while ADHD and control groups did not differ. Eichele et al. then conducted a post-hoc analysis to compare P3 amplitude between the children with TS divided into those with ADHD and those without ADHD, the children with ADHD and the control children. The aim of this analysis was to investigate whether the enhanced amplitude in TS was affected by co-occurring ADHD in the TS group. The results of this analysis revealed that the children with TS without ADHD showed significantly increased P3 amplitudes compared to the ADHD and control groups, while children with TS and co-occurring ADHD showed non-significant trends towards larger P3 amplitudes than ADHD and controls. Based on these findings, the authors concluded that children with TS, regardless of the presence of co-occurring ADHD, showed increased P3 amplitudes during the cognitive control task, which they suggested likely reflected adaptively enhanced engagement of attentional processes in TS.

A similar study by Lange et al. (2017) examined differences between adults with TS and control adults in amplitude of the P3 component measured during the Wisconsin Card Sorting Test, a neuropsychological task that is often used to assess cognitive flexibility. Compared to controls, the TS group showed significantly increased amplitude of the centro-parietal P3 component in response to cues indicating that a change in behavior would be required. The authors suggested that this finding reflected increased engagement of cognitive control resources in TS to enable flexible performance. Lange et al. then examined effects of co-occurring ADHD symptoms on this P3 component by (1) repeating the statistical comparison of TS and controls while covarying self-reported ADHD symptoms, and (2) examining correlations between self-reported ADHD symptoms and P3 amplitude in the TS group. The increased P3 amplitude to cues in TS remained significant while covarying ADHD symptoms and was not significantly correlated with ADHD symptoms. The authors did not discuss the implications of these additional analyses, but it might be inferred that co-occurring ADHD symptoms did not negatively affect the cue P3 amplitude increase in TS.

Finally, a more recent study by Loo et al. (2019) examined electrophysiological correlates of cognitive control in children with TS and control children during a paradigm in which participants were required to voluntarily control their eye blinks in response to visual cues. The children with TS showed stronger task-related modulations of oscillatory power in the alpha and gamma frequencies before, during and following the execution of blinks compared to the control children. Loo et al. suggested that these findings were indicative of heightened cognitive control processes involved in voluntary control of motor (blink) behaviors in TS. The authors then repeated the analyses while covarying ADHD symptoms and found that the findings of stronger oscillatory modulations in TS remained significant. These findings might suggest that co-occurring ADHD symptoms did not affect the electrophysiological correlates of cognitive control in the TS group.

In sum, the findings from Eichele et al. (2016), Lange et al. (2017) and Loo et al. (2019) suggest that increases in electrophysiological markers of cognitive control processes in TS, proposed to reflect compensatory enhancements in cognitive control, might not be detrimentally impacted by co-occurring ADHD symptoms. These findings appear to contrast with those from the studies using the four-group design, which showed widespread impairments in electrophysiological correlates of cognitive control associated with ADHD in TS (Shephard, Jackson, & Groom, 2016b; Yordanova et al., 2006). It should be noted, however, that there are some difficulties interpreting the findings from the Eichele et al. (2016), Lange et al. (2017) and Loo et al. (2019) studies. Concerning the Eichele et al. study, an alternative interpretation of the findings of the post-hoc analysis is that the children with TS and co-occurring ADHD did not show the same significant enhancement in P3 amplitude as children with TS without ADHD (the differences between TS+ADHD and ADHD/controls were non-significant trends, only). This pattern might indicate that the presence of ADHD in the TS+ADHD group undermined or reduced the superior attentional control in the TS group. The method of repeating analyses with ADHD symptoms included as a covariate used in the Lange et al. and Loo et al. studies is also problematic because levels of ADHD were not comparable across the TS and control groups (since ADHD symptoms are more common and severe in children with TS than control children). In their seminal article on the use and misuse of covariates, Miller and Chapman (2001) highlighted how systematic differences between groups in a covariate can bias main effects, either eliminating or inflating group differences (see also Schneider, Avivi-Reich, & Mozuraitis, 2015). Thus, it may be the case that the findings from the repeated analyses covarying ADHD symptoms were not valid reflections of the extent to which co-occurring ADHD symptoms affected the enhanced correlates of cognitive control in TS in the Lange et al. and Loo et al. studies. Finally, all three of these studies were conducted with modest samples of participants with TS ($N=25$ in Eichele et al.; $N=23$ in Lange et al. and Loo et al.) with small subsamples of individuals with TS and co-occurring ADHD ($n=14$ in Eichele et al.; $n=1$ in Lange et al.; $n=7$ in Loo et al.). Thus, low statistical power and variability in co-occurring ADHD symptoms may also have affected the analyses.

3.3 Effects of co-occurring ADHD on electrophysiological correlates of tic control

An important limitation of the work discussed above is that cognitive control was measured during experimental tasks that lack ecological validity. That is, the effects of ADHD on cognitive control processes in TS were examined in highly controlled laboratory settings and during computerized tasks that are designed to isolate specific aspects of cognitive control (e.g., inhibition on Nogo trials of the Go/Nogo task). It can therefore only be inferred from those studies how co-occurring ADHD symptoms might impact electrophysiological markers of cognitive control processes engaged during tic control in TS.

Tic control can be measured experimentally using the tic suppression paradigm (Conelea et al., 2018; Woods et al., 2008; Woods & Himle, 2004), in which individuals with TS are asked to alternate between periods of freely executing their tics and periods of voluntarily suppressing their tics. Electrophysiological activity can be compared between the free-tic and tic-suppression conditions to investigate electrophysiological correlates of tic control. Studies using this paradigm have shown that tic suppression appears to function similarly to the cognitive control processes measured with traditional experimental tasks such as the Go/Nogo and Flanker.

For instance, rewards have been shown to enhance tic suppression ability (Conelea et al., 2018) just as they do other cognitive control processes such as inhibition and performance monitoring (Botvinick & Braver, 2015). Performance on traditional cognitive control tasks has been shown to correlate with tic suppression ability during the tic suppression paradigm in some studies (Woods et al., 2008), although not in others (Specht et al., 2014). Further, similar patterns of electrophysiological activity have been reported during traditional cognitive control tasks and during tic suppression in TS (Serrien, Orth, Evans, Lees, & Brown, 2005). Specifically, Serrien et al. reported significant increases in fronto-central oscillatory alpha connectivity during motor response inhibition in the Go/Nogo task and during voluntary tic suppression in a small sample of adults with TS ($n = 9$).

As far as we are aware, only one published study (Morand-Beaulieu et al., 2021) has examined effects of co-occurring ADHD on electrophysiological activity during tic suppression in TS. The authors examined oscillatory activity during free-tic vs. tic-suppression conditions of the tic suppression paradigm in a large sample of 77 children with TS, of which $n = 26$ had co-occurring ADHD. Similar to Serrien et al.'s (2005) findings, Morand-Beaulieu et al. found increased oscillatory connectivity in the alpha frequency in a network of frontal, temporal, occipital and parietal regions during the tic suppression vs. free-tic conditions in the children with TS. The authors then examined effects of ADHD on this tic-suppression network by testing for statistical interactions between task condition (free-tic vs tic-suppression) and presence/absence of co-occurring ADHD. The interaction was non-significant, suggesting that ADHD did not affect electrophysiological correlates of tic suppression in the children with TS.

However, the authors did not examine main effects of ADHD on oscillatory activity across free-tic and tic-suppression conditions. This is important because ADHD is associated with altered alpha oscillations during both rest and task states (Cañigueral et al., 2021; Debnath, Miller, Morales, Seddio, & Fox, 2021; Shephard et al., 2019). Thus, in Morand-Beaulieu et al.'s sample, it may have been the case that co-occurring ADHD was associated with altered alpha connectivity during both free-tic and tic-suppression conditions in TS, which would not have been revealed by examination of the interaction between ADHD and task condition. Consistent with this suggestion, ADHD-related alterations in both spontaneous and cognitive control-related oscillatory activity were reported in TS +ADHD and ADHD in Yordanova et al. (2006). Further, a large-scale behavioral study reported that co-occurring ADHD symptoms were significantly associated with lower self-reported ability to suppress tics in a sample of 1032 children and adults with TS (Sambrani et al., 2016). Still, effects of ADHD on tic control have been found to be non-significant in behavioral studies using tic suppression paradigms (Conelea et al., 2018; Woods et al., 2008), supporting Morand-Beaulieu et al.'s finding that ADHD did not affect oscillatory networks involved in tic suppression in TS.

4. Conclusions and future directions

The studies reviewed in this chapter yield a somewhat conflicting picture of the effects of co-occurring ADHD on cognitive control in TS. On the one hand, several studies comparing cognitive control between participants with TS, TS+ADHD, ADHD and control participants without TS or ADHD provided evidence for impairing effects of co-occurring ADHD on a range of electrophysiological and behavioral correlates of cognitive control in individuals with TS (Greimel et al., 2011; (Shephard et al., 2016b); Sukhodolsky et al., 2010; Yordanova et al., 2006). On the other hand, studies that focused on specific cognitive control processes that were altered in TS provided evidence suggesting that co-occurring ADHD does not negatively impact electrophysiological correlates of cognitive control processes that are enhanced in TS (Eichele et al., 2016;Lange et al., 2017; Loo et al., 2019) or that are involved in tic suppression (Morand-Beaulieu et al., 2021). One potential explanation for these apparently contradictory findings is that ADHD does introduce broad impairments in cognitive control in TS, but compensatory enhancements in cognitive control related to the constant need to control tics in everyday life offset ADHD-related impairments in specific cognitive control processes involved in tic control.

While this explanation is appealing, several limitations of the field prohibit such conclusions based on the currently-available research. These limitations include the small number of studies conducted, the wide cross-study variability in methodological approach, experimental paradigms employed and electrophysiological measures examined, the lack of replication attempts across studies, and the small sample sizes included in individual studies (with the exception of the Morand-Beaulieu et al., 2021 and Sukhodolsky et al., 2010 studies). There are also interpretative difficulties for some of the findings reported in the literature, particularly those involving the use of co-occurring ADHD symptoms as covariates (Lange et al., 2017; Loo et al., 2019). It will be crucial for future research to attempt to address these limitations. For instance, future work could employ experimental tasks and measures that have been used in previous studies to enable analyses investigating the replicability of previously-reported effects of ADHD on cognitive control in TS. One advantage of EEG is that the experimental procedure is well-tolerated by participants (unlike the fMRI environment, for example). It should therefore be feasible for researchers to include a task that has previously been used (e.g., Morand-Beaulieu et al.'s tic suppression paradigm or the widely-used Go/Nogo task) in addition to any more novel experimental procedures in new studies.

Another key issue that should be investigated is heterogeneity in the effects of co-occurring ADHD on cognitive control in TS. ADHD is a highly heterogeneous neurodevelopmental condition with large inter-individual variability in the presence and form of neural (Wolfers et al., 2020) and cognitive (Fair, Bathula, Nikolas, & Nigg, 2012) alterations. In terms of cognitive control, some individuals with ADHD do not show impairments and, amongst those that do, the specific cognitive control processes affected varies considerably (Fair et al., 2012; Kofler et al., 2019). It is therefore unlikely that co-occurring ADHD symptoms would have the same effects on cognitive control processes in TS across different individuals. Future studies should directly investigate heterogeneity to understand the different cognitive control profiles associated with co-occurring ADHD that may affect individuals with TS. Such investigations necessitate larger participant samples than are currently the norm in this area of research. While it is challenging to collect large samples of clinical groups such as TS and ADHD, another approach is to conduct analyses on data pooled across different research studies. Data pooling requires that the same or similar experimental methods are used across research studies, which is another argument for replicating previously-used experimental paradigms and measures.

Finally, another important avenue for future research is the development and implementation of more naturalistic paradigms in which to measure the effects of ADHD on cognitive control over tics in TS. Tic severity and frequency are reported to increase under conditions of stress and emotional arousal (Conelea & Woods, 2008). These physiological states hinder cognitive control performance, including in individuals without cognitive control difficulties (Plessow, Kiesel, & Kirschbaum, 2012). Further, elevated stress responses and emotional dysregulation are strongly linked with ADHD (Angeli et al., 2018; Hirsch, Chavanon, Riechmann, & Christiansen, 2018; Shephard et al., 2021). Therefore, it is relevant to investigate tic control under naturalistic conditions that involve varying degrees of stress and emotion, rather than in controlled laboratory settings that minimize these physiological states. The development of mobile EEG in recent years has made data collection possible in real-life settings. For example, recent studies have successfully measured electrophysiological activity associated with socioemotional and attentional processing in children and adolescents during normal classwork at school (Bevilacqua et al., 2019; Prokofieva, Kostromina, Polevaia, & Fenouillet, 2019). This method could be used to measure effects of ADHD symptoms on cognitive control processes involved in tic control during naturalistic environments in TS.

References

American Psychiatric Association. (2013). *Diagnostic and statistical manual of mental disorders* (5th ed.). Arlington, VA: Author.

Angeli, E., Korpa, T., Johnson, E. O., Apostolakou, F., Papassotiriou, I., Chrousos, G. P., et al. (2018). Salivary cortisol and alpha-amylase diurnal profiles and stress reactivity in children with attention deficit hyperactivity disorder. *Psychoneuroendocrinology, 90*, 174–181.

Bellato, A., Norman, L., Idrees, I., Ogawa, C. Y., Waitt, A., Zuccolo, P. F., et al. (2021). A systematic review and meta-analysis of altered electrophysiological correlates of performance monitoring in obsessive-compulsive disorder (OCD), Gilles de la Tourette syndrome (GTS), attention-deficit/hyperactivity disorder (ADHD) and autism. *Neuroscience & Biobehavioral Reviews*. https://doi.org/10.1016/j.neubiorev.2021.10.018. In press.

Bevilacqua, D., Davidesco, I., Wan, L., Chaloner, K., Rowland, J., Ding, M., et al. (2019). Brain-to-brain synchrony and learning outcomes vary by student–teacher dynamics: Evidence from a real-world classroom electroencephalography study. *Journal of Cognitive Neuroscience, 31*(3), 401–411.

Botvinick, M., & Braver, T. (2015). Motivation and cognitive control: From behavior to neural mechanism. *Annual Review of Psychology, 66*, 83–113.

Botvinick, M. M., Braver, T. S., Barch, D. M., Carter, C. S., & Cohen, J. D. (2001). Conflict monitoring and cognitive control. *Psychological Review, 108*(3), 624.

Cañigueral, R., Palmer, J., Ashwood, K. L., Azadi, B., Asherson, P., Bolton, P. F., et al. (2021). Alpha oscillatory activity during attentional control in children with autism spectrum disorder (ASD), Attention-Deficit/Hyperactivity Disorder (ADHD), and ASD +ADHD. *Journal of Child Psychology and Psychiatry*. https://doi.org/10.1111/jcpp.13514. Available online ahead of print:.

Cavanagh, J. F., & Frank, M. J. (2014). Frontal theta as a mechanism for cognitive control. *Trends in Cognitive Sciences, 18*(8), 414–421.

Conelea, C. A., Wellen, B., Woods, D. W., Greene, D. J., Black, K. J., Specht, M., et al. (2018). Patterns and predictors of tic suppressibility in youth with tic disorders. *Frontiers in Psychiatry, 9*, 188.

Conelea, C. A., & Woods, D. W. (2008). The influence of contextual factors on tic expression in Tourette's syndrome: A review. *Journal of Psychosomatic Research, 65*(5), 487–496.

Cravedi, E., Deniau, E., Giannitelli, M., Xavier, J., Hartmann, A., & Cohen, D. (2017). Tourette syndrome and other neurodevelopmental disorders: A comprehensive review. *Child and Adolescent Psychiatry and Mental Health, 11*(1), 1–12.

Debnath, R., Miller, N. V., Morales, S., Seddio, K. R., & Fox, N. A. (2021). Investigating brain electrical activity and functional connectivity in adolescents with clinically elevated levels of ADHD symptoms in alpha frequency band. *Brain Research, 1750*, 147142.

Donchin, E., & Coles, M. G. (1988). Is the P300 component a manifestation of context updating? *Behavioral and Brain Sciences*, *11*(3), 357–374.
Donkers, F. C., & Van Boxtel, G. J. (2004). The N2 in go/no-go tasks reflects conflict monitoring not response inhibition. *Brain and Cognition*, *56*(2), 165–176.
Drury, H., Channon, S., Barrett, R., Young, M. B., Stern, J. S., Simmons, H., et al. (2012). Emotional processing and executive functioning in children and adults with Tourette's syndrome. *Child Neuropsychology*, *18*(3), 281–298.
Eddy, C. M., Cavanna, A. E., Gulisano, M., Calì, P., Robertson, M. M., & Rizzo, R. (2012). The effects of comorbid obsessive-compulsive disorder and attention-deficit hyperactivity disorder on quality of life in Tourette syndrome. *The Journal of Neuropsychiatry and Clinical Neurosciences*, *24*(4), 458–462.
Eichele, H., Eichele, T., Bjelland, I., Høvik, M. F., Sørensen, L., Van Wageningen, H., et al. (2016). Performance monitoring in medication-naïve children with Tourette syndrome. *Frontiers in Neuroscience*, *10*, 50.
Fair, D. A., Bathula, D., Nikolas, M. A., & Nigg, J. T. (2012). Distinct neuropsychological subgroups in typically developing youth inform heterogeneity in children with ADHD. *Proceedings of the National Academy of Sciences*, *109*(17), 6769–6774.
Falkenstein, M., Hoormann, J., Christ, S., & Hohnsbein, J. (2000). ERP components on reaction errors and their functional significance: A tutorial. *Biological Psychology*, *51*, 87–107.
Falkenstein, M., Hoormann, J., & Hohnsbein, J. (1999). ERP components in go/Nogo tasks and their relation to inhibition. *Acta Psychologica*, *101*(2–3), 267–291.
Gehring, W. J., Goss, B., Coles, M. G., Meyer, D. E., & Donchin, E. (1993). A neural system for error detection and compensation. *Psychological Science*, *4*(6), 385–390.
Gorman, D. A., Thompson, N., Plessen, K. J., Robertson, M. M., Leckman, J. F., & Peterson, B. S. (2010). Psychosocial outcome and psychiatric comorbidity in older adolescents with Tourette syndrome: Controlled study. *The British Journal of Psychiatry*, *197*(1), 36–44.
Greimel, E., Wanderer, S., Rothenberger, A., Herpertz-Dahlmann, B., Konrad, K., & Roessner, V. (2011). Attentional performance in children and adolescents with tic disorder and co-occurring attention-deficit/hyperactivity disorder: New insights from a 2 × 2 factorial design study. *Journal of Abnormal Child Psychology*, *39*(6), 819–828.
Groom, M. J., Bates, A. T., Jackson, G. M., Calton, T. G., Liddle, P. F., & Hollis, C. (2008). Event-related potentials in adolescents with schizophrenia and their siblings: A comparison with attention-deficit/hyperactivity disorder. *Biological Psychiatry*, *63*(8), 784–792.
Groom, M. J., Cahill, J. D., Bates, A. T., Jackson, G. M., Calton, T. G., Liddle, P. F., et al. (2010). Electrophysiological indices of abnormal error-processing in adolescents with attention deficit hyperactivity disorder (ADHD). *Journal of Child Psychology and Psychiatry*, *51*(1), 66–76.
Groom, M. J., & Cragg, L. (2015). Differential modulation of the N2 and P3 event-related potentials by response conflict and inhibition. *Brain and Cognition*, *97*, 1–9.
Hirsch, O., Chavanon, M., Riechmann, E., & Christiansen, H. (2018). Emotional dysregulation is a primary symptom in adult attention-deficit/hyperactivity disorder (ADHD). *Journal of Affective Disorders*, *232*, 41–47.
Hirschtritt, M. E., Lee, P. C., Pauls, D. L., Dion, Y., Grados, M. A., Illmann, C., et al. (2015). Lifetime prevalence, age of risk, and genetic relationships of comorbid psychiatric disorders in Tourette syndrome. *JAMA Psychiatry*, *72*(4), 325–333.
Houghton, G., & Tipper, S. P. (1996). Inhibitory mechanisms of neural and cognitive control: Applications to selective attention and sequential action. *Brain and Cognition*, *30*(1), 20–43.

Jackson, G. M., Draper, A., Dyke, K., Pépés, S. E., & Jackson, S. R. (2015). Inhibition, disinhibition, and the control of action in Tourette syndrome. *Trends in Cognitive Sciences*, *19*(11), 655–665.
Jackson, G. M., Mueller, S. C., Hambleton, K., & Hollis, C. P. (2007). Enhanced cognitive control in Tourette syndrome during task uncertainty. *Experimental Brain Research*, *182*(3), 357–364.
Jackson, S. R., Parkinson, A., Jung, J., Ryan, S. E., Morgan, P. S., Hollis, C., et al. (2011). Compensatory neural reorganization in Tourette syndrome. *Current Biology*, *21*(7), 580–585.
Kaiser, A., Aggensteiner, P. M., Baumeister, S., Holz, N. E., Banaschewski, T., & Brandeis, D. (2020). Earlier versus later cognitive event-related potentials (ERPs) in attention-deficit/hyperactivity disorder (ADHD): A meta-analysis. *Neuroscience & Biobehavioral Reviews*, *112*, 117–134.
Kofler, M. J., Irwin, L. N., Soto, E. F., Groves, N. B., Harmon, S. L., & Sarver, D. E. (2019). Executive functioning heterogeneity in pediatric ADHD. *Journal of Abnormal Child Psychology*, *47*(2), 273–286.
Kok, A. (2001). On the utility of P3 amplitude as a measure of processing capacity. *Psychophysiology*, *38*(3), 557–577.
Lange, F., Seer, C., Müller-Vahl, K., & Kopp, B. (2017). Cognitive flexibility and its electrophysiological correlates in Gilles de la Tourette syndrome. *Developmental Cognitive Neuroscience*, *27*, 78–90.
Lebowitz, E. R., Motlagh, M. G., Katsovich, L., King, R. A., Lombroso, P. J., Grantz, H., et al. (2012). Tourette syndrome in youth with and without obsessive compulsive disorder and attention deficit hyperactivity disorder. *European Child & Adolescent Psychiatry*, *21*(8), 451–457.
Loo, S. K., Miyakoshi, M., Tung, K., Lloyd, E., Salgari, G., Dillon, A., et al. (2019). Neural activation and connectivity during cued eye blinks in chronic tic disorders. *NeuroImage: Clinical*, *24*, 101956.
McGuire, J. F., Piacentini, J., Brennan, E. A., Lewin, A. B., Murphy, T. K., Small, B. J., et al. (2014). A meta-analysis of behavior therapy for Tourette syndrome. *Journal of Psychiatric Research*, *50*, 106–112.
McGuire, J. F., Sturm, A., Ricketts, E. J., Montalbano, G. E., Chang, S., Loo, S. K., et al. (2021). Cognitive control processes in behavior therapy for youth with Tourette's disorder. *Journal of Child Psychology and Psychiatry*. https://doi.org/10.1111/jcpp.13470. In press.
Miller, G. A., & Chapman, J. P. (2001). Misunderstanding analysis of covariance. *Journal of Abnormal Psychology*, *110*(1), 40.
Morand-Beaulieu, S., Wu, J., Mayes, L. C., Grantz, H., Leckman, J. F., Crowley, M. J., et al. (2021). Increased alpha-band connectivity during tic suppression in children with Tourette syndrome revealed by source EEG analyses. *Biological Psychiatry: Cognitive Neuroscience and Neuroimaging*. https://doi.org/10.1016/j.bpsc.2021.05.001. In press.
Mueller, S. C., Jackson, G. M., Dhalla, R., Datsopoulos, S., & Hollis, C. P. (2006). Enhanced cognitive control in young people with Tourette's syndrome. *Current Biology*, *16*(6), 570–573.
Pertermann, M., Bluschke, A., Roessner, V., & Beste, C. (2019). The modulation of neural noise underlies the effectiveness of methylphenidate treatment in attention-deficit/hyperactivity disorder. *Biological Psychiatry: Cognitive Neuroscience and Neuroimaging*, *4*(8), 743–750.
Plessow, F., Kiesel, A., & Kirschbaum, C. (2012). The stressed prefrontal cortex and goal-directed behaviour: Acute psychosocial stress impairs the flexible implementation of task goals. *Experimental Brain Research*, *216*(3), 397–408.

Prokofieva, V., Kostromina, S., Polevaia, S., & Fenouillet, F. (2019). Understanding emotion-related processes in classroom activities through functional measurements. *Frontiers in Psychology, 10*, 2263.

Ricketts, E. J., Wolicki, S. B., Danielson, M. L., Rozenman, M., McGuire, J. F., Piacentini, J., et al. (2021). Academic, interpersonal, recreational, and family impairment in children with Tourette syndrome and attention-deficit/hyperactivity disorder. *Child Psychiatry & Human Development*, 1–13.

Sambrani, T., Jakubovski, E., & Müller-Vahl, K. R. (2016). New insights into clinical characteristics of Gilles de la Tourette syndrome: Findings in 1032 patients from a single German center. *Frontiers in Neuroscience, 10*, 415.

Schneider, B. A., Avivi-Reich, M., & Mozuraitis, M. (2015). A cautionary note on the use of the analysis of covariance (ANCOVA) in classification designs with and without within-subject factors. *Frontiers in Psychology, 6*, 474.

Serrien, D. J., Orth, M., Evans, A. H., Lees, A. J., & Brown, P. (2005). Motor inhibition in patients with Gilles de la Tourette syndrome: Functional activation patterns as revealed by EEG coherence. *Brain, 128*(1), 116–125.

Shephard, E., Jackson, G. M., & Groom, M. J. (2016a). Electrophysiological correlates of reinforcement learning in young people with Tourette syndrome with and without co-occurring ADHD symptoms. *International Journal of Developmental Neuroscience, 51*, 17–27.

Shephard, E., Jackson, G. M., & Groom, M. J. (2016b). The effects of co-occurring ADHD symptoms on electrophysiological correlates of cognitive control in young people with Tourette syndrome. *Journal of Neuropsychology, 10*(2), 223–238.

Shephard, E., Tye, C., Ashwood, K. L., Azadi, B., Johnson, M. H., Charman, T., et al. (2019). Oscillatory neural networks underlying resting-state, attentional control and social cognition task conditions in children with ASD, ADHD and ASD+ADHD. *Cortex, 117*, 96–110.

Shephard, E., Zuccolo, P. F., Idrees, M. I., Godoy, M. P. B., Salomone, E., Ferrante, M. C., et al. (2021). Systematic review and Meta-analysis: The science of early-life precursors and interventions for attention-deficit/hyperactivity disorder. *Journal of the American Academy of Child & Adolescent Psychiatry*. https://doi.org/10.1016/j.jaac.2021.03.016. In press.

Specht, M. W., Nicotra, C. M., Kelly, L. M., Woods, D. W., Ricketts, E. J., Perry-Parrish, C., et al. (2014). A comparison of urge intensity and the probability of tic completion during tic freely and tic suppression conditions. *Behavior Modification, 38*(2), 297–318.

Sukhodolsky, D. G., Landeros-Weisenberger, A., Scahill, L., Leckman, J. F., & Schultz, R. T. (2010). Neuropsychological functioning in children with Tourette syndrome with and without attention-deficit/hyperactivity disorder. *Journal of the American Academy of Child & Adolescent Psychiatry, 49*(11), 1155–1164.

Tye, C., Asherson, P., Ashwood, K. L., Azadi, B., Bolton, P., & McLoughlin, G. (2014). Attention and inhibition in children with ASD, ADHD and co-morbid ASD +ADHD: An event-related potential study. *Psychological Medicine, 44*(5), 1101–1116.

Tye, C., Rijsdijk, F., & McLoughlin, G. (2014). Genetic overlap between ADHD symptoms and EEG theta power. *Brain and Cognition, 87*, 168–172.

Willcutt, E. G., Doyle, A. E., Nigg, J. T., Faraone, S. V., & Pennington, B. F. (2005). Validity of the executive function theory of attention-deficit/hyperactivity disorder: A meta-analytic review. *Biological Psychiatry, 57*(11), 1336–1346.

Wolfers, T., Beckmann, C. F., Hoogman, M., Buitelaar, J. K., Franke, B., & Marquand, A. F. (2020). Individual differences v. the average patient: Mapping the heterogeneity in ADHD using normative models. *Psychological Medicine, 50*(2), 314–323.

Woods, D. W., & Himle, M. B. (2004). Creating tic suppression: Comparing the effects of verbal instruction to differential reinforcement. *Journal of Applied Behavior Analysis*, *37*(3), 417–420.

Woods, D. W., Himle, M. B., Miltenberger, R. G., Carr, J. E., Osmon, D. C., Karsten, A. M., et al. (2008). Durability, negative impact, and neuropsychological predictors of tic suppression in children with chronic tic disorder. *Journal of Abnormal Child Psychology*, *36*(2), 237–245.

Yaniv, A., Benaroya-Milshtein, N., Steinberg, T., Ruhrman, D., Apter, A., & Lavidor, M. (2018). Executive control development in Tourette syndrome and its role in tic reduction. *Psychiatry Research*, *262*, 527–535.

Yordanova, J., Heinrich, H., Kolev, V., & Rothenberger, A. (2006). Increased event-related theta activity as a psychophysiological marker of comorbidity in children with tics and attention-deficit/hyperactivity disorders. *NeuroImage*, *32*(2), 940–955.

FSC
www.fsc.org
MIX
Papier | Fördert gute Waldnutzung
FSC® C083411

Kolibri 360 GmbH
Ferdinand-Jühlke-Straße 7
99095 Erfurt, Deutschland
produktsicherheit@kolibri360.de